HIGHER - ORDER SPECTRA ANALYSIS

PRENTICE HALL SIGNAL PROCESSING SERIES

Alan V. Oppenheim, Series Editor

ANDREWS AND HUNT *Digital Image Restoration*
BRIGHAM *The Fast Fourier Transform*
BRIGHAM *The Fast Fourier Transform and Its Applications*
BURDIC *Underwater Acoustic System Analysis, 2/E*
CASTLEMAN *Digital Image Processing*
COWAN AND GRANT *Adaptive Filters*
CROCHIERE AND RABINER *Multirate Digital Signal Processing*
DUDGEON AND MERSEREAU *Multidimensional Digital Signal Processing*
HAMMING *Digital Filters, 3/E*
HAYKIN, ED. *Advances in Spectrum Analysis and Array Processing, Vols. I & II*
HAYKIN, ED. *Array Signal Processing*
JAYANT AND NOLL *Digital Coding of Waveforms*
JOHNSON AND DUDGEON *Array Signal Processing: Concepts and Techniques*
KAY *Fundamentals of Statistical Signal Processing: Estimation Theory*
KAY *Modern Spectral Estimation*
KINO *Acoustic Waves: Devices, Imaging, and Analog Signal Processing*
LEA, ED. *Trends in Speech Recognition*
LIM *Two-Dimensional Signal and Image Processing*
LIM, ED. *Speech Enhancement*
LIM AND OPPENHEIM, EDS. *Advanced Topics in Signal Processing*
MARPLE *Digital Spectral Analysis with Applications*
MCCLELLAN AND RADER *Number Theory in Digital Signal Processing*
MENDEL *Lessons in Digital Estimation Theory*
NIKIAS AND PETROPULU *Higher-Order Spectra Analysis: A Nonlinear Signal Processing Framework*
OPPENHEIM, ED. *Applications of Digital Signal Processing*
OPPENHEIM AND NAWAB, EDS. *Symbolic and Knowledge-Based Signal Processing*
OPPENHEIM, WILLSKY, WITH YOUNG *Signals and Systems*
OPPENHEIM AND SCHAFER *Digital Signal Processing*
OPPENHEIM AND SCHAFER *Discrete-Time Signal Processing*
QUACKENBUSH ET AL. *Objective Measures of Speech Quality*
RABINER AND GOLD *Theory and Applications of Digital Signal Processing*
RABINER AND SCHAFER *Digital Processing of Speech Signals*
ROBINSON AND TREITEL *Geophysical Signal Analysis*
STEARNS AND DAVID *Signal Processing Algorithms*
STEARNS AND HUSH *Digital Signal Analysis, 2/E*
TRIBOLET *Seismic Applications of Homomorphic Signal Processing*
VAIDYANATHAN *Multirate Systems and Filter Banks*
WIDROW AND STEARNS *Adaptive Signal Processing*

HIGHER - ORDER SPECTRA ANALYSIS

A NONLINEAR SIGNAL PROCESSING FRAMEWORK

Chrysostomos L. Nikias
University of Southern California

Athina P. Petropulu
Drexel University

PTR Prentice Hall, Englewood Cliffs, New Jersey 07632

Library of Congress Cataloging-in-Publication Data

Nikias, Chrysostomos, L.
    Higher-order spectra analysis: a nonlinear signal processing framework / Chrysostomos L. Nikias, Athina P. Petropulu.
      p. cm. -- (Prentice Hall signal processing series)
    Includes bibliographical references and index.
    ISBN 0-13-678210-8
    1. Signal processing--Statistical methods. 2. Time-series analysis. 3. Spectrum analysis. I. Petropulu, Athina P. II. Title. III. Series.
TK5102.9.N54 1993   93-28
621.382'2--dc20   CIP

To: Niki, Georgiana, and Maria

To: Antonio and Artemis

Editorial/production supervision: bookworks
Manufacturing buyer: Mary McCartney
Cover Design: Wanda Lubelska

 © 1993 by PTR Prentice-Hall, Inc.
A Simon & Schuster Company
Englewood Cliffs, New Jersey 07632

All rights reserved. No part of this book may be reproduced, in any form or by any means, without permission in writing from the publisher.

The publisher offers discounts on this book when ordered in bulk quantities. For more information, contact:

Corporate Sales Department
PTR Prentice Hall
113 Sylvan Avenue
Englewood Cliffs, NJ 07632

Phone: 201-592-2863
Fax: 201-592-2249

Printed in the United States of America

10 9 8 7 6 5 4 3 2 1

ISBN   0-13-678210-8

Prentice-Hall International (UK) Limited, *London*
Prentice-Hall of Australia Pty. Limited, *Sydney*
Prentice-Hall Canada, Inc., *Toronto*
Prentice-Hall Hispanoamericana, S.A., *Mexico*
Prentice-Hall of India Private Limited, *New Delhi*
Prentice-Hall of Japan, Inc., *Tokyo*
Simon & Schuster Asia Pte. Ltd., *Singapore*
Editora Prentice-Hall do Brasil, Ltda., *Rio de Janeiro*

CONTENTS

PREFACE xvii

LIST OF SYMBOLS xxi

CHAPTER 1 INTRODUCTION 1
  1.1 POWER SPECTRUM 1
  1.2 WHY POLYSPECTRA IN SIGNAL PROCESSING? 2
  1.3 APPLICATIONS 4
  REFERENCES 4

CHAPTER 2 CUMULANT SPECTRA OF STOCHASTIC SIGNALS 7
  2.1 INTRODUCTION 7
  2.2 MOMENTS AND CUMULANTS 7
    2.2.1 Definitions 7
    2.2.2 Relationship Between Moments and Cumulants 9
    2.2.3 Properties of Moments and Cumulants 12
    2.2.4 Moments and Cumulants of Stationary Processes 15
    2.2.5 Variance, Skewness, and Kurtosis Measures 16
    2.2.6 Time-Reversible Process 17
    2.2.7 Cross-Cumulants 17
    2.2.8 Ergodicity and Moments 19
  2.3 CUMULANT SPECTRA 20
    2.3.1 Definition 20
    2.3.2 Alternative Definition 21
    2.3.3 Special Cases of Cumulant Spectra 21
    2.3.4 Variance, Skewness, and Kurtosis Measures 23
    2.3.5 Time-Reversible Processes 24
    2.3.6 Non-Gaussian White Noise Processes 24
    2.3.7 One-Dimensional Slices of Cumulants and their Cumulant Spectra 26
    2.3.8 Why Cumulant Spectra and not Moment Spectra 30
    2.3.9 The nth-Order Coherency Function 31

    2.3.10 Cross-Cumulant Spectra 31

    2.3.11 Linear Phase Shifts 33

    2.3.12 Complex Regression Coefficients 34

    2.3.13 Complex Processes 34

    2.3.14 The Wigner Bispectrum 36

2.4 CUMULANT SPECTRA OF NON-GAUSSIAN LINEAR
PROCESSES 37

    2.4.1 Cumulant Spectra of LTI Systems 37

    2.4.2 Cumulant Spectra of LTI Systems Driven by White Noise 38

    2.4.3 Nonminimum Phase LTI Systems 45

    2.4.4 Linear Processes with Gaussian and
Non-Gaussian Components 53

    2.4.5 Poisson Triggered Processes 56

    2.4.6 Regression Coefficients of Linear Processes 57

2.5 DETECTING AND IDENTIFYING NONLINEAR PROCESSES ... 58

    2.5.1 Sinusoidal Processes Applied to Nonlinear Systems 58

    2.5.2 Quadratic Phase Coupling 59

    2.5.3 Gaussian Processes Applied to Nonlinear Systems 61

    2.5.4 Linear Non-Gaussian versus Nonlinear Processes 64

2.6 SUMMARY 64

REFERENCES 66

CHAPTER 3 MOMENT SPECTRA OF DETERMINISTIC SIGNALS 71

3.1 INTRODUCTION 71

    3.1.1 Energy versus Power Deterministic Signals 71

    3.1.2 Fourier Analysis of Energy Signals 72

    3.1.3 Fourier Analysis of Periodic Power Signals 73

    3.1.4 Finite-Duration versus Periodic Signals 75

3.2 MOMENTS OF ENERGY SIGNALS 78

    3.2.1 Definition 78

    3.2.2 Properties 78

- 3.2.3 Special Cases 80
- 3.2.4 Cross-Moments 83
- 3.3 MOMENT SPECTRA OF APERIODIC ENERGY SIGNALS 84
  - 3.3.1 Definition 85
  - 3.3.2 An Alternative Definition 85
  - 3.3.3 Special Cases of Moment Spectra 88
  - 3.3.4 Scalar Measures of Energy Signals 89
  - 3.3.5 Moment Spectra and LTI System Relations 90
  - 3.3.6 Phase Conditions 94
  - 3.3.7 Cross-Moment Spectra 98
  - 3.3.8 The nth-Order Coherency of Energy Signals 99
- 3.4 MOMENTS OF POWER SIGNALS 100
  - 3.4.1 Definition 100
  - 3.4.2 Special Cases for Periodic Signals 101
  - 3.4.3 Cross-Moments of Periodic Signals 106
- 3.5 MOMENT SPECTRA OF PERIODIC POWER SIGNALS 107
  - 3.5.1 Definition 108
  - 3.5.2 An Alternative Definition 109
  - 3.5.3 Special Cases of Moment Spectra 110
  - 3.5.4 LTI System Relations 116
  - 3.5.5 Cross-Moment Spectra 117
  - 3.5.6 Higher-Order Coherency Function of Periodic Power Signals . 119
  - 3.5.7 Finite-Duration versus Periodic Signals:
    Moment Spectra Relationships 120
- 3.6 SUMMARY 121
- REFERENCES 121

CHAPTER 4 CONVENTIONAL METHODS FOR THE
ESTIMATION OF HIGHER-ORDER SPECTRA 123

- 4.1 INTRODUCTION 123
- 4.2 INDIRECT CLASS OF CONVENTIONAL METHODS 124

4.2.1 Estimation of Higher-Order Statistics 124

4.2.2 Estimation of Higher-Order Spectra 125

4.2.3 Window Functions 126

4.3 DIRECT CLASS OF CONVENTIONAL METHODS 132

4.3.1 The Higher-Order Periodogram 132

4.3.2 Direct Estimation of Moment Spectra 136

4.3.3 The Zurbenko Method for the Estimation
of Cumulant Spectra 138

4.4 COMPLEX DEMODULATES CLASS OF CONVENTIONAL
METHODS 141

4.5 STATISTICAL PROPERTIES OF CONVENTIONAL METHODS 142

4.5.1 Properties of Bispectral Estimates 142

4.5.2 Properties of Higher-Order Spectrum Estimation 145

4.6 TEST FOR ALIASING WITH THE BISPECTRUM 147

4.7 BISPECTRUM COMPUTATION ON POLAR RASTERS 149

4.7.1 Introduction 149

4.7.2 Problem Formulation 150

4.7.3 The Direct Approach (DA) 151

4.7.4 The Zero-Padding Approach (ZPA) 151

4.7.5 The "Super-FFT Approach" (SFA) 153

4.7.6 Trispectrum Computation on a Polar Raster 155

4.7.7 Extension to Stochastic Signals 158

4.8 SUMMARY 159

REFERENCES 159

CHAPTER 5 HIGHER-ORDER CEPSTRA (POLYCEPSTRA) 163

5.1 INTRODUCTION 163

5.2 THE COMPLEX CEPSTRUM OF DETERMINISTIC ENERGY
SIGNALS 164

5.2.1 Signal Assumptions 164

5.2.2 Definition 165

5.2.3 Properties 166

5.2.4 Computation and Phase Unwrapping	167

5.3 THE DIFFERENTIAL CEPSTRUM OF DETERMINISTIC ENERGY SIGNALS ... 169

5.3.1 Definition	169
5.3.2 Properties	170
5.3.3 Computation	171

5.4 THE POWER CEPSTRUM OF DETERMINISTIC AND STOCHASTIC SIGNALS ... 172

5.4.1 Definition	172
5.4.2 Properties	173
5.4.3 Computation	174

5.5 THE BICEPSTRUM AND TRICEPSTRUM OF DETERMINISTIC AND STOCHASTIC SIGNALS ... 177

5.5.1 Definition	179
5.5.2 Properties	181
5.5.3 Computation	182
5.5.4 The Tricepstrum	185

5.6 THE CEPSTRUM OF BICOHERENCY ... 187

5.6.1 Definition	189
5.6.2 Properties	189
5.6.3 Computation	190
5.6.4 Computation of $A^{(m)}, B^{(m)}$ from Second- and Third-Order Statistics	191

5.7 HIGHER-ORDER SPECTRUM FACTORIZATION AND INDEX FOR LINEARITY ... 191

5.8 INVERSE FILTER RECONSTRUCTION ... 193

5.9 THE CROSS-BICEPSTRUM OF DETERMINISTIC AND STOCHASTIC SIGNALS ... 200

5.9.1 Preliminaries	200
5.9.2 Signal Assumptions	203
5.9.3 Definition	203
5.9.4 Properties	204

5.9.5 Computation 205

5.10 ANALYTIC PERFORMANCE EVALUATION OF THE COMPLEX CEPSTRUM AND BICEPSTRUM 209

5.10.1 Preliminaries and Problem Definition 209

5.10.2 The Complex Cepstrum Evaluation 211

5.10.3 The Bicepstrum Evaluation 214

5.11 SUMMARY 215

REFERENCES 216

CHAPTER 6 NONPARAMETRIC METHODS FOR SIGNAL RECOVERY FROM HIGHER-ORDER SPECTRA 219

6.1 INTRODUCTION 219

6.2 MAGNITUDE AND PHASE ESTIMATION FROM HIGHER-ORDER SPECTRA 220

6.3 PHASE RECOVERY ALGORITHMS 220

6.3.1 The Brillinger Algorithm 221

6.3.2 The Lii-Rosenblatt Algorithm 222

6.3.3 The Bartelt-Lohman-Wirnitzer Algorithm 223

6.3.4 The Matsuoka-Ulrych Algorithm 224

6.3.5 Phase Reconstruction from a 45° Slice of the Bispectral Phase 228

6.4 PHASE AND MAGNITUDE ALGORITHMS BASED ON POLYSPECTRA/POLYCEPSTRA PROPERTIES 230

6.4.1 Preliminaries 230

6.4.2 Bispectrum Signal Reconstruction (BSR) 231

6.4.3 Two-Dimensional Phase Unwrapping 233

6.4.4 Duality Between Phase and Magnitude Recovery 234

6.4.5 Treating Singularities in the Vicinity of the Unit Circle 235

6.5 SIGNAL RECONSTRUCTION FROM ONLY THE PHASE OF THE BISPECTRUM 237

6.5.1 The Importance of Signal Recovery from Partial Information 237

6.5.2 Conditions for Signal Reconstruction from the Phase of the Bicepstrum 239

6.5.3 The Bispectrum Iterative Reconstruction Algorithm (BIRA)   240

6.6 SIGNAL RECONSTRUCTION FROM THE BISPECTRUM BASED ON THE METHOD OF POCS   249

6.7 BLIND DECONVOLUTION USING BIRA   253

    6.7.1 The Blind Deconvolution Scheme   254

    6.7.2 Computation of the Differences of Cepstral Coefficients   256

6.8 SUMMARY   259

REFERENCES   261

CHAPTER 7 PARAMETRIC METHODS FOR THE ESTIMATION OF HIGHER-ORDER SPECTRA   265

7.1 INTRODUCTION   265

7.2 MA METHODS   266

    7.2.1 Closed Form Solutions   267

    7.2.2 Optimization Methods   270

    7.2.3 Parameter Estimation Based on ARMA and AR Approximations   271

    7.2.4 The GM Method   276

7.3 NONCAUSAL AR METHODS   278

    7.3.1 The Huzii Method   278

    7.3.2 Cumulant Based Solutions   279

7.4 ARMA METHODS   282

    7.4.1 Estimation of the ARMA Parameters from the System Impulse Response   282

    7.4.2 Residual Time Series   284

    7.4.3 The Double $c(q,k)$ Algorithm   286

    7.4.4 The Q-Slice Algorithm   291

    7.4.5 Identification of ARMA Systems Via Second- and Fourth-Order Cumulant Matching   293

7.5 MODEL ORDER DETERMINATION   295

    7.5.1 MA Models   295

    7.5.2 AR Models   304

7.5.3 ARMA Models 307

7.6 SUMMARY 307

REFERENCES 308

CHAPTER 8 DIRECTION OF ARRIVAL ESTIMATION AND ANALYSIS OF TRANSIENT SIGNALS 313

8.1 INTRODUCTION 313

8.2 TIME DELAY ESTIMATION 314

    8.2.1 Problem Definition 314

    8.2.2 Bearing and Range Estimation 315

    8.2.3 Generalized Cross-Correlation (GCC) Method 317

    8.2.4 Conventional TDE Methods Based on Third-Order Statistics 322

    8.2.5 Parametric TDE Method in the Bispectrum Domain 327

    8.2.6 TDE Method Based on the Cepstrum of the Cross-Bispectrum 331

    8.2.7 TDE with the Mean Fourth-Cumulant (MFC) Criterion 332

8.3 BEARING ESTIMATION WITH HIGHER-ORDER STATISTICS 337

    8.3.1 Problem Formulation 339

    8.3.2 The Spatial Covariance Matrix 340

    8.3.3 Bearing Estimation Based on the Spatial Covariance Matrix . 342

    8.3.4 The Spatial Fourth-Order Cumulant Matrix and Bearing Estimation 346

    8.3.5 Bearing Estimation with Cross-Bispectrum Beamforming 352

    8.3.6 MUSIC-Like Method Based on Cumulants 361

8.4 PARAMETER ESTIMATION OF TRANSIENT SIGNALS 366

    8.4.1 Problem Formulation 366

    8.4.2 The Third-Order Recursion Equation 367

    8.4.3 Parameter Estimation Method 367

    8.4.4 Parameter Estimation with Fourth-Order Moments 370

8.5 DETECTION OF TRANSIENT SIGNALS 371

    8.5.1 Detection of Transient Signals Using the Bispectrum 373

    8.5.2 Signal Detection Using Matched Filter Operations 375

8.6 SUMMARY	377
REFERENCES	378

CHAPTER 9 ADAPTIVE FILTERING WITH HIGHER-ORDER STATISTICS ... 383

9.1 INTRODUCTION	383
9.2 ADAPTIVE FILTERING WITH SECOND-ORDER STATISTICS	384
9.2.1 The Steepest Descent Method	385
9.2.2 The LMS Algorithm	387
9.2.3 The RLS Algorithm	388
9.3 THE LMF AND RLF ALGORITHM	390
9.3.1 The LMF Algorithm	392
9.3.2 The RLF Algorithm	396
9.4 ADAPTIVE IIR ALGORITHMS BASED ON HOS	397
9.4.1 An Adaptive Algorithm for MA Processes Based on Third-Order Cumulants	397
9.4.2 Adaptive Algorithms for ARMA Processes	400
9.5 ADAPTIVE LATTICE LINEAR PREDICTION USING CUMULANTS	402
9.6 ADAPTIVE MA MODEL ESTIMATION VIA NONCAUSAL AR APPROXIMATIONS	406
9.7 ADAPTIVE TIME DELAY ESTIMATION	414
9.8 ADAPTIVE BLIND EQUALIZATION ALGORITHMS	418
9.8.1 The Bussgang Techniques	422
9.8.2 The Polyspectra Algorithms	427
9.9 SUMMARY	441
REFERENCES	443

CHAPTER 10 DETECTION AND CHARACTERIZATION OF NONLINEARITIES IN TIME SERIES ... 447

10.1 GENERAL VOLTERRA SYSTEMS	448
10.2 SECOND-ORDER VOLTERRA FILTERS	448

10.2.1 Identification Assuming Gaussian Input and Access
to Both Input and Output 449

10.2.2 Identification Assuming a Simple Linear Term and White
Unobservable Input 451

10.2.3 Identification Assuming General Non-Gaussian Input:
Three-Wave Coupling Equation 453

10.3 THE IDENTIFICATION OF A PARTICULAR NONLINEAR
TIME SERIES SYSTEM 455

10.4 QUADRATIC PHASE COUPLING 459

10.4.1 Conventional Techniques for Detection and
Quantification of Quadratic Phase Coupling 461

10.4.2 Parametric Methods for the Detection of
Phase Coupling 465

10.4.3 Time-Domain Approach to Quadratic Phase
Coupling Estimation 476

10.5 CUBIC PHASE COUPLING 482

10.6 SUMMARY 483

REFERENCES 484

CHAPTER 11 TIME-FREQUENCY DISTRIBUTIONS BASED ON
HIGHER-ORDER SPECTRA 487

11.1 INTRODUCTION 487

11.2 WIGNER-VILLE DISTRIBUTION (WD) 488

11.3 COHEN'S GENERAL CLASS OF TIME-FREQUENCY
DISTRIBUTIONS 491

11.3.1 Reduced Interference Distributions 492

11.4 WIGNER HIGHER-ORDER MOMENT SPECTRA:
CONTINUOUS CASE 494

11.4.1 Definition 495

11.4.2 Properties 496

11.4.3 General Class of Time-Frequency Higher-Order Spectra-Based
Distributions 501

11.4.4 Reduced Interference Higher-Order Distribution (RIHOD) . 502

11.5 WIGNER POLYSPECTRA: DISCRETE CASE 503

    11.5.1 Discrete Time WHOMS (DT-WHOMS) 504

    11.5.2 Discrete Frequency WHOMS (DF-WHOMS) 504

    11.5.3 Discrete Time and Frequency WHOMS (DTF-WHOMS) ... 505

    11.5.4 Computation of DTF-WHOMS 507

    11.5.5 Relation Between the WHOMS and the DTF-WHOMS
After Sampling 508

11.6 APPLICATIONS OF WIGNER HIGHER-ORDER SPECTRA ... 512

11.7 SUMMARY 520

REFERENCES 520

CHAPTER 12 CURRENT AND FUTURE TRENDS 525

    REFERENCES 526

    INDEX 529

PREFACE

Higher-Order Spectra (HOS) is a rapidly evolving signal analysis area with growing applications in science and engineering. In recent years, much interest by academic, industrial, and government laboratories has focused around the research and development of higher-order spectra techniques. Applications of HOS-based methods span a remarkably broad range that includes telecommunications, sonar, radar, geophysics, image processing, speech processing, biomedicine, oceanography, plasma physics, economic series, and fluid mechanics.

Although considerable work has been done in the area of power spectrum estimation over the past three decades, the information contained in the power spectrum is essentially that which is present in the second-order statistics (e.g., autocorrelation) of a signal; this would suffice for a complete statistical description of Gaussian processes only. Furthermore, second-order statistics are phase-blind and only describe linear mechanisms governing the process. However, there are signal processing situations in practice where signal analysts must look beyond the second-order statistics (power spectrum) of the process, into its higher-order statistics (or spectra), to extract phase information, as well as information due to deviations from Gaussianity and the presence of nonlinearities. Consequently, there are three main reasons for using higher-order spectra, such as the bispectrum and the trispectrum, in signal processing: to extract information due to deviations from Gaussianity, to recover the correct phase character of signals, and to detect and quantify nonlinearities in time series. Under broad signal and noise conditions, higher-order spectra also become high signal-to-noise ratio domains where detection, parameter estimation, and signal reconstruction can be performed.

The objective of this book is to cover the field of higher-order spectra from a nonlinear signal processing perspective and give an integrated account of the methods available in it. The intent is to introduce HOS-based techniques for all aspects of signal processing problems and to provide the reader with an understanding of the mathematical description of the techniques and their utility for signal processing applications. The book aims at an audience consisting of researchers and graduate students, as well as practicing engineers and scientists. This book has grown out of one decade of basic research and teaching in the field of higher-order spectra. Prerequisites for this book include an introductory course in digital signal processing and linear systems analysis, particularly in discrete systems with the use of z or Fourier transforms. We assume that the reader has a first-year graduate level academic background in probability theory and stochastic processes. Also, a course on modern power spectral analysis or correlation-based system identification would provide a helpful background.

Chapter 1 provides an introduction to higher-order spectra with an emphasis on motivations behind their utility in signal processing problems. The chapter briefly describes the developments in power spectrum estimation over the past three decades and points out the need for improvements using higher-order statistics. A higher-order spectra classification map is provided along with a summary of applications.

Chapter 2 presents the definitions and properties of moments, cumulants, and cumulant spectra of stochastic stationary signals. In fact, we start by providing the definitions and properties of moments and cumulants of a set of random variables, and by establishing their relationships. While not intending to be mathematically rigorous on several of the derivations of the properties of higher-order statistics, a hint is usually provided for a proof. A lot of the properties of cumulants and cumulant spectra described in this chapter serve as the basis for the signal processing methods presented in Chapters 4–11.

The definitions and properties of moments and higher-order moment spectra of deterministic signals is the subject of Chapter 3. Both classes of signals, namely energy and power, are treated in this chapter. Several of the examples described in Chapter 2 for stochastic processes are repeated in this chapter for "equivalent" deterministic signals.

Chapter 4 introduces conventional methods for the estimation of moments, cumulants, and higher-order spectra from finite lengths of stochastic or deterministic signals. Estimation techniques include direct, indirect, and complex demodulates approaches along with a description of their asymptotic normality properties. This chapter also describes a test for nonstationarity and FFT-based computation of the bispectrum and trispectrum on polar rasters.

The concept of polycepstra, that is, cepstra of higher-order spectra, is developed in Chapter 5. The definition, properties, and computation of the complex cepstrum, differential cepstrum, power cepstrum, bicepstrum, cepstrum of the bicoherence, and tricepstrum are described along with their interrelationships. Particular emphasis is placed on the description of methods for the computation of cepstra from higher-order statistics without the utility of phase unwrapping algorithms. Applications of polycepstra-based methods to signal recovery, blind system identification, and time delay estimation are discussed in the form of examples.

In Chapter 6, nonparametric methods for signal reconstruction from higher-order spectra are introduced. Reconstruction approaches include recursive and nonrecursive batch processing techniques as well as methods based on FFT algorithms. This chapter extends the discussion to signal recovery problems from only the phase of higher-order spectra and to blind deconvolution from multisensor observed signals.

In Chapter 7, we describe parametric methods for the estimation of higher-order spectra based on moving average (MA), noncausal autoregressive (AR), and ARMA models. Particular emphasis is placed on nonminimum phase system identification from output measurements only. Model order selection criteria based on higher-order statistics are also described.

The problems of direction of arrival estimation and analysis of transient signals with the utility of higher-order statistics are treated in Chapter 8. In particular, nonparametric and parametric time delay estimation methods are described along with high-resolution array signal processing techniques for estimating the number of sources and their bearings relative to an array configuration. A new bearing estimation method, which is based on the asymptotic normality properties of cross-bispectrum estimates and employs maximum likelihood theory, is also covered. The detection and parameter estimation of transient signals in noise using third- and fourth-order statistics is the last topic covered in this chapter.

Chapter 9 presents an overview of adaptive signal processing methods based on higher-order statistics. In this context, the LMF and the RLF algorithms are described, as well as techniques for adaptive system identification and time delay estimation. Adaptive systems based on both direct and lattice realization structures are considered. The last part of Chapter 9 focuses on blind equalization algorithms such as Bussgang techniques, polycepstra methods, and parametric approaches.

The detection and characterization of nonlinearities in time series with the utility of higher-order spectra is provided in Chapter 10. An introduction of Volterra systems leads to a discussion of quadratic filter identification techniques and their applications. Conventional and high-resolution parametric AR methods for the detection and characterization of quadratic phase coupling are also described. The more complicated cubic phase coupling phenomenon is briefly covered.

Finally, Chapter 11 describes time-frequency distributions based on higher-order spectra. An introduction to the Wigner-Ville distribution and Cohen's general class of time-frequency distributions leads to a description of Wigner Polyspectra for both the continuous and the discrete cases. An application of Wigner Polyspectra to the adaptive detection of transient signals in noise is also presented.

Chapter 12 brings the book to a close by providing an overview of current developments and future trends in terms of research and applications of higher-order spectra. Given the rapid expansion of the field, there is no doubt that some recently developed important new methods have not been described in the book. In particular the book does not cover higher-order spectra of multidimensional and multichannel processes and higher-order spectra of cyclo-stationary signals. It is our intent to include this material in a second edition of the book.

Many of the references cited at the end of each chapter are available in the short course manual by *C. L. Nikias and J. M. Mendel, Signal Processing with Higher-order Spectra, United Signal and Systems, (USS) Inc., 1990.* This volume provides a comprehensive bibliography on higher-order statistics, which contains more than 500 references. Many of the figures in the book were generated using *Hi-Spec*, USS's computer software on signal processing with HOS. We are therefore grateful to United Signals and Systems, Inc.

The authors appreciate the research support of the U.S. Government, in general, and in particular, of the Office of Naval Research which funded part of the research on which this book is partially based. We also wish to express our appreciation to those who generously read all or part of the book and have offered corrections and informative suggestions: Dana H. Brooks, Leon Cohen, Ted Dean, S. Dianat, Javier A. Fonollosa, Dimitrios Hatzinakos, Mysore R. Raghuveer, Andreas Spanias, and A. Swami. We acknowledge the many fruitful technical discussions related to topics of this book with Giorgios Giannakis, Melvin Hinich, Jerry M. Mendel, and John G. Proakis. Special thanks to Anna Patch and Gloria Proakis who provided much help in typing several chapters of the book. Finally, but most importantly, we are grateful to Mary and Nancy Toscano for their excellent job in typing and preparing the camera ready manuscript.

Chrysostomos L. Nikias
Rancho Palos Verdes, CA

Athina P. Petropulu
Philadelphia, PA

LIST OF SYMBOLS

SYMBOL	USE(S) IN THIS BOOK
$\phi(\omega_1,\ldots,\omega_n)$	joint characteristic function
$\tilde{\Psi}(\omega_1,\ldots,\omega_n)$	joint second characteristic function
m_n	moment of order n
c_n	cumulant of order n
$M_{om}[x_1,\ldots,x_n]$	joint moments
$c_{um}[x_1,\ldots,x_n]$	joint cumulants
$m_n^x(\tau_1,\ldots,\tau_{n-1})$	nth-order moment function
$c_n^x(\tau_1,\ldots,\tau_{n-1})$	nth-order cumulant function
γ_2^x	variance
γ_3^x	skewness
γ_4^x	kurtosis
$c_{xyz}(\tau_1,\tau_2)$ or $c_3^{xyz}(\tau_1,\tau_2)$	cross-cumulant function
$m_{xyz}(\tau_1,\tau_2)$ or $m_3^{xyz}(\tau_1,\tau_2)$	cross-moment function
$M_n^x(\omega_1,\ldots,\omega_n)$	nth-order moment spectrum
$C_n^x(\omega_1,\ldots,\omega_n)$	nth-order cumulant spectrum
$P_n^x(\omega_1,\ldots,\omega_n)$	nth-order coherency index
$C_{xyz}(\omega_1,\omega_2)$ or $C_3^{xyz}(\omega_1,\omega_2)$	cross-cumulant spectrum
$M_{xyz}(\omega_1,\omega_2)$ or $M_3^{xyz}(\omega_1,\omega_2)$	cross-moment spectrum
$H(\omega)$	Fourier transform
$|H(\omega)|$	Fourier magnitude
$\phi_h(\omega)$	Fourier phase
$H(z)$	Z transform
$|C_n^x(\omega_1,\ldots,\omega_n)|$	n-th order magnitude cumulant spectrum
$\Psi_n^x(\omega_1,\ldots,\omega_n)$	nth-order phase cumulant spectrum
E_x	total energy
P_x	average power
$\tilde{x}(k)$	periodic signal
$r_n^{\tilde{x}}(\tau_1,\ldots,\tau_{n-1})$	nth-order moment function of a periodic signal

$R_n^{\tilde{x}}(\lambda_1, \ldots, \lambda_{n-1})$	nth-order moment spectrum of a periodic signal
$R_{\tilde{x}_1 \cdots \tilde{x}_n}(\lambda_1, \ldots, \lambda_{n-1})$	nth-order cross-moment spectrum of a periodic signal
$w(u_1, \ldots, u_n)$	multidimensional window function
$c_h(m)$	complex cepstrum
$A^{(m)}, B^{(m)}$	cepstral coefficients or parameters
$d_h(m)$	differential cepstrum
$p_h(m)$	power cepstrum
$b_h(m, n)$	bicepstrum
$t_h(m, n, \ell)$	tricepstrum
$bic_h(m, n)$	cepstrum of bicoherency
$cb_{123}(m, n)$	cross-bicepstrum
$|C_3^x(\omega_1, \omega_2)|$	magnitude bispectrum
$\Psi_3^x(\omega_1, \omega_2)$	phase bispectrum
$\Psi_3^{xyz}(\omega_1, \omega_2)$	cross-bispectrum phase
$W_{1x}(t, f)$	Wigner distribution
$P_{GEN}(t, f)$	Cohen's general class of distribution
$P_{cw}(t, f)$	Choi-Williams distribution
$W_{nx}(t, f_1, \ldots, f_n)$	n-th order Wigner Higher-Order Moment Spectrum (WHOM)
$\text{RIHOD}_{nx}(t, f_1, \ldots, f_n)$	Reduced Interference Higher-Order Distribution

1

INTRODUCTION

For more than twenty-five years, significant contributions have been made in digital signal processing by combining ideas and methodologies from systems theory, statistics, numerical analysis, operations research, computer science, and integrated circuit technology. The objective in a signal processing problem is to process a finite number of data samples and extract important information which may be "hidden" in the data. This objective is usually achieved by combining the development of mathematical formulations with their algorithmic implementations (either in software or hardware) and their applications to real data. Various conflicting figures of merit are associated with digital signal processing techniques – namely, quality of the estimates, computational complexity, data throughput rate, cost of implementation, finite word-length effects, and structural properties.

1.1 POWER SPECTRUM

One frequently used digital signal processing technique has been the estimation of the power spectral density (PSD) or simply the power spectrum of discrete-time deterministic or stochastic processes. The past twenty years witnessed an expansion of new power spectrum estimation techniques which have proved essential to the creation of advanced communication, radar, sonar, speech, biomedical, geophysical, and imaging systems. The available power spectrum estimation techniques may be considered in a number of different categories; namely, conventional (or "Fourier type") methods, maximum-likelihood method of Capon with its modification, maximum-entropy and minimum cross-entropy methods, minimum energy methods based on autoregressive (AR), moving average (MA), and ARMA models, and harmonic decomposition methods such as Prony, Pisarenko, MUSIC, ESPRIT, and Singular Value Decomposition. Research in this area has also led to signal modeling, and to extensions to multidimensional, multichannel, and array processing problems. Each one of the aforementioned techniques has certain advantages and limitations not only in terms of estimation performance but also in terms of computational complexity. Therefore, depending on the signal environment, one has to choose the most appropriate. Today there are several textbooks on topics related to power spectrum estimation and its applications. See for example Marple [1987], Gardner [1988], Kay [1988], Childers [1978], Kesler [1986], Haykin [1983], and references therein.

In power spectrum estimation, the process under consideration is treated as a superposition of statistically uncorrelated harmonic components. The distribution of power among these frequency components is then estimated. As such, phase relations between frequency components are suppressed [Huber et al., 1971]. The information contained in the power spectrum is essentially present in the autocorrelation sequence; this would suffice for the complete statistical description of a Gaussian process of known mean. However, there are practical situations where we would have to look beyond the power spectrum or autocorrelation domain to obtain information regarding deviations from Gaussianess and presence of nonlinearities. Higher-order (greater than 2) spectra, also known as polyspectra defined in terms of higher-order statistics, do contain such information. Particular examples of higher-order spectra are the third-order spectrum, called the **bispectrum**, which is defined to be the Fourier transform of the third-order cumulant sequence, and the **trispectrum** (fourth-order spectrum), which is the Fourier transform of the fourth-order cumulant sequence of a stationary random process. Let us note that the **power spectrum** is, in fact, a member of the class of higher-order spectra; that is, it is a second-order spectrum [Rosenblatt, 1985], [Rao and Gabr, 1984], [Priestley, 1981].

1.2   WHY POLYSPECTRA IN SIGNAL PROCESSING?

Polyspectra consist of higher-order moment spectra and cumulant spectra and can be defined for both deterministic signals and random processes. As illustrated in this book, moment spectra can be very useful in the analysis of deterministic signals (transient and periodic) whereas cumulant spectra are of great importance in the analysis of stochastic signals. Figure 1.1 illustrates the two categories of polyspectra and the types of signals that can benefit from their use.

In general, there are three motivations behind the use of polyspectral analysis in signal processing [Nikias and Raghuveer, 1987]. These are: (1) to suppress Gaussian noise processes of unknown spectral characteristics in detection, parameter estimation and classification problems; the bispectrum also suppresses non-Gaussian noise with symmetric probability density function (pdf), (2) to reconstruct the phase and magnitude response of signals or systems, and (3) to detect and characterize nonlinearities in time series.

The first motivation is based on the property that for Gaussian processes only, all cumulant spectra of order greater than two are identically zero. If a non-Gaussian signal is received along with additive Gaussian noise, a transform to a higher-order cumulant domain will (in theory) eliminate the noise. Hence, in these signal processing settings, there will be certain advantages to detecting and/or estimating signal parameters from cumulant spectra of the observed data. In particular, we will show that cumulant spectra can become high signal-to-noise ratio (SNR) domains in which one may perform detection, parameter estimation, or even entire signal reconstruction. In chapters 5 to 11, several methods of parameter estimation in cumulant spectrum domains are presented, which may prove to be useful in practice.

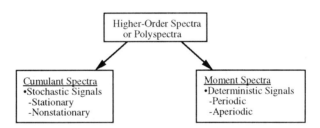

Figure 1.1 The polyspectra classification map

The second motivation is based on the fact that polyspectra (cumulant and moment) preserve the true phase character of signals. For modeling time series data in signal processing problems, second-order statistics are almost exclusively used because they are usually the result of least-squares optimization criteria. However, the autocorrelation domain suppresses phase information. An accurate phase reconstruction in the autocorrelation (or power spectrum) domain can only be achieved if the signal is minimum phase. On the other hand, nonminimum phase signal reconstruction or system identification can be achieved in higher-order spectrum domains due to the ability of polyspectra to preserve both magnitude and nonminimum phase information. Chapters 5 and 6 present non-parametric methods for nonminimum phase system identification based on polyspectra.

Finally, introduction of higher-order spectra is quite natural when we try to analyze the nonlinearity of a system operating under a random input. General relations for arbitrary stationary random data passing through arbitrary linear systems have been studied quite extensively for many years. In principle, most of these relations are based on power spectrum (or autocorrelation) matching criteria. On the other hand, general relations are not available for arbitrary stationary random data passing through arbitrary nonlinear systems. Instead, each type of nonlinearity has to be investigated as a special case [Schetzen, 1989]. Polyspectra can play a key role in detecting and characterizing the type of nonlinearity in a system from its output data. Chapter 10 is devoted to the discussion of signal processing methods for the detection and characterization of nonlinearities in time series using higher-order spectra.

1.3 APPLICATIONS

There are several papers that have been published over the past thirty years dealing with the applications of polyspectra in oceanography, geophysics, sonar, communications, biomedicine, speech processing, radioastronomy, image processing, fluid mechanics, economic time series, plasma physics, sunspot data, and so on. Procedures have been developed based on polyspectra for deconvolution (or equalization) and signal detection, for the identification of nonlinear, nonminimum phase, and spike-array type processes, for parameter estimation, and detection of quadratic phase coupling, and for detection of aliasing in discrete-time stochastic signals. A useful bibliography on higher-order spectra that covers the period 1950–1980 was provided by Tryon [1981]; see also the tutorial paper by Nikias and Raghuveer [1987] and references therein, as well as the review article by Mendel [1991]. A comprehensive bibliography on higher-order statistics (spectra) has been published by United Signals and Systems, Inc. [1992].

REFERENCES

Childers, D. G., **Modern Spectrum Analysis**, New York: IEEE Press, 1978.

Gardner, W. B., **Statistical Spectral Analysis**, Englewood Cliffs, NJ: Prentice Hall, 1988.

Haykin, S., **Nonlinear Methods of Spectral Analysis**, 2nd edition, Berlin, Germany: Springer-Verlag, 1983.

Huber, P. J., B. Kleiner, T. Gasser, and G. Dumermuth, "Statistical Methods for Investigating Phase Relations in Stationary Stochastic Processes," *IEEE Transactions Audio Electroacoustics*, **AU-19**, pp. 78-86, 1971.

Kay, S. M., **Modern Spectral Estimation**, Englewood Cliffs, NJ: Prentice Hall, 1988.

Kesler, S. B., **Modern Spectrum Analysis, II**, New York: IEEE Press, 1986.

Marple, S. L., Jr., **Digital Spectral Analysis with Applications**, Englewood Cliffs, NJ: Prentice Hall, 1987.

Mendel, J. M., "Tutorial on Higher-Order Statistics (Spectra) in Signal Processing and System Theory: Theoretical Results and Some Applications," *Proceedings IEEE*, **79**(3), pp. 278-305, March, 1991.

Nikias, C. L., and M. R. Raghuveer, "Bispectrum Estimation: A Digital Signal Processing Framework," *Proceedings IEEE*, **75**(7), pp. 869-891, July 1987.

Priestley, M. B., **Spectral Analysis and Time Series**, London: Academic, 1981.

Rao, T. S. and M. M. Gabr, "An Introduction to Bispectral Analysis and Bilinear Time Series," Lecture notes in **Statistics 24**, Berlin, Heidelberg: Springer-Verlag, 1984.

Rosenblatt, M., **Stationary Sequences and Random Fields**, Birkhauser, Boston: Birkhauser 1985.

Schetzen, M., **The Volterra and Wiener Theories on Nonlinear Systems**, updated edition, Malabar, FL: Krieger's, 1989.

Tryon, P. V., "The Bispectrum and Higher-Order Spectra: A Bibliography," *NBS Technical Note 1036*, 1981.

United Signals & Systems, Inc., **Comprehensive Bibliography on Higher-Order Statistics (Spectra)**, 1992.

2

CUMULANT SPECTRA OF STOCHASTIC SIGNALS

2.1 INTRODUCTION

In this chapter we are concerned primarily with the definitions and properties of moments, and cumulant spectra of stationary random processes. Although the value of stochastic signals at each instant of time is not known exactly, their higher-order statistics (moments and cumulants), when they exist, are multidimensional deterministic functions that possess special symmetry properties. We start by defining moments and a set of random variables, and by establishing their relationships. This is followed by the definition and properties of moments, and cumulant spectra of stationary random processes. Cumulant spectra of linear, non-Gaussian processes are then discussed, as well as their similarities and differences with cumulant spectra of nonlinear processes. Our primary goal in this chapter is to introduce all the important definitions and properties associated with polyspectra that can be found useful in applications of stochastic signal processing methods.

2.2 MOMENTS AND CUMULANTS

2.2.1 Definitions

Given a set of n real random variables $\{x_1, x_2, \ldots, x_n\}$, their joint **moments** of order $r = k_1 + k_2 + \ldots + k_n$ are given by [Papoulis, 1984]

$$\text{Mom } [x_1^{k_1}, x_2^{k_2}, \ldots, x_n^{k_n}] \triangleq E\{x_1^{k_1} x_2^{k_2} \ldots x_n^{k_n}\} =$$
$$= (-j)^r \left. \frac{\partial^r \Phi(\omega_1, \omega_2, \ldots, \omega_n)}{\partial \omega_1^{k_1} \partial \omega_2^{k_2} \ldots \partial \omega_n^{k_n}} \right|_{\omega_1 = \omega_2 = \cdots = \omega_n = 0} \quad (2.1)$$

where
$$\Phi(\omega_1, \omega_2, \ldots, \omega_n) \triangleq E\{\exp(j(\omega_1 x_1 + \omega_2 x_2 + \cdots + \omega_n x_n))\}$$
is their joint characteristic function. $E\{\cdot\}$ denotes the expectation operation. For example, for two random variables $\{x_1, x_2\}$, we have the second-order moments $\text{Mom}[x_1, x_2] = E\{x_1 \cdot x_2\}$, Mom $[x_1^2] = E\{x_1^2\}$ and $\text{Mom}[x_2^2] = E\{x_2^2\}$.

Another form of the joint characteristic function is defined as the natural logarithm [Papoulis, 1984] of $\Phi(\omega_1, \omega_2, \ldots, \omega_n)$; i.e.,

$$\tilde{\Psi}(\omega_1, \omega_2, \ldots, \omega_n) \triangleq \ln[\Phi(\omega_1, \omega_2, \ldots, \omega_n)]. \tag{2.2}$$

The joint **cumulants** (also called **semi-invariants**) of order r, $\text{Cum}[x_1^{k_1}, x_2^{k_2}, \ldots, x_n^{k_n}]$, of the same set of random variables, are defined as the coefficients in the Taylor expansion of the second characteristic function about zero [Shiryaev, 1960; 1963; Brillinger, 1965; Rosenblatt, 1983; 1985]; i.e.,

$$\text{Cum}[x_1^{k_1}, x_2^{k_2}, \ldots, x_n^{k_n}] \triangleq (-j)^r \frac{\partial^r \tilde{\Psi}(\omega_1, \omega_2, \ldots, \omega_n)}{\partial \omega_1^{k_1} \partial \omega_2^{k_2} \ldots \partial \omega_n^{k_n}}\bigg|_{\omega_1 = \omega_2 = \ldots = \omega_n = 0}. \tag{2.3}$$

Thus, the joint can be expressed in terms of the joint moments of a set of random variables. For example, the moments

$$\begin{aligned} m_1 &= \text{Mom}[x_1] = E\{x_1\} & m_2 &= \text{Mom}[x_1, x_1] = E\{x_1^2\} \\ m_3 &= \text{Mom}[x_1, x_1, x_1] = E\{x_1^3\} & m_4 &= \text{Mom}[x_1, x_1, x_1, x_1] = E\{x_1^4\} \end{aligned}$$

of the random variable $\{x_1\}$ are related to its cumulants by

$$\begin{aligned} c_1 &= \text{Cum}[x_1] = m_1 & c_2 &= \text{Cum}[x_1, x_1] = m_2 - m_1^2 \\ c_3 &= \text{Cum}[x_1, x_1, x_1] = m_3 - 3m_2 m_1 + 2m_1^3 & & \\ c_4 &= \text{Cum}[x_1, x_1, x_1, x_1] = m_4 - 4m_3 m_1 - 3m_2^2 + 12m_2 m_1^2 - 6m_1^4. \end{aligned} \tag{2.4}$$

These relationships can be verified by substituting

$$\Phi(\omega_1) = 1 + j\omega_1 m_1 - \frac{\omega_1^2}{2!} m_2 \cdots + \frac{(j\omega_1)^k}{k!} m_k + \cdots$$

into (2.1), (2.2), (2.3) and working out differentiations about zero. If $E\{x_1\} = m_1 = 0$, it follows that $c_2 = m_2$, $c_3 = m_3$, and $c_4 = m_4 - 3m_2^2$.

Example 2.1

Consider the three symmetric probability density functions (*pdfs*) shown in Figure

2.1; namely, Laplace, Gaussian, and Uniform. Their moments m_n of order $n = 1, 2, 3, 4$ can be generated from [Papoulis, 1984]

$$m_n = \int_{-\infty}^{+\infty} x^n f(x) \, dx$$

where $f(x)$ is the probability density function (*pdf*). From (2.1), we compute the characteristic function

$$\Phi(\omega) = \int_{-\infty}^{+\infty} \exp(j\omega x) f(x) \, dx.$$

The cumulants c_n, $n = 1, 2, 3, 4$ of the *pdf*s follow easily from the moments in (2.4). Figure 2.1 also illustrates the moments and cumulants of the *pdf*s from order first to fourth. Let us note that for the symmetric *pdf*s all m_n and c_n for n odd are identical to zero and that for the Gaussian case all cumulants c_n of order greater than second ($n > 2$) are also zero. □

Example 2.2

Figure 2.2 illustrates three nonsymmetric *pdf*s; i.e., Exponential, Rayleigh, and K-distribution [Watts, 1985], as well as their moments and for orders $n = 1, 2, 3, 4$. □

2.2.2 Relationship Between Moments and Cumulants

The general relationship between moments of $\{x_1, x_2, \ldots, x_n\}$ and joint cumulants $\text{Cum}[x_1, x_2, \ldots, x_n]$ of order $r = n$ is given by [Leonov and Shiryaev, 1959; Brillinger, 1965; Rosenblatt, 1985]

$$\text{Cum}[x_1, x_2, \ldots, x_n] = \sum (-1)^{p-1}(p-1)! \cdot E\{\prod_{i \in s_1} x_i\} \cdot E\{\prod_{i \in s_2} x_i\} \cdots E\{\prod_{i \in s_p} x_i\} \quad (2.5)$$

where the summation extends over all partitions $(s_1, s_2, \ldots, s_p)$, $p = 1, 2, \ldots, n$, of the set of integers $(1, 2, \ldots, n)$. For example, the set of integers $(1, 2, 3)$ can be partitioned into

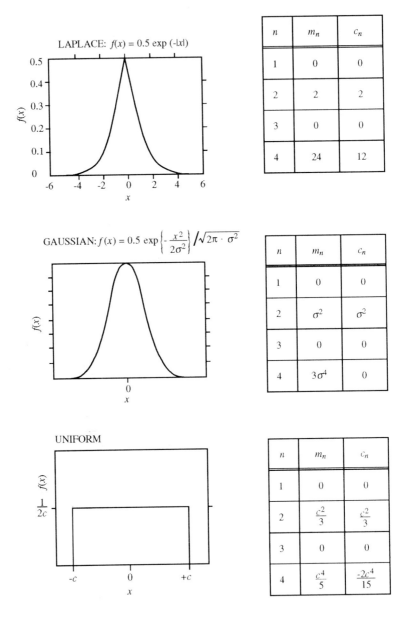

Figure 2.1 The nth-order moments and cumulants for $n = 1, 2, 3, 4$ of the Laplace, Gaussian, and Uniform Probability Density Functions (pdfs)

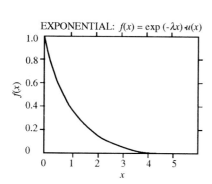
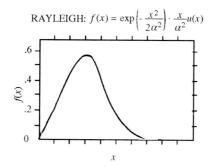
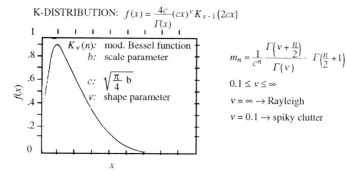

Figure 2.2 The nth-order moments and cumulants for $n = 1, 2, 3, 4$ of Exponential, Rayleigh, and K-distribution pdfs

$$\begin{aligned}
p=1 \quad & s_1 = \{1,2,3\} \\
p=2 \quad & s_1 = \{1\}, \quad & s_2 = \{2,3\} \\
& s_1 = \{2\}, \quad & s_2 = \{1,3\} \\
& s_1 = \{3\}, \quad & s_2 = \{1,2\} \\
p=3 \quad & s_1 = \{1\}, \quad & s_2 = \{2\}, \quad & s_3 = \{3\}
\end{aligned}$$

and therefore (2.5) becomes

$$\begin{aligned}
\operatorname{Cum}[x_1, x_2, x_3] = & E\{x_1 x_2 x_3\} - E\{x_1\} \cdot E\{x_2 x_3\} - E\{x_2\} \cdot E\{x_1 x_3\} \\
& - E\{x_3\} \cdot E\{x_1 x_2\} + 2E\{x_1\} \cdot E\{x_2\} \cdot E\{x_3\}.
\end{aligned} \quad (2.6)$$

Clearly, (2.6) is identical to c_3 of (2.4) for $x_1 = x_2 = x_3$. On the other hand, if we are given $\{x_1, x_2, x_3, x_4\}$, then all possible partitions of the set of integers $(1,2,3,4)$ is given in Table 2.1. As such, (2.5) takes the form shown in (2.7):

$$\begin{aligned}
\operatorname{Cum}[x_1, x_2, x_3, x_4] = & E\{x_1 x_2 x_3 x_4\} - E\{x_1 x_2\} \cdot E\{x_3 x_4\} \\
& - E\{x_1 x_3\} \cdot E\{x_2 x_4\} - E\{x_1 x_4\} \cdot E\{x_2 x_3\} \\
& - E\{x_1\} \cdot E\{x_2 x_3 x_4\} - E\{x_2\} \cdot E\{x_1 x_3 x_4\} \\
& - E\{x_3\} \cdot E\{x_1 x_2 x_4\} - E\{x_4\} \cdot E\{x_1 x_2 x_3\} \\
& + 2E\{x_1 x_2\} \cdot E\{x_3\} \cdot E\{x_4\} \\
& + 2E\{x_1 x_3\} \cdot E\{x_2\} \cdot E\{x_4\} + 2E\{x_1 x_4\} \cdot E\{x_2\} \cdot E\{x_3\} \\
& + 2E\{x_2 x_4\} \cdot E\{x_1\} \cdot E\{x_3\} + 2E\{x_3 x_4\} \cdot E\{x_1\} \cdot E\{x_2\} \\
& + 2E\{x_2 x_3\} \cdot E\{x_1\} \cdot E\{x_4\} \\
& - 6E\{x_1\} \cdot E\{x_2\} \cdot E\{x_3\} \cdot E\{x_4\}.
\end{aligned} \quad (2.7)$$

Two important observations can be made from (2.7). First, (2.7) becomes identical to c_4 of (2.4) if we assume $x_1 = x_2 = x_3 = x_4$. Second, if the random variables have zero-mean (i.e., $E\{x_i\} = 0$, $i = 1, 2, 3, 4$), then (2.7) turns out to be the well-known expression

$$\begin{aligned}
\operatorname{Cum}[x_1, x_2, x_3, x_4] = & E\{x_1 x_2 x_3 x_4\} - E\{x_1 x_2\} \cdot E\{x_3 x_4\} \\
& - E\{x_1 x_3\} \cdot E\{x_2 x_4\} - E\{x_1 x_4\} \cdot E\{x_2 x_3\}
\end{aligned} \quad (2.8)$$

The relationship (2.5) implies that **the computation of joint of order r requires knowledge of all moments up to order r.**

2.2.3 Properties of Moments and Cumulants

The properties of moments and cumulants may be summarized as follows [Shiryaev, 1960; 1963; Sinai, 1963; Brillinger, 1965; Rosenblatt, 1983; 1985; Brillinger and Rosenblatt, 1967; Rao and Gabr, 1984]:

TABLE 2.1 ALL POSSIBLE PARTITIONS OF SET $\{1,2,3,4\}$ NEEDED FOR COMPUTATION OF $\text{Cum}[x_1, x_2, x_3, x_4]$.

p \ Partitions	s_1	s_2	s_3	s_4
1	1,2,3,4	ϕ	ϕ	ϕ
	1,2	3,4	ϕ	ϕ
	1,3	2,4	ϕ	ϕ
	1,4	2,3	ϕ	ϕ
2	1	2,3,4	ϕ	ϕ
	2	1,3,4	ϕ	ϕ
	3	1,2,4	ϕ	ϕ
	4	1,2,3	ϕ	ϕ
	1,2	3	4	ϕ
	1,3	2	4	ϕ
	1,4	2	3	ϕ
3	2,4	1	3	ϕ
	3,4	1	2	ϕ
	2,3	1	4	ϕ
4	1	2	3	4

1. $\text{Mom}[a_1 x_1, a_2 x_2, \ldots, a_n x_n] = a_1 \ldots a_n \text{Mom}[x_1, \ldots, x_n]$ and
   $\text{Cum}[a_1 x_1, a_2 x_2, \ldots, a_n x_n] = a_1 \ldots a_n \text{Cum}[x_1, \ldots, x_n]$
   where $(a_1, a_2, \ldots, a_n)$ are constants. This follows directly from (2.1) and (2.5).

2. Moments and cumulants are symmetric functions in their arguments, e.g.,
   $\text{Mom}[x_1, x_2, x_3] = \text{Mom}[x_2, x_1, x_3] = \text{Mom}[x_3, x_2, x_1]$, and so on.

3. If the random variables $\{x_1, x_2, \ldots, x_n\}$ can be divided into any two or more groups which are statistically independent, their nth-order cumulant is identical to zero; i.e., $\text{Cum}[x_1, x_2, \ldots, x_n] = 0$ whereas, in general, $\text{Mom}[x_1, x_2, \ldots, x_n] \neq 0$. For example, if the two independent groups are $\{x_1, x_2 \ldots, x_\lambda\}$ and $\{x_{\lambda+1}, \ldots, x_n\}$, then their joint characteristic function is $\Phi(\omega_1, \omega_2, \ldots, \omega_n) = \Phi_1(\omega_1, \ldots, \omega_\lambda) \cdot \Phi_2(\omega_{\lambda+1}, \ldots, \omega_n)$. On the other hand, their joint second characteristic function is $\tilde{\Psi}(\omega_1, \omega_2, \ldots, \omega_n) = \tilde{\Psi}_1(\omega_1, \ldots, \omega_\lambda) + \tilde{\Psi}_2(\omega_{\lambda+1}, \ldots, \omega_n)$. The proof of this property easily follows if we substitute $\tilde{\Psi}(\omega_1, \ldots, \omega_n)$ and $\Phi(\omega_1, \ldots, \omega_n)$ into (2.3) and (2.1), respectively.

4. If the sets of random variables $\{x_1, x_2, \ldots, x_n\}$ and $\{y_1, y_2, \ldots, y_n\}$ are independent, then

$$\text{Cum}[x_1 + y_1, x_2 + y_2, \ldots, x_n + y_n] = \text{Cum}[x_1, \ldots, x_n] + \text{Cum}[y_1, \ldots, y_n]$$

whereas in general

$$\text{Mom}[x_1 + y_1, \ldots, x_n + y_n] = E\{(x_1 + y_1)(x_2 + y_2) \cdots (x_n + y_n)\}$$
$$\neq \text{Mom}[x_1, \ldots, x_n] + \text{Mom}[y_1, \ldots, y_n].$$

However, for the random variables $\{y_1, x_1, x_2, \ldots, x_n\}$ we have that

$$\text{Cum}[x_1 + y_1, x_2, \ldots, x_n] = \text{Cum}[x_1, x_2, \ldots, x_n] + \text{Cum}[y_1, x_2, \ldots, x_n]$$

and

$$\text{Mom}[x_1 + y_1, x_2, \ldots, x_n] = \text{Mom}[x_1, x_2, \ldots, x_n] + \text{Mom}[y_1, x_2, \ldots, x_n].$$

5. If the set of random variables $\{x_1, \ldots, x_n\}$ is jointly Gaussian, then all the information about their distribution is contained in the moments of order $n \leq 2$. Therefore, all moments of order greater than two $(n > 2)$ have no new information to provide. This leads to the fact that all joint cumulants of order $n > 2$ are identical to zero for Gaussian random vectors. Hence, the cumulants of order greater than two, in some sense, measure the non-Gaussian nature (or nonnormality) of a time series.

Example 2.3

Consider the random variables

$$z_i = y_i + x_i, \quad i = 1, 2, 3$$

where the joint probability density function of $\{y_1, y_2, y_3\}$ is non-Gaussian and $\{x_1, x_2, x_3\}$ is jointly Gaussian and independent from $\{y_1, y_2, y_3\}$. Let us also assume that $E\{y_i\} \neq 0$, $E\{x_i\} \neq 0$ for $i = 1, 2, 3$. From properties (4), (5) of cumulants, it follows that

$$\text{Cum}[z_1, z_2, z_3] = \text{Cum}[y_1, y_2, y_3]$$

because $\text{Cum}[x_1, x_2, x_3] = 0$. On the other hand,

$$\begin{aligned}
\text{Mom}[z_1, z_2, z_3] &= E\{z_1 \cdot z_2 \cdot z_3\} = E\{(y_1 + x_1)(y_2 + x_2)(y_3 + x_3)\} \\
&= \text{Mom}[y_1, y_2, y_3] + \text{Mom}[x_1, x_2, x_3] \\
&+ \text{Mom}[y_1, x_2, y_3] + \text{Mom}[x_1, y_2, y_3] + \text{Mom}[x_1, x_2, y_3] \\
&+ \text{Mom}[y_1, y_2, x_3] + \text{Mom}[y_1, x_2, x_3] + \text{Mom}[x_1, y_2, x_3].
\end{aligned}$$

We see that if $E\{y_i\} = E\{x_i\} = 0$ for all i, then

$$\text{Mom}[z_1, z_2, z_3] = \text{Mom}[y_1, y_2, y_3] + \text{Mom}[x_1, x_2, x_3].$$

This simple example demonstrates one of the key motivations behind the use of cumulants in signal processing problems; namely, their ability to suppress noise (x_i) when it is additive Gaussian. □

2.2.4 Moments and Cumulants of Stationary Processes

If $\{X(k)\}$, $k = 0, \pm 1, \pm 2, \pm 3, \ldots$ is a real stationary random process and its moments up to order n exist, then

$$\text{Mom}[X(k), X(k+\tau_1), \ldots, X(k+\tau_{n-1})] = E\{X(k) \cdot X(k+\tau_1) \cdots X(k+\tau_{n-1})\}$$

will depend only on the time differences $\tau_1, \tau_2, \ldots, \tau_{n-1}$, $\tau_i = 0, \pm 1, \pm 2, \ldots$ for all i. We now write the moments of a stationary random process as:

$$m_n^x(\tau_1, \tau_2, \ldots, \tau_{n-1}) \triangleq E\{X(k) \cdot X(k+\tau_1) \cdots X(k+\tau_{n-1})\}. \quad (2.9)$$

Similarly, the nth-order cumulants of $\{X(k)\}$ are $(n-1)$-dimensional functions which we now write in the form:

$$c_n^x(\tau_1, \tau_2, \ldots, \tau_{n-1}) \triangleq \text{Cum}[X(k), X(k+\tau_1), \ldots, X(k+\tau_{n-1})]. \quad (2.10)$$

Combining (2.5), (2.9), and (2.10), we obtain the following relationships between moment and cumulant sequences of $X(k)$:

1st-order cumulants:

$$c_1^x = m_1^x = E\{X(k)\} \quad \text{(mean value)} \quad (2.11)$$

2nd-order cumulants:

$$\begin{aligned} c_2^x(\tau_1) &= m_2^x(\tau_1) - (m_1^x)^2 \quad \text{(covariance sequence)} \\ &= m_2^x(-\tau_1) - (m_1^x)^2 = c_2^x(-\tau_1) \end{aligned} \quad (2.12)$$

where $m_2^x(\tau_1)$ is the autocorrelation sequence.

3rd-order cumulants:

$$c_3^x(\tau_1, \tau_2) = m_3^x(\tau_1, \tau_2) - m_1^x[m_2^x(\tau_1) + m_2^x(\tau_2) + m_2^x(\tau_2 - \tau_1)] + 2(m_1^x)^3 \quad (2.13)$$

where $m_3^x(\tau_1, \tau_2)$ is the 3rd-order moment sequence. This follows if we combine (2.6) and (2.10).

4th-order cumulants: Combining (2.7) and (2.10), we get

$$\begin{aligned}
c_4^x(\tau_1, \tau_2, \tau_3) &= m_4^x(\tau_1, \tau_2, \tau_3) - m_2^x(\tau_1) \cdot m_2^x(\tau_3 - \tau_2) - m_2^x(\tau_2) \cdot m_2^x(\tau_3 - \tau_1) \\
&- m_2^x(\tau_3) \cdot m_2^x(\tau_2 - \tau_1) - m_1^x[m_3^x(\tau_2 - \tau_1, \tau_3 - \tau_1) \\
&+ m_3^x(\tau_2, \tau_3) + m_3^x(\tau_2, \tau_4) + m_3^x(\tau_1, \tau_2)] \\
&+ (m_1^x)^2[m_2^x(\tau_1) + m_2^x(\tau_2) + m_2^x(\tau_3) + m_2^x(\tau_3 - \tau_1) + m_2^x(\tau_3 - \tau_2) \\
&+ m_2^x(\tau_2 - \tau_1)] - 6(m_1^x)^4.
\end{aligned}$$

(2.14)

If the process $\{X(k)\}$ is zero-mean ($m_1^x = 0$), it follows from (2.12) and (2.13) that the second- and third-order cumulants are identical to the second- and third-order moments, respectively. However, to generate the fourth-order, we need knowledge of the fourth-order and second-order moments in (2.14).

The nth-order cumulant function of a non-Gaussian stationary random process $X(k)$ can be written as (for $n = 3, 4$ only):

$$c_n^x(\tau_1, \tau_2, \ldots, \tau_{n-1}) = m_n^x(\tau_1, \tau_2, \ldots, \tau_{n-1}) - m_n^G(\tau_1, \tau_2, \ldots, \tau_{n-1})$$

where $m_n^x(\tau_1, \ldots, \tau_{n-1})$ is the nth-order moment function of $X(k)$ and $m_n^G(\tau_1, \tau_2, \ldots, \tau_{n-1})$ is the nth-order moment function of an equivalent Gaussian process that has the same mean value and autocorrelation sequence as $X(k)$. Clearly, if $X(k)$ is Gaussian, $m_n^x(\tau_1, \ldots, \tau_{n-1}) = m_n^G(\tau_1, \ldots, \tau_{n-1})$ and thus $c_n^x(\tau_1, \ldots, \tau_{n-1}) = 0$. Note, however, that this is only true of orders $n = 3$ and 4.

2.2.5 Variance, Skewness, and Kurtosis Measures

By putting $\tau_1 = \tau_2 = \tau_3 = 0$ in (2.12), (2.13), (2.14) and assuming $m_1^x = 0$ we get

$$\begin{aligned}
\gamma_2^x &= E\{X(k)^2\} = c_2^x(0) & \text{(variance)} \\
\gamma_3^x &= E\{X^3(k)\} = c_3^x(0,0) & \text{(skewness)} \\
\gamma_4^x &= E\{X^4(k)\} - 3[\gamma_2^x]^2 = c_4^x(0,0,0) & \text{(kurtosis)}.
\end{aligned}$$

(2.15)

Normalized kurtosis is defined as $\gamma_4^x/[\gamma_2^x]^2$. Equation (2.15) gives the variance, skewness, and kurtosis measures in terms of cumulant lags.

2.2.6 Time-Reversible Process

A process $\{X(k)\}$ is said to be time-reversible if the probability structure of $\{X(-k)\}$ is the same as that of $\{X(k)\}$. This implies that

$$c_n^x(\tau_1, \tau_2, \ldots, \tau_{n-1}) = c_n^x(-\tau_1, -\tau_2, \ldots, -\tau_{n-1}) \quad (2.16)$$

for all integer values of $\tau_1, \tau_2, \ldots, \tau_{n-1}$. If the nth-order cumulants of a process satisfy (2.16), the process is time-reversible [Rao and Gabr, 1984]. Clearly, Gaussian processes ($n = 2$) are time-reversible due to the symmetry property of the covariance function; i.e., $c_2^x(\tau_1) = c_2^x(-\tau_1)$. Remember that higher-order ($n > 2$) cumulants of a Gaussian process are all zero.

2.2.7 Cross-Cumulants

Suppose we are given stationary real random processes $\{X(k)\}$, $\{Y(k)\}$, $\{Z(k)\}$, $k = 0, \pm 1, \pm 2, \ldots$. Their cross-cumulants may be defined as follows.

2nd-order cross-cumulants

$$\begin{aligned} c_{xy}(\tau_1) &= \text{Cum}[X(k), Y(k+\tau_1)] \\ c_{yz}(\tau_1) &= \text{Cum}[Y(k), Z(k+\tau_1)]. \end{aligned} \quad \text{(cross – covariances)} \quad (2.17)$$

Thus, if the processes are zero-mean, then $c_{xy}(\tau_1) = E\{X(k)Y(k+\tau_1)\}$ and $c_{yz}(\tau_1) = E\{Y(k)Z(k+\tau_1)\}$.

3rd-order cross-cumulants

$$\begin{aligned} c_{xyz}(\tau_1, \tau_2) &= \text{Cum}[X(k), Y(k+\tau_1), Z(k+\tau_2)] \\ &= E\{(X(k) - m_x)(Y(k+\tau_1) - m_y)(Z(k+\tau_2) - m_z)\} \end{aligned} \quad (2.18)$$

where $m_x = E\{X(k)\}$, $m_y = E\{Y(k)\}$, and $m_z = E\{Z(k)\}$. For zero-mean processes we have

$$c_{xyz}(\tau_1, \tau_2) = E\{X(k)Y(k+\tau_1)Z(k+\tau_2)\}.$$

Additional cross-cumulants are, for example

$$c_{xyy}(\tau_1,\tau_2) = \text{Cum}[X(k), Y(k+\tau_1), Y(k+\tau_2)]$$
$$c_{xyx}(\tau_1,\tau_2) = \text{Cum}[X(k), Y(k+\tau_1), X(k+\tau_2)] \quad (2.19)$$

So, the nth-order cross-cumulant sequence of stationary processes $\{X_i(k)\}$, $i = 1, 2, \ldots, n$ is defined as

$$c_{x_1 x_2 \ldots x_n}(\tau_1, \tau_2, \ldots, \tau_{n-1}) \triangleq \text{Cum}[X_1(k), X_2(k+\tau_1), \ldots, X_n(k+\tau_{n-1})]. \quad (2.20)$$

These quantities become useful in nonlinear system identification problems when we have access to input and output measurements [Brillinger, 1965; Brillinger and Rosenblatt, 1967; Brillinger, 1977]. Essentially, we will use $c_{x_1,x_2,\ldots,x_n}(\tau_1,\ldots,\tau_{n-1})$ to test for the nonlinearity of a function of order $n - 1$.

Example 2.4

Consider the narrow-band process

$$Z(k) = X(k) \cos(\omega_o k) + Y(k) \sin(\omega_o k)$$

where $X(k), Y(k)$ are independent stationary random processes with $E\{X(k)\} = E\{Y(k)\} = 0$, $m_2^x(\tau) = E\{X(k)X(k+\tau)\} = m_2^y(\tau)$, and $m_3^x(\tau_1,\tau_2) = E\{X(k)X(k+\tau_1)X(k+\tau_2)\} = m_3^y(\tau_1,\tau_2)$.

We now evaluate the autocorrelation and third-order moment sequence of $Z(k)$. The second-order moment is:

$$\begin{aligned}\text{Mom}[Z(k), Z(k+\tau)] &= E\{(X(k)\cos(\omega_o k) + Y(k)\sin(\omega_o k)) \cdot \\ &\quad (X(k+\tau)\cos(\omega_o(k+\tau)) + Y(k+\tau)\sin(\omega_o(k+\tau)))\} = \\ &= m_2^x(\tau) \cos(\omega_o \tau) = m_2^z(\tau)\end{aligned}$$

which is independent of k. Thus, $\{Z(k)\}$ is a wide-sense stationary random process. However, the third-order moments are

$$\text{Mom}[Z(k), Z(k+\tau_1), Z(k+\tau_2)]$$
$$= m_3^x(\tau_1, \tau_2)[\cos(\omega_o k)\cos(\omega_o(k+\tau_1))\cos(\omega_o(k+\tau_2))$$
$$+ \sin(\omega_o k)\sin(\omega_o(k+\tau_1))\sin(\omega_o(k+\tau_2))]$$

and the quantities in square brackets are dependent on k for τ_1, τ_2. Hence $\{Z(k)\}$ is nonstationary in its third-order statistics. □

2.2.8 Ergodicity and Moments

According to Papoulis [1984], a process $\{X(k)\}$ is ergodic in the most general form if, with probability one, all its moments can be determined from a single observation. In other words, the expected values $E\{\cdot\}$ (or ensemble averages) can be replaced by time averages; i.e.,

$$E\{X(k) \cdot X(k+\tau_1) \cdots X(k+\tau_{n-1})\} = <X(k) \cdots X(k+\tau_{n-1})> = \lim_{M \to \infty} \frac{1}{2M+1} \sum_{k=-M}^{+M} X(k) \, X(k+\tau_1) \cdots X(k+\tau_{n-1}) \quad (2.21)$$

where $<\cdot>$ is the time-average operator which has the same properties as the ensemble average operation $E\{\cdot\}$ if the process is ergodic [Sinai, 1963].

We see from (2.21) that time-averages of higher-order moments are functions of infinitely many random variables and, therefore, can be viewed as random variables themselves. What ergodicity implies is that the time averages of all possible sample sequences are equal to the same constant which, in turn, equals the ensemble average. Clearly, a process might be ergodic for certain higher-order moments and not for others [Papoulis, 1984].

We shall not discuss here the various criteria for ergodicity related to time averages of higher-order moments. Throughout this chapter we assume that if the process is ergodic, then (2.21) holds for all orders up to n. This implies that nth-order cumulants exist and can be generated from (2.5).

In practice, when we are given a finite length single realization of an ergodic process, i.e., $X(k)$, $k = -M, \ldots, 0, \ldots, +M$, we cannot compute the limits of (2.21) but the **estimates**

$$<X(k) \cdots X(k+\tau_{n-1})>_M = \frac{1}{2M+1} \sum_{k=-M}^{+M} X(k) \cdots X(k+\tau_{n-1}). \quad (2.22)$$

The estimation of higher-order moments and thus of a stochastic process is a problem of statistics which will be examined in Chapter 4.

2.3 CUMULANT SPECTRA

Suppose that the process $\{X(k)\}$, $k = 0, \pm1, \pm2, \ldots$ is real, strictly stationary, with nth-order cumulant sequence $c_n^x(\tau_1, \ldots, \tau_{n-1})$ defined by (2.10).

2.3.1 Definition

Assuming that the cumulant sequence satisfies the condition

$$\sum_{\tau_1=-\infty}^{+\infty} \cdots \sum_{\tau_{n-1}=-\infty}^{+\infty} \left| c_n^x(\tau_1, \ldots, \tau_{n-1}) \right| < \infty,$$

or the condition

$$\sum_{\tau_1=-\infty}^{+\infty} \cdots \sum_{\tau_{n-1}=-\infty}^{+\infty} (1+|\tau_j|) \left| c_n^x(\tau_1, \ldots, \tau_{n-1}) \right| < \infty. \quad (2.23)$$

for $j = 1, 2, \ldots, n-1$, the nth-order cumulant spectrum $C_n^x(\omega_1, \ldots, \omega_{n-1})$ of $\{X(k)\}$ exists, is continuous, and is defined as the $(n-1)$-dimensional Fourier transform of the nth-order cumulant sequence; e.g., [Brillinger, 1965; Rosenblatt, 1983; 1985]. Note that (2.23) describe the usual conditions for a Fourier transform to be well defined. The nth-order cumulant spectrum is thus defined:

$$C_n^x(\omega_1, \omega_2, \ldots, \omega_{n-1}) = \sum_{\tau_1=-\infty}^{+\infty} \cdots \sum_{\tau_{n-1}=-\infty}^{+\infty} c_n^x(\tau_1, \tau_2, \ldots, \tau_{n-1})$$

$$\exp\{-j(\omega_1\tau_1 + \omega_2\tau_2 + \cdots + \omega_{n-1}\tau_{n-1})\} \quad (2.24)$$

$|\omega_i| \leq \pi$ for $i = 1, 2, \ldots, n-1$ and $|\omega_1 + \omega_2 + \ldots + \omega_{n-1}| \leq \pi$.

In general, $C_n^x(\omega_1, \omega_2, \ldots, \omega_{n-1})$ is complex, i.e., it has magnitude and phase

$$C_n^x(\omega_1, \ldots, \omega_{n-1}) = \left| C_n^x(\omega_1, \ldots, \omega_{n-1}) \right| \exp\{j\Psi_n^x(\omega_1, \ldots, \omega_{n-1})\}. \quad (2.25)$$

The cumulant spectrum is also periodic with period 2π, i.e.,

$$C_n^x(\omega_1, \ldots, \omega_{n-1}) = C_n^x(\omega_1 + 2\pi, \ldots, \omega_{n-1} + 2\pi).$$

The notion of considering a spectral representation for a cumulant function as shown in (2.24) (cumulant spectrum) is acknowledged to be due to Kolmogorov [Shiryaev,

1960; 1963]. The term "higher-order spectrum" is due to Brillinger [1965] and Akaike [1966]. The term "polyspectra" is due to Brillinger [1965].

2.3.2 Alternative Definition

The physical significance of cumulant spectra becomes apparent when expressed in terms of the components $dZ(\omega)$ of the Fourier-Stieltjes representation of $\{X(k)\}$ (Cramer spectral representation) [Rosenblatt, 1983; 1985].

$$X(k) = \frac{1}{2\pi} \int_{-\infty}^{+\infty} \exp\{j\omega k\}\, dZ(\omega) \qquad (2.26)$$

for all k, where

$$E\{dZ(\omega)\} = 0$$

$$\mathrm{Cum}[dZ(\omega_1), dZ(\omega_2), \ldots, dZ(\omega_n)] = \begin{cases} C_n^x(\omega_1, \ldots, \omega_{n-1}) d\omega_1 \cdots d\omega_{n-1}, \text{ for} \\ \omega_1 + \omega_2 + \ldots + \ldots \omega_{n-1} + \omega_n = 0 \\ 0, \text{ for} \\ \omega_1 + \omega_2 + \ldots + \omega_{n-1} + \omega_n \neq 0. \end{cases} \qquad (2.27)$$

It is therefore apparent that the cumulant spectrum of order n represents the cumulant contribution of n Fourier components, the sum of whose frequencies equal zero. Although the cumulant spectrum of order n is a function of $n-1$ variables $\omega_1, \omega_2, \ldots, \omega_{n-1}$, it should be kept in mind that there is a hidden variable $\omega_n = -\omega_1 - \ldots - \omega_{n-1}$ in (2.27) [Rosenblatt, 1983].

2.3.3 Special Cases of Cumulant Spectra

The power spectrum, bispectrum, and trispectrum are special cases of the nth-order cumulant spectrum defined by (2.24) [Brillinger and Rosenblatt, 1967 a and b].

Power Spectrum: n=2

$$C_2^x(\omega) = \sum_{\tau=-\infty}^{+\infty} c_2^x(\tau) \exp\{-j(\omega\tau)\}, \qquad (2.28)$$

$|\omega| \leq \pi$ where $c_2^x(\tau)$ is the covariance sequence of $\{X(k)\}$ given by (2.12). If the process $\{X(k)\}$ is zero-mean, then (2.28) becomes the Wiener-Khintchine identity.

From (2.12) and (2.28) we have

$$c_2^x(\tau) = c_2^x(-\tau)$$
$$C_2^x(\omega) = C_2^x(-\omega) \qquad (2.29)$$
$$C_2^x(\omega) \geq 0 \quad \text{(real, nonnegative function)}$$

Bispectrum: n=3

$$C_3^x(\omega_1,\omega_2) = \sum_{\tau_1=-\infty}^{+\infty} \sum_{\tau_2=-\infty}^{+\infty} c_3^x(\tau_1,\tau_2) \exp\{-j(\omega_1\tau_1 + \omega_2\tau_2)\} \qquad (2.30)$$

$$|\omega_1| \leq \pi, |\omega_2| \leq \pi, |\omega_1 + \omega_2| \leq \pi$$

where $c_3^x(\tau_1,\tau_2)$ is the third-order cumulant sequence of $\{X(k)\}$ described by (2.13). Important symmetry conditions follow from the properties of moments and (2.13):

$$\begin{aligned} c_3^x(\tau_1,\tau_2) &= c_3^x(\tau_2,\tau_1) = c_3^x(-\tau_2,\tau_1-\tau_2) \\ &= c_3^x(\tau_2-\tau_1,-\tau_1) = c_3^x(\tau_1-\tau_2,-\tau_2) \\ &= c_3^x(-\tau_1,\tau_2-\tau_1). \end{aligned} \qquad (2.31)$$

As a consequence, knowing the third-order cumulants in any of the six sectors, I through VI, shown in Figure 2.3(a), would enable us to find the entire third-order cumulant sequence. These sectors include their boundaries so that, for example, sector I is an infinite wedge bounded by the lines $\tau_1 = 0$, and $\tau_1 = \tau_2$; $\tau_1, \tau_2 \geq 0$.

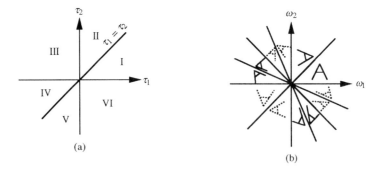

Figure 2.3 (a) Symmetry regions of third-order moments. (b) Symmetry regions of the bispectrum.

The definition of the bispectrum in (2.30) and the properties of third-order cumulants in (2.31) give

$$\begin{aligned} C_3^x(\omega_1,\omega_2) &= C_3^x(\omega_2,\omega_1) \\ &= C_3^{x*}(-\omega_2,-\omega_1) = C_3^x(-\omega_1-\omega_2,\omega_2) \\ &= C_3^x(\omega_1,-\omega_1-\omega_2) = C_3^x(-\omega_1-\omega_2,\omega_1) \\ &= C_3^x(\omega_2,-\omega_1-\omega_2). \end{aligned} \qquad (2.32)$$

Thus knowledge of the bispectrum in the triangular region $\omega_2 \geq 0$, $\omega_1 \geq \omega_2$, $\omega_1 + \omega_2 \leq \pi$ shown in Figure 2.3(b) is enough for a complete description of the bispectrum. For real processes, the bispectrum has 12 symmetry regions.

Trispectrum: n=4

$$C_4^x(\omega_1,\omega_2,\omega_3) = \sum_{\tau_1=-\infty}^{+\infty}\sum_{\tau_2=-\infty}^{+\infty}\sum_{\tau_3=-\infty}^{+\infty} c_4^x(\tau_1,\tau_2,\tau_3)\cdot\exp\{-j(\omega_1\tau_1+\omega_2\tau_2+\omega_3\tau_3)\} \qquad (2.33)$$

$$|\omega_1| \leq \pi, |\omega_2| \leq \pi, |\omega_3| \leq \pi, |\omega_1+\omega_2+\omega_3| \leq \pi$$

where $c_4^x(\tau_1,\tau_2,\tau_3)$ is the fourth-order cumulant sequence given by (2.14).

From the definition (2.14) of fourth-order cumulants, a lot of symmetry properties can be derived for the trispectrum, similar to those given in (2.32) for the bispectrum. For example, since the moments and are symmetric functions in their arguments, we have

$$\begin{aligned} C_4^x(\omega_1,\omega_2,\omega_3) &= C_4^x(\omega_2,\omega_1,\omega_3) = C_4^x(\omega_3,\omega_2,\omega_1) \\ &= C_4^x(\omega_1,\omega_3,\omega_2) = C_4^x(\omega_2,\omega_3,\omega_1) \\ &= C_4^x(\omega_3,\omega_1,\omega_2) = \text{ etc.} \end{aligned} \qquad (2.34)$$

Pflug et al. [1992] point out that the trispectrum of real processes has 96 symmetry regions.

2.3.4 Variance, Skewness, and Kurtosis Measures

Inverse Fourier transformation on (2.24) yields

$$c_n^x(\tau_1, \tau_2, \ldots, \tau_{n-1}) = \frac{1}{(2\pi)^{y-1}} \int_{-\pi}^{+\pi} \int_{-\pi}^{+\pi} \cdots \int_{-\pi}^{+\pi} C_n^x(\omega_1, \ldots, \omega_{n-1})$$
$$\exp\{j(\omega_1\tau_1 + \ldots + \omega_{n-1}\tau_{n-1})\}$$
$$d\omega_1 \cdots d\omega_{n-1}.$$
(2.35)

By choosing $n = 2, 3, 4$ and setting $(\tau_i) = 0$, $i = 1, 2, \ldots, n-1$, we get

$$c_2^x(0) = \frac{1}{2\pi} \int_{-\pi}^{+\pi} C_2^x(\omega) d\omega \quad \text{(variance } \gamma_2^x\text{)}$$

$$c_3^x(0,0) = \frac{1}{(2\pi)^2} \int_{-\pi}^{+\pi} \int_{-\pi}^{+\pi} C_3^x(\omega_1, \omega_2) d\omega_1 d\omega_2 \quad \text{(skewness } \gamma_3^x\text{)}$$

$$c_4^x(0,0,0) = \frac{1}{(2\pi)^3} \int_{-\pi}^{+\pi} \int_{-\pi}^{+\pi} \int_{-\pi}^{+\pi} C_4^x(\omega_1, \omega_2, \omega_3) d\omega_1 d\omega_2 d\omega_3 \quad \text{(kurtosis } \gamma_4^x\text{)}$$

(2.36)

which are the measures also described in (2.15). However, the measures in (2.36) are expressed in terms of cumulant spectra.

2.3.5 Time-Reversible Processes

An alternative way to (2.16) of testing whether a given stationary random process $\{X(k)\}$ is time-reversible is by examining the imaginary part of its cumulant spectra [Rao and Gabr, 1984]. From the condition of time-reversibility (2.16) and the definition of cumulant spectra (2.24), it follows that a process $\{X(k)\}$ is said to be time-reversible if the imaginary parts of all its cumulant spectra are identically zero. Since a Gaussian process has all its cumulant spectra of order $n > 2$ equal to zero and the imaginary part of its 2nd-order spectrum (power spectrum) is also zero, we conclude that a Gaussian process is time-reversible.

2.3.6 Non-Gaussian White Noise Processes

If $\{W(k)\}$ is a stationary non-Gaussian process with $E\{W(k)\} = 0$ and with nth-order cumulant sequence

$$c_n^w(\tau_1, \tau_2, \ldots, \tau_{n-1}) = \mathrm{Cum}[W(k), W(k+\tau_1), \ldots, W(k+\tau_{n-1})]$$
$$= \gamma_n^w \, \delta(\tau_1, \tau_2, \ldots, \tau_{n-1}) \qquad (2.37)$$

where γ_n^w is constant and $\delta(\tau_1, \tau_2, \ldots, \tau_{n-1})$ is the $(n-1)$-dimensional Kronecker delta function, then $\{W(k)\}$ is said to be white of order n [Brillinger, 1965; Giannakis and Mendel, 1989]. Of course, one need not assume all γ_n^w are finite. The cumulants γ_n^w cannot all be zero (assuming they exist) for $n \geq 2$. Combining (2.37) and (2.24) we obtain

$$C_n^w(\omega_1, \ldots, \omega_{n-1}) = \cdot \gamma_n^w \qquad (2.38)$$

which is a flat spectrum for all frequencies. Hence, consideration of (2.36), (2.37), and (2.38) leads to the following important special cases of white noise higher-order spectra.

$$\begin{aligned} C_2^w(\omega) &= \gamma_2^w \quad \text{(Power Spectrum)} \\ C_3^w(\omega_1, \omega_2) &= \gamma_3^w \quad \text{(Bispectrum)} \\ C_4^w(\omega_1, \omega_2, \omega_3) &= \gamma_4^w \quad \text{(Trispectrum)} \end{aligned} \qquad (2.39)$$

where γ_2^w is the variance, γ_3^w - the skewness, and γ_4^w - the kurtosis of $\{W(k)\}$.

Example 2.5

Consider the simple example [Lii and Rosenblatt, 1988]

$$X(k) = W(k) - W(k-1), \quad k = 0, \pm 1, \pm 2, \ldots$$

where $\{W(k)\}$ are independent, identically distributed (i.i.d.) with $E\{W(k)\} = 0$, $E\{W^2(k)\} = 1$ and $E\{W^3(k)\} = 1$. The covariance sequence of $\{X(k)\}$ is given by

$$\begin{aligned} c_2^x(\tau) = m_2^x(\tau) &= E\{X(k)X(k+\tau)\} \\ &= E\{(W(k) - W(k-1))(W(k+\tau) - W(k+\tau-1))\} \\ &= 2\delta(\tau) - \delta(\tau-1) - \delta(\tau+1) \end{aligned}$$

where $\delta(\tau)$ is the Kronecker delta function. Thus,

$$c_2^x(\tau) = \begin{cases} 2, & \tau = 0 \\ -1, & \tau = 1, \tau = -1 \\ 0, & \text{otherwise.} \end{cases}$$

On the other hand, the third-order cumulants of $\{X(k)\}$ are computed from

$$\begin{aligned} c_3^x(\tau_1,\tau_2) = m_3^x(\tau_1,\tau_2) &= E\{X(k)X(k+\tau_1)X(k+\tau_2)\} \\ &= E\{(W(k)-W(k-1))(W(k+\tau_1)-W(k+\tau_1-1)) \\ & \quad (W(k+\tau_2)-W(k+\tau_2-1))\} \end{aligned}$$

which leads to

$$\begin{aligned} c_3^x(\tau_1,\tau_2) = & - \delta(\tau_1-1,\tau_2) + \delta(\tau_1-1,\tau_2-1) - \delta(\tau_1,\tau_2-1) \\ & + \delta(\tau_1+1,\tau_2) + \delta(\tau_1+1,\tau_2+1) - \delta(\tau_1,\tau_2+1). \end{aligned}$$

Figure 2.4 illustrates the covariance and third-order cumulant sequences of $\{X(k)\}$. It is important to note that although the skewness $c_3^x(0,0,0) = \gamma_3^x = 0$, the third-order cumulants $c_3^x(\tau_1,\tau_2)$ are generally different from zero.

The power spectrum of the random process is given by

$$\begin{aligned} C_2^x(\omega) &= \sum_{\tau=-1}^{+1} c_2^x(\tau)\exp(-j\omega\tau) \\ &= (2 - 2\cos\omega) \end{aligned}$$

whereas its bispectrum is given by

$$\begin{aligned} C_3^x(\omega_1,\omega_2) &= \left(-e^{-j(\omega_1)} + e^{-j(\omega_1+\omega_2)} - e^{-j(\omega_2)} + e^{+j(\omega_1)} \right. \\ &\qquad \left. - e^{+j(\omega_1+\omega_2)} + e^{+j(\omega_2)}\right) \\ &= \left(2j\sin\omega_1 + 2j\sin\omega_2 - 2j\sin(\omega_1+\omega_2)\right) \\ &= 2j\left(\sin\omega_1 + \sin\omega_2 - \sin(\omega_1+\omega_2)\right). \end{aligned}$$

We observe that the real part of the bispectrum is zero and the imaginary part equals $2(\sin\omega_1 + \sin\omega_2 - \sin(\omega_1+\omega_2))$. Figure 2.4 also illustrates the power spectrum and the bispectrum of $\{X(k)\}$. This example illustrates that **zero skewness does not necessarily imply zero bispectrum** because the skewness of a signal contributes to the real part of the bispectrum only. □

2.3.7 One-Dimensional Slices of Cumulants and their Cumulant Spectra

Since higher-order cumulant spectra are multidimensional functions, their computation may be impractical in some applications due to excessive number crunching. Nagata [1970] suggested the use of certain 1-d slices of multidimensional cumulant sequences, and their 1-d Fourier transforms, as ways of extracting useful information from higher-order statistics (or moments) of non-Gaussian stationary processes.

Consider a non-Gaussian process $\{X(k)\}$ with third-order cumulants given by (2.13); i.e.,

$$c_3^x(\tau_1, \tau_2) = \text{Cum}\{X(k), X(k+\tau_1), X(k+\tau_2)\}. \tag{2.40}$$

One-dimensional slices of $c_3^x(\tau_1, \tau_2)$ can be defined as shown in (2.41):

$$\begin{aligned} r_{2,1}^x(\tau) &\triangleq \text{Cum}\{X(k), X(k), X(k+\tau)\} = c_3^x(0, \tau) \\ r_{1,2}^x(\tau) &\triangleq \text{Cum}\{X(k), X(k+\tau), X(k+\tau)\} = c_3^x(\tau, \tau) \\ r_{1,2}(\tau) &\triangleq r_{1,2}(-\tau) \end{aligned} \tag{2.41}$$

which represent two straight lines with slopes 90° and 45°, respectively. Furthermore, we can define

$$\begin{aligned} s_{2,1}^x(\tau) &\triangleq \frac{1}{2}[r_{2,1}^x(\tau) + r_{1,2}^x(\tau)], \\ q_{2,1}^x(\tau) &\triangleq \frac{1}{2}[r_{2,1}^x(\tau) - r_{1,2}^x(\tau)] \end{aligned} \tag{2.42}$$

which correspond to even and odd functions, respectively. If we define as 1-d spectrum

$$R_{2,1}^x(\omega) \triangleq \sum_{\tau=-\infty}^{+\infty} r_{2,1}^x(\tau) \exp(-j\omega\tau) \tag{2.43}$$

it follows from (2.42) and (2.43) that

$$R_{2,1}^x(\omega) = \sum_{\tau=-\infty}^{+\infty} \{s_{2,1}^x(\tau)\cos(\omega\tau) - jq_{2,1}^x(\tau)\sin(\omega\tau)\}. \tag{2.44}$$

Since $s_{2,1}^x(0) = \gamma_3^x$ and $q_{2,1}^x(0) = 0$, the *effective contribution to the skewness comes only from the real and symmetrical part*. Furthermore, from (2.41), (2.43), and (2.30), we obtain the relation between $R_{2,1}^x(\omega)$ and the bispectrum $C_3^x(\omega_1, \omega_2)$; viz.

$$R_{2,1}^x(\omega) = \frac{1}{2\pi} \int_{-\pi}^{+\pi} C_3^x(\omega, \sigma) d\sigma, \tag{2.45}$$

which represents the integrated bispectrum along a frequency line. Nagata [1970] points out that because the real part of $C_3^x(\omega_1, \omega_2)$ is not positive definite, even if $R_{2,1}^x(\omega_0)$ is very small, we cannot conclude that the value of $R_{2,1}^x(\omega)$ at $\omega = \omega_0$

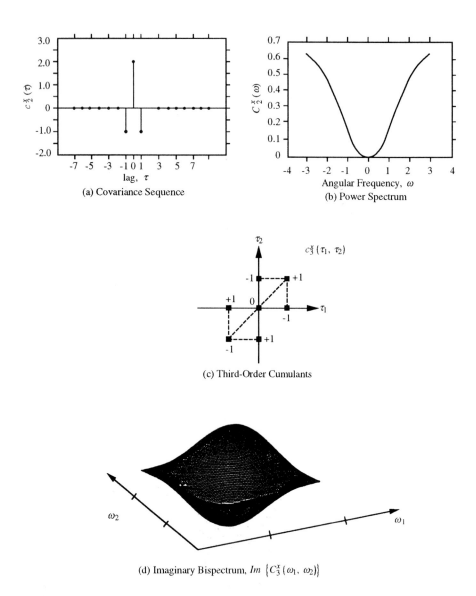

Figure 2.4 (a) The covariance sequence, (b) power spectrum, (c) third-order cumulants, and (d) bispectrum of the random process $X(k) = W(k) - W(k-1)$. Note that the real bispectrum is zero.

does not play a major role in determining the skewness. Gurbatov et al. [1987] point out that the imaginary part of $R_{2,1}^x(\omega)$ corresponds to "energy transfer" in the spectrum. A positive value of $I_m[R_{2,1}^x(\omega_o)]$ indicates energy transfer to the frequency $\omega = \omega_o$, and a negative value indicates the leakage of energy from ω_o.

If we now consider the fourth-order cumulants of $\{X(k)\}$ given by (2.14), we can define the following 1-d slices:

$$\begin{aligned}
r_{3,1}(\tau) &\triangleq \mathrm{Cum}[X(k), X(k), X(k), X(k+\tau)] = c_4^x(0,0,\tau) = c_4^x(0,\tau,0) = c_4^x(\tau,0,0) \\
r_{2,2}(\tau) &\triangleq \mathrm{Cum}[X(k), X(k), X(k+\tau), X(k+\tau)] = c_4^x(0,\tau,\tau) = c_4^x(\tau,0,\tau) \\
&= c_4^x(\tau,\tau,0) \\
r_{1,3}(\tau) &\triangleq \mathrm{Cum}[X(k), X(k+\tau), X(k+\tau), X(k+\tau)] = c_4^x(\tau,\tau,\tau)
\end{aligned} \quad (2.46)$$

and their functions

$$\begin{aligned}
s_{3,1}^x(\tau) &= \frac{1}{2}[r_{3,1}(\tau) + r_{1,3}(\tau)] \quad \text{(even)} \\
q_{3,1}^x(\tau) &= \frac{1}{2}[r_{3,1}(\tau) - r_{1,3}(\tau)] \quad \text{(odd)} \\
s_{2,2}^x(\tau) &= r_{2,2}(\tau). \quad \text{(even)}
\end{aligned} \quad (2.47)$$

Useful 1-d spectra based on 4-th order cumulants (2.46) are the following:

$$\begin{aligned}
R_{3,1}^x(\omega) &= \sum_{\tau=-\infty}^{+\infty} r_{3,1}^x(\tau)\exp(-j\omega\tau) \\
&= \sum_{\tau=-\infty}^{+\infty} \{s_{3,1}^x(\tau)\cos(\omega\tau) + jq_{3,1}^x(\tau)\sin(\omega\tau)\}
\end{aligned} \quad (2.48)$$

and

$$\begin{aligned}
R_{2,2}^x(\omega) &= \sum_{\tau=-\infty}^{+\infty} r_{2,2}^x(\tau)\exp(-j\omega\tau) \\
&= \sum_{\tau=-\infty}^{+\infty} s_{2,2}^x(\tau)\cos(\omega\tau).
\end{aligned} \quad (2.49)$$

Since $s_{2,2}^x(\tau)$ is an even sequence, its resulting spectrum $R_{2,2}^x(\omega)$ is real.

Combining (2.48), (2.46), and (2.33), we obtain the relation between $R_{3,1}^x(\omega)$ and $C_4^x(\omega_1, \omega_2, \omega_3)$. That is

$$R_{3,1}^x(\omega) = \frac{1}{(2\pi)^2} \int_{-\pi}^{+\pi} \int_{-\pi}^{+\pi} C_4^x(\omega, \sigma_1, \sigma_2)\, d\sigma_1\, d\sigma_2 \quad (2.50)$$

where $C_4^x(\omega, \sigma_1, \sigma_2)$ can be replaced by $C_4^x(\sigma_1, \omega, \sigma_2)$ or $C_4^x(\sigma_1, \sigma_2, \omega)$. Because $q_{3,1}^x(0) = 0$, **the net contribution to kurtosis comes from $s_{3,1}^x(\tau)$ and thus** from the real part of $R_{3,1}^x(\omega)$. From (2.49), (2.46), and (2.33) we obtain

$$R_{2,2}^x(\omega_1+\omega_2) = \frac{1}{2}\int_{-\pi}^{+\pi}\int_{-\pi}^{+\pi} C_4^x(\omega_1,\omega_2,\omega_3)\, d(\omega_1-\omega_2)\, d\omega_3. \qquad (2.51)$$

Therefore, $R_{2,2}^x(\omega_o)$ represents the integrated trispectrum on the plane $\omega_1+\omega_2=\omega_o$ or equivalently on the planes $\omega_1+\omega_3=\omega_o$ or $\omega_2+\omega_3=\omega_o$.

Example 2.6

Let us consider the process $\{X(k)\}$ described in Example 2.5. One-dimensional slices of its third-order cumulant sequence (shown in Figure 2.4) are obtained by utilizing (2.41); i.e.,

$$\begin{aligned} r_{2,1}(\tau) &= -\delta(\tau)+\delta(\tau+1) \\ r_{1,2}(\tau) &= \delta(\tau)-\delta(\tau+1). \end{aligned}$$

From these expressions and (2.42) we obtain

$$\begin{aligned} s_{2,1}^x(\tau) &= 0 \\ q_{2,1}^x(\tau) &= -\delta(\tau-1)+\delta(\tau+1) \end{aligned}$$

which is consistent with the fact that the real part of the bispectrum of $\{X(k)\}$ is zero. Applying to this example equation (2.43) or (2.44) we obtain

$$\begin{aligned} R_{2,1}^x(\omega) &= -j(-\sin(\omega))+\sin(-\omega) \\ &= 2j\sin\omega. \end{aligned}$$

This result is verified if we substitute in (2.45) the bispectrum of $\{X(k)\}$; i.e.,

$$C_3^x(\omega,\sigma) = 2j\bigl(\sin\omega+\sin\sigma-\sin(\omega+\sigma)\bigr)$$

and perform the integration

$$\begin{aligned} R_{2,1}^x(\omega) &= \frac{1}{2\pi}\int_{-\pi}^{+\pi} C_3^x(\omega,\sigma)\, d\sigma \\ &= 2j\sin\omega. \end{aligned}$$

Figure 2.5 illustrates the 1-d cumulant slices and $R_{2,1}^x(\omega)$ of the process $\{X(k)\}$. □

2.3.8 Why Cumulant Spectra and not Moment Spectra?

Cumulant spectra can be found more useful in the processing of random signals than moment spectra. The reason is threefold: (a) cumulant spectra of order $n>2$

are zero if the process is Gaussian and nonzero cumulant spectra provide a measure of extent of non-Gaussianity; (b) cumulants provide a suitable measure of extent of statistical dependence in time series; (c) the cumulant spectrum of the sum of two independent, nonzero mean, stationary random processes equals the sum of their individual cumulant spectra. However, this latter property does not hold in the case of moment spectra. Finally, Brillinger [1965] points out that ergodicity assumptions are met more easily in estimating cumulants rather than moments.

2.3.9 The nth-Order Coherency Function

A normalized cumulant spectrum or the nth-order coherency index is a function that combines two completely different entities, namely, the cumulant spectrum of order n, $C_n^x(\omega_1,\ldots,\omega_{n-1})$ and the power spectrum $C_2^x(\omega)$ of a process. The nth-order coherency index is defined as

$$P_n^x(\omega_1,\omega_2,\ldots,\omega_{n-1}) \triangleq \frac{C_n^x(\omega_1,\omega_2,\ldots,\omega_{n-1})}{\left[C_2^x(\omega_1) \cdot C_2^x(\omega_2) \cdots C_2^x(\omega_{n-1}) \cdot C_2^x(\omega_1+\omega_2+\cdots+\omega_{n-1})\right]^{\frac{1}{2}}}. \quad (2.52)$$

The third-order ($n = 3$) coherency index is also called bicoherency (normalized bispectrum) [Hasselman et al., 1963; Raghuveer and Nikias, 1985]. The nth-order coherency index is very useful for the detection and characterization of non-linearities in time series via phase relations of their harmonic components. Also, the nth-order coherency index becomes useful in studying the phase response of non-Gaussian linear processes, i.e., processes whose spectra are modeled by the same linear filter. The magnitude of the nth-order coherency, $|P_n^x(\omega_1,\ldots,\omega_{n-1})|$, is called the coherence index.

2.3.10 Cross-Cumulant Spectra

The cross-cumulant spectra are defined as the multidimensional Fourier transforms of the corresponding cross-cumulants. Formation of the Fourier transform of the relationship (2.20) gives

$$C_{x_1 x_2 \ldots x_n}(\omega_1,\omega_2,\ldots,\omega_{n-1}) \triangleq \sum_{\tau_1=-\infty}^{+\infty} \cdots \sum_{\tau_{n-1}=-\infty}^{+\infty} c_{x_1 \ldots x_n}(\tau_1,\tau_2,\ldots,\tau_{n-1})\exp\{-j(\omega_1\tau_1+\omega_2\tau_2+\cdots+\omega_{n-1}\tau_{n-1})\} \quad (2.53)$$

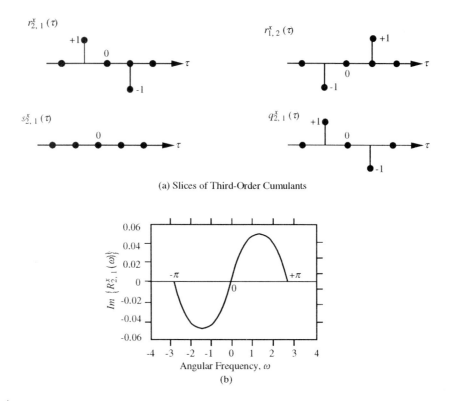

Figure 2.5 Random process $X(k) = W(k) - W(k-1)$ described in Example (2.6) (a) 1-d slices of its third-order cumulants, (b) the corresponding 1-d spectrum.

which is the nth-order cross-cumulant spectrum of processes $\{X_i(k)\}$, $i = 1, 2, \ldots, n$. The summability of the cross-cumulant sequence is assumed. For example, combining (2.18) and (2.53) we get a cross-bispectrum of $\{X(k)\}$ and $\{Y(k)\}$; i.e., [Brillinger, 1977]

$$C_{xyy}(\omega_1, \omega_2) = \sum_{\tau_1=-\infty}^{+\infty} \sum_{\tau_2=-\infty}^{+\infty} c_{xyy}(\tau_1, \tau_2) \exp\{-j(\omega_1 \tau_1 + \omega_2 \tau_2)\}. \qquad (2.54)$$

On the other hand, combining (2.17) and (2.53) we obtain

$$C_{xy}(\omega) = \sum_{\tau=-\infty}^{+\infty} c_{xy}(\tau) \exp\{-j\omega\tau\} \qquad (2.55)$$

which is the cross-spectrum (2nd-order) between $\{X(k)\}$ and $\{Y(k)\}$. Akaike [1966] defines as "mixed spectrum" the quantity

$$B_{xyy}(\omega;\sigma) \triangleq \cdot \sum_{\tau=-\infty}^{+\infty} c_{yyx}(\sigma,\tau)\exp\{-j\omega\tau\} \qquad (2.56)$$

as it relates to a spectral function mixed in time and frequency. The "mixed spectrum" $B_{xyy}(\omega;\sigma)$ gives the cross-spectrum between $\{X(k)\}$ and $\{Y(k)Y(k+\tau)\}$ under the assumption of being stationary. The cross-bispectrum $C_{yyx}(\omega_1,\omega_2)$ and the mixed spectrum $B_{yyx}(\omega;\sigma)$ satisfy the relationship [Akaike, 1966]

$$C_{yyx}(\omega_1,\omega_2) = \sum_{\sigma=-\infty}^{+\infty} B_{yyx}(\omega_2;\sigma)\exp\{-j\omega,\sigma\}. \qquad (2.57)$$

Extensions of the "mixed spectrum" definition (2.56) to higher-order cases will be straightforward.

2.3.11 Linear Phase Shifts

Consider a zero-mean stationary random process $\{X(k)\}$ with finite moments up to order n. Let us form a new process $Y(k) = X(k - D)$ where D is a constant integer. From (2.5), (2.9), and (2.10) we conclude that

$$\begin{aligned}\text{Cum}[X(k), X(k+\tau_1),&\ldots,X(k+\tau_{n-1})] = \\ \text{Cum}[Y(k), Y(k+\tau_1),&\ldots,Y(k+\tau_{n-1})] \\ = c_n^x(\tau_1,\tau_2,&\ldots,\tau_{n-1})\end{aligned} \qquad (2.58)$$

which implies that processes $\{X(k)\}$, $\{Y(k)\}$ have identical cumulant spectra. In other words, **cumulant spectra suppress linear phase shifts**.

On the other hand, if we form the signals $X_i(k) = X(k)$ for all $i \neq 2$, $X_2(k) = Y(k) = X(k-D)$, and generate the cross-cumulants

$$c_{xyx\ldots x}(\tau_1,\tau_2,\ldots,\tau_{n-1}) = \text{Cum}[X(k), Y(k+\tau_1), X(k+\tau_2),\ldots,X(k+\tau_{n-1})]$$

we obtain

$$c_{xyx\ldots x}(\tau_1,\tau_2,\ldots,\tau_{n-1}) = c_n^x(\tau_1,\tau_2 - D,\tau_3,\ldots,\tau_{n-1}). \qquad (2.59)$$

Combining (2.53) and (2.59), we obtain

$$C_{xyx\ldots x}(\omega_1,\omega_2,\ldots,\omega_{n-1}) = C_n^x(\omega_1,\omega_2,\ldots,\omega_{n-1}) \cdot \exp\{j\omega_2 D\}. \qquad (2.60)$$

From (2.59) and (2.60), it is apparent that cross-cumulants and their corresponding cross-spectra do preserve linear phase shifts. Specifically, the cross-cumulant spectrum between $\{X(k)\}$ and $\{X(k-D)\}$ equals the cumulant spectrum of $\{X(k)\}$ times a linear phase shift component with slope determined by D. Extensions of (2.59) to higher-order cross-cumulant spectra are straightforward. This key property of cross-cumulants will be considered further later in this book in the study of time delay estimation and array processing problems (Chapter 8).

2.3.12 Complex Regression Coefficients

The complex "regression" coefficient, $R(\omega)$, of a process $\{X(k)\}$ on the process $\{Y(k)\}$, is defined by Brillinger [1977] as

$$R(\omega) \triangleq \frac{C_{xy}(\omega)}{C_{xx}(\omega)} \qquad (2.61)$$

where $C_{xy}(\omega)$, $C_{xx}(\omega)$ are the cross-spectrum and power spectrum, respectively. Higher-order complex "regression" coefficients may be defined as

$$R_1(\omega_1, \omega_2) \triangleq \frac{C_{xxy}(\omega_1, \omega_2)}{C_{xy}(\omega_1) C_{xy}(\omega_2)} \qquad (2.62)$$

and

$$R_2(\omega_1, \omega_2) \triangleq \frac{C_{yxx}(\omega_1, \omega_2)}{C_{xxx}(\omega_1, \omega_2)}. \qquad (2.63)$$

Regression coefficients (2.61), (2.63) become useful, as we see later, in linear and non-linear system identification problems using cumulant spectra of input/output measurements [Rosenblatt, 1985].

2.3.13 Complex Processes

If a given process $\{X(k)\}$ is complex, its nth-order cumulant sequence has more than one definition depending on where we place the conjugation "*" operation. For example, 3rd-order cumulant sequences may be defined as

$$\begin{aligned} c_3^{(1)}(\tau_1, \tau_2) &\triangleq \mathrm{Cum}[X(k), X(k+\tau_1), X(k+\tau_2)] \\ c_3^{(2)}(\tau_1, \tau_2) &\triangleq \mathrm{Cum}[X(k), X^*(k+\tau_1), X(k+\tau_2)] \\ c_3^{(3)}(\tau_1, \tau_2) &\triangleq \mathrm{Cum}[X(k), X^*(k+\tau_1), X^*(k+\tau_2)] \end{aligned} \qquad (2.64)$$

etc., and their bispectra will follow from (2.30). Each one of these bispectrum functions will be different from the others. In general, there are 2^n different nth-order complex cumulant definitions.

Example 2.7

Let us consider the process

$$X(k) = a \exp\{j\omega_o k\}$$

where ω_o is constant and a is a random variable with $E\{a\} = 0$, $E\{a^2\} = Q$, $E\{a^3\} = 0$ and $E\{a^4\} = \mu$. Its second-order moments are given by (2.9).

$$\begin{aligned} \text{Mom}[X(k), X^*(k+\tau)] &= E\{X(k)X^*(k+\tau)\} \\ &= Q \exp\{j(-\omega_o \tau)\} \\ &= c_2^x(\tau), \end{aligned} \quad (2.65)$$

which implies that the process is wide-sense stationary.

Two of the sixteen possible definitions of the fourth-order moments of $\{X(k)\}$ are

$$\begin{aligned} \text{Mom}[X(k), X^*(k+\tau_1), X(k+\tau_2), X^*(k+\tau_3)] &= \mu \exp\{j\omega_o(\tau_2 - \tau_1 - \tau_3)\} \\ &= m_4^x(\tau_1, \tau_2, \tau_3) \end{aligned} \quad (2.66)$$

and

$$\begin{aligned} \text{Mom} \ [X(k), X(k+\tau_1), X(k+\tau_2), X^*(k+\tau_3)] &= \\ = \mu \exp\{j(2\omega_o k)\} \cdot \exp\{j\omega_o(\tau_2 + \tau_1 - \tau_3)\} &= \\ m_4^x(k; \tau_1, \tau_2, \tau_3). \end{aligned}$$

Here we see two different fourth-order moment sequences of the process where only the first one is stationary. Combining (2.65) and (2.66) with (2.14), we obtain

$$c_4^x(\tau_1, \tau_2, \tau_3) = (\mu - 3Q^2) \exp\{j\omega_o(\tau_2 - \tau_1 - \tau_3)\}. \quad (2.67)$$

Combining (2.46), (2.47), and (2.67), we can get the following $1-d$ slices of $c_4^x(\tau_1, \tau_2, \tau_3)$:

$$\begin{aligned} r_{3,1}^x(\tau) &= \gamma \exp\{j(-\omega_o \tau)\} \\ r_{1,3}^x(\tau) &= \gamma \exp\{j(-\omega_o \tau)\} \\ r_{2,2}^x(\tau) &= \gamma \end{aligned} \quad (2.68)$$

where $\gamma = \mu - 3Q^2$. If we substitute (2.68) into (2.47), we obtain

$$s_{3,1}^x(\tau) = \gamma \exp\{j(-\omega_0 \tau)\}$$
$$q_{3,1}^x(\tau) = 0$$
$$s_{2,2}^x(\tau) = \gamma.$$

Substituting these results into (2.48), we see that

$$R_{3,1}^x(\omega) = \frac{\gamma}{Q} \cdot C_2^x(\omega)$$

where $C_2^x(\omega)$ is the power spectrum of the process. □

2.3.14 The Wigner Bispectrum

Gerr [1988] introduced the third-order Wigner time-frequency distribution or Wigner bispectrum which is a mixed time-frequency representation that extends the standard Wigner distribution in the same way that the bispectrum extends the power spectrum. The Wigner time-frequency distribution of a real-valued signal $\{X(k)\}$ may be defined as

$$W_2(t,\omega) = \int X(t+a(\tau))X(t+b(\tau)) \exp\{-j\omega\tau\} d\tau \qquad (2.69)$$

where $a(\tau) \triangleq \tau/2$, $b(\tau) \triangleq \tau/2$. If $\{X(t)\}$ is a zero-mean stationary random process, then

$$\begin{aligned} E\{W_2(t,\omega)\} &= \int c_2^x(\tau) \exp\{-j\omega\tau\} d\tau \\ &= C_2^x(\omega) \end{aligned} \qquad (2.70)$$

where $c_2^x(\tau)$ is the covariance function and $C_2^x(\omega)$ is the power spectral density of $\{X(t)\}$. Analogous to (2.69), the Wigner bispectrum is defined as [Gerr, 1988]

$$W_3(t,\omega_1,\omega_2) \triangleq \iint X(t+a(\tau_1,\tau_2))X(t+b(\tau_1,\tau_2))X(t+c(\tau_1,\tau_2)) \\ \cdot \exp\{-j(\omega_1+\omega_2)\tau_1 - j\omega_2\tau_2\} d\tau_1 d\tau_2 \qquad (2.71)$$

where the lag functions are given by

$$a(\tau_1,\tau_2) = -\frac{2}{3}\tau_1 - \frac{1}{3}\tau_2$$
$$b(\tau_1,\tau_2) = \frac{1}{3}\tau_1 - \frac{1}{3}\tau_2$$
$$c(\tau_1,\tau_2) = \frac{1}{3}\tau_1 + \frac{2}{3}\tau_2.$$

It is easy to see from (2.71) that

$$\begin{aligned} E\{W_3(t,\omega_1,\omega_2)\} &= \iint c_3^x(\tau_1, \tau_1+\tau_2) \cdot \exp\{-j(\omega_1+\omega_2)\tau_1 - j\omega_2\tau_2\} d\tau_1 d\tau_2 \\ &= C_3^x(\omega_1,\omega_2) \end{aligned}$$
(2.72)

where $c_3^x(\tau_1,\tau_2)$, $C_3^x(\omega_1,\omega_2)$ are the third-order cumulants and bispectrum of $\{X(t)\}$, respectively.

The Wigner bispectrum may prove useful for extracting time-varying phase information or phase coupling between frequency components, as well as for detection, parameter estimation, and classification of deterministic signals in stochastic noise.

2.4 CUMULANT SPECTRA OF NON-GAUSSIAN LINEAR PROCESSES

Let $\{X(k)\}$ be a zero-mean non-Gaussian process with all its moments finite up to order nth (stationary to order n) and with cumulant spectrum $C_n^x(\omega_1, \omega_2, \ldots, \omega_{n-1})$. Assume that $\{X(k)\}$ is the input to a linear time-invariant (LTI) system described by

$$Y(k) = \sum_{i=-\infty}^{+\infty} h(k-i)\, X(i) \tag{2.73}$$

where

$$h(k) \triangleq \frac{1}{2\pi} \int_{-\pi}^{+\pi} H(\omega) \exp\{j\omega k\} d\omega \tag{2.74}$$

is the impulse response of the system and $H(\omega)$ its frequency response function. The LTI system is assumed to be stable; i.e., its impulse response is absolutely summable

$$\sum_{k=-\infty}^{+\infty} |h(k)| < \infty.$$

2.4.1 Cumulant Spectra of LTI Systems

Brillinger and Rosenblatt [1967 a and b] established that the nth-order cumulant spectra of the input $\{X(k)\}$ and output $\{Y(k)\}$ are related by

$$\begin{aligned} C_n^y(\omega_1,\ldots,\omega_{n-1}) = H(\omega_1) \cdot H(\omega_2) \cdots H(\omega_{n-1}) &\cdot H^*(\omega_1 + \cdots + \omega_{n-1}) \\ &\cdot C_n^x(\omega_1,\ldots,\omega_{n-1}). \end{aligned}$$
(2.75)

If we express the frequency response of the LTI system as

$$H(\omega) = |H(\omega)| \cdot \exp\{j\phi_h(\omega)\} \tag{2.76}$$

and

$$C_n^x(\omega_1,\ldots,\omega_{n-1}) = |C_n^x(\omega_1,\ldots,\omega_{n-1})| \exp\{j\Psi_n^x(\omega_1,\ldots,\omega_{n-1})\} \tag{2.77}$$

then (2.75) can be written as

$$|C_n^y(\omega_1,\ldots,\omega_{n-1})| = |H(\omega_1)|\cdots|H(\omega_{n-1})| \cdot |H^*(\omega_1+\cdots+\omega_{n-1})| \\ \cdot |C_n^x(\omega_1,\ldots,\omega_{n-1})| \tag{2.78}$$

and

$$\Psi_n^y(\omega_1,\ldots,\omega_{n-1}) = \phi_h(\omega_1) + \phi_h(\omega_2) + \cdots + \phi_h(\omega_{n-1}) - \\ \phi_h(\omega_1+\omega_2+\cdots+\omega_{n-1}) + \Psi_n^x(\omega_1,\ldots,\omega_{n-1}). \tag{2.79}$$

Hence, the output cumulant spectra of order $n > 2$ carry phase information about the LTI system $H(\omega)$.

2.4.2 Cumulant Spectra of LTI Systems Driven by White Noise

In the special case where $\{X(k)\}$ is non-Gaussian white, we obtain from (2.39) and (2.75) [Rosenblatt, 1980]

$$C_n^y(\omega_1,\ldots,\omega_{n-1}) = \gamma_n^x H(\omega_1) \cdot H(\omega_2) \cdots H(\omega_{n-1}) H^*(\omega_1+\cdots+\omega_{n-1}). \tag{2.80}$$

where γ_n^x is defined in (2.38). Important special cases are:

Power Spectrum (n=2):

$$C_2^y(\omega) = \gamma_2^x \cdot |H(\omega)|^2 \tag{2.81}$$

where γ_2^x is the variance of the input white sequence. $C_2^y(\omega)$ is real and nonnegative. Since $|H(\omega)|^2 = H(\omega)H^*(\omega) = |H(\omega)|\exp\{j\phi\} \cdot |H(\omega)| \cdot \exp\{-j\phi\}$, we see that nonminimum phase information is lost for $n = 2$.

Bispectrum (n=3):

From (2.39) and (2.80), it follows that

$$|C_3^y(\omega_1,\omega_2)| = |\gamma_3^x| \cdot |H(\omega_1)| \cdot |H(\omega_2)| \cdot |H(\omega_1+\omega_2)| \tag{2.82}$$

and
$$\Psi_3^y(\omega_1,\omega_2) = \phi_h(\omega_1) + \phi_h(\omega_2) - \phi_h(\omega_1 + \omega_2) \tag{2.83}$$

where γ_3^x is the skewness of the input process.

Trispectrum (n=4):

Similarly,
$$|C_4^y(\omega_1,\omega_2,\omega_3)| = |\gamma_4^x| \cdot |H(\omega_1)| \cdot |H(\omega_2)| \cdot |H(\omega_3)| \cdot |H(\omega_1+\omega_2+\omega_3)| \tag{2.84}$$

and
$$\Psi_4^y(\omega_1,\omega_2,\omega_3) = \phi_h(\omega_1) + \phi_h(\omega_2) + \phi_h(\omega_3) - \phi_h(\omega_1+\omega_2+\omega_3) \tag{2.85}$$

where γ_4^x is the kurtosis of the input white process.

The cumulant spectrum of order n described by (2.80) is related to that of order $(n-1)$ by the following simple identity:

$$C_n^y(\omega_1,\ldots,\omega_{n-2},0) = C_{n-1}^y(\omega_1,\ldots,\omega_{n-2}) \cdot H(0) \cdot \frac{\gamma_n^y}{\gamma_{n-1}^y}. \tag{2.86}$$

Thus, the power spectrum of non-Gaussian linear processes may be reconstructed from its bispectrum up to a constant term (provided $H(0) \neq 0$); i.e.,

$$C_3^y(\omega,0) = C_2^y(\omega) \cdot H(0) \cdot \frac{\gamma_3^y}{\gamma_2^y}. \tag{2.87}$$

Since $\{X(k)\}$ is non-Gaussian with all its moments finite, it follows that for some order n, $\gamma_n \neq 0$. However, the LTI system with frequency transfer function $H(\omega)$ may be identified from the cumulant spectrum $C_n^y(\omega_1,\omega_2,\ldots,\omega_{n-1})$ for $n > 2$ up to a sign and except possibly for an additive linear phase term [Rosenblatt, 1980; 1985]. From the convolutional relationship (2.73), we see that one can change the sign of all the $h(k)$'s and all the $X(i)$'s without altering the observed process $Y(k)$. Further, from identity (2.58), we observe that cumulants and therefore cumulant spectra suppress linear phase terms.

In Chapters 5–7, we show that LTI system identification may be performed from the power spectrum and one of the nth-order ($n > 2$) cumulant spectra of the output $\{Y(k)\}$ if $\gamma_n^x \neq 0$, or directly from the nth-order cumulant spectrum. Finally, if the input $\{X(k)\}$ is stationary Gaussian, then $C_n^x(\omega_1,\ldots,\omega_{n-1}) = 0$ and hence $C_n^y(\omega_1,\ldots,\omega_{n-1}) = 0$ for all $n > 2$. Thus, only the power spectrum $C_2^y(\omega)$ is different from zero. Consequently, nonminimum phase systems cannot be identified from their output only if they are driven by white Gaussian stationary processes. Recall that a minimum phase system has all its poles and zeros inside the unit circle in the z-domain. A maximum phase system has all its poles and zeros outside the unit circle. A stable nonminimum phase system has all its poles inside the unit circle and its zeros inside and outside the unit circle (i.e., anywhere in the z-domain).

Example 2.8

Suppose we have a first-order FIR system with impulse response

$$h(k) = \delta(k) - a\ \delta(k-1)$$

and frequency transfer function

$$H(\omega) = 1 - a\ e^{-j\omega}$$

which is driven by zero-mean non-Gaussian white noise $\{X(k)\}$ with variance γ_2^x and skewness γ_3^x. The output of the system is given by

$$Y(k) = \sum_{i=0}^{1} h(i)\ X(k-i) =$$

$$X(k) - a\ X(k-1).$$

The power spectrum and bispectrum of $\{Y(k)\}$ follow from (2.81), (2.82), and (2.83);

$$\begin{aligned} C_2^y(\omega) &= \gamma_2^x \cdot |1 - a\ e^{-j\omega}|^2 \\ &= \gamma_2^x \cdot (1 + a^2 - 2a\ \cos\omega) \end{aligned} \quad \text{(power spectrum)}$$

and

$$\begin{aligned} C_3^y(\omega_1,\omega_2) &= \gamma_3^x \cdot (1 - ae^{-j\omega_1})(1 - ae^{-j\omega_2})(1 - ae^{j(\omega_1+\omega_2)}) \\ &= \gamma_3^x \cdot \{(1 - a^3) + a^2\ e(\omega_1,\omega_2) - a\ e^*(\omega_1,\omega_2)\} \end{aligned} \quad \text{(bispectrum)}$$

where $e(\omega_1,\omega_2) = e^{j\omega_1} + e^{j\omega_2} + e^{-j(\omega_1+\omega_2)}$.

Calculating

$$C_3^y(\omega,0) = \gamma_3^x\{(1-a^3) + a^2 e(\omega,0) - ae^*(\omega,0)\}$$
$$= \gamma_3^x \cdot \{(1-a^3) + a^2(1 + e^{j\omega} + e^{-j\omega}) - a(1 + e^{-j\omega} + e^{j\omega})\}$$
$$= \gamma_3^x\{(1-a^3+a^2-a) + (a^2-a)2\cos\omega\}$$
$$= \gamma_3^x \cdot H(0)\{1 + a^2 - 2a\cos\omega\}$$
$$= \frac{\gamma_3^x}{\gamma_2^x} \cdot H(0) C_2^y(\omega)$$

we can verify (2.87).

The power spectrum and magnitude and phase bispectra of $\{Y(k)\}$ are illustrated in Figure 2.6 for $a = 0.5, 2,$ and 1. □

Example 2.9

Consider a fourth-order autoregressive (AR) LTI system with frequency response function [Raghuveer and Nikias, 1985]

$$H(\omega) = \frac{1}{\sum_{i=0}^{4} a_i \exp\{j\omega i\}}$$

where $a_0 = 1$, $a_1 = 0.1$, $a_2 = 0.2238$, $a_3 = 0.084$, and $a_4 = 0.0294$. Let us assume that the system is driven by a non-Gaussian zero-mean stationary random process $\{X(k)\}$ with power spectrum

$$C_2^x(\omega) = |1 - 0.5\, e^{-j\omega}|^2$$

and bispectrum

$$C_3^x(\omega_1,\omega_2) = 0.875 + 0.25\, e(\omega_1,\omega_2) - 0.5\, e^*(\omega_1,\omega_2)$$

where $e(\omega_1,\omega_2)$ was defined in Example 2.8.

The power spectrum and bispectrum of the system output $\{Y(k)\}$ follows from (2.75).

These are

$$C_2^y(\omega) = |H(\omega)|^2 \cdot C_2^x(\omega)$$

$$= \frac{|1 - 0.5\, e^{-j\omega}|^2}{\left|\sum_{i=o}^{4} a_i\, e^{-j(\omega i)}\right|^2}$$

and

Figure 2.6 The power spectrum and bispectrum of the output of FIR system described in Example 2.8 for parameter values (a) $a = 0.5$, (b) $a = 2$, and (c) $a = 1$.

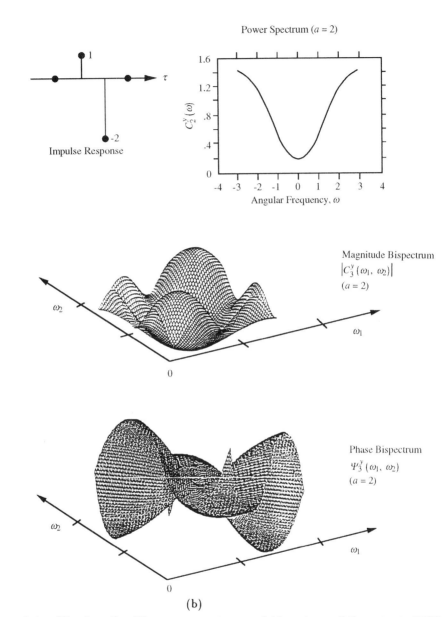

Figure 2.6 (Continued) The power spectrum and bispectrum of the output of FIR system described in Example 2.8 for parameter values (a) $a = 0.5$, (b) $a = 2$, and (c) $a = 1$.

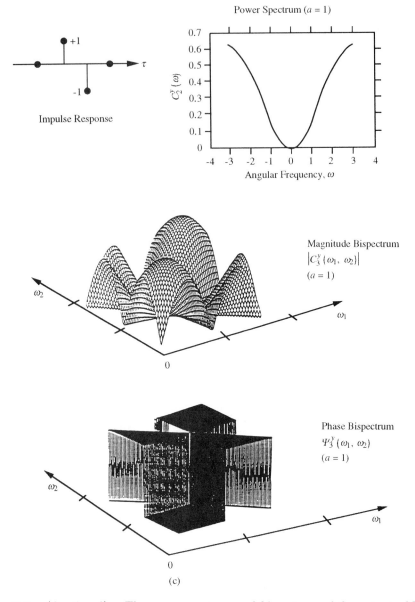

Figure 2.6 (Continued) The power spectrum and bispectrum of the output of FIR system described in Example 2.8 for parameter values (a) $a = 0.5$, (b) $a = 2$, and (c) $a = 1$.

$$C_3^y(\omega_1,\omega_2) = H(\omega_1)\,H(\omega_2)\,H^*(\omega_1+\omega_2)\cdot C_3^x(\omega_1,\omega_2)$$

$$= \frac{(0.875 + 0.25\,e(\omega_1,\omega_2) - 0.5\,e^*(\omega_1,\omega_2))}{\left(\sum_{i=0}^{4} e^{-j\omega_1 i}\right)\left(\sum_{\lambda=0}^{4} a_\lambda e^{-j\omega_2 \lambda}\right)\left(\sum_{\ell=0}^{4} a_\ell e^{+j(\omega_1+\omega_2)\ell}\right)}$$

$$= \frac{0.875 + 0.25 e(\omega_1,\omega_2) - 0.5 e^*(\omega_1,\omega_2)}{\sum_{i=0}^{4}\sum_{\lambda=0}^{4}\sum_{\ell=0}^{4} a_i a_\lambda a_\ell e^{-j(\omega_1 i + \omega_2 \lambda - \omega_1 \ell - \omega_2 \ell)}}$$

The bispectrum $C_3^y(\omega_1,\omega_2)$ (magnitude and phase) is illustrated in Figure 2.7. □

2.4.3 Nonminimum Phase LTI Systems

What makes higher-order moment and cumulant spectra of order $n > 2$ useful in nonminimum phase system identification is their ability to retain phase information (except, of course, for the linear phase term). We demonstrate this fundamental property of polyspectra with a simple example. Consider a second-order finite impulse response (FIR) system [also called moving average (MA)] driven by zero-mean, non-Gaussian white noise $\{W(k)\}$. Depending on the location of the system's zeros relative to the unit circle, we may have the following cases [Nikias and Raghuveer, 1987]:

Minimum Phase MA System:

$$Y_1(k) = W(k) - (a+b)W(k-1) + ab\,W(k-2)$$
or
$$H_1(z) = (1 - az^{-1})(1 - bz^{-1}) \tag{2.88}$$

where $Y_1(k)$ is the system output, $H_1(z)$ is the MA transfer function and $0 < a < 1$, $0 < b < 1$. This is a case where both zeros lie inside the unit circle.

Maximum Phase MA System:

$$Y_2(k) = W(k) - (a+b)W(k+1) + ab\,W(k+2) \tag{2.89}$$

or

$$H_2(z) = (1 - az)(1 - bz).$$

Both zeros of the system lie outside the unit circle at $(1/a)$ and $(1/b)$, where $0 < a < 1$, $0 < b < 1$.

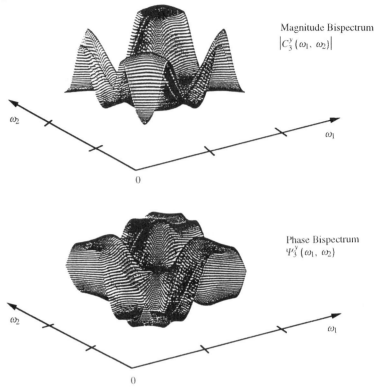

Figure 2.7 Bispectrum of the output of an AR system driven by colored non-Gaussian noise (Example 2.9)

Nonminimum or Mixed Phase MA System:

$$Y_3(k) = -a\,W(k+1) + (1+ab)\,W(k) - b\,W(k-1)$$
or
$$H_3(z) = (1-az)(1-bz^{-1}) \tag{2.90}$$

where one of the zeros is inside the unit circle at (b) and the other one outside at $(1/a)$, where $0 < a < 1$, $0 < b < 1$.

The output sequences $\{Y_1(k)\}$, $\{Y_2(k)\}$, $\{Y_3(k)\}$ have identical autocorrelations given by

$$c_2^y(\tau) = E\{Y_1(k)\,Y_1(k+\tau)\} = E\{Y_2(k)\,Y_2(k+\tau)\}$$
$$= E\{Y_3(k)\,Y_3(k+\tau)\}$$

or
$$\begin{aligned} c_2^y(0) &= 1 + a^2b^2 + (a+b)^2 \\ c_2^y(1) &= -(a+b)(1+ab) \\ c_2^y(2) &= ab \\ c_2^y(\tau) &= 0 \quad \text{for} \quad \tau > 2, \end{aligned} \quad (2.91)$$

which implies, of course, that they have identical power spectra; i.e.,

$$C_2^y(\omega) = \gamma_2^w \ |H_1(z)|^2 \ = \gamma_2^w \ |H_2(z)|^2 = \gamma_2^w \ |H_3(z)|^2 \quad (2.92)$$

for $z = \exp\{j\omega\}$. FIR systems $H_1(z), H_2(z)$, and $H_3(z)$ are said to be spectrally equivalent.

This is an expected result because the magnitude squared of a MA transfer function does not recognize a zero (z_o) from its reciprocal $(1/z_o^*)$. On the other hand, the output sequences $\{Y_1(k)\}, \{Y_2(k)\}, \{Y_3(k)\}$ have different nth-order cumulants and thus different cumulant spectra. Table 2.2 illustrates the third-order cumulants of each of the output sequences. From this table, it is apparent that the bispectrum of the system output is capable of "seeing" its true phase character, unlike the power spectrum that "sees" the system as always being minimum phase.

Let us consider now the general case of a real autoregressive moving average ARMA (p,q) process $\{Y(k)\}$ described by the equation

$$\sum_{i=0}^{p} a_i \ Y(k-i) = \sum_{l=0}^{q} b_l \ W(k-l), \quad a_o = 1 \quad (2.93)$$

where $\{W(k)\}$ is non-Gaussian, white and $Y(k)$ is independent of $W(k)$ for $k < m$. Since $\{W(k)\}$ is also nth-order stationary, it follows that $\{Y(k)\}$ is nth-order stationary, assuming it is a stable ARMA model. The transfer function of the ARMA model is given by

$$H(z) = B(z) \Big/ A(z)$$

where

$$\begin{aligned} A(z) &= \sum_{i=0}^{p} a_i \ z^{-i} \\ B(z) &= \sum_{l=0}^{q} b_l \ z^{-l}. \end{aligned} \quad (2.94)$$

The nth-order cumulant spectrum of $Y(k)$ is given by (2.80) where $H(\omega) = H(z)|_{z=\exp j\omega}$.

If $W(k)$ is Gaussian (and hence $\{Y(k)\}$), then any real zero (z_r) of $H(z)$ can be replaced by its inverse $(1/z_r)$ and pairs of non-zero conjugate roots (z_o) by their paired conjugated inverses $(1/z_o^*)$ without changing the power spectrum of the output $\{Y(k)\}$ and therefore its probability structure. For example, with real distinct roots there are 2^{p+q} ways of choosing the roots without changing the power spectrum $C_2^y(\omega)$; i.e., resulting in the same autocorrelation sequence. This implies that there are 2^{p+q} ways of choosing the phase $\phi_h(\omega)$ in (2.76) without changing the magnitude $|H(\omega)|$. For example, if $p = 0$ and $q = 2$ (second-order MA model), then there are four ways of choosing the roots without changing the output autocorrelation sequence. These are illustrated in Table 2.2.

In the case of non-Gaussian structure satisfying (2.93), each one of the 2^{p+q} zero patterns corresponds to a different process [Rosenblatt, 1980]. Thus, the true system transfer function can be estimated from cumulant spectra ($n > 2$) if $\gamma_n^w \neq 0$ for some n. Methods for nonminimum phase system identification when the input is inaccessible that utilize cumulants of the system output only are discussed extensively in Chapters 5, 6, and 7.

Example 2.10

Let us consider the following 6th-order MA systems

$$\begin{aligned}
H_1(z) &= (1 - 0.85z^{-1})(1 - 1.2z^{-1} + 0.45z^{-2})(1 - 0.869z^{-1}) \\
&\quad (1 + 1.1z^{-1} + 0.617z^{-2}) \\
H_2(z) &= (1 - 0.85z^{-1})(1 - 1.2z^{-1} + 0.45z^{-2})(1 - 0.869z) \\
&\quad (1 + 1.1z + 0.617z^2) \\
H_3(z) &= (1 - 0.85z^{-1})(1 - 1.2z^1 + 0.45z^2)(1 - 0.869z)(1 + 1.1z + 0.617z^2)
\end{aligned}$$

that have identical magnitude response but different phase responses. The first system is minimum phase, whereas the other two are nonminimum phase systems. Figure 2.8 illustrates the zero locations, impulse responses, magnitude and phase responses of $H_1(z)$, $H_2(z)$, and $H_3(z)$.

Assume that each of the MA systems is driven by non-Gaussian, zero-mean, white noise with variance $\gamma_2^w = 1$ and skewness $\gamma_3^w = 1$. The power spectrum of the systems' output follows from (2.81), (2.82), and (2.83);

$$C_2^y(\omega) = |H_1(\omega)|^2 = |H_2(\omega)|^2 = \frac{1}{2\pi}|H_3(\omega)|^2.$$

Using (2.82), the magnitude bispectra are

TABLE 2.2 MINIMUM, MAXIMUM, AND NONMINIMUM PHASE PROCESSES WITH IDENTICAL POWER SPECTRA (OR AUTOCORRELATIONS), $0 < a < 1, 0 < b < 1$.

	Minimum Phase	Maximum Phase	Nonminimum Phase I	Nonminimum Phase II
$H(z)$	$(1-az^{-1})(1-bz^{-1})$	$(1-az)(1-bz)$	$(1-az)(1-bz^{-1})$	$(1-az^{-1})(1-bz)$
$Y(k)$	$W(k)-(a+b)W(k-1)$ $+abW(k-2)$	$W(k)-(a+b)W(k+1)$ $+abW(k+2)$	$-aW(k+1)+(1+ab)W(k)$ $-bW(k-1)$	$-bW(k+1)+(1+ab)W(k)$ $-aW(k-1)$
Zero Location	$\lvert z\rvert=1$, (a,b)	$\lvert z\rvert=1$, $\left(\frac{1}{a},\frac{1}{b}\right)$	$\lvert z\rvert=1$, $\left(\frac{1}{a},b\right)$	$\lvert z\rvert=1$, $\left(a,\frac{1}{b}\right)$
Third-Order Cumulants $c_3^y(0,0)$ $c_3^y(1,1)$ $c_3^y(2,2)$ $c_3^y(1,0)$ $c_3^y(2,0)$ $c_3^y(2,1)$	$1-(a+b)^3+a^3b^3$ $(a+b)^2-(a+b)a^2b^2$ a^2b^2 $-(a+b)+ab(a+b)^2$ ab $-(a+b)ab$	$1-(a+b)^3+a^3b^3$ $-(a+b)^2+ab(a+b)^2$ ab $(a+b)^2-(a^2+b^2)$ a^2b^2 $-(a+b)ab$	$(1+ab)^3-a^3-b^3$ $-a(1+ab)^2+(1+ab)^2b$ $-a^2b$ $a^2(1+ab)-(1+ab)^2b$ $-a^2b$ $ab(1+ab)$	$(1+ab)^3-a^3-b^3$ $-b(1+ab)^2+(1+ab)a^2b$ $-ba^2$ $b^2(1+ab)-(1+ab)^2a$ $-b^2a$ $ab(1+ab)$
Auto-correlations $c_2^y(0)$ $c_2^y(1)$ $c_2^y(2)$	$1+a^2b^2+(a+b)^2$ $-(a+b)(1+ab)$ ab			

$$W(k) \longrightarrow \boxed{H(z)} \longrightarrow Y(k)$$

$$|C_3^{y_1}(\omega_1,\omega_2)| = |C_3^{y_2}(\omega_1,\omega_2)| = |C_3^{y_3}(\omega_1,\omega_2)|$$
$$= |H_1(\omega_1)|\,|H(\omega_2)||H(\omega_1+\omega_2)|,$$

and from (2.83) the phase bispectra are given by

$$\Psi_3^{y_1}(\omega_1,\omega_2) = \phi_{h_1}(\omega_1) + \phi_{h_1}(\omega_2) - \phi_{h_1}(\omega_1+\omega_2),$$
$$\Psi_3^{y_2}(\omega_1,\omega_2) = \phi_{h_2}(\omega_1) + \phi_{h_2}(\omega_2) - \phi_{h_2}(\omega_1+\omega_2),$$
$$\Psi_3^{y_3}(\omega_1,\omega_2) = \phi_{h_3}(\omega_1) + \phi_{h_3}(\omega_2) - \phi_{h_3}(\omega_1+\omega_2).$$

Figure 2.8 also illustrates the power spectrum and magnitude and phase bispectra, and it clearly demonstrates the ability of the bispectrum to preserve correct phase information. □

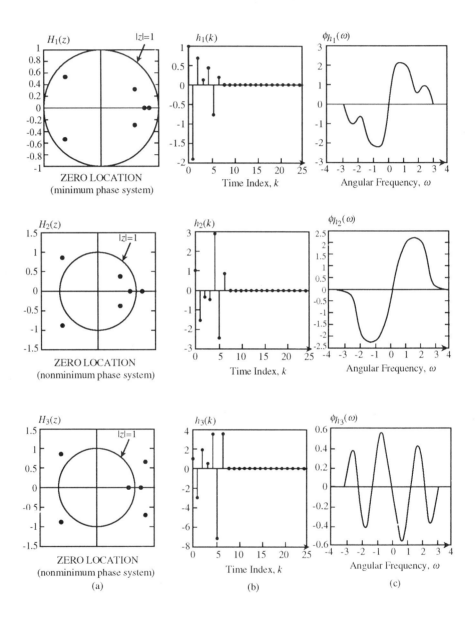

Figure 2.8 MA(6) LTI systems of Example 2.10: (a) zero location, (b) impulse response, (c) phase response, (d) magnitude response and power spectrum, (e) magnitude bispectrum, (f) phase bispectrum.

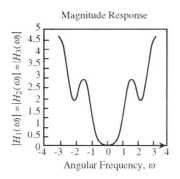
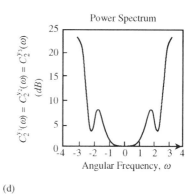

(d)

$$\left|C_3^{y_1}(\omega_1,\omega_2)\right| = \left|C_3^{y_2}(\omega_1,\omega_2)\right| = \left|C_3^{y_3}(\omega_1,\omega_2)\right|$$

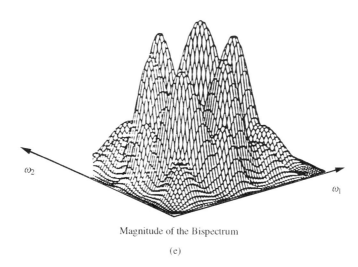

Magnitude of the Bispectrum

(e)

Figure 2.8 (Continued) MA(6) LTI systems of Example 2.10: (a) zero location, (b) impulse response, (c) phase response, (d) magnitude response and power spectrum, (e) magnitude bispectrum, (f) phase bispectrum.

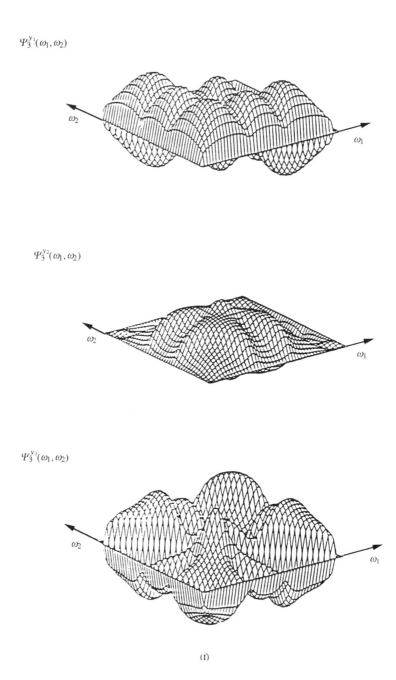

(f)

Figure 2.8 (Continued) MA(6) LTI systems of Example 2.10: (a) zero location, (b) impulse response, (c) phase response, (d) magnitude response and power spectrum, (e) magnitude bispectrum, (f) phase bispectrum.

2.4.4 Linear Processes with Gaussian and Non-Gaussian Components

There are signal processing problems where the observed signal consists of a non-Gaussian stationary random process embedded in additive Gaussian noise. For example, digital communication signals are usually assumed to follow a discrete symmetric distribution; they traverse a linear channel with additive white Gaussian noise (AWGN). Consequently, there will be certain advantages if we estimate signal parameters in cumulant spectrum domains because the additive Gaussian noise will be suppressed (in theory). Consider the process $S(k)$ shown in Figure 2.9, which is the sum of three processes; viz:

$$S(k) = X(k) + Y(k) + Z(k) \qquad (2.95)$$

where $\{X(k)\}$ is an AR(p_1) Gaussian process shown in (2.96), generated by

$$X(k) = -\sum_{i=1}^{p_1} a_i\, X(k-i) + E(k). \qquad (2.96)$$

$\{E(k)\}$ is a Gaussian, white, zero-mean process with variance $\gamma_2^E \neq 0$; $\{Y(k)\}$ is a MA(q) non-Gaussian process generated by

$$Y(k) = \sum_{i=0}^{q} b_i\, N(k-i) \qquad (2.97)$$

where $\{N(k)\}$ is a non-Gaussian, white, zero-mean noise process with non-zero variance ($\gamma_2^N \neq 0$), skewness, ($\gamma_3^N \neq 0$) and, kurtosis ($\gamma_4^N \neq 0$). $\{Z(k)\}$ is an AR(p_2) non-Gaussian process generated by

$$Z(k) = -\sum_{i=1}^{p_2} c_i\, Z(k-i) + W(k) \qquad (2.98)$$

where $\{W(k)\}$ is also non-Gaussian, white, zero-mean noise with non-zero variance ($\gamma_2^W \neq 0$), zero skewness ($\gamma_3^W = 0$), and non-zero kurtosis ($\gamma_4^W \neq 0$). The noises $\{E(k)\}$, $\{N(k)\}$, and $\{W(k)\}$ are assumed to be independent, which implies that $\{X(k)\}$, $\{Y(k)\}$, and $\{Z(k)\}$ are statistically independent.

From (2.95) and the assumptions made above, we have that the cumulant spectrum of order n of $\{S(k)\}$ is the sum of the individual cumulant spectra, i.e.,

$$C_n^s(\omega_1, \ldots, \omega_{n-1}) = C_n^x(\omega_1, \ldots, \omega_{n-1}) + C_n^y(\omega_1, \ldots, \omega_{n-1}) \\ + C_n^z(\omega_1, \ldots, \omega_{n-1}). \qquad (2.99)$$

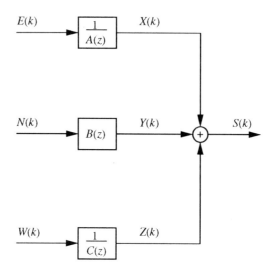

Figure 2.9 The output process $S(k)$ has ARMA power spectrum, MA bispectrum, and ARMA trispectrum. Input processes $\{E(k)\}, \{N(k)\}$, and $\{W(k)\}$ are assumed to be independent white noise processes.

Important special cases follow.

Power Spectrum (n=2):

$$C_2^s(\omega) = C_2^x(\omega) + C_2^y(\omega) + C_2^z(\omega) \tag{2.100}$$

where

$$C_2^x(\omega) = \gamma_2^E \cdot \frac{1}{|A(\omega)|^2}; A(\omega) = \sum_{i=0}^{P_1} a_i e^{-j\omega i}$$

$$C_2^y(\omega) = \gamma_2^N \cdot |B(\omega)|^2; B(\omega) = \sum_{i=0}^{q} b_i e^{-j\omega i}$$

and

$$C_2^z(\omega) = \gamma_2^W \cdot \frac{1}{|C(\omega)|^2}; C(\omega) = \sum_{i=0}^{P_2} c_i e^{-ij\omega i}.$$

By substituting the individual power spectra into (2.100) we obtain

$$C_2^s(\omega) = \left\{ \frac{\gamma_2^E \cdot |C(\omega)|^2 + \gamma_2^N |A(\omega)|^2 \cdot |B(\omega)|^2 \cdot |C(\omega)|^2 + \gamma_2^W |A(\omega)|^2}{|A(\omega)|^2 \cdot |C(\omega)|^2} \right\} \quad (2.101)$$

which is an ARMA power spectrum because it can be written as a ratio of two polynomials.

Bispectrum (n=3):

Since the process $\{X(k)\}$ is Gaussian, all its cumulant spectra of order $n > 2$ are zero. Furthermore, since $\{W(k)\}$ has zero skewness, from (2.98) we have that $\{Z(k)\}$ has zero bispectrum. Therefore, (2.99) becomes

$$\begin{aligned} C_3^s(\omega_1, \omega_2) &= C_3^y(\omega_1, \omega_2) \\ &= \gamma_3^N \cdot B(\omega_1) B(\omega_2) B^*(\omega_1 + \omega_2); \end{aligned} \quad (2.102)$$

i.e., the process $\{S(k)\}$ has MA bispectrum which is identical to that of $\{Y(k)\}$. Processes $\{X(k)\}$ and $\{Z(k)\}$ do not contribute any information to the bispectrum (or third-order cumulants) of $\{S(k)\}$.

Trispectrum (n=4):

In this case only the Gaussian process $\{X(k)\}$ does not contribute to the trispectrum of (2.99), and therefore

$$C_4^s(\omega_1, \omega_2, \omega_3) = C_4^y(\omega_1, \omega_2, \omega_3) + C_4^z(\omega_1, \omega_2, \omega_3) \quad (2.103)$$

where

$$C_4^y(\omega_1, \omega_2, \omega_3) = \gamma_4^N B(\omega_1) \cdot B(\omega_2) \cdot B(\omega_3) \cdot B^*(\omega_1 + \omega_2 + \omega_3)$$

and

$$C_4^z(\omega_1, \omega_2, \omega_3) = \gamma_4^N \cdot \frac{1}{C(\omega_1) C(\omega_2) C(\omega_3) C^*(\omega_1 + \omega_2 + \omega_3)}.$$

We see that $\{S(k)\}$ has an ARMA trispectrum. However, the ARMA model that describes the trispectrum of $\{S(k)\}$ is **different** from the ARMA model that describes its power spectrum.

This example, which is also illustrated in Figure 2.9, clearly shows that a linear model that describes the power spectrum of a process may in general be different from one that describes its bispectrum and different from the model that describes its trispectrum. In this specific example, the autocorrelation of $\{S(k)\}$ may be described exactly by an ARMA model, its third-order cumulants by an MA model, and its fourth-order cumulants by a different ARMA model. More examples that address this issue can be found in Hasselman et al. [1963] and Raghuveer and Nikias [1985].

Conversely, suppose that a stationary process has power spectrum $C_2(\omega)$, bispectrum $C_3(\omega_1, \omega_2)$, and trispectrum $C_4(\omega_1, \omega_2, \omega_3)$. Then finding a LTI system with transfer function $H(\omega)$ to match the power spectrum amounts to solving the spectral factorization problem

$$C_2(\omega) = |H(\omega)|^2, \qquad (2.104)$$

whereas finding a LTI system with transfer function $G(\omega)$ to match the bispectrum is equivalent to solving [Raghuveer and Nikias, 1985]

$$C_3(\omega_1, \omega_2) = G(\omega_1)G(\omega_2)G^*(\omega_1 + \omega_2). \qquad (2.105)$$

Similarly, in order to find $F(\omega)$ to match the trispectrum, we must solve

$$C_4(\omega_1, \omega_2, \omega_3) = F(\omega_1)F(\omega_2)F(\omega_3)F^*(\omega_1 + \omega_2 + \omega_3). \qquad (2.106)$$

The solution to (2.105) or (2.106), when it exists, is generally different from that of (2.104). Also, the solution to (2.105) will generally be different from that of (2.106). The existence of higher-order spectrum factorization, as well as algorithms for testing factorization, have been established by Tekalp and Erdem [1989] and Raghuveer and Dianat [1989].

2.4.5 Poisson Triggered Processes

Huber et al. [1971] studied the bispectrum of a Poisson triggered process $\{X(k)\}$ described by

$$X(k) = \sum_m h(k - T_m) \qquad (2.107)$$

where $\ldots, T_{-1}, T_0, T_1, \ldots$ are the times of events of a Poisson process, with $E\{T_{m+1} - T_m\} = \mu$. Assuming that $\{h(k)\}$ is deterministic and is related to $H(\omega)$ as shown in (2.74), the power spectrum and the bispectrum of the process $\{X(k)\}$ are

$$C_2^x(\omega) = 1|H(\omega)|^2 \qquad (2.108)$$

$$C_3^x(\omega_1, \omega_2) = \frac{1}{\mu} H(\omega_1)H(\omega_2)H^*(\omega_1 + \omega_2). \qquad (2.109)$$

Comparing (2.108) with (2.81), and (2.109) with (2.82) and (2.83), we note that the Poisson triggered process $\{X(k)\}$ has essentially the same power spectrum and bispectrum as the linear process (2.73). From (2.108), and (2.109), we also see that

$$\begin{aligned}C_2^x(\omega) &= C_3^x(\omega,0) \cdot \frac{1}{H(0)} \\ &= C_3^x(0,\omega) \cdot \frac{1}{H(0)},\end{aligned} \quad (2.110)$$

which implies that the power spectrum may be reconstructed from the bispectrum, provided $H(0) \neq 0$.

When the process $\{X(k)\}$ of (2.107) acts as additive noise (impulsive or shapes of spikes) to a stationary Gaussian process $\{Y(k)\}$; i.e.,

$$Z(k) = Y(k) + X(k) \quad (2.111)$$

where $\{Y(k)\}$ and $\{X(k)\}$ are statistically independent, then the power spectrum and bispectrum of $\{Z(k)\}$ are given by

$$C_2^z(\omega) = C_2^y(\omega) + C_2^x(\omega) \quad (2.112)$$

and

$$C_3^z(\omega_1,\omega_2) = C_3^x(\omega_1,\omega_2). \quad (2.113)$$

Using (2.110) and (2.113) we can reconstruct the power spectrum, $C_2^x(\omega)$, from the bispectrum and use it in (2.112) to calculate $C_2^y(\omega)$ from $C_2^z(\omega)$.

2.4.6 Regression Coefficients of Linear Processes

If we assume a LTI system described by (2.73) or (2.74) where $\{X(k)\}$ denotes its input and $\{Y(k)\}$ the output, then the complex regression coefficient $R(\omega)$ in (2.61) takes the form [Brillinger, 1977; Rosenblatt, 1985]

$$R(\omega) = H(\omega) \quad (2.114)$$

because $C_{xy}(\omega) = H(\omega) \cdot C_{xx}(\omega)$. In other words, $R(\omega)$ provides the transfer function of the best linear filter for predicting the sequence $\{Y(k)\}$ from $\{X(k)\}$. The transfer function $H(\omega)$ may also be obtained by using higher-order regression coefficients. For example, from (2.63) and (2.75) it follows that

$$R_2(\omega_1,\omega_2) = H(\omega_1) \quad (2.115)$$

because $C_{yxx}(\omega_1,\omega_2) = H(\omega_1)C_{xxx}(\omega_1,\omega_2)$. The importance of $R_2(\omega_1,\omega_2)$ shows in linear system identification when the cumulant spectra and cross-polyspectra of input and output measurements are being used. Further, regression coefficients $R(\omega)$ and $R_2(\omega_1,\omega_2)$ may serve as the basis for time delay estimation methods with two sensor measurements.

2.5 DETECTING AND IDENTIFYING NONLINEAR PROCESSES

Higher-order moment and cumulant spectra or polyspectra provide a means of detecting and quantifying nonlinearities in stochastic signals. These stochastic signals usually arise when a nonlinear system operates under a random input. General relations for arbitrary stationary random data passing through arbitrary linear systems have been studied quite extensively for many years. In principle, most of these relations are based on autocorrelation, power spectrum, or cross-correlation matching criteria. On the other hand, general relations are not available for arbitrary stationary random data passing through arbitrary nonlinear systems. Instead, each type of nonlinearity has been investigated as a special case [Akaike, 1964; 1966]. Polyspectra could play a key role in detecting and characterizing the type of nonlinearity in a system from its output sequence. In addition, cross-polyspectra may be used for nonlinear system identification from observations of input and output data.

2.5.1 Sinusoidal Processes Applied to Nonlinear Systems

Consider a linear time-invariant (LTI) system as shown in Figure 2.10(a) with input

$$X(k) = \sum_m A_m \exp\{j(\lambda_m k + \phi_m)\} \qquad (2.116)$$

where $\{\phi_m\}$ are independent, identically distributed (i.i.d.) random variables, uniformly distributed over $[-\pi, +\pi]$. The output of the LTI system, $\{Y(k)\}$, is given by

$$Y(k) = \sum_m B_m \exp\{j(\lambda_m k + \theta_m)\}. \qquad (2.117)$$

The output is also a superposition of sinusoidal signals with the same frequencies as the input sinusoids. If we now pass the same process $\{X(k)\}$ through a nonlinear system of a Volterra type, then the output $Z(k)$ will be given by

$$Z(k) = \sum_{i_1=0}^{M-1} \cdots \sum_{i_N=0}^{M-1} h_{12\ldots N}(i_1, i_2, \ldots, i_N) X(k-i_1) X(k-i_2) \ldots X(k-i_N) \qquad (2.118)$$

Chapter 2	59

which will contain sinusoidal signals of the form

$$\exp\left\{j\left[\left(\sum_i \lambda_i\right)k + \left(\sum_i \phi_i\right)\right]\right\}. \tag{2.119}$$

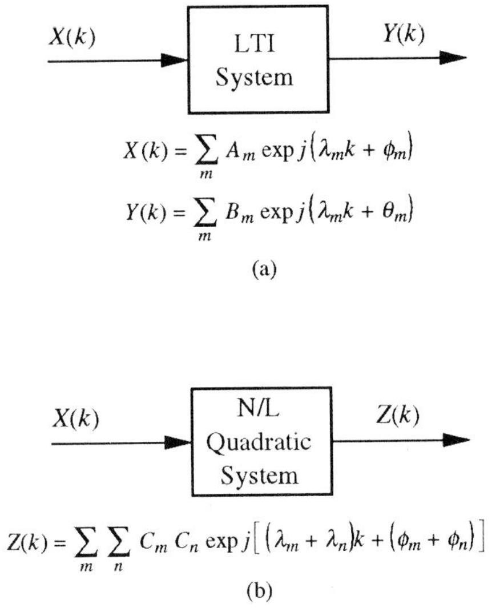

Figure 2.10 (a) Output of a linear time-invariant (LTI) system to sinusoidal inputs; (b) Output of a quadratic system to sinusoidal inputs.

We see that in the case of a nonlinear system the sinusoidal outputs are frequency and phase related to the input [Brillinger, 1965; 1977]. Consider a second-order Volterra filter (quadratic) in (2.118). Then $Z(k)$ will take the form [Figure 2.10(b)]

$$Z(k) = \sum_m \sum_n C_m C_n \exp\{j[(\lambda_m + \lambda_n)k + \phi_m + \phi_n)]\}. \tag{2.120}$$

Notice that while the third-order cumulants (bispectrum) of $Y(k)$ in (2.117) are identical to zero, the bispectrum of $Z(k)$ in (2.120) will not vanish because the third-order cumulants $c_3^z(\tau_1, \tau_2)$ are different from zero. Thus a nonzero bispectrum in the output of a system with input $X(k)$ will imply the presence of a quadratic nonlinearity in the system.

2.5.2 Quadratic Phase Coupling

There are situations in practice in which the interaction between two harmonic components causes contribution to the power at their sum and/or difference frequencies. For example, suppose that (2.116) takes the form

$$X(k) = A_1 \cos(\lambda_1 k + \theta_1) + A_2 \cos(\lambda_2 k + \theta_2)$$

and the nonlinear system in (2.118) is a simple quadratic of the form

$$Z(k) = X(k) + \varepsilon X^2(k) \qquad (2.121)$$

where ε is a nonzero constant. The signal $Z(k)$ contains cosinusoidal terms in $(\lambda_1, \theta_1), (\lambda_2, \theta_2), (2\lambda_1, 2\theta_1), (2\lambda_2, 2\theta_2), (\lambda_1 + \lambda_2, \theta_1 + \theta_2)$, and $(\lambda_1 - \lambda_2, \theta_1 - \theta_2)$. Such a phenomenon, which gives rise to certain phase relations of the same type as the frequency relations, is called **quadratic phase coupling** [Hasselman et al., 1963; Brillinger and Tukey, 1984; Kim and Powers, 1978; Raghuveer and Nikias, 1985]. In certain applications it is necessary to determine if peaks at harmonically related positions in the power spectrum are, in fact, phase coupled. Since the power spectrum suppresses all phase relations, it cannot provide the answer. The third-order cumulants (the bispectrum), however, are capable of detecting and characterizing quadratic phase coupling. This is best illustrated in the following example.

Example 2.11

Consider the signals

$$X_1(k) = \cos(\lambda_1 k + \theta_1) + \cos(\lambda_2 k + \theta_2) + \cos(\lambda_3 k + \theta_3) \qquad (2.122)$$

and

$$X_2(k) = \cos(\lambda_1 k + \theta_1) + \cos(\lambda_2 k + \theta_2) + \cos(\lambda_3 k + (\theta_1 + \theta_2)) \qquad (2.123)$$

where $\lambda_3 = \lambda_1 + \lambda_2$; i.e., $(\lambda_1, \lambda_2, \lambda_3)$ are said to be harmonically related and $\theta_1, \theta_2, \theta_3$ are independent random variables uniformly distributed between $[-\pi, +\pi]$. From (2.122) it is apparent that λ_3 is an independent harmonic component because θ_3 is an independent random-phase variable. On the other hand, λ_3 of $X_2(k)$ in (2.123) is a result of phase coupling between λ_1 and λ_2. From (2.122) and (2.123) we see that $E\{X_1(k)\} = E\{X_2(k)\} = 0$. Thus from (2.122) and (2.123), we have that the autocorrelation sequences of the signals are given by

$$c_2^{x_1}(\tau_1) = c_2^{x_2}(\tau_1) = \frac{1}{2}\{\cos(\lambda_1 \tau_1) + \cos(\lambda_2 \tau_1) + \cos(\lambda_3 \tau_1)\}; \qquad (2.124)$$

i.e., $\{X_1(k)\}$ and $\{X_2(k)\}$ have identical power spectra consisting of impulses at λ_1, λ_2 and $\lambda_3 = \lambda_1 + \lambda_2$ [Figure 2.11(a)].

On the other hand, the third-order cumulants of the signals can be easily obtained [Raghuveer and Nikias, 1985]

$$\begin{aligned}
c_3^{x_1}(\tau_1, \tau_2) &\equiv 0 \\
c_3^{x_2}(\tau_1, \tau_2) &= \frac{1}{4}\{\cos(\lambda_2\tau_1 + \lambda_1\tau_2) + \cos(\lambda_3\tau_1 - \lambda_1\tau_2) \\
&\quad + \cos(\lambda_1\tau_1 + \lambda_2\tau_2) + \cos(\lambda_3\tau_1 - \lambda_2\tau_2) \\
&\quad + \cos(\lambda_1\tau_1 - \lambda_3\tau_2) + \cos(\lambda_2\tau_1 - \lambda_3\tau_2)\}.
\end{aligned} \quad (2.125)$$

Therefore, the bispectrum of $\{X_1(k)\}$ is identically zero whereas the bispectrum of $\{X_2(k)\}$ shows an impulse in the triangular region $\omega_2 \geq 0$, $\omega_1 \geq \omega_2$, $\omega_1 + \omega_2 \leq \pi$ (Figure 2.11(b)). The impulse is located at (λ_1, λ_2) if $\lambda_1 \geq \lambda_2$.

Harmonically related peaks in the power spectrum (e.g., $\lambda_1+\lambda_2, \lambda_1-\lambda_2, 2\lambda_1, 2\lambda_2$) are necessary conditions for the presence of quadratic nonlinearities in the data. However, the power spectrum suppresses phase relations of harmonic components and therefore fails to discriminate $\{X_1(k)\}$ from $\{X_2(k)\}$, as shown in Figure 2.11(a). On the other hand, we have demonstrated above that the bispectrum does preserve quadratic phase relations, and therefore becomes useful for extracting them quantitatively. We elaborate more on quadratic, as well as cubic phase coupling and their applications, in Chapter 10. □

2.5.3 Gaussian Processes Applied to Nonlinear Systems

Let $\{X(k)\}$ be a Gaussian stationary process with zero-mean, autocorrelation $\{c_2^x(\tau)\}$, and power spectrum $C_2^x(\omega)$. Let

$$Y(k) = G[X(k)] \quad (2.126)$$

where $G[\cdot]$ is a nonlinear function. If $G[\cdot]$ is odd, $G(-x) = -G(x)$, the bispectrum of $\{Y(k)\}$ vanishes [Huber et al., 1971]. Brillinger [1965] points out that many nonlinear functional relationships may be approximated by relationships of the form

$$G(x) = x + \varepsilon x^{n-1} \quad (2.127)$$

where ε is small such that ε^3 is negligible. Combining (2.126) with (2.127), we obtain

$$Y(k) = X(k) + \varepsilon X(k)^{n-1}. \quad (2.128)$$

Let us note that when $n = 3$, (2.128) becomes identical to (2.121). For the special case of a quadratic system $(n = 3)$, the autocorrelation sequence and the power spectrum of $\{Y(k)\}$ are given by

$$\begin{aligned} m_2^y(\tau) &= m_2^x(\tau) + 2\varepsilon^2 [m_2^x(\tau)]^2 + \varepsilon^2 [m_2^x(0)]^2 \\ M_2^y(\omega) &= M_2^x(\omega) + 2\varepsilon^2 \cdot \int M_2^x(\lambda) M_2^x(\omega - \lambda) d\lambda + \varepsilon^2 [m_2^x(0)]^2 \delta(\omega). \end{aligned} \quad (2.129)$$

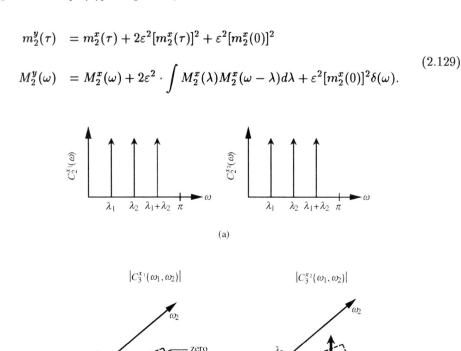

Figure 2.11 Quadratic Phase coupling (a) power spectra of $X_1(k), X_2(k)$ given by (2.122), (2.123), respectively, (b) their magnitude bispectra.

Its third-order moments and third-order moment spectrum are given by

$$\begin{aligned} m_3^y(\tau_1, \tau_2) &= 2\varepsilon\{m_2^x(\tau_1)m_2^x(\tau_2) \quad +m_2^x(\tau_1)m_2^x(\tau_2 - \tau_1) \\ &\qquad\qquad\qquad\qquad\quad +m_2^x(\tau_2)m_2^x(\tau_2 - \tau_1)\} \\ &\quad + \varepsilon\{m_2^x(\tau_1) + m_2^x(\tau_2) \quad +m_2^x(\tau_2 - \tau_1)\} \cdot m_2^x(0) \\ M_3^y(\omega_1, \omega_2) &= 2\varepsilon\{M_2^x(\omega_1)M_2^x(\omega_2) \quad +M_2^x(\omega_1)M_2^x(\omega_1 + \omega_2) \\ &\qquad\qquad\qquad\qquad\quad +M_2^x(\omega_2)M_2^x(\omega_1 + \omega_2)\} \\ &\quad + \varepsilon\{M_2^x(\omega_1)\delta(\omega_2) \quad +M_2^x(\omega_2)\delta(\omega_1) \\ &\qquad\qquad\qquad\qquad\quad +M_2^x(\omega_1)\delta(\omega_1 + \omega_2)\}m_2^x(0). \end{aligned} \quad (2.130)$$

Knowing the power spectrum, $M_2^x(\omega)$, of the input Gaussian process and computing the third-order moment spectrum of the output, $M_3^y(\omega_1, \omega_2)$, we can easily estimate the parameter ε of the quadratic system using identity (2.130). If $\varepsilon = 0$, then $M_2^y(\omega) = M_2^x(\omega)$ and $M_3^y(\omega_1, \omega_2) \equiv 0$. From (2.130) we also observe that

$$M_3^y(\omega, 0) = M_2^x(\omega)\{4\varepsilon M_2^x(0) + 2\varepsilon M_2^x(\omega)\} + \\ + \varepsilon\{M_2^x(\omega)\delta(0) + M_2^x(0)\delta(\omega) + M_2^x(\omega)\delta(\omega)\}m_2^x(0) \quad (2.131)$$

which implies that the power spectrum of $\{X(k)\}$ cannot be easily identified from the bispectrum of $\{Y(k)\}$. Thus, for linear non-Gaussian processes identity (2.87) provides an easy way of reconstructing the power spectrum of the process from its bispectrum. This is not the case, however, for nonlinear processes.

Brillinger [1965] provides the general relationship between the $(n-1)$th-order polyspectra of $Y(k)$ and the power spectrum of $X(k)$. Gaussian random inputs to nonlinear systems result in non-Gaussian outputs with some of the polyspectra components being nonzero (depending on the degrees of nonlinearities in the system). It is, therefore, natural to expect that nonzero output polyspectra may be expressed in terms of the power spectrum of the Gaussian input and the parameters of the nonlinear system. Furthermore, general system identification procedures may be devised based on the output polyspectra and input power spectrum. Some of these methods are discussed in Chapter 10.

Example 2.12

Consider a zero-mean, Gaussian, white process $\{X(k)\}$ with $E\{X^2(k)\} = Q$ as input to a square-law device of the form described by (2.128) with $n = 3$ and $\varepsilon = 0.01$. From (2.129) it follows that the autocorrelation and 2nd-order moment spectrum of the output are respectively given by

$$m_2^y(\tau) = Q(1 + 2\varepsilon^2 Q)\delta(\tau) + \varepsilon^2 Q^2$$
$$M_2^y(\omega) = Q(1 + 2\varepsilon^2 Q) + \varepsilon^2 Q^2 \delta(\omega).$$

Third-order moments and moment spectrum of the output are given by (2.130); i.e.,

$$m_3^y(\tau_1, \tau_2) = 2\varepsilon Q^2 \{\delta(\tau_1)\delta(\tau_2) + \delta(\tau_1)\delta(\tau_2 - \tau_1) + \delta(\tau_2)\delta(\tau_2 - \tau_1)\} + \\ \varepsilon Q^2\{\delta(\tau_1) + \delta(\tau_2) + \delta(\tau_2 - \tau_1)\}$$

and

$$M_3^y(\omega_1, \omega_2) = 6\varepsilon Q^2 + \varepsilon Q^2\{\delta(\omega_2) + \delta(\omega_1) + \delta(\omega_1 + \omega_2)\}$$

Figure 2.12 illustrates the moments and moment spectra of this example. The purpose of this example was to demonstrate the relationship between second- and third-order moments of the output with the variance of the input to a square-law device. □

2.5.4 Linear Non-Gaussian versus Nonlinear Processes

There are situations in practice where it is necessary to be able to distinguish between a linear non-Gaussian process and a nonlinear process. Assuming that both linear and nonlinear processes have nonzero nth-order cumulants, a good test statistic may be chosen to be the nth-order coherence function (normalized magnitude cumulant spectrum) defined in (2.52) [Hinich, 1982; Rao and Gabr, 1984]. If the non-Gaussian process is linear, then its nth-order cumulant spectrum, $C_n^y(\omega_1,\ldots,\omega_{n-1})$, is given by (2.80), and its power spectrum $C_2^y(\omega)$ by (2.81). Thus, the nth-order coherency function takes the form

$$P_n^y(\omega_1,\ldots,\omega_{n-1}) = \frac{\gamma_n^x}{(\gamma_2^x)^{n/2}} \frac{H(\omega_1)H(\omega_2)\cdots H(\omega_{n-1})\cdot H^*(\omega_1+\cdots+\omega_{n-1})}{|H(\omega_1)||H(\omega_2)|\cdots|H(\omega_{n-1})|\cdot|H^*(\omega_1+\cdots+\omega_{n-1})|}. \tag{2.132}$$

Consequently, its magnitude response or nth-order coherence index

$$|P_n^y(\omega_1,\ldots,\omega_{n-1})| = \frac{\gamma_n^x}{(\gamma_2^x)^{n/2}} \tag{2.133}$$

is constant at all frequencies. If the linear process is Gaussian, then $\gamma_n^x = 0$ for all $n > 2$ and $|P_n^y(\omega_1,\ldots,\omega_{n-1})| \equiv 0$. On the other hand, if the value of the nth-order coherence index depends upon the frequencies $\omega_1,\omega_2,\ldots,\omega_{n-1}$, then the process $\{Y(k)\}$ is nonlinear. Examples of nonlinear processes with frequency dependent bicoherency, $P_3^y(\omega_1,\omega_2)$, include $\{X_2(k)\}$ of (2.123) and $\{Y(k)\}$ of (2.128). In chapters 4, 5, and 10, we examine detection schemes based on (2.132) that distinguish linear non-Gaussian from nonlinear processes for finite segments of data.

2.6 SUMMARY

In this chapter, we introduced the definitions and properties of cumulants and cumulant spectra of stochastic signals. In particular, we introduced moments, and cumulant spectra of stochastic signals, as well as cumulant spectra of non-Gaussian linear processes. Finally, we discussed detection and identification of nonlinear processes with cumulant spectra.

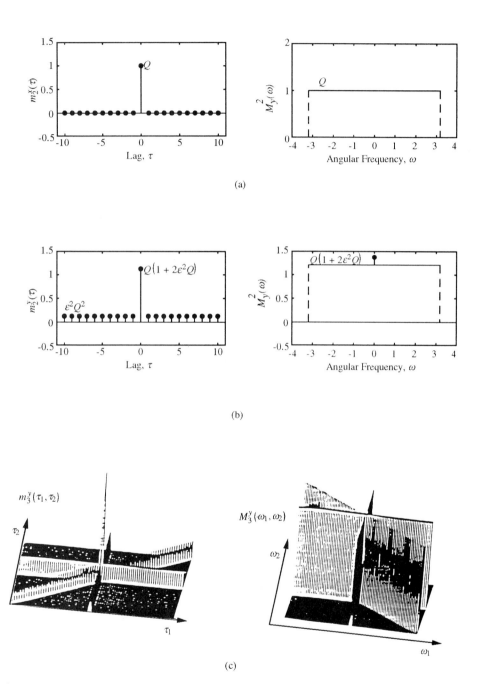

Figure 2.12 White Gaussian input to a square-law device with $\varepsilon = 0.01$. (a) 2nd-order statistics of input process, (b) 2nd-order statistics of output process, (c) 3rd-order statistics of output process.

The first tutorial paper on polyspectra was published in the statistics literature by Brillinger [1965]. The definitions and properties of cumulant spectra are treated in Brillinger [1965], Brillinger and Rosenblatt [1967], Rosenblatt [1985], and Rao and Gabr [1984]. The first engineering tutorial paper on higher-order spectra, emphasizing digital signal processing applications, was published by Nikias and Raghuveer [1987]. The papers by Brillinger [1977], Hasselman et al. [1963], Huber et al. [1971], Kim and Powers [1978], and Raghuveer and Nikias [1985] provide discussions on quadratic phase coupling and detection/identification of nonlinearities. Linear non-Gaussian processes and their higher-order spectral properties are discussed in the papers by Hinich [1982], Lii and Rosenblatt [1988], and Rosenblatt [1980; 1983]. The polyspectra factorization problem can be found in Raghuveer and Nikias [1985], Raghuveer and Dianat [1989], and Tekalp and Erdem [1989]. Finally, definitions of time-frequency bispectra are given in Akaike [1966] and Gerr [1988].

REFERENCES

Akaike, H., Notes on Lectures at the Institute of Statistical Mathematics, *The Statistical Theory of Spectral Estimation*, pp. 1–130, 1964.

Akaike, H., "Note on Higher-Order Spectra," *Ann. Inst. Statist. Math.*, **18**, pp. 123–126, 1966.

Brillinger, D. R., "An Introduction to Polyspectra," *Ann. Math. Statist.*, **36**, pp. 1351–1374, 1965.

Brillinger, D. R., "The Identification of a Particular Nonlinear Time Series System," *Biometrika*, **65**, pp. 509–515, 1977.

Brillinger, D. R., and M. Rosenblatt, "Asymptotic Theory of kth-Order Spectra," in **Spectral Analysis of Time Series**, B. Harris (ed.), pp. 153–188, New York: Wiley, 1967a.

Brillinger, D. R., and M. Rosenblatt, "Computation and Interpretation of kth-Order Spectra," in **Spectral Analysis of Time Series**, B. Harris (ed.), pp. 189–232, New York: Wiley, 1967b.

Brillinger, D. R., and J. W. Tukey, "Spectrum Analysis in the Presence of Noise: Some Issues and Examples" in **The Collected Works of John W. Tukey, II**, Time Series, 1965–1984, D. R. Brillinger (ed.), Belmont, CA: Wadsworth, 1984.

Gerr, N. L., "Introducing a Third-Order Wigner Distribution", *Proc. IEEE*, **76**(3), pp. 290–292, March, 1988.

Giannakis, G. B., and J. M. Mendel, "Identification of Non-Minimum Phase Systems Using Higher-Order Statistics," *IEEE Trans. on Acoustics, Speech and Signal Processing*, **37**, pp. 360–377, March, 1989.

Gurbatov, S. N., A. N. Malakhov, and N. V. Pronchatov-Rubtson, "Applications of Higher-Order Spectra in Problems of the Diagnosis of Strong Acoustic Noise," *Sov. Phys. Acoust.*, **35**(5), pp. 549-550, September-October 1987.

Hasselman, K., W. Munk, and G. MacDonald, "Bispectra of Ocean Waves," in **Time Series Analysis**, M. Rosenblatt (ed.), pp. 125–130, New York: Wiley, pp. 125-130, 1963.

Hinich, M. J., "Testing for Gaussianity and Linearity of a Stationary Time Series," *J. Time Series Analysis*, **3**, pp. 169–176, 1982.

Huber, P. J., B. Kleiner, T. Gasser, and G. Dumermuth, "Statistical Methods for Investigating Phase Relations in Stationary Stochastic Processes", *IEEE Trans. Audio Electroacoust.*, **AU-19**, pp. 78–86, 1971.

Kim, Y. C., and E. J. Powers, "Digital Bispectral Analysis of Self-Excited Fluctuations Spectra," *Phys. Fluids*, **21**(8), pp. 1452–1453, 1978.

Leonov, V. P., and A. N. Shiryaev, "On the Technique of Computing Semi-Invariants," *Theory of Probability Applications*, No. 4, 319–329 (English translation), 1959.

Lii, K. S., and M. R. Rosenblatt, "Estimation and Deconvolution when the Transfer Function has Zeros," *J. Theoretical Probability*, **1**(1), pp. 93–113, January, 1988.

Nagata, Y., "Lag Joint Probability, Higher-Order Covariance Function and Higher-Order Spectrum," *Bulletin de la Société franco-japonaise d'oceanographie*, **8**(2), pp. 78–94, May 1970.

Nikias, C. L., and M. R. Raghuveer, "Bispectrum Estimation: A Digital Signal Processing Framework," *Proc. IEEE*, **75**(7), pp. 869–891, July, 1987.

Papoulis, A., **Probability Random Variables and Stochastic Processes**, New York: McGraw-Hill, 1984.

Pflug, A. L., G. E. Ioup, J. W. Ioup and R. L. Field, "Properties of Higher-Order Correlations and Spectra for Band Limited, Deterministic Transients," *J. Acoust. Soc. Am.*, **91**(2), pp. 975–988, February, 1992.

Raghuveer, M. R., and S. Dianat, "Polyspectra Factorization," *Proceedings ICASSP'89*, pp. 2322–2324, Scotland, U.K., May, 1989.

Raghuveer, M. R., and C. L. Nikias, "Bispectrum Estimation: A Parametric Approach," *IEEE Acoustics, Speech, and Signal Processing*, **ASSP-33**, pp. 1213–1230, 1985.

Rao, T. S., and M. M. Gabr, **An Introduction to Bispectral Analysis and Bilinear Time Series Models**, Lecture Notes in Statistics, **24**, Berlin, Heidelberg: Springer-Verlag, 1984.

Rosenblatt, M., "Linear Processes and Bispectra," *J. Appl. Prob.*, **17**, pp. 265–270, 1980.

Rosenblatt, M., "Cumulants and Cumulant Spectra," in **Handbook of Statistics**, **3**, pp. 369–382, D. R. Brillinger and P. R. Krishnaiah (eds.), Elsevier Science Publishers B. V., 1983.

Rosenblatt, M., **Stationary Sequences and Random Fields**, Boston: Birkhauser, 1985.

Shiryaev, A. N., "Some Problems in the Spectral Theory of Higher-Order Moments I," *Theory Probl. Appl.*, **5**, pp. 265–284, 1960.

Shiryaev, A. N., "On Conditions for Ergodicity of Stationary Processes in Terms of Higher-order Moments," *Theory Probl. Appl.*, **8**, pp. 436–439, 1963.

Sinai, Y. G., "On Higher-Order Spectral Measures of Ergodic Stationary Processes," *Theory Prob. Appl.*, **8**, pp. 429–436, 1963.

Tekalp, A. M., and A. T. Erdem, "Higher-Order Factorization in One and Two Dimensions with Applications in Signal Modeling and Nonminimum Phase System Identification," *IEEE Trans. Acoustics, Speech, and Signal Processing*, **ASSP-37**(10), pp. 1537–1549, October 1989.

Watts, S., "Radar Detection Prediction in Sea Clutter Using the Compound k-Distribution Model," *IEEE Proceedings*, **132**(7), pp. 613-620, December, 1985.

3

MOMENT SPECTRA OF DETERMINISTIC SIGNALS

3.1 INTRODUCTION

Many examples exist in signal processing applications where the value of a signal is known at every instant of time. Such examples include finite duration signals whose value is known to be zero outside the signal duration interval. This class of signals is said to be deterministic, in contrast to stochastic signals (random processes), whose value is not known exactly at each instant of time. The definitions and properties of moments, cumulants, and cumulant spectra of stochastic signals were discussed in Chapter 2. In this chapter we introduce the definitions and properties of higher-order moments and moment spectra of deterministic signals. In particular, the moments and moment spectra of both energy and power deterministic signals are discussed, and several examples are given. We consider only moments in our discussion because there is no clear advantage to using cumulants for the analysis of deterministic signals.

3.1.1 Energy versus Power Deterministic Signals

Signals in general may be classified in accordance with their energy and power content. On one hand, finite energy signals correspond to deterministic transients and have average power equal to zero. Power signals, on the other hand, have average power greater than zero (hence infinite energy) and correspond to deterministic aperiodic or periodic signals and stationary random processes. An excellent treatment of energy and power signals and their "dichotomies" is given by Lob [1985].

If $\{x(k)\}$, $k = 0, \pm 1, \pm 2, \ldots$ is a real deterministic signal, its instantaneous power is given by $|x(k)|^2$ and its total energy is given by

$$E_x = \sum_{k=-\infty}^{+\infty} |x(k)|^2. \tag{3.1}$$

Similarly, the average power in $\{x(k)\}$ is given by

$$P_x = \lim_{M \to \infty} \frac{1}{2M+1} \sum_{k=-M}^{M} |x(k)|^2 \tag{3.2}$$

or

$$P_x = <|x(k)|^2>$$

where $<\cdot>$ corresponds to the time-averaging operation. Throughout this chapter we assume that the sampling period equals one.

Deterministic transients (energy signals) have finite total energy and hence zero average power ($E_x < \infty$, $P_x = 0$), whereas power signals have total average power greater than zero and infinite total energy ($P_x > 0$, $E_x = \infty$). For example, all periodic signals, $x(k) = x(k+N)$, are power signals; the unit step function, i.e., $u(k) = 1$ for $k \geq 0$, and $u(k) = 0$ for $k < 0$, is an aperiodic power signal. On the other hand, finite-duration signals, exponential decays, or damped sinusoidal signals are energy signals. Lob [1985] gives several examples of energy and power signals including certain signals which belong to neither class, such as the unit ramp signal; i.e., $r(k) = k$ for $k \geq 0$ and $r(k) = 0$ for $k < 0$.

3.1.2 Fourier Analysis of Energy Signals

Let $\{x(k)\}$, $k = 0, \pm 1, \pm 2, \ldots$ be a finite-energy deterministic transient signal. From (3.1) and (3.2) it follows that $E_x < \infty$ and $P_x = 0$. The Fourier transform of the signal $x(k)$ is defined as

$$X(\omega) \triangleq \sum_{k=-\infty}^{+\infty} x(k)\, e^{-j\omega k} \tag{3.3}$$

and its inverse Fourier transform as

$$x(k) \triangleq \frac{1}{2\pi} \int_{-\pi}^{+\pi} X(\omega) e^{j\omega k} d\omega. \tag{3.4}$$

Since the signal is assumed to have finite energy, representation (3.3) always exists and the sum converges so that the mean-square error is zero, according to the Riesz-Fisher theorem [Gårding and Hormander, 1988]. The $X(\omega)$ is a periodic function of ω with period 2π and is generally complex. That is

$$X(\omega) = X_R(\omega) + jX_I(\omega) \qquad (3.5)$$

where $X_R(\omega)$ denotes the real part of $X(\omega)$, and $X_I(\omega)$ denotes the imaginary part, or in terms of magnitude and phase

$$X(\omega) = |X(\omega)|\exp\{j\phi_x(\omega)\}. \qquad (3.6)$$

If $\{x(k)\}$ is real, its Fourier transform is conjugate symmetric; i.e., $X(\omega) = X^*(-\omega)$. It then follows that

$$X_R(\omega) = X_R(-\omega) \qquad \text{(even function)} \qquad (3.7)$$

$$X_I(\omega) = -X_I(-\omega) \qquad \text{(odd function)}. \qquad (3.8)$$

Also,

$$|X(\omega)| = |X(-\omega)| \qquad \text{(even function)} \qquad (3.9)$$

and

$$\phi_x(\omega) = -\phi_x(-\omega) \qquad \text{(odd function)}. \qquad (3.10)$$

If the signal (real or complex) is even, that is, it satisfies $x(k) = x^*(-k)$, it follows from (3.3) that $X(\omega) = X^*(\omega)$, which implies that the Fourier transform of the signal is a real function. Consequently, even signals have zero-phase response; i.e., $\phi_x(\omega) = 0$ for $|\omega| \leq \pi$.

Parseval's relation for $\{x(k)\}$ states that

$$E_x = \sum_{k=-\infty}^{+\infty} |x(k)|^2 = \frac{1}{2\pi} \int_{-\pi}^{+\pi} |X(\omega)|^2 \, d\omega, \qquad (3.11)$$

which provides an alternative expression of the total energy E_x of the signal in terms of its Fourier transform. Oppenheim and Schafer [1989] discuss in detail the properties of the Fourier transform of energy signals.

3.1.3 Fourier Analysis of Periodic Power Signals

Consider a periodic sequence $\{\tilde{x}(k)\}$ with period N so that $\tilde{x}(k) = \tilde{x}(k+N)$ for all integer values of k. This and other periodic sequences have total average power

$P_x > 0$ and infinite total energy. The periodic sequence can be represented exactly as a weighted sum of complex exponential sequences with frequencies that are integer multiples of the fundamental frequency $2\pi/N$. However, there are only N distinct complex exponentials having a frequency that is an integer multiple of $2\pi/N$. Thus, the Fourier series representation of a periodic sequence $\{\tilde{x}(k)\}$ has the form [Oppenheim and Schafer, 1989]

$$\tilde{x}(k) = \frac{1}{N} \sum_{\lambda=0}^{N-1} \tilde{X}(\lambda) \, e^{j(2\pi/N)k\lambda}, \text{ where } k = 0, 1, \ldots, N-1 \qquad (3.12)$$

$$\tilde{X}(\lambda) = \sum_{k=0}^{N-1} \tilde{x}(k) \, e^{-j(2\pi/N)k\lambda}, \lambda = 0, 1, \ldots, N-1. \qquad (3.13)$$

Clearly, the Fourier series sequence $\{\tilde{X}(\lambda)\}$ is generally complex and periodic with period N; i.e., $\tilde{X}(\lambda) = \tilde{X}(\lambda + N)$ for any integer value of λ.

We note from (3.2) that the average power in the periodic sequence $\{\tilde{x}(k)\}$ is given by

$$P_{\tilde{x}} = \lim_{M \to \infty} \frac{1}{2M+1} \sum_{k=-M}^{M} |\tilde{x}(k)|^2 = \frac{1}{N} \sum_{k=J}^{J+N-1} |\tilde{x}(k)|^2 \qquad (3.14.1)$$

or

$$\tilde{x} = <|\tilde{x}(k)|^2>_N \qquad (3.14.2)$$

where J, the starting point of the summation, is arbitrary and $< \cdot >_N$ is the time-average operator for periodic sequences. The time-average operator $< \cdot >$ defined in (3.2.2) for power signals becomes $< \cdot >_N$ when the signal is periodic with period N. Consequently, (3.12) can be written as

$$\tilde{x}(k) = <\tilde{X}(\lambda) \, e^{j(2\pi/N)k\lambda}>_N . \qquad (3.15)$$

From (3.13) we observe that for real $\{\tilde{x}(k)\}$ the $\tilde{X}(0)$ is real and that $\tilde{X}(\lambda)$ possesses conjugate (Hermitian) symmetry. In other words, if the periodic signal is real, with Fourier series representation

$$\tilde{X}(\lambda) = |\tilde{X}(\lambda)| \exp\{j\phi_{\tilde{x}}(\lambda)\}, \lambda = 0, 1, \ldots, N-1, \qquad (3.16)$$

it follows that

$$|\tilde{X}(\lambda)| = |\tilde{X}(-\lambda)| \qquad \text{(even sequence)} \qquad (3.17)$$

and

$$\phi_{\tilde{x}} = -\phi_{\tilde{x}}(-\lambda) \qquad \text{(odd sequence)}. \qquad (3.18)$$

Parseval's relation for periodic sequences of period N takes the form

$$\sum_{k=o}^{N-1} |\tilde{x}(k)|^2 = \frac{1}{N} \sum_{\lambda=0}^{N-1} |\tilde{X}(\lambda)|^2, \qquad (3.19)$$

or by combining (3.14.2) and (3.19),

$$P_{\tilde{x}} = <|\tilde{x}(k)|^2>_N = \frac{1}{N} <|\tilde{X}(\lambda)|^2>_N . \qquad (3.20)$$

3.1.4 Finite-Duration Versus Periodic Signals

In the previous two sections we considered the Fourier analysis of energy signals (deterministic transients), and that of a special subclass of power signals, namely, periodic sequences. An important subclass of energy signals is that of finite-duration sequences whose value is **known** to be zero outside of their duration interval. The purpose of this section is to show the relationship between the Fourier representations of finite-duration and periodic sequences, and to establish under what assumptions these relationships hold.

There are many examples in signal processing applications where we are given a signal $\{x(k)\}$ whose value is known only in the duration interval $k = 0, 1, \ldots, N-1$. The value of the signal is unknown outside its observation interval, i.e., for $k = -\infty, \ldots, -1, N, \ldots, +\infty$. Thus we cannot classify the signal as either an energy or a power signal unless we make certain assumptions about its unavailable data samples. Two possibilities, which by no means are the only ones, assume that the unavailable data is either zero or is a periodic repetition of the available data segment. That is

$$y(k) = \begin{cases} x(k), & k = 0, 1, \ldots, N-1 \\ 0, & \text{otherwise} \end{cases} \quad \begin{array}{l}(\text{finite} - \text{duration}) \\ (E_y < \infty, P_y = 0)\end{array} \qquad (3.21)$$

or

$$\tilde{z}(k) = \sum_{r=-\infty}^{+\infty} x(k+rN) \qquad \begin{array}{l}(\text{periodic}), \\ (0 < P_z < \infty, E_z = \infty)\end{array} \qquad (3.22)$$

Clearly, $\{y(k)\}$ is a deterministic transient signal with Fourier transform obtained from (3.3) and written as

$$Y(\omega) = \sum_{k=0}^{N-1} x(k)\, e^{-j\omega k}. \qquad (3.23)$$

Also, from (3.11) and (3.21) we have

$$E_y = \sum_{k=-\infty}^{+\infty} |y(k)|^2 = \sum_{k=0}^{N-1} |x(k)|^2 = \frac{1}{2\pi} \int_{-\pi}^{+\pi} |Y(\omega)|^2 d\omega. \qquad (3.24)$$

On the other hand, the signal $\{\tilde{z}(k)\}$ is periodic with period N, i.e., $\tilde{z}(k) = \tilde{z}(k+N)$. Thus, its Fourier series representation is obtained from (3.13) and written as

$$\tilde{Z}(\lambda) = \sum_{k=0}^{N-1} x(k) \, e^{-j(2\pi/N)k\lambda}, \lambda = 0, 1, \ldots, N-1. \tag{3.25}$$

Furthermore, Parseval's identity (3.20) takes the form

$$P_{\tilde{z}} = \frac{1}{N} \sum_{k=0}^{N-1} |\tilde{z}(k)|^2 = \frac{1}{N} \sum_{k=0}^{N-1} |x(k)|^2 = \frac{1}{N^2} \sum_{\lambda=0}^{N-1} |\tilde{Z}(\lambda)|^2. \tag{3.26}$$

Combining (3.25), (3.23), and (3.15), we obtain [Oppenheim and Schafer, 1989]

$$Y(\omega) = \sum_{\lambda=0}^{N-1} \tilde{Z}(\lambda) \, \Phi\left(\omega - \frac{2\pi}{N}\lambda\right) \tag{3.27}$$

where

$$\Phi(\omega) = \frac{\sin(\omega N/2)}{N \sin(\omega/2)} \cdot \exp\{-j\omega(N-1)/2\} \cdot \exp\{j\pi\lambda(1 - (1/N))\}. \tag{3.28}$$

Since the function $\Phi(\omega)$ has the property

$$\Phi\left(\frac{2\pi}{N}\lambda\right) = \begin{cases} 0, & \lambda = 1, 2, \ldots, N-1 \\ 1, & \lambda = 0, \end{cases}$$

it follows from (3.27) that

$$Y\left(\omega = \frac{2\pi}{N}\lambda\right) = \tilde{Z}(\lambda), \quad \lambda = 0, 1, \ldots, N-1. \tag{3.29}$$

If we substitute (3.27) into (3.24), we obtain

$$E_y = \sum_{\lambda_1=0}^{N-1} \sum_{\lambda_2=0}^{N-1} \tilde{Z}(\lambda_1) \tilde{Z}^*(\lambda_2) \frac{1}{2\pi} \int_{-\pi}^{+\pi} \Phi\left(\omega - \frac{2\pi}{N}\lambda_1\right) \Phi^*\left(\omega - \frac{2\pi}{N}\lambda_2\right) d\omega$$

$$= \frac{1}{N} \sum_{\lambda=0}^{N-1} |\tilde{Z}(\lambda)|^2 = N \cdot P_{\tilde{z}} \tag{3.30}$$

which is identical to (3.26).

The results of this section suggest that if we choose to treat $\{x(k)\}$ $k = 0, 1, \ldots, N-1$ as one period of a periodic sequence $\{\tilde{z}(k)\}$, the Fourier transform of the equivalent finite-duration sequence $\{y(k)\}$ can easily be reconstructed utilizing the interpolation formula (3.27). On the other hand, if we choose to view $\{x(k)\}$, $k = 0, 1, \ldots, N-1$ as a finite duration sequence $\{y(k)\}$ whose value is zero outside the duration interval, the Fourier series representation of the equivalent periodic sequence $\{\tilde{z}(k)\}$ can be obtained by sampling $Y(\omega)$ at every $(2\pi/N)$ points, by utilizing (3.29). Further discussion and examples on the subject can be found in Oppenheim and Schafer [1989] and Lob [1985].

Example 3.1

Consider the signal $x(k) = A \exp\{j\omega_o k\}$ where A and ω_o are constants. From the energy definition (3.1) we have

$$E_x = \sum_{k=-\infty}^{+\infty} |A \exp\{j\omega_o k\}|^2$$

$$= \sum_{k=-\infty}^{+\infty} |A|^2 = \infty,$$

which implies that $x(k)$ is not an energy signal. The average power of the signal is given by (3.2). That is

$$P_x = \lim_{M \to \infty} \frac{1}{2M+1} \sum_{k=-M}^{+M} |A \exp\{j\omega_o k\}|^2$$

$$= <|A|^2> = \lim_{M \to \infty} \frac{1}{2M+1} \cdot (2M+1)|A|^2$$

$$= A^2 < \infty.$$

Consequently, the signal is a power signal. If f_o is a rational number ($\omega_o = 2\pi f_o$), the signal $x(k)$ is periodic. Otherwise, $x(k)$ will be an aperiodic power signal. □

Example 3.2

Consider the aperiodic sequence

$$y(k) = \begin{cases} 1, & k = 0, 1, 2 \\ 0, & \text{otherwise} \end{cases}$$

and an equivalent periodic sequence

$$\tilde{z}(k) = \sum_{r=-\infty}^{+\infty} x(k + r \cdot N)$$

where $N = 6$ and
$$x(k) = \begin{cases} 1, & k = 0, 1, 2 \\ 0, & k = 3, 4, 5 \end{cases}$$

Both $y(k)$ and $\tilde{z}(k)$ are illustrated in Figure 3.1. From (3.23), we obtain

$$Y(\omega) = e^{-j\omega} \cdot \frac{\sin(3\omega/2)}{\sin(\omega/2)}; \quad |\omega| \leq \pi,$$

whereas from (3.25) we have

$$\tilde{Z}(\lambda) = e^{-j\frac{\pi\lambda}{3}} \cdot \frac{\sin(\pi\lambda/2)}{\sin(\pi\lambda/6)}, \lambda = 0, 1, 2, 3, 4, 5.$$

It can easily be verified that (3.27) and (3.29) hold for this example. $Y(\omega)$ and $\tilde{Z}(\lambda)$ are also illustrated in Figure 3.1. □

3.2 MOMENTS OF ENERGY SIGNALS

In this section we will introduce the definition and properties of higher-order moments of energy signals (deterministic transients).

3.2.1 Definition

Let us consider a finite-energy real signal $\{x(k)\}$, $k = 0, \pm 1, \pm 2, \ldots$ and assume that its moments (or multiple correlations) exist. Then the nth-order moments of energy signals are $(n-1)$-dimensional functions defined by

$$m_n^x(\tau_1, \ldots, \tau_{n-1}) \triangleq \sum_{k=-\infty}^{+\infty} x(k)x(k + \tau_1) \cdots x(k + \tau_{n-1}) \tag{3.31}$$

where $\tau_i = 0, \pm 1, \pm 2, \ldots$ for all values of i. These moments are numerical measures of the degree of similarity between a signal and a product of delayed or advanced versions of itself. For example, the value of the nth-order moment function at the origin, i.e., $m_n^x(0, \ldots, 0) = \sum_{k=-\infty}^{+\infty} x(k)x^{n-1}(k)$, is the "inner product" or "cross-energy" of $\{x(k)\}$ and $\{x^{n-1}(k)\}$.

3.2.2 Properties

Some of the properties we saw in Chapter 2 for moments of stationary random processes also apply for moments of deterministic energy signals. From (3.31) we note:

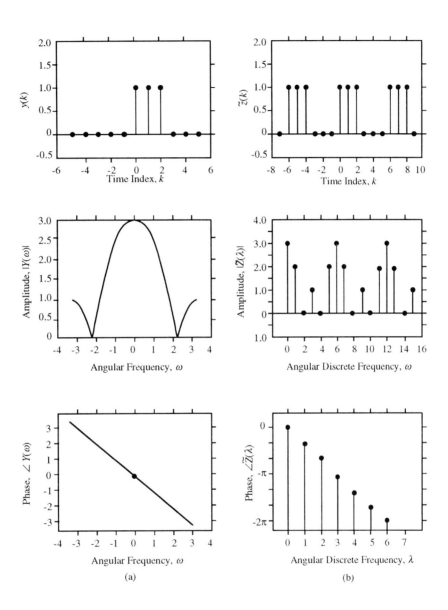

Figure 3.1 Time sequence, magnitude, and phase response of (a) a aperiodic sequence, (b) a periodic equivalent sequence.

1. If $x(k) = ax_1(k)$ where a is an arbitrary constant, then $m_n^x(\tau_1, \ldots, \tau_{n-1}) = a^n\, m_n^{x_1}(\tau_1, \ldots, \tau_{n-1})$.

2. If $x(k) = x_1(k) + x_2(k)$ then, in general, $m_n^x(\tau_1, \ldots, \tau_{n-1}) \neq m_n^{x_1}(\tau_1, \ldots, \tau_{n-1}) + m_n^{x_2}(\tau_1, \ldots, \tau_{n-1})$.

3. Moments are symmetric functions in their arguments. For example, $m_n^x(\tau_1, \tau_2, \ldots, \tau_{n-1}) = m_n^x(\tau_2, \tau_1, \ldots, \tau_{n-1}) = m_n^x(\tau_{n-1}, \tau_2, \ldots, \tau_1)$, etc.

4. If the signal $\{x(k)\}$ is even, i.e., $x(k) = x(-k)$, its nth-order moments satisfy the symmetry condition

$$m_n^x(\tau_1, \tau_2, \ldots, \tau_{n-1}) = m_n^x(-\tau_1, -\tau_2, \ldots, -\tau_{n-1}) \qquad (3.32)$$

for all $n \geq 2$. According to definition (2.15) in Chapter 2, a sequence that satisfies (3.32) is said to be "time reversible."

3.2.3 Special Cases

Important special cases of the definition (3.31) are the following:

First-Order Moment:

$$m_1^x = \sum_{k=-\infty}^{+\infty} x(k) \qquad \text{(also known as the "mean" value)}. \qquad (3.33)$$

We know that m_1^x exists because we assumed that $\{x(k)\}$ is a finite-energy signal.

Second-Order Moments:

$$m_2^x(\tau_1) = \sum_{k=-\infty}^{+\infty} x(k)\, x(k+\tau_1) \qquad \text{(autocorrelation sequence)}. \qquad (3.34)$$

Clearly, $m_2^x(0)$ is the total energy E_x in the real signal $\{x(k)\}$. The autocorrelation sequence is always even; i.e., $m_2^x(\tau_1) = m_2^x(-\tau_1)$ for all integer values of τ_1. $m_2^x(\tau_1)$ may also be seen as the **linear** convolution between $\{x(k)\}$ and $\{x(-k)\}$.

Third-Order Moments:

$$m_3^x(\tau_1, \tau_2) = \sum_{k=-\infty}^{+\infty} x(k) x(k+\tau_1) x(k+\tau_2) \qquad \text{(triple correlation)}. \qquad (3.35)$$

The triple correlation function satisfies the symmetry conditions of (2.31) in its arguments that correspond to third-order cumulants of stationary random processes. From (3.35) we note that $m_3^x(0,0)$ is a measure of skewness, or symmetry of the signal.

Fourth-Order Moments:

$$m_4^x(\tau_1, \tau_2, \tau_3) = \sum_{k=-\infty}^{+\infty} x(k)x(k+\tau_1)x(k+\tau_2)x(k+\tau_3). \qquad (3.36)$$

It is interesting to note that $m_3^x(0,\tau_2)$, $m_3^x(\tau_1,0)$, and $m_3^x(\tau,\tau)$ provide the cross-correlation sequence between $\{x^2(k)\}$ and $\{x(k)\}$, whereas $m_4^x(0,0,\tau_3)$, $m_4^x(\tau_1,0,0)$, $m_4^x(0,\tau_2,0)$, and $m_4^x(\tau,\tau,\tau)$ correspond to the cross-correlation between $\{x^3(k)\}$ and $\{x(k)\}$ [Nagata, 1978].

Example 3.3

Let us consider the delta function $\delta(k)$ which is defined by

$$\delta(k) \triangleq \begin{cases} 1, & k = 0 \\ 0, & k \neq 0. \end{cases} \qquad (3.37)$$

If we substitute $\delta(k)$ for $x(k)$ in (3.31), we obtain

$$m_n^\delta(\tau_1, \ldots, \tau_{n-1}) = \delta(\tau_1)\delta(\tau_2)\ldots\delta(\tau_{n-1}), \qquad (3.38)$$

which implies that all higher-order moments of $\delta(k)$ are multidimensional Kronecker delta functions; i.e., their value equals one at the origin and zero elsewhere. □

Example 3.4

Compute the moments of order one through four of the sequence $x(k) = a^k u(k)$ where $-1 < a < +1$ and $u(k)$ is the unit step function. Using (3.31) we obtain

$$\begin{aligned} m_1^x &= \frac{1}{1-a} \quad \text{(mean value)}, \\ m_2^x(\tau) &= \frac{a^{|\tau|}}{1-a^2} \quad \text{(autocorrelation)}, \end{aligned} \qquad (3.39)$$

where $m_2^x(0) = 1/(1-a^2)$ is the total energy of the signal, and

$$m_3^x(\tau_1, \tau_2) = \begin{cases} \dfrac{a^{(\tau_1+\tau_2)}}{1-a^3}, & \text{for } \tau_1 > 0, \tau_2 > 0, \\ \dfrac{a^{(\tau_1+\tau_2)}}{1-a^3} \cdot a^{-3\min(\tau_1,\tau_2)}, & \text{otherwise.} \end{cases} \quad (3.40)$$

The fourth-order moments of $x(k)$ for $\tau = \tau_1 = \tau_2 = \tau_3$ are given by

$$m_4^x(\tau, \tau, \tau) = \begin{cases} \dfrac{a^{|\tau|}}{1-a^4}, & \text{for } \tau < 0 \\ \dfrac{a^{3\tau}}{1-a^4}, & \text{for } \tau \geq 0. \end{cases} \quad (3.41)$$

Comparing (3.41) with (3.39), we see that

$$m_4^x(\tau, \tau, \tau) = \begin{cases} m_2^x(\tau)\dfrac{1}{1+a^2}, & \text{for } \tau < 0 \\ m_2^x(\tau)\dfrac{a^{2\tau}}{1+a^2}, & \text{for } \tau \geq 0. \end{cases}$$

Figure 3.2 illustrates the exponential sequence and its higher-order moments. □

Example 3.5

Consider the finite-duration sequence

$$s(k) = \begin{cases} 1, & k = 0 \\ -1, & k = 1 \\ 0, & \text{otherwise,} \end{cases} \quad (3.42)$$

and compute its moments of orders one through four. Using (3.31), we obtain

$$m_1^s = 0,$$

$$m_2^s(\tau) = \begin{cases} 2, & \tau_1 = 0 \\ -1, & \tau_1 = 1, -1 \\ 0, & \text{otherwise,} \end{cases}$$

$$m_3^s(\tau_1, \tau_2) = \begin{cases} 0, & \tau_1 = \tau_2 = 0 \\ 1, & (\tau_1 = \tau_2 = 1), (\tau_1 = -1, \tau_2 = 0), (\tau_1 = 0, \tau_2 = -1) \\ -1, & (\tau_1 = 1, \tau_2 = 0), (\tau_1 = 0, \tau_2 = 1), (\tau_1 = \tau_2 = -1) \\ 0, & \text{otherwise,} \end{cases}$$

$$m_4^s(\tau, \tau, \tau) = m_2^s(\tau).$$

$$(3.43)$$

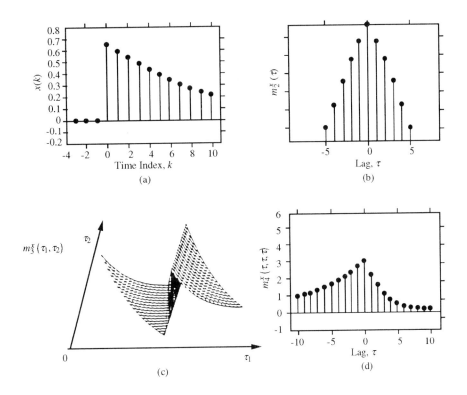

Figure 3.2 The higher-order moments of the sequence $x(k) = a^k u(k)$: (a) the exponential sequence, (b) its autocorrelation sequence, (c) its third-order moments, and (d) its fourth-order moments along a diagonal straight line ($\tau = \tau_1 = \tau_2 = \tau_3$).

The fourth-order moments were computed along a straight line $\tau_1 = \tau_2 = \tau_3 = \tau$. It is interesting to note that although $m_3^s(0,0) = 0$, the third-order moments of $s(k)$ are not all identical to zero. □

3.2.4 Cross-Moments

Suppose that we are given the real energy signals $\{x_i(k)\}$, $i = 1, 2, \ldots, n$. Their nth-order cross-moments are defined as

$$m_{x_1 x_2 \ldots x_n}(\tau_1, \tau_2, \ldots, \tau_{n-1}) \triangleq \sum_{k=-\infty}^{+\infty} x_1(k) x_2(k+\tau_1) \cdots x_n(k+\tau_{n-1}). \quad (3.44)$$

So, for the special case of two signals ($n = 2$), the above expression defines the cross-correlation function, $m_{x_1 x_2}(\tau)$, of $x_1(k)$ and $x_2(k)$. Furthermore, if $x_i(k) = x(k)$ for all i, (3.44) becomes (3.31). Equation (3.44) is the counterpart to (2.9), but for deterministic energy signals. If we consider $x_1(k) = y_1(k) + y_2(k)$, it follows from (3.44) that

$$m_{x_1x_2\ldots x_n}(\tau_1,\ldots,\tau_{n-1}) = m_{y_1x_2\ldots x_n}(\tau_1,\ldots,\tau_{n-1}) + m_{y_2x_2\ldots x_n}(\tau_1,\ldots,\tau_{n-1}).$$

Example 3.6

Determine the cross-correlation $m_{x_1x_2}(\tau)$, as well as third-order cross-moments $m_{x_1x_2x_1}(\tau_1,\tau_2)$, $m_{x_1x_1x_2}(\tau_1,\tau_2)$, and $m_{x_1x_2x_2}(\tau_1,\tau_2)$ of the sequences $x_1(k) = \delta(k)$ and $x_2(k) = \delta(k-D)$, where $\delta(k)$ is the unit impulse function. Applying (3.44) to this example, we obtain

$$\begin{aligned} m_{x_1x_2}(\tau) &= \sum_{k=-\infty}^{+\infty} \delta(k)\,\delta(k-D+\tau) \\ &= \delta(\tau - D) \end{aligned}$$

whereas

$$\begin{aligned} m_{x_1x_2x_1}(\tau_1,\tau_2) &= \sum_{k=-\infty}^{+\infty} \delta(k)\,\delta(k-D+\tau_1)\,\delta(k+\tau_2) \\ &= \delta(\tau_2)\,\delta(\tau_1 - D), \end{aligned}$$

$$\begin{aligned} m_{x_1x_1x_2}(\tau_1,\tau_2) &= \sum_{k=-\infty}^{+\infty} \delta(k)\,\delta(k+\tau_1)\,\delta(k-D+\tau_2) \\ &= \delta(\tau_1)\,\delta(\tau_2 - D), \end{aligned}$$

$$\begin{aligned} m_{x_1x_2x_2}(\tau_1,\tau_2) &= \sum_{k=-\infty}^{+\infty} \delta(k)\,\delta(k-D+\tau_1)\,\delta(k-D+\tau_2) \\ &= \delta(\tau_1 - D)\,\delta(\tau_2 - D). \end{aligned}$$

We observe that the time delay D between these two sequences is preserved not only by their cross-correlation sequence but also by their third-order cross-moments sequence. Figure 3.3 illustrates the sequences and their cross-moments assuming $D > 0$. □

3.3 MOMENT SPECTRA OF APERIODIC ENERGY SIGNALS

Let $\{x(k)\}$, $k = 0, \pm 1, \pm 2, \ldots$ be a real energy signal with nth-order moments,

$m_n^x(\tau_1, \ldots, \tau_{n-1})$.

3.3.1 Definition

The nth-order moment spectrum of $\{x(k)\}$ is defined as

$$M_n^x(\omega_1, \omega_2, \ldots, \omega_{n-1}) \triangleq \sum_{\tau_1=-\infty}^{+\infty} \cdots \sum_{\tau_{n-1}=-\infty}^{+\infty} m_n^x(\tau_1, \ldots, \tau_{n-1}) \quad (3.45)$$
$$\cdot \exp\{-j(\omega_1 \tau_1 + \omega_2 \tau_2 + \cdots + \omega_{n-1} \tau_{n-1})\}$$

where $|\omega_i| \leq \pi$ for $i = 1, 2, \ldots, n-1$ and $|\omega_1 + \omega_2 + \cdots + \omega_{n-1}| \leq \pi$. The moment spectra of energy signals are continuous in frequencies $\omega_1, \ldots, \omega_{n-1}$. Clearly, $M_n^x(\omega_1, \ldots, \omega_{n-1})$ is generally complex for $n > 2$, and periodic with period 2π. We often work with the magnitude and phase response of the nth-order moment spectrum rather than with its real and imaginary parts. Then,

$$M_n^x(\omega_1, \ldots, \omega_{n-1}) = |M_n^x(\omega_1, \ldots, \omega_{n-1})| \cdot \exp\{j\Psi_n^x(\omega_1, \ldots, \omega_{n-1})\}. \quad (3.46)$$

3.3.2 An Alternative Definition

The nth-order moment spectrum can be easily expressed in terms of the Fourier transform of the energy signal $\{x(k)\}$. Substituting (3.31) into (3.45) and utilizing (3.3), after some algebra we obtain

$$M_n^x(\omega_1, \ldots, \omega_{n-1}) = \sum_{k=-\infty}^{+\infty} x(k) \left(\sum_{\tau_1=-\infty}^{+\infty} x(k+\tau_1) e^{-j\omega_1 \tau_1} \right) \cdots$$
$$\left(\sum_{\tau_{n-1}=-\infty}^{+\infty} x(k+\tau_{n-1}) e^{-j\omega_{n-1} \tau_{n-1}} \right)$$
$$= \sum_{k=-\infty}^{+\infty} x(k) \cdot e^{j(\omega_1 + \cdots + \omega_{n-1})k} \cdot X(\omega_1) X(\omega_2) \cdots X(\omega_{n-1})$$

or
$$M_n^x(\omega_1, \ldots, \omega_{n-1}) = X(\omega_1) \cdot X(\omega_2) \cdots X(\omega_{n-1}) X^*(\omega_1 + \cdots + \omega_{n-1}). \quad (3.47)$$

Thus, the nth-order moment spectrum can be obtained directly from the Fourier transform, $X(\omega)$, of the energy signal. If we recall equation (2.80), we see that it is equivalent (to within a constant term) to (3.47). In terms of Fourier magnitude

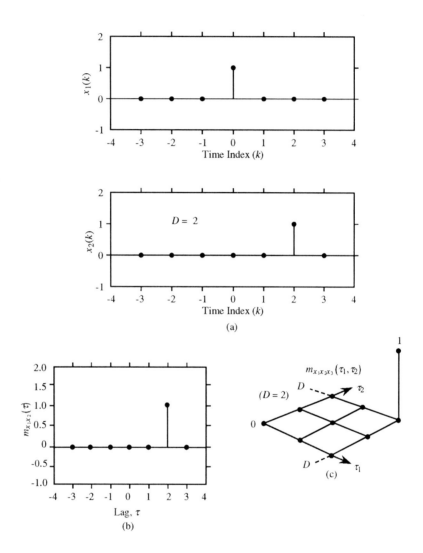

Figure 3.3 Cross-moments of two sequences $x_1(k), x_2(k)$ where $x_2(k) = x_1(k-D)$. (a) The time sequences, (b) their cross-correlations, (c) their third-order cross-moments.

and phase, we have

$$|M_n^x(\omega_1,\ldots,\omega_{n-1})| = |X(\omega_1)|\cdots|X(\omega_{n-1})|\cdot|X(\omega_1+\cdots+\omega_{n-1})| \quad (3.48)$$

and

$$\Psi_n^x(\omega_1,\ldots,\omega_{n-1}) = \phi_x(\omega_1)+\phi_x(\omega_2)+\cdots+\phi_x(\omega_{n-1})-\phi_x(\omega_1+\cdots+\omega_{n-1}). \quad (3.49)$$

Lob [1985] provides a signal analysis map for energy signals illustrating the relationships between time, Fourier, and power spectrum domains. A generalization of this map that includes moments and moment spectra is given in Figure 3.4.

From (3.47) we see that the nth-order moment spectrum is related to that of order $n-1$ by

$$M_n^x(\omega_1,\ldots,\omega_{n-2},0) = X(0)M_{n-1}^x(\omega_1,\ldots,\omega_{n-2}), \text{ etc.} \quad (3.50)$$

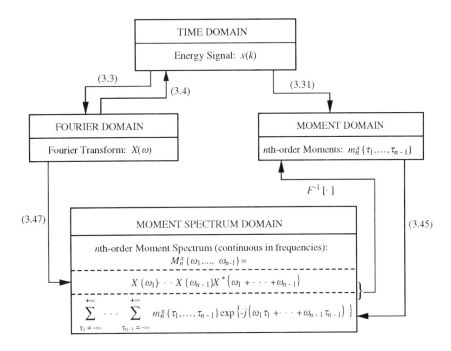

Figure 3.4 The signal analysis map of energy signals illustrating the relationships between time, Fourier, moment, and moment spectrum domains. $F^{-1}[\cdot]$ denotes the multidimensional inverse Fourier transform.

Identity (3.50) is equivalent to (2.86) of cumulant spectra of linear non-Gaussian processes.

3.3.3 Special Cases of Moment Spectra

Important special cases of moment spectra include those of orders $n = 2, 3$ and 4. The term bispectrum (trispectrum) was used in Chapter 2 to denote a third-order (fourth-order) cumulant spectrum of a stationary random process. However, several authors have also used the term bispectrum (trispectrum) in the case of deterministic signals to denote a third-order (fourth-order) moment spectrum because of the equivalence between the definitions (3.47) and (2.80) [Lohmann and Wirnitzer, 1984; Dainty and Northcott, 1986].

Energy Spectrum: n=2

$$M_2^x(\omega) = X(\omega)X^*(\omega) \tag{3.51}$$

or from (3.48) and (2.49) we have

$$M_2^x(\omega) = |X(\omega)|^2 \tag{3.52}$$

and

$$\Psi_2^x(\omega) = 0, \quad \text{for all } \omega \tag{3.53}$$

The energy spectrum is a real, nonnegative even function that suppresses phase information. We show in Chapter 6 that only minimum phase or maximum phase signals can be uniquely recovered (to within a sign and a delay) from the energy spectrum [Hayes, 1987].

Third-Order Moment Spectrum: n=3 (bispectrum)

From (3.47), it follows that

$$M_3^x(\omega_1, \omega_2) = X(\omega_1)X(\omega_2)X^*(\omega_1 + \omega_2) \tag{3.54}$$

or in terms of magnitude and phase

$$|M_3^x(\omega_1, \omega_2)| = |X(\omega_1)||X(\omega_2)||X(\omega_1 + \omega_2)|, \tag{3.55}$$

$$\Psi_3^x(\omega_1, \omega_2) = \phi_x(\omega_1) + \phi_x(\omega_2) - \phi_x(\omega_1 + \omega_2). \tag{3.56}$$

The third-order moment spectrum is generally a complex function that satisfies the symmetry properties shown by (2.32) in Chapter 2. The distinct difference between third-order moment spectra and energy spectra becomes apparent when we compare (3.53) with (3.56), that is, the third-order moment spectrum preserves phase information. Thus, whereas we can use the bispectrum to recover magnitude and phase information, the energy spectrum ($n = 2$) retains magnitude information but can only identify the phase as minimum.

Fourth-Order Moment Spectrum: n=4 (trispectrum)

Similarly, we obtain

$$M_4^x(\omega_1, \omega_2, \omega_3) = X(\omega_1)X(\omega_2)X(\omega_3)X^*(\omega_1 + \omega_2 + \omega_3) \qquad (3.57)$$

and

$$|M_4^x(\omega_1, \omega_2, \omega_3)| = |X(\omega_1)||X(\omega_2)|X(\omega_3)||X(\omega_1 + \omega_2 + \omega_3)| \qquad (3.58)$$

$$\Psi_4^x(\omega_1, \omega_2, \omega_3) = \phi_x(\omega_1) + \phi_x(\omega_2) + \phi(\omega_3) - \phi_x(\omega_1 + \omega_2 + \omega_3). \qquad (3.59)$$

The $n = 4$ moment spectrum satisfies many symmetry properties, such as those described by (2.34) in Chapter 2. For nonzero mean energy signals, there is no clear advantage to using the trispectrum instead of the bispectrum because both magnitude and phase information of the signal can be recovered from its bispectrum. However, for stationary random processes, we saw in Chapter 2 that phase and magnitude recovery can only be achieved in the trispectrum domain if the process has a symmetric probability density function.

A special case of (3.50) relates moment spectra of order $n = 3, 4$ with the energy spectrum, that is

$$M_3^x(\omega, 0) = M_3^x(0, \omega) = X(0)M_2^x(\omega) \qquad (3.60)$$

and

$$M_4^x(\omega, 0, 0) = M_4^x(0, \omega, 0) = M_4^x(0, 0, \omega) = X^2(0)M_2^x(\omega). \qquad (3.61)$$

3.3.4 Scalar Measures of Energy Signals

In Chapter 2 we saw three important scalar measures of stationary random processes and their relationship to cumulant spectra; namely, the variance, skewness, and kurtosis measures defined by (2.36). In a similar manner, we can define equivalent measures for energy signals as follows:

$$m_2^x(0) = \sum_{k=-\infty}^{+\infty} |x(k)|^2 = \frac{1}{2\pi} \int_{-\pi}^{+\pi} M_2^x(\omega) d\omega$$

(Energy or "variance" scalar measure),

$$m_3^x(0,0) = \sum_{k=-\infty}^{+\infty} x^3(k) = \frac{1}{(2\pi)^2} \int_{-\pi}^{+\pi} \int_{-\pi}^{+\pi} M_3^x(\omega_1,\omega_2) d\omega_1 d\omega_2$$

("Skewness" scalar measure),

and

$$m_4^x(0,0,0) = \sum_{k=-\infty}^{+\infty} |x(k)|^4 = \frac{1}{(2\pi)^3} \int_{-\pi}^{+\pi} \int_{-\pi}^{+\pi} M_4^x(\omega_1,\omega_2,\omega_3)(d\omega_1 d\omega_2 d\omega_3)$$

("Kurtosis" scalar measure). (3.62)

The third-order scalar measure can take positive or negative values, whereas the second- and fourth-order measures cannot be negative.

3.3.5 Moment Spectra and LTI System Relations

Let $\{h(k)\}$ be the impulse response of a stable linear time-invariant (LTI) system with frequency transfer function $H(\omega)$. That is

$$H(\omega) = \sum_{k=-\infty}^{+\infty} h(k) \exp\{-j\omega k\}. \tag{3.63}$$

Let $\{x(k)\}$ be a deterministic transient signal passing through the LTI system and let $\{y(k)\}$ be its output. The time-domain and frequency-domain relationships between input and output signals are respectively given by

$$y(k) = \sum_{n=-\infty}^{+\infty} h(n)\, x(k-n) \qquad \text{(time domain)} \tag{3.64}$$

and

$$Y(\omega) = H(\omega)\, X(\omega) \qquad \text{(frequency domain)}. \tag{3.65}$$

From (3.47) we have that the nth-order moment spectrum of the output, $M_n^y(\omega_1,\ldots,\omega_{n-1})$, is given by

$$M_n^y(\omega_1,\ldots,\omega_{n-1}) = Y(\omega_1)Y(\omega_2)\cdots Y^*(\omega_1+\cdots+\omega_{n-1})$$

and thus from (3.65) we obtain

$$M_n^y(\omega_1,\ldots,\omega_{n-1}) = M_n^h(\omega_1,\ldots,\omega_n) \cdot M_n^x(\omega_1,\cdots,\omega_n) \qquad (3.66)$$

where

$$M_n^h(\omega_1,\ldots,\omega_{n-1}) = H(\omega_1)\cdots H(\omega_{n-1})H^*(\omega_1+\cdots+\omega_{n-1}) \qquad (3.67)$$

and $M_n^x(\omega_1,\ldots,\omega_{n-1}) = X(\omega_1)\cdots X(\omega_{n-1})X^*(\omega_1+\cdots+\omega_{n-1})$ are the nth-order moment spectra of the impulse response of the LTI system and its input signal, respectively. The nth-order moments of the input and output signals are related through

$$m_n^y(\tau_1,\ldots,\tau_{n-1}) = m_n^h(\tau_1,\ldots,\tau_{n-1}) * m_n^x(\tau_1,\ldots,\tau_{n-1}) \qquad (3.68)$$

where "*" denotes the $(n-1)$-dimensional linear convolution operation.

Example 3.7

The nth-order moments of the impulse sequence $\delta(k)$ were computed in Example 3.3 and found to be a $(n-1)$-dimensional Kronecker delta function. Hence, the nth-order moment spectrum of $\delta(k)$ follows from (3.45) to be

$$M_n^\delta(\omega_1,\ldots,\omega_{n-1}) = 1 \qquad (3.69)$$

for all values of $\omega_1,\ldots,\omega_{n-1}$, that is, it is a flat hyperplane with magnitude one in the $(n-1)$-dimensional space. □

Example 3.8

Consider the sequence $x(k) = a^k u(k)$ whose moments of orders $n = 1,2,3,4$ are given by (3.39)–(3.41) and are illustrated in Figure 3.2. One way of computing moment spectra is by using (3.45). An easier way, however, is to compute the Fourier transform of $x(k)$,

$$X(\omega) = \frac{1}{1 - ae^{-j\omega}} \qquad (3.70)$$

and then utilize (3.47). It follows that

$$M_2^x(\omega) = \frac{1}{1 + a^2 - 2a \cos\omega} \tag{3.71}$$

and

$$M_3^x(\omega_1, \omega_2) = \frac{1}{1 - a^3 - ae(\omega_1, \omega_2) + a^2 e^*(\omega_1, \omega_2)}, \tag{3.72}$$

where $e(\omega_1, \omega_2) = \exp(-j\omega_1) + \exp(-j\omega_2) + \exp(j(\omega_1 + \omega_2))$. Figure 3.5 illustrates the energy spectrum, and the third-order moment spectrum of the exponential sequence. □

Example 3.9

We continue Example 3.5 to compute moment spectra of the finite duration sequence $s(k)$ given by (3.42). Its Fourier transform, energy spectrum, and bispectrum follow from (3.3) and (3.47). These are

$$\begin{aligned} S(\omega) &= 1 - e^{-j\omega} \\ M_2^x(\omega) &= 2 - 2\cos\omega \\ M_3^s(\omega_1, \omega_2) &= 2j[\sin\omega_1 + \sin\omega_2 - \sin(\omega_1 + \omega_2)]. \end{aligned} \tag{3.73}$$

Since $M_3^s(\omega, 0) = S(0) M_2^s(\omega)$ and $S(0) = 0$, we observe that this is an example where the energy spectrum of the signal cannot be reconstructed from its bispectrum. On the other hand, if we compute the fourth-order moment spectrum (trispectrum)

$$M_4^s(\omega_1, \omega_2, \omega_3) = S(\omega_1) S(\omega_2) S(\omega_3) S^*(\omega_1 + \omega_2 + \omega_3) \tag{3.74}$$

we observe that for $\omega_2 = -\omega_3$ and $\omega_1 = \omega_2 = \omega$ we have

$$\begin{aligned} M_4^s(\omega, \omega, -\omega) &= S(\omega) S(\omega) S(-\omega) S^*(\omega) \\ &= [S(\omega) S^*(\omega)][S(\omega) S(-\omega)] \\ &= M_2^s(\omega) M_2^s(\omega), \end{aligned} \tag{3.75}$$

which implies that the energy spectrum can be recovered from the trispectrum. □

Example 3.10

Consider the complex deterministic transient signal

$$x(k) = A\, e^{(-a + j\omega_0)k}\, u(k), \quad k = 0, 1, 2, \ldots \tag{3.76}$$

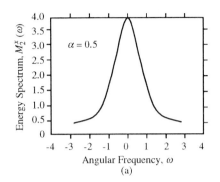

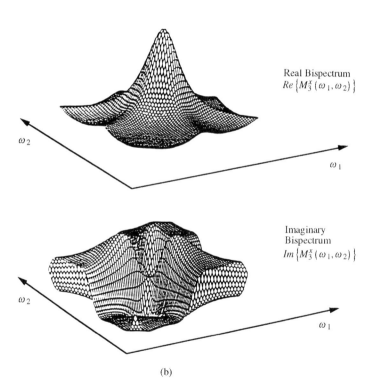

Figure 3.5 Moment spectra of the exponential sequence $x(k) = a^k u(k)$: (a) energy spectrum and (b) third-order moment spectrum (bispectrum). The autocorrelation and third-order moment sequence of $x(k)$ are illustrated in Figure 3.2.

where A is complex, $a > 0$, and ω_o is a constant frequency. The total energy and Fourier transform of the signal are

$$E_x = |A|^2 \Big/ [1 - e^{-2a}], \text{ and}$$
$$X(\omega) = A \Big/ [1 - p\, e^{-j\omega}], \tag{3.77}$$

where $p = \exp\{-a + j\omega_0\}$. From (3.51) it follows that

$$M_2^x(\omega) = \frac{|A|^2}{1 + e^{-2a} - 2e^{-a}\cos[\omega - \omega_0]}. \tag{3.78}$$

The third-order moment spectrum of the signal follows from (3.54). That is

$$M_3^x(\omega_1, \omega_2) = \frac{|A|^2 \cdot A}{\left(1 - pe^{j\omega_1}\right)\left(1 - pe^{-j\omega_2}\right)\left(1 - p^*e^{+j(\omega_1+\omega_2)}\right)}. \tag{3.79}$$

Figure 3.6 illustrates the energy spectrum and third-order moment spectrum of the signal $x(k)$. □

3.3.6 Phase Conditions

Let $\{h(k)\}$ be a deterministic transient signal with z transform $H(z)$. In general, $H(z)$ is a rational function of the form

$$H(z) = B(z)/A(z) \tag{3.80}$$

where

$$B(z) = \sum_{i=0}^{q} b_i\, z^{-i}$$

$$A(z) = \sum_{i=0}^{p} a_i\, z^{-i}.$$

It is also assumed that there is no pole-zero cancellation.

If the signal $\{h(k)\}$ with z-transform $H(z)$ has all its singularities (poles and zeros) inside the unit circle, it is called a **minimum phase signal**. It corresponds to the requirement that $h(k)$ is a causal and stable sequence. However, not all causal signals are minimum phase [Oppenheim and Schafer, 1989]. On the other hand, if all the singularities of $H(z)$ lie outside the unit circle, the signal $h(k)$ is called **maximum phase**. Finally, if some of its singularities lie inside the unit circle and the rest outside, the signal is called **mixed phase** or **nonminimum phase**.

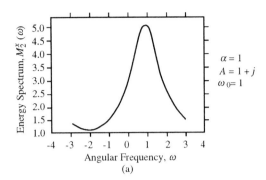
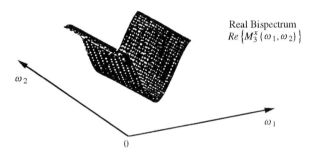
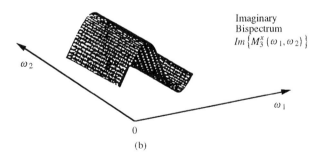

Figure 3.6 Moment spectra of a complex deterministic transient described in Example 3.10: (a) its energy spectrum, (b) real and imaginary parts of its bispectrum.

If $B(z)$ and $A(z)$ [with $z = \exp(j\omega)$] have the same phase response, but different magnitudes, then $h(k)$ is a **zero-phase** signal [Hayes, 1987]. By definition, a zero-phase signal must have a Fourier transform which is real, i.e., $h(k) = h^*(-k)$. Therefore, the singularities of $H(z)$ occur in conjugate reciprocal pairs so that if $H(z)$ has a zero (pole) at $z = r_o$ then $H(z)$ must also have a zero (pole) at $z = 1/r_o^*$. For example, it follows from (3.34) that autocorrelation sequences are always zero-phase signals.

Let us consider the finite-duration sequences $\{x(k)\}, \{y(k)\},$ and $\{z(k)\}$ illustrated in Figure 3.7. Their Fourier transforms are given by

$$X(\omega) = (1 - ae^{-j\omega})(1 - be^{-j\omega})$$

$$Y(\omega) = (e^{-j\omega} - a)(e^{-j\omega} - b)$$

and

$$Z(\omega) = (e^{-j\omega} - a)(1 - be^{-j\omega}). \qquad (3.81)$$

We observe that $\{x(k)\}$ is minimum phase, $\{y(k)\}$ is maximum phase, and $\{z(k)\}$ is nonminimum phase. Furthermore, we also observe that these three sequences have **identical** magnitude responses:

$$\begin{aligned} |X(\omega)| &= |Y(\omega)| = |Z(\omega)| \\ &= [1 + a^2 - 2a\cos\omega]^{\frac{1}{2}} \cdot [1 + b^2 - 2b\cos\omega]^{1/2}. \end{aligned} \qquad (3.82)$$

However, their phase responses are different:

$$\begin{aligned} \phi_x(\omega) &= \tan^{-1}\left[\frac{\sin\omega}{\frac{1}{a} - \cos\omega}\right] + \tan^{-1}\left[\frac{\sin\omega}{\frac{1}{b} - \cos\omega}\right] \\ \phi_y(\omega) &= \tan^{-1}\left[\frac{-\sin\omega}{\cos\omega - a}\right] + \tan^{-1}\left[\frac{-\sin\omega}{\cos\omega - b}\right] \qquad (3.83) \\ \phi_z(\omega) &= \tan^{-1}\left[\frac{-\sin\omega}{\cos\omega - a}\right] + \tan^{-1}\left[\frac{\sin\omega}{\frac{1}{b} - \cos\omega}\right]. \end{aligned}$$

Figure 3.7 illustrates the magnitude and phase response of each sequence. As a result of (3.82) and (3.83), the three sequences have identical energy spectra (or autocorrelations), but different bispectra. The autocorrelations and third-order moments of these sequences are shown in Table 2.2 in Chapter 2.

The sequences $\{x(k)\}, \{y(k)\}, \{z(k)\}$ have the same total energy; i.e.,

$$E = 1 + (a + b)^2 + (ab)^2.$$

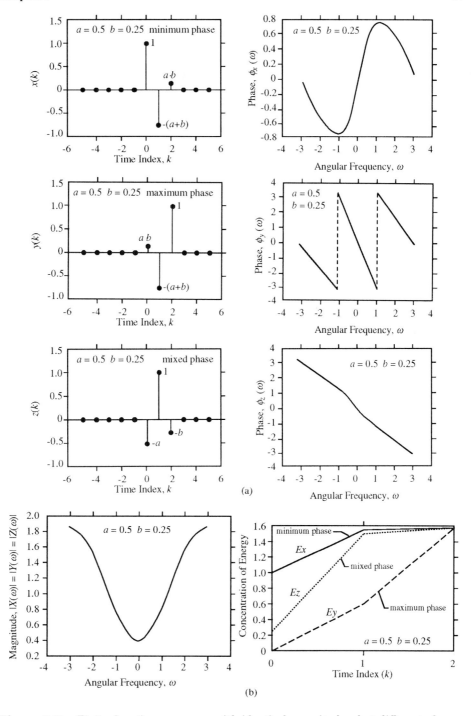

Figure 3.7 Finite-duration sequences with identical magnitudes, but different phases. (a) time sequences and their corresponding Fourier phase response, (b) their Fourier magnitude response and energy concentrations.

However, Oppenheim and Schafer [1989] point out that the concentration of energy around the time axis is different for each sequence. If we define

$$E_h(m) \triangleq \sum_{k=0}^{m} |h(k)|^2 \qquad (3.84)$$

as the energy contributed by the first $m+1$ samples, we note that

$$E_y(m) \le E_z(m) \le E_x(m).$$

These quantities are also plotted in Figure 3.7. We need to remind the reader that the latter result is true in general for cumulative energy transfer: $E_{MAX} \le E_{MIX} \le E_{MIN}$.

3.3.7 Cross-Moment Spectra

Given the real energy signals $\{x_i(k)\}$, $i = 1, 2, \ldots, n$ and their nth-order cross-moments $m_{x_1 x_2 \cdots x_n}(\tau_1, \ldots, \tau_{n-1})$ shown by (3.44), their nth-order cross-moment spectrum is defined as

$$M_{x_1 \ldots x_n}(\omega_1, \ldots, \omega_{n-1}) = \sum_{\tau_1=-\infty}^{+\infty} \cdots \sum_{\tau_{n-1}=-\infty}^{+\infty} m_{x_1 \ldots x_n}(\tau_1, \ldots, \tau_{n-1}) \cdot \\ \exp[-j \cdot \{\omega_1 \tau_1 + \cdots + \omega_{n-1} \tau_{n-1}\}]. \qquad (3.85)$$

If $\{X_i(\omega)\}$ corresponds to the Fourier transform of the ith energy signal, then an alternative definition of the nth-order cross-moment spectrum is given by

$$M_{x_1 \ldots x_n}(\omega_1, \ldots, \omega_{n-1}) = X_2(\omega_1) \cdot X_3(\omega_2) \cdots X_n(\omega_{n-1}) \cdot X_1^*(\omega_1 + \cdots + \omega_{n-1}). \qquad (3.86)$$

Special cases widely used in practice include the second-order cross-moment spectrum

$$M_{x_1 x_2}(\omega) = X_2(\omega) X_1^*(\omega), \qquad (3.87)$$

and the third-order cross-moment spectrum

$$M_{x_1 x_2 x_3}(\omega_1, \omega_2) = X_2(\omega_1) X_3(\omega_2) X_1^*(\omega_1 + \omega_2). \qquad (3.88)$$

Suppose we are given two signals, $x_1(k)$ and $x_2(k) = x_1(k-D)$ where D is an integer. Then the cross-moment spectrum of this sequence with a delayed version of itself is given by

$$M_{x_1 x_2}(\omega) = X_1(\omega) \, e^{-j\omega D} \, X_1^*(\omega).$$

If we normalize the cross-spectrum by the energy spectrum of $x_1(k)$, we obtain

$$I_2(\omega) = \frac{M_{x_1x_2}(\omega)}{M_2^{x_1}(\omega)} = e^{-j\omega D}. \tag{3.89}$$

Similarly, if we generate the third-order cross-moment spectrum $M_{x_1x_2x_1}(\omega_1,\omega_2)$ and normalize it by the third-order moment spectrum $M_3^{x_1}(\omega_1,\omega_2)$, we obtain

$$I_3(\omega_1,\omega_2) = \frac{M_{x_1x_2x_1}(\omega_1,\omega_2)}{M_3^{x_1}(\omega_1,\omega_2)} = e^{-j\omega_1 D}. \tag{3.90}$$

Hence, the time delay information between the two sequences can be recovered by both second- and third-order cross-moment spectra. Several estimation methods that are based on (3.89) and (3.90) are discussed in Chapter 8.

Example 3.11

We continue Example 3.6 to generate cross-moment spectra of the sequences $x_1(k) = \delta(k)$ and $x_2(k) = \delta(k - D)$. Since $X_1(\omega) = 1$ and $X_2(\omega) = e^{-j\omega D}$, it follows from (3.87) and (3.98) that

$$\begin{aligned}
M_{x_1x_2}(\omega) &= e^{-j\omega D}, \\
M_{x_1x_2x_1}(\omega_1,\omega_2) &= e^{-j\omega_1 D}, \\
M_{x_1x_1x_2}(\omega_1,\omega_2) &= e^{-j\omega_2 D}, \quad \text{and} \\
M_{x_1x_2x_2}(\omega_1,\omega_2) &= e^{-j\omega_1 D}\, e^{-j\omega_2 D}.
\end{aligned}$$

□

3.3.8 The nth-Order Coherency of Energy Signals

The nth-order coherency of energy signals is defined in terms of their nth-order moment spectrum, $M_n^x(\omega_1,\ldots,\omega_{n-1})$, and energy spectrum, $M_2^x(\omega)$, as follows:

$$E_n^x(\omega_1,\ldots,\omega_{n-1}) \triangleq \frac{M_n^x(\omega_1,\omega_2,\ldots,\omega_{n-1})}{[M_2^x(\omega_1)\cdot M_2^x(\omega_2)\cdots M_2^x(\omega_{n-1})M_2^x(\omega_1+\cdots+\omega_{n-1})]^{\frac{1}{2}}}. \tag{3.91}$$

Using (3.46), (3.48), (3.49), and (3.52), we obtain

$$\begin{aligned}
E_n^x(\omega_1,\ldots,\omega_{n-1}) &= \exp\{j\Psi_n^x(\omega_1,\ldots,\omega_{n-1})\} \\
&= \exp\{j[\phi_x(\omega_1)+\cdots+\phi_x(\omega_{n-1})-\phi_x(\omega_1+\cdots+\omega_{n-1})]\}.
\end{aligned} \tag{3.92}$$

Thus, the nth-order coherency of energy signals has constant magnitude equal to one and phase identical to the phase of the nth-order moment spectrum. Therefore, an alternative definition to (3.91) is

$$E_n^x(\omega_1,\ldots,\omega_{n-1}) = \left[\frac{M_n^x(\omega_1,\ldots,\omega_{n-1})}{M_n^{x*}(\omega_1,\ldots,\omega_{n-1})}\right]^{1/2}. \tag{3.93}$$

Comparing (3.91) with (2.132), we see that the nth-order coherency of energy signals has the same properties as those of **linear** non-Gaussian processes.

3.4 MOMENTS OF POWER SIGNALS

In the previous two sections we discussed the moments and moment spectra of deterministic transient (energy) signals. In those sections we were concerned with both finite- and infinite-duration energy signals. An important conclusion was that moment spectra of energy signals are **continuous** functions in frequencies. In the next sections we present the definition, properties, and examples of higher-order moments of deterministic power signals. The general class of power signals includes the periodic and aperiodic deterministic power signals, and the stationary random processes [Lob, 1985]. Since the case of random processes was discussed in Chapter 2, we shall treat here the class of deterministic power signals as a whole, with emphasis on the important subclass of periodic signals.

3.4.1 Definition

Let us consider a real power signal $\{x(k)\}$, $k = 0, \pm 1, \pm 2, \ldots$ and assume that its higher-order moments exist. Then the nth-order moments of the power signal are defined by [Nagata, 1978; Lohmann and Wirnitzer, 1984; Dainty and Northcott, 1986]

$$r_n^x(\tau_1,\ldots,\tau_{n-1}) \stackrel{\Delta}{=} <x(k)x(k+\tau_1)\cdots x(k+\tau_{n-1})> \tag{3.94}$$

where $< \cdot >$ corresponds to the time-averaging operation defined by (3.2). If the power signal is periodic with period N, i.e., $\tilde{x}(k) = \tilde{x}(k+N)$, then (3.94) becomes

$$\begin{aligned} r_n^{\tilde{x}}(\tau_1,\ldots,\tau_{n-1}) &= <\tilde{x}(k)\tilde{x}(k+\tau_1)\cdots\tilde{x}(k+\tau_{n-1})>_N \\ &= \frac{1}{N}\sum_{k=J}^{J+N-1}\tilde{x}(k)\tilde{x}(k+\tau_1)\cdots\tilde{x}(k+\tau_{n-1}). \end{aligned} \tag{3.95}$$

All the properties defined earlier for moments of energy signals also apply here for $r_n^x(\tau_1,\ldots,\tau_{n-1})$ and $r_n^{\tilde{x}}(\tau_1,\ldots,\tau_{n-1})$. An additional property of $r_n^{\tilde{x}}(\tau_1,\ldots,\tau_{n-1})$ which has no counterpart with moments of energy signals is the fact that

$$r_n^{\tilde{x}}(\tau_1,\ldots,\tau_{n-1}) = r_n^{\tilde{x}}(\tau_1 + N,\ldots,\tau_{n-1} + N); \qquad (3.96)$$

i.e., the nth-order moment sequence of a periodic signal with period N is, in turn, periodic with the same period.

3.4.2 Special Cases for Periodic Signals

We concentrate on the periodic signals and consider as special cases their moments of orders $n = 1, 2, 3,$ and 4. These are:

First-Order Moments:

$$\begin{aligned} r_1^{\tilde{x}} &= <\tilde{x}(k)>_N \text{ (mean value)} \\ &= \frac{1}{N}\sum_{k=J}^{J+N-1}\tilde{x}(k). \end{aligned} \qquad (3.97)$$

Second-Order Moments:

$$\begin{aligned} r_2^{\tilde{x}}(\tau_1) &= <\tilde{x}(k)\tilde{x}(k+\tau_1)>_N \quad \text{(autocorrelation)} \\ &= \frac{1}{N}\sum_{k=J}^{J+N-1}\tilde{x}(k)\tilde{x}(k+\tau_1). \end{aligned} \qquad (3.98)$$

At $\tau_1 = 0$ the autocorrelation equals the total average power in the signal. The $r_2^{\tilde{x}}(\tau_1)$ is an even function that satisfies $r_2^{\tilde{x}}(0) \geq |r_2^{\tilde{x}}(\tau_1)|$ for any τ_1. Equation (3.98) may also be seen (to within a constant) as the **periodic** convolution of $\{\tilde{x}(k)\}$ and $\{\tilde{x}(-k)\}$. That is

$$r_2^{\tilde{x}}(\tau_1) = \frac{1}{N}\tilde{x}(k) \circledast \tilde{x}(-k) \qquad (3.99)$$

where "$\circledast$" denotes periodic convolution. Clearly, $r_2^{\tilde{x}}(\tau_1) = r_2^{\tilde{x}}(\tau_1 + mN)$ for any integer m.

Third-Order Moments:

$$\begin{aligned} r_3^{\tilde{x}}(\tau_1, \tau_2) &= <\tilde{x}(k)\tilde{x}(k+\tau_1)\tilde{x}(k+\tau_2)>_N \\ &= \frac{1}{N}\sum_{k=J}^{J+N-1}\tilde{x}(k)\tilde{x}(k+\tau_1)\tilde{x}(k+\tau_2). \end{aligned} \quad (3.100)$$

We observe that $r_3^{\tilde{x}}(0,0)$ corresponds to a measure of skewness of the data samples within a period. Nagata [1978] defines the one-dimensional third-order moment sequence $r_3^{\tilde{x}}(0,\tau_2)$, which is essentially the periodic convolution between $\{\tilde{x}^2(k)\}$ and $\{\tilde{x}(-k)\}$; i.e.,

$$r_3^{\tilde{x}}(0,\tau_2) = \frac{1}{N}\tilde{x}^2(k) \circledast \tilde{x}(-k).$$

Also,

$$r_3^{\tilde{x}}(\tau,\tau) = r_3^{\tilde{x}}(0,\tau).$$

Fourth-Order Moments:

$$\begin{aligned} r_4^{\tilde{x}}(\tau_1,\tau_2,\tau_3) &=<\tilde{x}(k)\tilde{x}(k+\tau_1)\tilde{x}(k+\tau_2)\tilde{x}(k+\tau_3)>_N \\ &= \frac{1}{N}\sum_{k=J}^{J+N-1}\tilde{x}(k)\tilde{x}(k+\tau_1)\tilde{x}(k+\tau_2)\tilde{x}(k+\tau_3). \end{aligned} \quad (3.101)$$

One-dimensional slices of $r_4^{\tilde{x}}(\tau_1,\tau_2,\tau_3)$ are:

$$\begin{aligned} r_4^{\tilde{x}}(\tau,\tau,\tau) &= \frac{1}{N}\tilde{x}(k) \circledast \tilde{x}^3(-k) \\ r_4^{\tilde{x}}(0,\tau,\tau) &= \frac{1}{N}\tilde{x}^2(k) \circledast \tilde{x}^2(-k). \end{aligned} \quad (3.102)$$

Example 3.12

Consider the impulse train sequence of period $N = 1$ defined by

$$\tilde{\delta}(k) \triangleq A \sum_{l=-\infty}^{+\infty} \delta(k+l) \quad (3.103)$$

where $\delta(k)$ is the unit impulse sequence. Its higher-order moments follow from (3.97)–(3.101) [Nagata, 1978]:

$$\begin{aligned} r_1^{\tilde{\delta}} &= A \\ r_2^{\tilde{\delta}}(\tau) &= A^2 \quad \text{for all} \quad \tau \\ r_3^{\tilde{\delta}}(\tau_1,\tau_2) &= A^3, \quad \text{for all} \quad (\tau_1,\tau_2), \quad \text{and} \\ r_4^{\tilde{\delta}}(\tau_1,\tau_2,\tau_3) &= A^4, \quad \text{for all} \quad (\tau_1,\tau_2,\tau_3). \end{aligned} \quad (3.104)$$

Chapter 3   103

These moments are also periodic sequences with period $N = 1$.   □

Example 3.13

Figure 3.8(a) illustrates the periodic sequence

$$\tilde{s}(k) = \cos\left(\frac{2\pi}{N}k + \phi\right), \tag{3.105}$$

which has period N. Combining (3.97) and (3.105), we get $r_1^{\tilde{s}} = 0$, i.e., $\{\tilde{s}(k)\}$ is a zero-mean sequence. The autocorrelation, third- and fourth-order moments of $\tilde{s}(k)$ follow from (3.98)–(3.101) and (3.105):

$$\begin{aligned} r_2^{\tilde{s}}(\tau) &= \frac{1}{N}\sum_{k=0}^{N-1} \cos\left(\frac{2\pi}{N}k + \phi\right)\cos\left(\frac{2\pi}{N}(k+\tau) + \phi\right) \\ &= \frac{1}{2}\cos\left(\frac{2\pi}{N}\tau\right), \end{aligned} \tag{3.106}$$

$$r_3^{\tilde{s}}(\tau_1, \tau_2) \equiv 0, \tag{3.107}$$

and

$$\begin{aligned} r_4^{\tilde{s}}(\tau_1, \tau_2, \tau_3) &= \frac{1}{16}\left\{\cos\left[\frac{2\pi}{N}(\tau_1 + \tau_2 - \tau_3)\right] + \cos\left[\frac{2\pi}{N}(\tau_1 - \tau_2 - \tau_3)\right]\right. \\ &\left. + \cos\left[\frac{2\pi}{N}(\tau_1 - \tau_2 + \tau_3)\right]\right\} \end{aligned} \tag{3.108}$$

The autocorrelation and slices of fourth-order moments are illustrated in Figure 3.8(b),(c). This example illustrates that a deterministic cosine wave has third-order moments equal to zero.   □

Example 3.14

Let us now consider the same periodic sequence $\tilde{\delta}(k)$ as that from Example 3.13, but with a non-zero constant dc value being added to the data; i.e.,

$$\tilde{x}(k) = \cos\left(\frac{2\pi}{N}k + \phi\right) + \tilde{\delta}(k) \tag{3.109}$$

where $\tilde{\delta}(k)$ is the impulse train sequence defined in (3.103). The signal $\tilde{x}(k)$ is periodic with period N because $\tilde{\delta}(k)$ may be viewed as periodic with the same period, hence $\tilde{x}(k) = \tilde{x}(k+N)$. This sequence is shown in Figure 3.9(a). Following similar steps as before, we obtain the moments of $\tilde{x}(k)$ for orders $n = 1, 2, 3$:

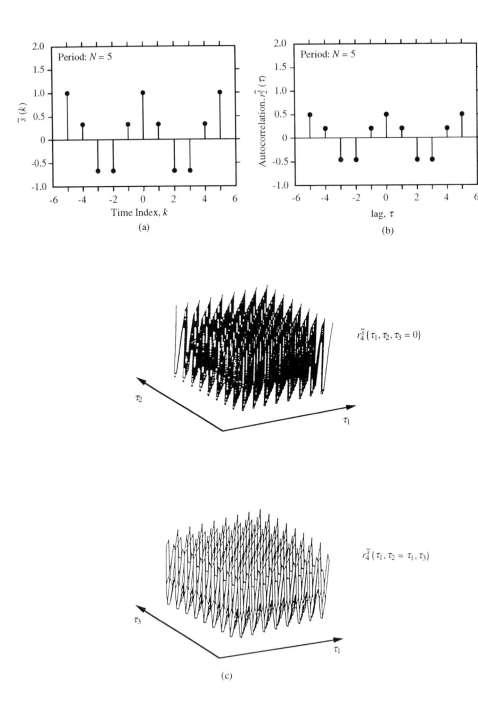

Figure 3.8 Higher-order moments of the periodic sequence $\tilde{s}(k) = \left(\frac{2\pi}{N}k + \phi\right)$. (a) the time sequence, (b) its autocorrelation sequence, (c) slices of its fourth-order moment sequence. Note that its third-order moments are identically zero.

$$r_1^{\tilde{x}} = A$$
$$r_2^{\tilde{x}}(\tau) = A^2 + \frac{1}{2}\cos\frac{2\pi}{N}\tau \qquad (3.110)$$
$$r_3^{\tilde{x}}(\tau_1,\tau_2) = A^3 + \frac{A}{2}\left\{\cos\frac{2\pi}{N}\tau_1 + \cos\frac{2\pi}{N}\tau_2 + \cos\frac{2\pi}{N}(\tau_1-\tau_2)\right\}.$$

□

The autocorrelation and third-order moments are illustrated in Figure 3.9(b),(c). Lohmann and Wirnitzer [1984] point out that what we actually learn from the last two examples is that third-order moment computation might be a very sensitive test for detecting the presence of a bias (e.g., dc value) in sinusoidal periodic sequences.

Example 3.15

Consider the following periodic sequences with period N

$$\tilde{y}(k) = 2\cos\left(\frac{2\pi}{N}k + \phi\right) + \cos\left(\frac{4\pi}{N}k + 2\phi\right)$$

and

$$\tilde{z}(k) = 2\cos\left(\frac{2\pi}{N}k + \phi\right) + \cos\left(\frac{4\pi}{N}k + 2\phi_1\right) \qquad (3.111)$$

where $\Delta\phi = 2\phi - 2\phi_1 \neq 0$. If $\Delta\phi = 0$, then $\tilde{y}(k) \equiv \tilde{z}(k)$. These sequences have identical autocorrelations but different third-order moments; viz:

$$r_3^{\tilde{y}}(\tau_1,\tau_2) = \frac{1}{2}\left\{\cos\frac{2\pi}{N}(\tau_1+\tau_2) + \cos\frac{2\pi}{N}(\tau_1-2\tau_2) + \cos\frac{2\pi}{N}(2\tau_1-\tau_2)\right\}, \qquad (3.112)$$

$$r_3^{\tilde{z}}(\tau_1,\tau_2) = \frac{1}{2}\left\{\cos\left[\frac{2\pi}{N}(\tau_1+\tau_2) + \Delta\phi\right] + \cos\left[\frac{2\pi}{N}(\tau_1-2\tau_2) + \Delta\phi\right] + \cos\left[\frac{2\pi}{N}(\tau_2-2\tau_1) + \Delta\phi\right]\right\}. \qquad (3.113)$$

We therefore conclude that $\tilde{y}(k)$ can be discriminated from $\tilde{z}(k)$ in the third-order moment domain. For example, $r_3^{\tilde{y}}(0,0) = 3/2$ whereas $r_3^{\tilde{z}}(0,0) = \frac{3}{2}\cos\Delta\phi$. This simple example demonstrates the utility of third-order moments for detecting weak quadratic nonlinearities of a system from its output sequence [Lohmann and

Wirnitzer, 1984; Sato et al., 1977]. □

Example 3.16

Let us consider the sawtooth periodic sequence [Lohmann and Wirnitzer, 1984] of the form

$$\tilde{i}(k) = \begin{cases} 0, & k = 0 \\ 0.5, & k = 1 \\ 1, & k = 2 \end{cases}$$

where $\tilde{i}(k) = \tilde{i}(k+3m)$ where m is any integer. We calculate the following moments:

$$r_1^{\tilde{i}} = 0.5$$

$$r_2^{\tilde{i}}(\tau) = \begin{cases} (5/4) \cdot \frac{1}{3}, & \tau = 0 \\ (0.5)\frac{1}{3}, & \tau = 1 \\ (0.5)\frac{1}{3}, & \tau = 2 \\ r_2^{\tilde{i}}(\tau + 3), & \text{otherwise} \end{cases}$$

$$r_3^{\tilde{i}}(\tau_1, \tau_2) = \begin{cases} (9/8)\frac{1}{3}, & \tau_1 = \tau_2 = 0 \\ (1/4)\frac{1}{3}, & \tau_1 = 1, \tau_2 = 0 \\ (1/2)\frac{1}{3}, & \tau_1 = 2, \tau_2 = 0 \\ (1/2) \cdot \frac{1}{3}, & \tau_1 = \tau_2 = 1 \\ (1/4) \cdot \frac{1}{3}, & \tau_1 = \tau_2 = 2 \\ 0, & \tau_1 = 2, \tau_2 = 1, \end{cases}$$

which are illustrated in Figure 3.10. Note that $r_2^{\tilde{i}}(\tau)$ and $r_3^{\tilde{i}}(\tau_1, \tau_2)$ are also periodic because $\tilde{i}(k)$ is periodic. □

3.4.3 Cross-Moments of Periodic Signals

If we are given the periodic signals $\{\tilde{x}_i(k)\}$, $i = 1, 2, \ldots, n$ with period N, i.e., $\tilde{x}_i(k) = \tilde{x}_i(k + N)$ for all k and i indices, the nth-order cross-moments are defined as

$$r_{\tilde{x}_1\tilde{x}_2\cdots\tilde{x}_n}(\tau_1,\tau_2,\ldots,\tau_{n-1}) = <\tilde{x}_1(k)\tilde{x}_2(k+\tau_1)\cdots\tilde{x}_n(k+\tau_{n-1})>_N$$
$$= \frac{1}{N}\sum_{k=J}^{J+N-1}\tilde{x}_1(k)\tilde{x}_2(k+\tau_1)\cdots\tilde{x}_n(k+\tau_{n-1}). \quad (3.114)$$

The function $r_{\tilde{x}_1\ldots\tilde{x}_n}(\tau_1,\ldots,\tau_{n-1})$ is periodic with period N.

3.5 MOMENT SPECTRA OF PERIODIC POWER SIGNALS

Let $\{\tilde{x}(k)\}$, $k = 0, \pm 1, \pm 2, \ldots$ be a real periodic signal with period N whose higher-order moments, $r_n^{\tilde{x}}(\tau_1,\ldots,\tau_{n-1})$, exist.

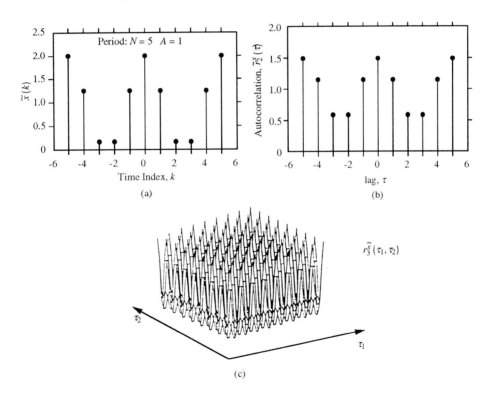

Figure 3.9 Higher-order moments of $\left\{\tilde{x}(k) = 1 + \cos\left(\frac{2\pi}{N}k + \phi\right)\right\}$. (a) Time sequence, (b) its autocorrelation, (c) its third-order moment.

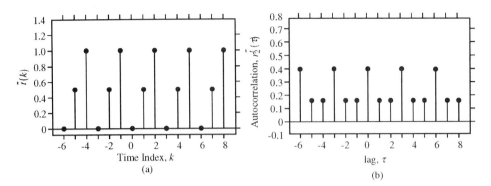

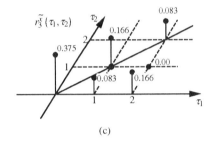

Figure 3.10 Sawtooth Sequence: (a) time domain, (b) autocorrelation, (c) third-order moment sequence.

3.5.1 Definition

The nth-order moment spectrum of $\{\tilde{x}(k)\}$ is defined as

$$R_n^{\tilde{x}}(\lambda_1, \lambda_2, \ldots, \lambda_{n-1}) \triangleq \sum_{\tau_1=J}^{J+N-1} \cdots \sum_{\tau_{n-1}=J}^{J+N-1} r_n^{\tilde{x}}(\tau_1, \ldots, \tau_{n-1}) \cdot \exp\left\{-j\frac{2\pi}{N}(\tau_1\lambda_1 + \cdots + \tau_{n-1}\lambda_{n-1})\right\} \quad (3.115)$$

where $\frac{2\pi}{N}\lambda_i$, $i = 1, 2, \ldots, N-1$ are discrete frequencies, $\lambda_i = 0, 1, \ldots, N-1$, and $R_n^{\tilde{x}}(\lambda_1, \ldots, \lambda_{n-1})$ is generally complex for $n > 2$ and multi-dimensionally periodic with periods N. The nth-order moment spectrum can be written as

$$R_n^{\tilde{x}}(\lambda_1, \ldots, \lambda_{n-1}) = \left|R_n^{\tilde{x}}(\lambda_1, \ldots, \lambda_{n-1})\right| \cdot \exp\{j\Psi_n^{\tilde{x}}(\lambda_1, \ldots, \lambda_{n-1})\}. \quad (3.116)$$

The higher-order moment spectra of periodic signals are **discrete functions** in their frequencies $\frac{2\pi}{N}\lambda_1, \ldots, \frac{2\pi}{N}\lambda_{n-1}$.

3.5.2 An Alternative Definition

As in the case of moment spectra of energy signals, we can define moment spectra of periodic signals directly in terms of their Fourier series representation. If we substitute (3.12) into (3.95) and assume $J = 0$ (without loss of generality), we get

$$\begin{aligned}
r_n^{\tilde{x}}(\tau_1,\ldots,\tau_{n-1}) &= \frac{1}{N} \sum_{k=0}^{N-1} \tilde{x}(k) \left(\frac{1}{N} \sum_{\lambda_1=0}^{N-1} \tilde{X}(\lambda_1) \exp\left[j\frac{2\pi}{N}\lambda_1(k+\tau_1)\right] \right) \cdots \\
&\qquad \left(\frac{1}{N} \sum_{\lambda_{n-1}=0}^{N-1} \tilde{X}(\lambda_{n-1}) \exp\left[j\frac{2\pi}{N}\lambda_{n-1}(k+\tau_{n-1})\right] \right) \\
&= \frac{1}{N^n} \sum_{\lambda_1=0}^{N-1} \cdots \sum_{\lambda_{n-1}=0}^{N-1} \tilde{X}(\lambda_1)\cdots\tilde{X}(\lambda_{n-1}) \\
&\qquad \exp\left[j\frac{2\pi}{N}(\lambda_1\tau_1 + \cdots + \lambda_{n-1}\tau_{n-1})\right] \\
&= \frac{1}{N^n} \sum_{\lambda_1=0}^{N-1} \cdots \left(\sum_{k=0}^{N-1} \tilde{x}(k)\exp\left[j\frac{2\pi}{N}k(\lambda_1 + \cdots + \lambda_{n-1})\right] \right)
\end{aligned}$$

or

$$\begin{aligned}
r_n^{\tilde{x}}(\tau_1,\ldots,\tau_{n-1}) &= \frac{1}{N^n} \sum_{\lambda_1=0}^{N-1} \cdots \sum_{\lambda_{n-1}=0}^{N-1} \tilde{X}(\lambda_1)\cdots\tilde{X}(\lambda_{n-1})\tilde{X}^*(\lambda_1+\cdots+\lambda_{n-1}) \\
&\qquad \cdot \exp\left[j\frac{2\pi}{N}(\lambda_1\tau_1 + \cdots + \lambda_{n-1}\tau_{n-1})\right]
\end{aligned} \quad (3.117)$$

Substituting (3.117) into (3.115), we obtain

$$R_n^{\tilde{x}}(\lambda_1,\lambda_2,\ldots,\lambda_{n-1}) \triangleq \frac{1}{N}\tilde{X}(\lambda_1)\tilde{X}(\lambda_2)\cdots\tilde{X}(\lambda_{n-1})\tilde{X}^*(\lambda_1+\cdots+\lambda_{n-1}). \quad (3.118)$$

Combining (3.116) with (3.118), we obtain the nth-order magnitude and phase moment spectra, i.e.,

$$|R_n^{\tilde{x}}(\lambda_1,\ldots,\lambda_{n-1})| = \frac{1}{N}|\tilde{X}(\lambda_1)|\cdots|\tilde{X}(\lambda_{n-1})||\tilde{X}(\lambda_1+\cdots+\lambda_{n-1})| \quad (3.119)$$

and

$$\Psi_n^{\tilde{x}}(\lambda_1,\ldots,\lambda_{n-1}) = \phi_{\tilde{x}}(\lambda_1) + \cdots + \phi_{\tilde{x}}(\lambda_{n-1}) - \phi_{\tilde{x}}(\lambda_1+\cdots+\lambda_{n-1}) \quad (3.120)$$

Moment spectra of periodic signals may easily be computed utilizing a fast Fourier transform (FFT) algorithm in (3.13) and then employing (3.118). Figure 3.11 illustrates the signal analysis map of periodic signals.

The $(n-1)$-order moment spectrum of a periodic signal is related to that of order n by

$$\begin{aligned} R^{\tilde{x}}_{n-1}(\lambda_1,\ldots,\lambda_{n-2}) &= R^{\tilde{x}}_n(\lambda_1,\ldots,\lambda_{n-2},0) \Big/ \tilde{X}(0) \\ &= R^{\tilde{x}}_n(\lambda_1,\ldots,\lambda_{n-3},0,\lambda_{n-2}) \Big/ \tilde{X}(0), \quad \text{etc.} \end{aligned} \quad (3.121)$$

3.5.3 Special Cases of Moment Spectra

The power spectrum, third- and fourth-order moment spectra of periodic signals are the special cases widely used in practice. The following spectra are derived using (3.118)–(3.120):

Power Spectrum: n=2

$$R^{\tilde{x}}_2(\lambda) = \frac{1}{N}|\tilde{X}(\lambda)|^2$$

and

$$\Psi^{\tilde{x}}_2(\lambda) = 0, \quad \text{for} \quad \lambda = 0, 1, 2, \ldots, N-1. \quad (3.122)$$

So, the power spectrum of periodic sequences is not only a real, nonnegative, and even function, but also is discrete with frequencies $\frac{2\pi}{N}\lambda$. The average power of the signal, $P_{\tilde{x}}$, is related to the power spectrum by

$$P_{\tilde{x}} = r^{\tilde{x}}_2(0) = \frac{1}{N}\sum_{\lambda=0}^{N-1} R^{\tilde{x}}_2(\lambda). \quad (3.123)$$

Third-Order Moment Spectrum: n=3

The magnitude bispectrum is

$$|R^{\tilde{x}}_3(\lambda_1,\lambda_2)| = \frac{1}{N}|\tilde{X}(\lambda_1)|\,|\tilde{X}(\lambda_2)|\,|\tilde{X}(\lambda_1+\lambda_2)|, \quad (3.124)$$

whereas the phase bispectrum is

$$\Psi^{\tilde{x}}_3(\lambda_1,\lambda_2) = \phi_{\tilde{x}}(\lambda_1) + \phi_{\tilde{x}}(\lambda_2) - \phi_{\tilde{x}}(\lambda_1+\lambda_2), \quad (3.125)$$

(for $\lambda_1,\lambda_2 = 0,1,\ldots,N-1$). The third-order scalar measure $r^{\tilde{x}}_3(0,0)$ of a periodic signal is given by (3.100) for $\tau_1 = \tau_2 = 0$:

$$r^{\tilde{x}}_3(0,0) = \frac{1}{N}\sum_{k=0}^{N-1} \tilde{x}^3(k) = \frac{1}{N^2}\sum_{\lambda_1=0}^{N-1}\sum_{\lambda_2=0}^{N-1} R^{\tilde{x}}_3(\lambda_1,\lambda_2). \quad (3.126)$$

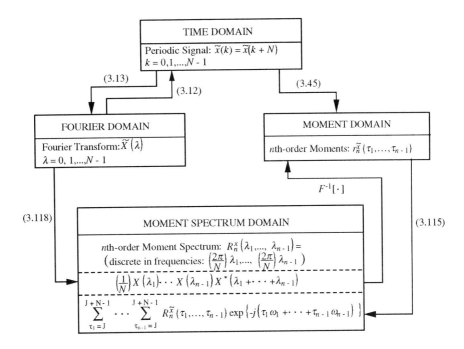

Figure 3.11 The signal analysis map of periodic signals illustrating the relationships between time, Fourier, moment, and moment spectra domains. $F^{-1}[\cdot]$ denotes multidimensional inverse Fourier transform.

Fourth-Order Moment Spectrum: n=4

$$R_4^{\tilde{x}}(\lambda_1, \lambda_2, \lambda_3) = \frac{1}{N}\tilde{X}(\lambda_1)\tilde{X}(\lambda_2)\tilde{X}(\lambda_3)\tilde{X}^*(\lambda_1 + \lambda_2 + \lambda_3) \qquad (3.127)$$

for $(\lambda_1, \lambda_2, \lambda_3 = 0, 1, \ldots, N-1)$. From (3.121), it follows that

$$\begin{aligned}
R_4^{\tilde{x}}(\lambda_1, \lambda_2, 0) &= \tilde{X}(0) \cdot R_3^{\tilde{x}}(\lambda_1, \lambda_2), \\
R_4^{\tilde{x}}(\lambda_1, 0, 0) &= \tilde{X}^2(0) \cdot R_2^{\tilde{x}}(\lambda_1), \\
R_4^{\tilde{x}}(\lambda_1, -\lambda_1, \lambda_3) &= (R_2^{\tilde{x}}(\lambda_1))^2 \cdot R_2^{\tilde{x}}(\lambda_3).
\end{aligned} \qquad (3.128)$$

Thus, the fourth-order moment spectrum contains the second- and third-order moment spectra of the signal. The "kurtosis" measure, $r_4^{\tilde{x}}(0,0,0)$, of a periodic signal is given by

$$r_4^{\tilde{x}}(0,0,0) = \frac{1}{N}\sum_{k=0}^{N-1}\tilde{x}^4(k) = \frac{1}{N^3}\sum_{\lambda_1=0}^{N-1}\sum_{\lambda_2=0}^{N-1}\sum_{\lambda_3=0}^{N-1} R_4^{\tilde{x}}(\lambda_1, \lambda_2, \lambda_3). \qquad (3.129)$$

Example 3.17

We compute moment spectra of the impulse train sequence, $\tilde{\delta}(k)$, described in Example 3.12. Combining (3.104) and (3.115), we obtain

$$R_2^{\tilde{\delta}}(\lambda) = r_2^{\tilde{\delta}}(\tau = 0) = A^2 \text{ for } \lambda = 0,$$

$$R_3^{\tilde{\delta}}(\lambda_1, \lambda_2) = r_3^{\tilde{\delta}}(\tau_1 = 0, \tau_2 = 0) = A^3 \text{ for } (\lambda_1, \lambda_2) = (0,0)$$

and

$$R_4^{\tilde{\delta}}(\lambda_1, \lambda_2, \lambda_3) = r_4^{\tilde{\delta}}(\tau_1 = 0, \tau_2 = 0, \tau_3 = 0) = A^4 \text{ for } (\lambda_1, \lambda_2, \lambda_3) = (0,0,0).$$

□

Example 3.18

The higher-order moments of the periodic sequence $\tilde{s}(k) = \cos(\frac{2\pi}{N} k + \phi)$ were computed in Example 3.13 (see (3.106)–(3.108)) and illustrated in Figure 3.8. One way of computing its higher-order moment spectra is by using (3.115). Another way, however, is by the use of the alternative definition (3.118) which requires knowledge of the Fourier series representation $\tilde{S}(\lambda)$. That is obtained by combining (3.105) and (3.13), and is given by

$$\tilde{S}(\lambda) = \sum_{k=0}^{N-1} \frac{1}{2} \left\{ e^{j(\frac{2\pi}{N} k + \phi)} + e^{-j(\frac{2\pi}{N} k + \phi)} \right\} \cdot e^{-j\frac{2\pi}{N} k \lambda}$$

$$= \frac{e^{j\phi}}{2} \sum_{k=0}^{N-1} e^{-j\frac{2\pi}{N} k(\lambda-1)} + \frac{e^{-j\phi}}{2} \sum_{k=0}^{N-1} e^{-j\frac{2\pi}{N} k(\lambda+1)}$$

$$= \begin{cases} \frac{N}{2} e^{j\phi}, & \lambda = 1 \\ \frac{N}{2} e^{-j\phi}, & \lambda = N-1 \\ 0, & \lambda = 0, 2, 3, \ldots, N-2. \end{cases} \quad (3.130)$$

The power spectrum, bispectrum, and trispectrum of $\tilde{s}(k)$ are respectively given by

$$R_2^{\tilde{s}}(\lambda) = \frac{1}{N} \tilde{S}(\lambda) \tilde{S}^*(\lambda)$$

$$= \begin{cases} \frac{N}{4}, & \lambda = 1 \text{ and } \lambda = N-1 \\ 0, & \lambda = 0, 2, 3, \ldots, N-2, \end{cases} \quad (3.131)$$

$$R_3^{\tilde{s}}(\lambda_1,\lambda_2) = \frac{1}{N}\tilde{S}(\lambda_1)\tilde{S}(\lambda_2)\tilde{S}^*(\lambda_1+\lambda_2)$$
$$= 0, \quad \text{for all } (\lambda_1,\lambda_2), \tag{3.132}$$

and

$$R_4^{\tilde{s}}(\lambda_1,\lambda_2,\lambda_3) = \frac{1}{N}\tilde{S}(\lambda_1)\tilde{S}(\lambda_2)\tilde{S}(\lambda_3)\tilde{S}^*(\lambda_1+\lambda_2+\lambda_3)$$

$$= \begin{cases} \dfrac{N^3}{16}, & \text{for } \lambda_1 = \lambda_2 = 1 \text{ and } \lambda_3 = N-1 \\ & \text{for } \lambda_1 = \lambda_3 = 1 \text{ and } \lambda_2 = N-1 \\ & \text{for } \lambda_2 = \lambda_3 = 1 \text{ and } \lambda_1 = N-1 \\ 0, & \text{otherwise.} \end{cases} \tag{3.133}$$

Note that $\tilde{S}(\lambda) = \tilde{S}(\lambda+N)$ and hence $\tilde{S}(N+1) = \tilde{S}(1)$. Figure 3.12 illustrates the Fourier series representation, power spectrum, and trispectrum of $\tilde{s}(k)$. □

Example 3.19

In Example (3.14) we computed the autocorrelation and third-order moment sequences of a cosine wave of the form $\tilde{x}(k) = \cos(\frac{2\pi}{N}k + \phi) + \tilde{\delta}(k)$ (see also Figure 3.9). The Fourier series representation of this sequence is

$$\tilde{X}(\lambda) = \frac{e^{j\phi}}{2}\sum_{k=0}^{N-1}e^{-j\frac{2\pi}{N}k(\lambda-1)} + \frac{e^{-j\phi}}{2}\sum_{k=0}^{N-1}e^{-j\frac{2\pi}{N}k(\lambda+1)}$$

$$+ A\sum_{k=0}^{N-1}e^{-j\frac{2\pi}{N}k\lambda}$$

Hence,

$$\tilde{X}(\lambda) = \begin{cases} \dfrac{N}{2}e^{j\phi}, & \lambda = 1 \\ \dfrac{N}{2}e^{-j\phi}, & \lambda = N-1 \\ NA, & \lambda = 0 \\ 0, & \lambda = 2, 3, \ldots, N-2. \end{cases} \tag{3.134}$$

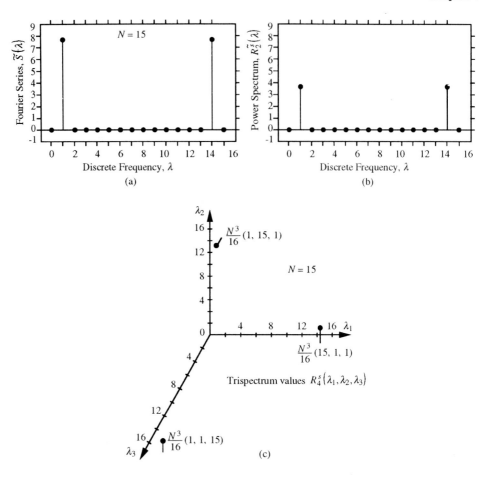

Figure 3.12 Moment spectra of periodic signal $\tilde{s}(k) = \cos\left(\frac{2\pi}{N}k + \phi\right)$. (a) Its Fourier series representation, (b) power spectrum, and (c) trispectrum. The bispectrum of the signal is equal to zero.

The power spectrum is given by

$$R_2^{\tilde{x}}(\lambda) = \frac{1}{N}\tilde{X}(\lambda)\tilde{X}^*(\lambda)$$
$$= \begin{cases} \dfrac{N}{4}, & \lambda = 1, N-1 \\ NA^2, & \lambda = 0 \\ 0, & \text{otherwise,} \end{cases} \qquad (3.135)$$

whereas the third-order moment spectrum (bispectrum) is given by

$$R_3^{\tilde{x}}(\lambda_1, \lambda_2) = \frac{1}{N}\tilde{X}(\lambda_1)\tilde{X}(\lambda_2)\tilde{X}^*(\lambda_1+\lambda_2)$$
$$= \begin{cases} N^2 A^3, & \text{for } (\lambda_1,\lambda_2)=(0,0) \\ \dfrac{N^2 A}{4}, & \text{for } \{(\lambda_1,\lambda_2):(0,1),(0,N-1),(1,0),(1,N-1),(N-1,0),(N-1,1)\} \\ 0, & \text{otherwise.} \end{cases}$$
(3.136)

Figure 3.13 illustrates $\tilde{X}(\lambda), R_2^{\tilde{x}}(\lambda)$ and $R_3^{\tilde{x}}(\lambda_1, \lambda_2)$. If we apply (3.123) and (3.125) in our example, we obtain

$$P_{\tilde{x}} = \frac{1}{N}\sum_{\lambda=0}^{N-1} R_2^{\tilde{x}}(\lambda) = \frac{1}{N}\left\{NA^2 + \frac{2N}{4}\right\}$$
$$= A^2 + \frac{1}{2},$$

which is identical to $r_2^{\tilde{x}}(0)$ in (3.110), and

$$r_3^{\tilde{x}}(0,0) = \frac{1}{N^2}\sum_{\lambda_1=0}^{N-1}\sum_{\lambda_2=0}^{N-1} R_3^{\tilde{x}}(\lambda_1,\lambda_2)$$
$$= \frac{1}{N^2}\left[N^2 A^3 + \frac{6N^2 A}{4}\right]$$
$$= A^3 + \frac{3}{2}A$$

which may also be obtained directly from (3.110). □

Example 3.20

We compute moment spectra of the sawtooth periodic sequence described in Example 3.16. Its Fourier series representation is given by

$$\tilde{T}(\lambda) = e^{-j\frac{2\pi}{3}\lambda}\left\{\frac{1}{2} + e^{-j\frac{2\pi}{3}\lambda}\right\} \quad \text{for} \quad \lambda=0,1,2. \tag{3.137}$$

Combining (3.137) and (3.118), we obtain the power spectrum

$$R_2^{\tilde{t}}(\lambda) = \frac{1}{3}\tilde{T}(\lambda)\,\tilde{T}^*(\lambda)$$
$$= \frac{1}{3}\left\{\frac{5}{4} + \cos\frac{2\pi}{3}\lambda\right\} \quad \text{for} \quad \lambda=0,1,2.$$
(3.138)

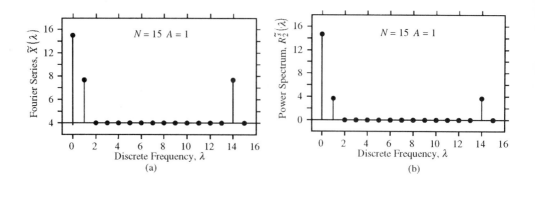
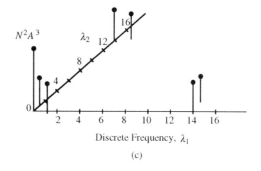

Figure 3.13 Moment spectra of the periodic signal $\tilde{x}(k) = \cos\left(\frac{2\pi}{N}k + \phi\right) + \tilde{\delta}(k)$. (a) Its Fourier series representation, (b) power spectrum, and (c) bispectrum.

The third-order moment spectrum (bispectrum) is given by

$$R_3^{\tilde{x}}(\lambda_1, \lambda_2) = \frac{1}{3}\left\{\frac{9}{8} + \frac{1}{4}\left(e^{-j\frac{2\pi}{3}\lambda_2} + e^{-j\frac{2\pi}{3}\lambda_1}\right)\right.$$
$$+ \frac{1}{2}\left(e^{j\frac{2\pi}{3}\lambda_2} + e^{j\frac{2\pi}{3}\lambda_1}\right) \qquad (3.139)$$
$$\left. + \frac{1}{2}e^{-j\frac{2\pi}{3}(\lambda_1+\lambda_2)} + \frac{1}{4}e^{j\frac{2\pi}{3}(\lambda_1+\lambda_2)}\right\}$$

for $\lambda_1, \lambda_2 = 0, 1, 2$. $R_2^{\tilde{x}}(\lambda)$ and $R_3^{\tilde{x}}(\lambda_1, \lambda_2)$ are illustrated in Figure 3.14. □

3.5.4 LTI System Relations

Suppose we are given the stable LTI system described by (3.63). Let us also assume that the input to the system, $\tilde{x}(k)$, is periodic with period N. Since

$$\tilde{x}(k) = \frac{1}{N}\sum_{\lambda=0}^{N-1}\tilde{X}(\lambda)e^{j\frac{2\pi}{N}k\lambda},$$

the output, $\tilde{y}(k)$, of the system is also periodic with the same period and is given by

$$\tilde{y}(k) = \frac{1}{N}\sum_{\lambda=0}^{N-1}\tilde{X}(\lambda)H\left(\frac{2\pi\lambda}{N}\right)\cdot e^{j\frac{2\pi}{N}k\lambda} \qquad (3.140)$$

where $H\left(\frac{2\pi\lambda}{N}\right)$ is the frequency transfer function of the system computed at frequencies $\omega_\lambda = \frac{2\pi}{N}\lambda$, $\lambda = 0, 1, \ldots, N-1$. Hence, the Fourier series coefficients of the system response, $\tilde{Y}(\lambda)$, are related to those of the input by

$$\tilde{Y}(\lambda) = \tilde{X}(\lambda)\cdot H\left(\frac{2\pi}{N}\lambda\right), \quad \lambda = 0, 1, \ldots, N-1. \qquad (3.141)$$

From (3.118) we have that the nth-order moment spectrum of the system response is given by

$$R_n^{\tilde{y}}(\lambda_1,\ldots,\lambda_{n-1}) = \frac{1}{N}\tilde{Y}(\lambda_1)\cdots\tilde{Y}(\lambda_{n-1})\tilde{Y}^*(\lambda_1+\cdots+\lambda_{n-1}). \qquad (3.142)$$

Combining (3.141) and (3.142), we obtain

$$R_n^{\tilde{y}}(\lambda_1,\ldots,\lambda_{n-1}) = H\left(\frac{2\pi}{N}\lambda_1\right)\cdots H\left(\frac{2\pi}{N}\lambda_{n-1}\right)\ H^*\left(\frac{2\pi}{N}(\lambda_1+\cdots+\lambda_{n-1})\right)$$
$$\cdot R_n^{\tilde{x}}(\lambda_1,\ldots,\lambda_{n-1}), \qquad (3.143)$$

which relates the nth-order moment spectra of the system input and output.

3.5.5 Cross-Moment Spectra

The nth-order cross-moment spectrum of a set of periodic signals $\{\tilde{x}_i(k)\}, i = 1, 2, \ldots, n$ with period N is defined as

$$R_{\tilde{x}_1\cdots\tilde{x}_n}(\lambda_1,\ldots,\lambda_{n-1}) \triangleq \sum_{\tau_1=J}^{J+N-1}\cdots\sum_{\tau_{n-1}=J}^{J+N-1} r_{\tilde{x}_1\cdots\tilde{x}_n}(\tau_1,\ldots,\tau_{n-1})\cdot$$
$$\exp\left\{-j\frac{2\pi}{N}(\lambda_1\tau_1+\cdots+\lambda_{n-1}\tau_{n-1})\right\} \qquad (3.144)$$

where $r_{\tilde{x}_1\cdots\tilde{x}_n}(\tau_1,\ldots,\tau_{n-1})$ is the nth-order cross-moment sequence. Substituting (3.114) into (3.144), we obtain, after some algebra, an alternative definition directly

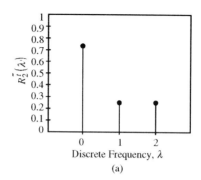

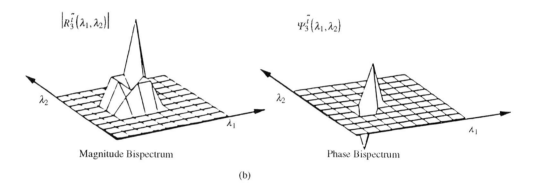

Figure 3.14 Moment spectra of sawtooth periodic sequence. (a) power spectrum, and (b) magnitude and phase bispectrum.

in terms of Fourier series coefficients:

$$R_{\tilde{x}_1\cdots\tilde{x}_n}(\lambda_1,\ldots,\lambda_{n-1}) = \frac{1}{N}\tilde{X}_2(\lambda_1)\cdots\tilde{X}_n(\lambda_{n-1})\cdot \tilde{X}_1^*(\lambda_1+\cdots+\lambda_{n-1}). \quad (3.145)$$

Hence, the second-order cross-spectrum of two periodic signals with the same period is given by

$$R_{\tilde{x}_1\tilde{x}_2}(\lambda_1) = \frac{1}{N}\tilde{X}_2(\lambda_1)\cdot \tilde{X}_1^*(\lambda_1), \quad (3.146)$$

whereas the cross-bispectrum of three periodic signals takes the form

$$R_{\tilde{x}_1\tilde{x}_2\tilde{x}_3}(\lambda_1,\lambda_2) = \frac{1}{N}\tilde{X}_2(\lambda_1)\tilde{X}_3(\lambda_2)\tilde{X}_1^*(\lambda_1+\lambda_2). \quad (3.147)$$

Example 3.21

Consider the periodic signals $\tilde{x}_1(k) = \cos\left(\frac{2\pi}{N}k + \phi\right)$ and $\tilde{x}_2(k) = \cos\left(\frac{2\pi}{N}k + \phi\right) +$

$\tilde{\delta}(k)$ of which the Fourier series coefficients are given by (3.130) and (3.134), respectively. The second-order cross-spectrum of these signals is given by

$$R_{\tilde{x}_1\tilde{x}_2}(\lambda) = \frac{1}{N}X_2(\lambda)X_1^*(\lambda)$$

$$= \begin{cases} 0, & \lambda = 0 \\ \dfrac{N}{4}, & \lambda = 1 \text{ and } \lambda = N-1 \\ 0, & \text{otherwise,} \end{cases}$$

whereas their cross-bispectrum is given by

$$R_{\tilde{x}_1\tilde{x}_2\tilde{x}_2}(\lambda_1,\lambda_2) = \frac{1}{N}X_2(\lambda_1)X_2(\lambda_2)X_1^*(\lambda_1+\lambda_2)$$

$$= \begin{cases} \dfrac{N^2 A}{4}e^{j2\phi}, & \{(\lambda_1,\lambda_2):(0,1),(1,0)\} \\ \dfrac{N^2 A}{4}, & \{(\lambda_1,\lambda_2):(0,N-1),(N-1,0)\} \\ 0, & \text{otherwise.} \end{cases}$$

However, if we choose to define the cross-bispectrum of the two periodic signals as

$$R_{\tilde{x}_2\tilde{x}_2\tilde{x}_1}(\lambda_1,\lambda_2) = \frac{1}{N}\tilde{X}_2(\lambda_1)\tilde{X}_1(\lambda_2)\tilde{X}_2^*(\lambda_1+\lambda_2)$$

$$= \begin{cases} \dfrac{N^2 A}{4}, & \{(\lambda_1,\lambda_2):(0,1),(0,N-1),(1,N-1),(N-1,1)\} \\ 0, & \text{otherwise,} \end{cases}$$

we observe that $R_{\tilde{x}_2\tilde{x}_2\tilde{x}_1}(\lambda_1,\lambda_2)$ is different from $R_{\tilde{x}_1\tilde{x}_2\tilde{x}_2}(\lambda_1,\lambda_2)$. □

3.5.6 Higher-Order Coherency Function of Periodic Power Signals

The nth-order coherency function of a periodic signal may be defined in terms of the phase of its Fourier series coefficients as follows:

$$\begin{aligned}P_n^{\tilde{x}}(\lambda_1,\ldots,\lambda_{n-1}) &= \exp\{\Psi_n^{\tilde{x}}(\lambda_1,\ldots,\lambda_{n-1})\} \\ &= \exp\{\phi_{\tilde{x}}(\lambda_1)+\cdots+\phi_{\tilde{x}}(\lambda_{n-1})-\phi_{\tilde{x}}(\lambda_1+\cdots+\lambda_{n-1})\},\end{aligned}$$
(3.148)

which is essentially the phase of its nth-order moment spectrum.

3.5.7 Finite-Duration versus Periodic Signals: Moment Spectra Relationships

In section 3.1.4 we established the relationships between the Fourier series representation $\tilde{Z}(\lambda)$ of a periodic sequence $\tilde{z}(k) = \tilde{z}(k+N)$ and the Fourier transform $Y(\omega)$ of a finite-duration sequence $y(k)$ of length N. One period of the periodic sequence was identical to the finite-duration sequence. The purpose of this section is to establish similar relationships between the moment spectra of sequences $y(k)$ and $\tilde{z}(k)$.

From (3.47), we see that the nth-order moment spectrum of $y(k)$ is given by

$$M_n^y(\omega_1, \ldots, \omega_{n-1}) = Y(\omega_1)Y(\omega_2)\cdots Y(\omega_{n-1})Y^*(\omega_1 + \cdots + \omega_{n-1}). \quad (3.149)$$

Similarly, from (3.118) we obtain

$$R_n^{\tilde{z}}(\lambda_1, \ldots, \lambda_{n-1}) = \frac{1}{N}\tilde{Z}(\lambda_1)\cdots\tilde{Z}(\lambda_{n-1})\tilde{Z}^*(\lambda_1 + \cdots + \lambda_{n-1}). \quad (3.150)$$

Substitution of (3.29) into (3.150) yields

$$R_n^{\tilde{z}}(\lambda_1, \ldots, \lambda_{n-1}) = \frac{1}{N}Y\left(\frac{2\pi}{N}\lambda_1\right)\cdots Y\left(\frac{2\pi}{N}\lambda_{n-1}\right)\cdot Y^*\left(\frac{2\pi}{N}(\lambda_1 + \cdots + \lambda_{n-1})\right). \quad (3.151)$$

Combining (3.149) and (3.151), we obtain

$$R_n^{\tilde{z}}(\lambda_1, \ldots, \lambda_{n-1}) = \frac{1}{N}\cdot M_n^y\left(\frac{2\pi}{N}\lambda_1, \ldots, \frac{2\pi}{N}\lambda_{n-1}\right). \quad (3.152)$$

Thus, if we know the moment spectrum of a finite-duration sequence, the moment spectrum of its equivalent periodic sequence is given by (3.152).

On the other hand, if we are given $R_n^{\tilde{z}}(\lambda_1, \ldots, \lambda_{n-1})$, the moment spectrum $M_n^y(\omega_1, \ldots, \omega_{n-1})$ of the equivalent finite duration sequence can be computed using the following identity:

$$\begin{aligned}M_n^y(\omega_1, \ldots, \omega_{n-1}) = &\frac{1}{N}\sum_{\lambda_1=0}^{N-1}\cdots\sum_{\lambda_{n-2}=0}^{N-1}\sum_{\lambda_{n-1}=0}^{N-1}\\ R_n^{\tilde{z}}(\lambda_2, \ldots, \lambda_{n-1}, -\lambda_1 - \cdots - \lambda_{n-1})&\cdot F\left(\frac{\omega_1 - 2\pi(-\lambda_1 - \cdots - \lambda_{n-1})}{2N}\right)\\ &\cdot F\left(\frac{\omega_2 - 2\pi\lambda_2}{2N}\right)\cdots F\left(\frac{\omega_{n-1} - 2\pi\lambda_{n-1}}{2N}\right)\end{aligned} \quad (3.153)$$

where
$$F(\omega) = \frac{\sin[\omega(1-2N)]}{\sin\omega}. \tag{3.154}$$

Notice that
$$R_n^{\tilde{z}}(\lambda_2,\ldots,\lambda_{n-1},-\lambda_1-\cdots-\lambda_{n-1}) = \tilde{Z}(\lambda_2)\cdots\tilde{Z}(\lambda_{n-1})$$
$$\cdot \tilde{Z}(-\lambda_1-\cdots-\lambda_{n-1})\cdot\tilde{Z}^*(-\lambda_1). \tag{3.155}$$

3.6 SUMMARY

In this chapter, we introduced the definitions and properties of moments and moment spectra of energy and power deterministic signals. Many examples were provided to establish the duality between moments and moment spectra of energy and power signals. Within the class of power signals, special emphasis was given to the moment spectra of periodic sequences.

A tutorial paper on triple correlations and the bispectrum of deterministic signals was published by Lohmann and Wirnitzer [1984]. Bispectra of special periodic sequences have been studied by Nagata [1978]. An exposition of energy and power signals and their "dichotomies" is given by Lob [1985]. The papers by Dainty and Northcott [1986] and Sato et al. [1977] discuss specific applications of bispectra of deterministic signals.

REFERENCES

Dainty, J. C., and M. J. Northcott, "Imaging a Randomly Translating Object at Low Light Levels Using the Triple Correlation," *Optics Communications*, **58**(1), pp. 11–14, Elsevier Science Publishers B. V., 1986.

Gårding, L., and L. Hörmander (eds.), **Marcel Riesz Collected Papers**, Berlin Heidelbert: Springer-Verlag, 1988.

Hayes, M. H., "The Unique Reconstruction of Multi-dimensional Sequences from Fourier Transform Magnitude or Phase," in **Image Recovery: Theory and Application**, pp. 195-230, H. Stark (ed.), Orlando, Florida: Academic Press, 1987.

Lob, W. H., "Signal Analysis and AM Theory," **Lecture Notes of Signal Analysis**, Boston: Northeastern University Press, 1985.

Lohmann, A. W., and B. Wirnitzer, "Triple Correlations," *Proceedings IEEE*, **72** (7), pp. 889–901, July 1984.

Nagata, Y., "Bispectra of Spike-Array Type Time Series and Their Application to the Analysis of Oceanic Microstructures," *J. Oceanographical Society of Japan*, **34**, pp. 204–216, 1978.

Oppenheim, A. V., and R. W. Schafer, **Discrete-Time Signal Processing**, Englewood Cliffs, NJ: Prentice Hall, 1989.

Sato, T., K. Sasaki, and Y. Nakamura, "Real-Time Bispectral Analysis of Gear Noise and Its Application to Contactless Diagnosis", *J. Acoust. Soc. Amer.*, **62**(2), p. 404, 1977.

4

CONVENTIONAL METHODS FOR THE ESTIMATION OF HIGHER-ORDER SPECTRA

4.1 INTRODUCTION

Two equivalent definitions of cumulant spectra were introduced in Chapter 2 for stochastic signals; e.g., equations (2.24) and (2.27). Similarly, definitions of moment spectra for energy signals were given by (3.45) and (3.47), and definitions for periodic signals by (3.115) and (3.118). All these definitions (except for periodic signals) are based on the assumptions that an infinite sequence of time samples, or higher-order statistics, are available and that the higher-order statistics are known exactly. Furthermore, in the case of periodic signals, the definitions require that the fundamental period of the signal is known exactly. The problem met in practice, however, is one of estimating the cumulant or moment spectra of a process when a finite set of observation measurements is available. There are two chief approaches that can be used to estimate higher-order spectra; namely, the conventional ("Fourier" type) and the parametric approach, which is based on autoregressive (AR), moving average (MA), ARMA or Volterra models. The purpose of this chapter is to discuss the conventional methods, as well as their statistical properties, and computational complexity.

The conventional methods may be classified into the following three classes: (1) The **Indirect class** of techniques which are approximations of the definitions given by (2.24) or (3.45), (2) The **Direct class** of techniques, which approximate definitions (2.27) or (3.47); and (3) The **Complex demodulates class** of techniques. While the conventional methods are straightforward and their implementation can be based on fast Fourier transform (FFT) algorithms, limitations on statistical variance of the estimates and frequency resolution may impose severe limits on their applicability.

4.2 INDIRECT CLASS OF CONVENTIONAL METHODS

Each of the methods considered here estimates first the higher-order statistics (cumulants or moments) from a finite length record of data, then generates higher-order spectral estimates using multidimensional window functions. Hence, it is similar to the estimation of the power spectrum using the Blackman-Tuckey method [Marple, 1987; Kay, 1988].

4.2.1 Estimation of Higher-Order Statistics

Let $\{X(1), X(2), \ldots, X(N)\}$ be the given data set which may well represent a realization from a strictly stationary random process or a deterministic sequence. Then, we have the following [Brillinger and Rosenblatt, 1967; Nikias and Raghuveer, 1987; Rao and Gabr, 1988]:

1. Segment the data into K records of M samples each; i.e., $N = K \cdot M$. However, if the data samples correspond to a deterministic energy signal, then the data segmentation is inappropriate. Also, if the process is deterministic periodic, then M should be equal to the period of the signal or multiple integers of the period.
2. Subtract the average value of each record (this is optional for deterministic signals).
3. Assuming that $\{x^{(i)}(k), k = 0, 1, \ldots, M-1\}$ is the data per segment ($i = 1, 2, \ldots, K$), natural estimates of higher-order moments are given by

$$m_n^{(i)}(\tau_1, \ldots, \tau_{n-1}) = \frac{1}{M} \sum_{k=s_1}^{s_2} x^{(i)}(k) x^{(i)}(k+\tau_1) \cdots x^{(i)}(k+\tau_{n-1}) \quad (4.1)$$

where $n = 2, 3, \ldots, i = 1, 2, \ldots, K, \tau_k = 0, \pm 1, \pm 2, \ldots, s_1 = \max(0, -\tau_1, \ldots, -\tau_{n-1}), s_2 = \min(M-1, M-1-\tau_1, \ldots, M-1-\tau_{n-1})$, and $|\tau_k| \leq L_n$. Note that L_n determines the region of support of the estimated nth-order moment function.
4. The average over all segments is given by

$$\hat{m}_n^x(\tau_1, \ldots, \tau_{n-1}) = \frac{1}{K} \sum_{i=1}^{K} m_n^{(i)}(\tau_1, \ldots, \tau_{n-1}), \quad (4.2)$$

$$n = 2, 3, \ldots, \quad |\tau_k| \leq L_n.$$

The $\hat{m}_n^x(\tau_1, \ldots, \tau_{n-1})$ is considered the usual estimate for $m_n^x(\tau_1, \ldots, \tau_{n-1})$. If the signal is deterministic, then $\hat{m}_n^x(\tau_1, \ldots, \tau_{n-1}) = m_n^{(1)}(\tau_1, \ldots, \tau_{n-1})$; i.e., $K = 1$ and thus $N = M$.

5. For stochastic signals, generate the cumulants $\hat{c}_n^x(\tau_1, \ldots, \tau_{n-1})$ using (2.5). If the average value of each record is subtracted (i.e., so each can be considered as a zero-mean signal), it follows from (2.12)–(2.14) that

$$\begin{aligned}
\hat{c}_2^x(\tau_1) &= \hat{m}_2^x(\tau_1) \\
\hat{c}_3^x(\tau_1, \tau_2) &= \hat{m}_3^x(\tau_1, \tau_2) \\
\hat{c}_4^x(\tau_1, \tau_2, \tau_3) &= \hat{m}_4^x(\tau_1, \tau_2, \tau_3) - \hat{m}_2^x(\tau_1)\hat{m}_2^x(\tau_3 - \tau_2) - \\
&\quad \hat{m}_2^x(\tau_2) \cdot \hat{m}_2^x(\tau_3 - \tau_1) - \hat{m}_2^x(\tau_3) \cdot \hat{m}_2^x(\tau_2 - \tau_1),
\end{aligned} \quad (4.3)$$

where $|\tau_k| \leq L_n \quad k = 1, 2, 3$.

Rosenblatt and Van Ness [1965] point out that there are two customary requirements made of such estimates: (i) the estimate should be asymptotically unbiased, and (ii) the variance of the estimate should go to zero as $K \to \infty$, or as $N \to \infty$ for $K = 1$. If the stationary random process, $\{X(k)\}$, satisfies certain mixing conditions, the estimates of higher-order statistics given above have both properties [Rosenblatt and Van Ness, 1965; Rosenblatt, 1985].

4.2.2 Estimation of Higher-Order Spectra

The nth-order moment spectrum estimate is given by

$$\begin{aligned}
\hat{M}_n^x(\omega_1, \ldots, \omega_{n-1}) &= \\
= \sum_{\tau_1 = -L_n}^{L_n} \cdots &\sum_{\tau_{n-1} = -L_n}^{L_n} \hat{m}_n^x(\tau_1, \ldots, \tau_{n-1}) \cdot \\
\cdot w(\tau_1 \Delta_n, \ldots, &\tau_{n-1} \Delta_n) \cdot \exp\{j(\omega_1 \tau_1 + \cdots + \omega_{n-1} \tau_{n-1})\}]
\end{aligned} \quad (4.4)$$

while the cumulant spectrum estimate is given by

$$\begin{aligned}
\hat{C}_n^x(\omega_1, \ldots, \omega_{n-1}) &= \sum_{\tau = -L_n}^{L_n} \cdots \sum_{\tau_{n-1} = -L_n}^{L_n} \hat{c}_n^x(\tau_1, \ldots, \tau_{n-1}) \\
\cdot w(\tau_1 \Delta_n, \ldots, \tau_{n-1} \Delta_n) &\cdot \exp\{-j(\omega_1 \tau_1 + \cdots + \omega_{n-1} \tau_{n-1})\}]
\end{aligned} \quad (4.5)$$

where $w(u_1, \cdots, u_{n-1})$ is a continuous window function of bounded support and Δ_n is the bandwidth usually taken $\Delta_n \triangleq 1/L_n$. Notice that for $K = 1$, the L_n and M are chosen such that $(L_n^2/M) \to 0$ as $L_n \to \infty$, $M \to \infty$ [Rao and Gabr, 1984]. The computational complexity associated with higher-order spectrum estimates (4.4) or (4.5) can be reduced substantially if the symmetry properties of higher-order statistics are taken into account during their calculation by (4.1), and if the symmetry properties of higher-order spectra are incorporated in the computations of (4.4) or (4.5).

4.2.3 Window Functions

As in the case of conventional power spectrum estimation, to get smooth estimates, suitable window functions should be used. The window function, $w(\tau_1 \Delta_n, \ldots, \tau_{n-1} \Delta_n)$ for higher-order spectrum estimation should satisfy the following properties: [Rosenblatt and Van Ness, 1965; Sasaki, et al. 1975; Rao and Gabr, 1984]:

1. Possess symmetry properties of higher-order moments or cumulants. For example, in the case of bispectrum estimation of real signals, the window function should satisfy the symmetries $w(m,n) = w(n,m) = w(-m, n-m) = w(m-n, -n)$.

2. Be zero outside the region of support of estimated higher-order statistics; i.e., $w(\tau_1 \Delta_n, \ldots, \tau_{n-1} \Delta_n) = 0$ for $|\tau_i| > L_n$, $i = 1, 2, \ldots, n-1$.

3. Be equal to one at the origin; i.e., $w(0, 0, \ldots, 0) = 1$ (normalizing condition).

4. Have a real nonnegative Fourier transform; i.e., $W(\omega_1, \ldots, \omega_{n-1}) \geq 0$ for $|\omega_i| \leq \pi$, $i = 1, 2, \ldots, n-1$. The window function should also have finite energy.

A class of functions which satisfies these properties can easily be generated using standard one-dimensional lag windows. That is,

$$w(\tau_1 \Delta_n, \ldots, \tau_{n-1} \Delta_n) = d(\tau_1 \Delta_n) \cdots d(\tau_{n-1} \Delta_n) \cdot d((\tau_1 + \ldots + \tau_{n-1}) \Delta_n), \quad (4.6)$$

where $d(\tau \Delta_n)$ is a one-dimensional function with properties

$$\begin{aligned} d(\tau \Delta_n) &= d(-\tau \Delta_n) \\ d(\tau \Delta_n) &= 0, \quad \text{for} \quad |\tau| > L_n \\ d(0) &= 1 \\ D(\omega) &\geq 0, \quad \text{for} \quad |\omega| \leq \pi. \end{aligned} \quad (4.7)$$

However, not all one-dimensional windows satisfy constraint $D(\omega) \geq 0$ for all ω. For example, the Hanning window has negative sidelobes in the frequency domain. Sasaki et al. [1975] showed several different windows that satisfy the properties of (4.7). Table 4.1 defines one-dimensional (1-d) window functions that satisfy (4.7) and can be used in (4.6) to generate multidimensional windows for higher-order spectrum estimation [Rao and Gabr, 1984; Sasaki et al., 1975]. Figure 4.1 illustrates the 1-d windows and their Fourier transforms.

TABLE 4.1 DEFINITIONS OF ONE-DIMENSIONAL WINDOW FUNCTIONS SUITABLE FOR GENERATION OF MULTIDIMENSIONAL WINDOWS FOR HIGHER-ORDER SPECTRUM ESTIMATION

Name	Mathematical Description
Daniell	$d(u) = \begin{cases} \frac{\sin(\pi u)}{\pi u}, & \|u\| \leq 1 \\ 0, & \text{otherwise} \end{cases}$
Hamming	$d(u) = \begin{cases} 0.54 + 0.46\cos(\pi u), & \|u\| \leq 1 \\ 0, & \text{otherwise} \end{cases}$
Parzen	$d(u) = \begin{cases} 1 - 6u^2 + 6\|u\|^3, & \|u\| \leq 0.5 \\ 2(1-\|u\|)^3, & 0.5 \leq \|u\| \leq 1 \\ 0, & \text{otherwise} \end{cases}$
Priestley	$d(u) = \begin{cases} \frac{3}{(\pi u)^2} \cdot [\frac{\sin(\pi u)}{\pi u} - \cos(\pi u)], & \|u\| \leq 1 \\ 0, & \text{otherwise} \end{cases}$
Sasaki	$d(u) = \begin{cases} \frac{1}{\pi}\|\sin(\pi u)\| + (1 - \|u\|)\cos(\pi u), & \|u\| \leq 1 \\ 0, & \text{otherwise} \end{cases}$

$u = \tau_i \Delta_n$, where $\Delta_n = \frac{1}{L_n}$, and τ_i take only integer values.

Two window functions that do not belong to the separable class described by (4.6) are the uniform window in the frequency domain [Nikias and Raghuveer, 1987] and the mean-squared error (MSE) optimal window derived by Rao and Gabr [1984]. The uniform window is defined by:

$$W(\omega_1, \ldots, \omega_{n-1}) = \begin{cases} \frac{4}{3}\left(\frac{\pi}{\Omega_o}\right), & |\omega| < \Omega_o = \alpha_o \cdot \Delta_n \\ 0, & |\omega| > \Omega_o \end{cases} \quad (4.8)$$

where $|\omega| = \max[|\omega_1|, |\omega_2|, \ldots, |\omega_{n-1}|, |\omega_1 + \ldots + \omega_{n-1}|]$, α_o is a constant parameter and $\Delta_n = 1/L_n$. In the case of the bispectrum, this window is distributed uniformly in its hexagonal frequency region of support. On the other hand, the MSE optimal window for bispectrum estimation is defined by [Rao and Gabr, 1984]:

$$w(u_1, u_2) = \frac{8}{7\pi^3}\left\{g(u_1, u_2) + g(-u_1, u_2 - u_1) + g(u_1 - u_2, -u_2)\right\}, \quad (4.9.1)$$

where

$$g(u_1, u_2) = \frac{2u_1^2 + 2u_2^2 + u_1 u_2}{\pi u_1^3 u_2^3}\cos[(u_2 - u_1)\pi] - \frac{u_2 - u_1}{u_1^2 u_2^2} \cdot \sin[(u_2 - u_1)\pi], \quad (4.9.2)$$

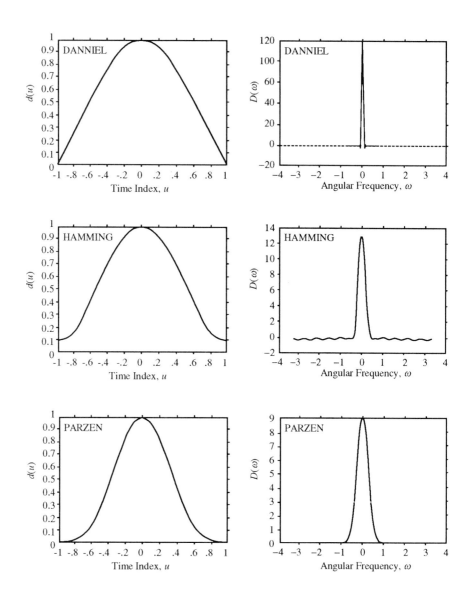

Figure 4.1 Illustration of window functions and their Fourier transforms. The mathematical descriptions of these window functions are defined in Table 4.1.

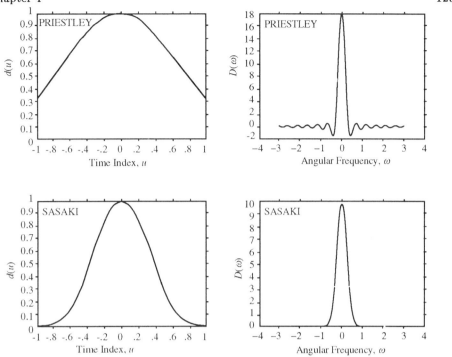

Figure 4.1 (Continued) Illustration of window functions and their Fourier transforms. The mathematical descriptions of these window functions are defined in Table 4.1.

$$|u_1| \leq 1, \quad |u_2| \leq 1.$$

Notice that $u_1 = \tau_1 \Delta_2$ and $u_2 = \tau_2 \Delta_2$, where $\Delta_2 = 1/L_2$.

Figure 4.2 illustrates the optimal MSE window (4.9), as well as two-dimensional windows generated by (4.6) using the expressions of Table 4.1. These window functions can be evaluated in terms of bispectrum bias supremum (J), bispectrum variance (V), and mean-squared error (MSE) between the true and estimated bispectrum. Rao and Gabr [1984] showed that the MSE is directly proportional to an index of efficiency, E, defined by $E \triangleq V \cdot B$, where

$$B \triangleq -\frac{1}{(2\pi)^2} \int_{-\pi}^{+\pi} \int_{-\pi}^{+\pi} \omega_1 \omega_2 W(\omega_1, \omega_2) d\omega_1 d\omega_2, \quad \text{and} \quad (4.10)$$

$$V \triangleq \frac{1}{(2\pi)^2} \int_{-\pi}^{+\pi} \int_{-\pi}^{+\pi} |W(\omega_1, \omega_2)|^2 \, d\omega_1 d\omega_2. \quad (4.11)$$

Sasaki et al. [1975] demonstrated that the bispectrum variance is approximately proportional to the index V, and that the bispectrum bias supremum is proportional to the index J, where

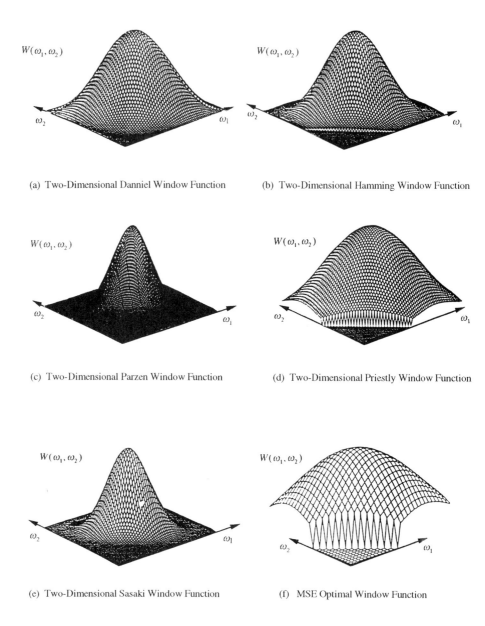

Figure 4.2 Two-dimensional windows for bispectrum estimation: (a) Daniell, (b) Hamming, (c) Parzen, (d) Priestley, (e) Sasaki, and (f) MSE optimal.

TABLE 4.2 VALUES OF INDEXES J, V, B, AND E OF TWO-DIMENSIONAL WINDOWS FOR BISPECTRUM ESTIMATION

Window	J	V	B	MSE(E)
Daniell	99468.50	0.1199	8990.00	1078.50
Parzen	8392.43	0.0409	1324.78	54.20
Hamming	60664.80	0.9067	6261.80	567.76
Priestley	288002.12	0.2032	10909.30	2216.91
Sasaki	1315.20	0.0486	2007.43	97.29
MSE Optimal	2220.74	0.0691	458.69	31.68

$$J \triangleq \frac{1}{(2\pi)^2} \int_{-\pi}^{+\pi} \int_{-\pi}^{+\pi} (\omega_1 - \omega_2)^2 W(\omega_1, \omega_2) d\omega_1 d\omega_2. \qquad (4.12)$$

In Table 4.2, the values of J, V, B, and E are given for the MSE optimal window, and the product windows are illustrated in Figure 4.2. As expected, the MSE optimal window has the smallest E value. The Sasaki window has the smallest J value because it was originally derived to minimize the bispectrum bias supremum [Sasaki et al., 1975]. (The Parzen window, on the other hand, has the smallest V value.)

Example 4.1

Consider the real discrete process

$$X(k) = \sum_{i=1}^{6} \cos(\omega_i k + \phi_i); \quad k = 0, 1, \ldots, N-1, \qquad (4.13)$$

where $\omega_i = 2\pi f_i$, $f_3 = f_1 + f_2$ and $f_6 = f_4 + f_5$, $\phi_3 = \phi_1 + \phi_2$, $\phi_6 = \phi_4 + \phi_5$, and $\phi_1, \phi_2, \phi_4, \phi_5$ are independent and uniformly distributed random variables on $[0, 2\pi]$. Let $f_1 = 0.12$Hz, $f_2 = 0.30$Hz, $f_4 = 0.19Hz$, and $f_5 = 0.17$Hz. Consequently, $f_3 = 0.42$Hz and $f_6 = 0.36$Hz. The length of the data is taken to be N=8192. We obtain K=32 independent records of the process with each record containing M=256 samples (the records are independent because ϕ_1, ϕ_2, ϕ_4, and ϕ_5 were generated afresh for each record). The average value of each record is computed and subtracted. The power spectrum and bispectrum of the process are estimated via (4.1)–(4.5), using $L_2 = 128$ and $L_3 = 64$, respectively. Figure 4.3 illustrates the estimated power spectra (indirect method) utilizing rectangular,

Parzen, and Sasaki 1-d windows. Figure 4.4 demonstrates the magnitude bispectrum estimates of the process obtained by the indirect method with rectangular, Parzen, and Sasaki 2-d windows. From this figure, it is apparent that peaks occur at (0.12Hz, 0.30Hz) and (0.19Hz, 0.17Hz). □

Example 4.2

Consider the deterministic transient signal

$$X(k) = \alpha_1^k \cos(\omega_1 k) + \alpha_2^k \cos(\omega_2 k), \quad k = 0, 1, \ldots, N-1, \qquad (4.14)$$

where $\alpha_1 = 0.8, \omega_1 = 2\pi f_1$ with $f_1 = 0.2Hz, \alpha_2 = 0.6, \omega_2 = 2\pi f_2$ with $f_2 = 0.3$Hz and N=64 samples. Since this is a deterministic energy signal, data segmentation is inappropriate. Figure 4.5 illustrates power spectrum estimates of the signal utilizing $L_2 = 15$ covariance lags and two different windows; i.e., Parzen and Sasaki. On the other hand, Figure 4.6 illustrates bispectrum estimates of the signal with Parzen and Sasaki 2-d windows. The window size of the estimated third-order moments was chosen to be $L_3 = 10$. □

4.3 DIRECT CLASS OF CONVENTIONAL METHODS

The definitions of polyspectra, as given by (2.27) or (3.47), provide a way to estimate them. If a finite length record of data has been observed, its Fourier transform coefficients are estimated and their averaged triple (quadruple) products are then estimates of its bispectrum (trispectrum). Hence, the direct class of methods for higher-order spectrum estimation are similar to the "averaged periodogram" or Welch method for power spectrum estimation [Marple, 1987; Kay, 1988].

4.3.1 The Higher-Order Periodogram

Let $\{X(k)\}, k = 0, 1, \ldots, N-1$ be a real-valued stationary time series with zero-mean. Let the discrete Fourier transform of $\{X(k)\}$ be

$$F_x(\omega_\lambda) = T \cdot \sum_{k=0}^{N-1} X(k) \exp\{-j\omega_\lambda k\}, \qquad (4.15)$$

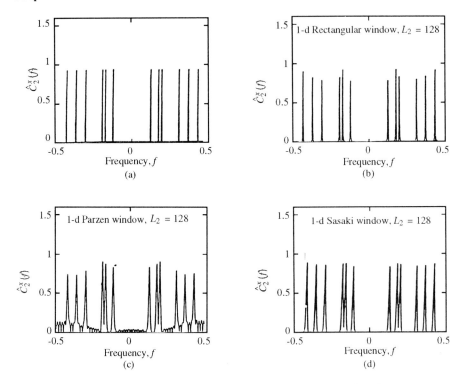

Figure 4.3 (a) The true spectrum of the process described by (4.13). Estimated power spectra by the indirect method using: (b) the rectangular window, (c) the Parzen window, and (d) the Sasaki window.

where $\omega_\lambda = \frac{2\pi}{N}\lambda$ for $\lambda = 0, 1, \ldots, N-1$ and T is the sampling period; if the mean of $\{X(k)\}$ is not removed, then set $F_x(0) = 0$. The higher-order periodogram is then defined as [Helland, Itsweire, and Lii, 1985]:

$$M_n^x(\omega_{\lambda_1}, \ldots, \omega_{\lambda_{n-1}}) = \frac{1}{NT} \cdot F_x(\omega_{\lambda_1}) \cdots F_x(\omega_{\lambda_{n-1}}) \cdot F_x^*(\omega_{\lambda_1} + \cdots + \omega_{\lambda_{n-1}}). \quad (4.16)$$

Brillinger and Rosenblatt [1967] point out that for large N, $\hat{M}_n^x(\omega_{\lambda_1}, \ldots, \omega_{\lambda_{n-1}})$ has the statistical expectation

$$E\left\{\hat{M}_n^x(\omega_{\lambda_1}, \ldots, \omega_{\lambda_{n-1}})\right\} \approx M_n^x(\omega_{\lambda_1}, \ldots, \omega_{\lambda_{n-1}}), \quad (4.17)$$

and the asymptotic variance

$$\begin{aligned} \text{var} \ & \left\{Re[\hat{M}_n^x(\omega_{\lambda_1}, \ldots, \omega_{\lambda_{n-1}})]\right\} \\ & \cong \text{var}\left\{Im[\hat{M}_n^x(\omega_{\lambda_1}, \ldots, \omega_{\lambda_{n-1}})]\right\} \\ & \approx NT \cdot M_2^x(\omega_{\lambda_1}) \cdots M_2^x(\omega_{\lambda_{n-1}}) \cdot M_2^x(\omega_{\lambda_1} + \cdots + \omega_{\lambda_{n-1}}) \end{aligned} \quad (4.18)$$

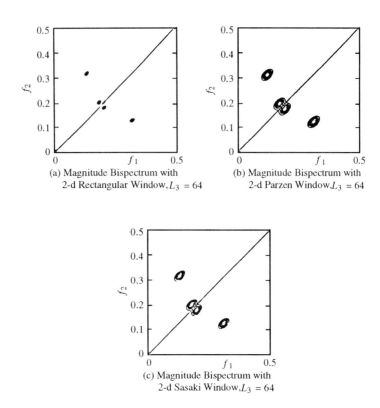

Figure 4.4 The magnitude bispectrum of the process described by (4.13), estimated by the indirect method: with (a) the rectangular window, (b) the Parzen window, and (c) the Sasaki window.

where $M_2^x(\omega_\lambda)$ is the **true power spectrum** of $\{X(k)\}$, and $M_n^x(\omega_{\lambda_1}, \ldots, \omega_{\lambda_{n-1}})$ its **true nth-order moment spectrum**. Clearly, the higher-order periodogram estimate (4.16) is asymptotically unbiased, while its variance is proportional to the product of the power spectra. Furthermore, as N becomes larger, the variance of the estimate increases. Consequently, *the higher-order periodogram estimate is inconsistent.*

There are two main approaches we can use to reduce estimation variance, namely, smoothing the higher-order periodogram over neighboring frequencies, or averaging periodogram estimates over disjoint time blocks (records). We describe in the next section the direct method that essentially combines the aforementioned two approaches.

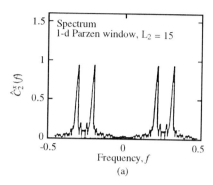
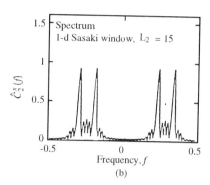

Figure 4.5 Power spectrum estimates of the deterministic energy signal (4.14) from the indirect method using (a) the Parzen window and (b) the Sasaki window.

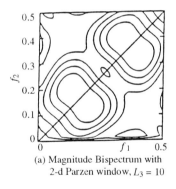
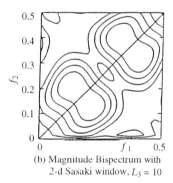

Figure 4.6 Magnitude bispectrum estimates of the deterministic energy signal described by (4.14).

4.3.2 Direct Estimation of Moment Spectra

Let $\{X(1), \cdots, X(N)\}$ be the available set of observations of a stationary stochastic or deterministic sequence. Let us assume that the sampling period is $T = 1$, and $\Delta_n \triangleq 1/N_n$ is the required spacing between frequency samples in the higher-order spectrum domain along horizontal or vertical directions. Finally, assume that the higher-order spectrum is to be estimated for frequencies between 0 and 0.5 with a bandwidth Δ_n. The direct method proceeds as follows [Huber et al., 1971; Lii and Helland, 1981; Hinich and Clay, 1968]:

1. Segment the data into K records of M samples each, i.e., $N = K \cdot M$ and subtract the average value of each segment. If the signal is deterministic, then use only one record ($N = M$). If necessary, add zeros at the end of each record to obtain a convenient length M for a FFT algorithm.

2. Assuming that $\{x^{(i)}(k)\}, k = 0, 1, \ldots, M-1$ is the data of segment $\{i\}$, generate the DFT coefficients

$$F_x^{(i)}(\lambda) = \sum_{k=0}^{M-1} x^{(i)}(k) \exp\{-j\frac{2\pi}{M}k\lambda\}, \qquad (4.19)$$

$$\lambda = 0, 1, \ldots, M-1 \quad \text{and} \quad i = 1, 2, \ldots, K$$

3. In general, the relationship between N_n and M is given by $M = M_n \times N_n$ where M_n is a positive odd integer; e.g., $M_n = 2J_n + 1$. In other words, M_n determines the size of higher-order spectrum smoothing over neighboring frequencies. Since M is even and M_n is odd, we compromise on the value of N_n (closest integer). Estimate the nth-order moment spectrum, $\hat{M}_n^{(i)}(\lambda_1, \ldots, \lambda_{n-1})$, by *frequency-domain averaging*

$$\begin{aligned}\hat{M}_n^{(i)}&(\lambda_1, \ldots, \lambda_{n-1}) = \\ &= \frac{1}{\Delta_n^{n-1}} \sum_{k_1=-J_n}^{J_n} \cdots \sum_{k_{n-1}=-J_n}^{J_n} F_x^{(i)}(\lambda_1 + k_1) \cdots F_x^{(i)}(\lambda_{n-1} + k_{n-1}) \cdot \\ &\quad F_x^{(i)*}(\lambda_1 + \cdots + \lambda_{n-1} + k_1 + \cdots + k_{n-1}), \; i = 1, 2, \ldots, K\end{aligned}$$
(4.20)

over a 2-d rectangular window of size $M_n \times M_n$ [Huber et al., 1971]. For the special case where no averaging is performed in the frequency domain, $M_n = 1 (J_n = 0)$ and, therefore,

$$\hat{M}_n^{(i)}(\lambda_1, \ldots, \lambda_{n-1}) = \frac{1}{\Delta_n^{n-1}} \cdot F_x^{(i)}(\lambda_1) \cdots F_x^{(i)}(\lambda_{n-1}) F_x^{(i)^*}(\lambda_1 + \cdots + \lambda_{n-1}), \quad (4.21)$$

$$i = 1, 2, \ldots, K$$

4. Finally, the nth-order moment spectrum of the given data is obtained by averaging over the K pieces:

$$\hat{M}_n^x(\omega_1, \ldots, \omega_{n-1}) = \frac{1}{K} \sum_{i=1}^{K} \hat{M}_n^{(i)}(\omega_1, \ldots, \omega_{n-1}), \quad (4.22)$$

where $\omega_j \triangleq (2\pi\Delta_n)\lambda_j$, and $j = 1, 2, \ldots, n-1$.

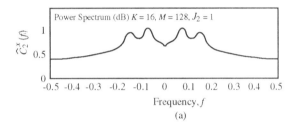

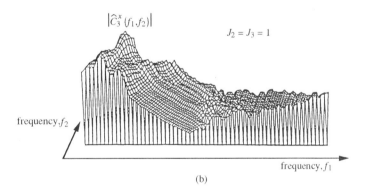

Figure 4.7 Direct conventional estimates of (a) power spectrum and (b) magnitude bispectrum for the AR(4) process described by (4.23) using $K = 16$, $M = 128$, and $J_2 = J_3 = 1$.

Example 4.3

Consider the purely autoregressive process

$$X(k+4) + \alpha_{4,1}X(k+3) + \alpha_{4,2}X(k+2) + \alpha_{4,3}X(k+1) + \alpha_{4,4}X(k) = W(k), \quad (4.23)$$

where $\alpha_{4,1} = -2.623$, $\alpha_{4,2} = 3.516$, $\alpha_{4,3} = -2.518$, and $\alpha_{4,4} = 0.922$. The driving noise process $\{W(k)\}$ is non-Gaussian (one-sided exponential) and white, with zero-mean and nonzero variance and skewness. It was generated by the GGEXN subroutine of the IMSL library with skewness $E\{W^3(k)\} = 2$. Figure 4.7 illustrates the power spectrum and magnitude bispectrum of $\{X(k)\}$ estimated by (4.19)-(4.22) using K=16 records of M=128 samples each and $J_2 = J_3 = 1$. In addition, Figures 4.8 and 4.9 show the estimated functions for (K=16, M=128, $J_2 = J_3 = 3$) and (K=32, M=128, $J_2 = J_3 = 1$), respectively. From these figures, it appears that for this specific example the more segments there are available and larger J (size of frequency domain smoothing window) is, the smoother are the estimates of power spectrum and magnitude bispectrum. □

4.3.3 The Zurbenko Method for the Estimation of Cumulant Spectra

According to Zurbenko [1986], the direct and indirect class of conventional methods described in the previous sections have an important drawback; namely, the impossibility of obtaining a cumulant spectrum estimate for all values of the argument. From (2.27) we see that nth-order cumulant spectra are defined at all points which correspond to $\omega_1 + \omega_2 + \cdots + \omega_n = 0$ (ω_n is the hidden variable). However, the direct and indirect class of methods yield very unstable estimates about manifolds of which their arguments satisfy the relation $\omega_{k_1} + \cdots + \omega_{k_p} = 0$, $1 \leq k_i \leq n$, $i = 1, 2, \ldots, p$ for $p < n$ [Zurbenko, 1986]. For example, let us assume that $C_3^x(\omega_1, \omega_2)$ is the bispectrum and $C_2^x(\omega)$ is the power spectrum of a linear non-Gaussian process and that $C_2^x(\omega) = C_3^x(\omega, -\omega)$. Using the conventional methods described earlier we can obtain an asymptotically unbiased and consistent estimate of the bispectrum, $C_3^x(\omega_1, \omega_2)$, for all ω_1, ω_2 except $\omega_1 + \omega_2 = 0$. The power spectrum function obtained from $\hat{C}_2^x(\omega) = \hat{C}_3^x(\omega, -\omega)$ will be *biased*. This will be true for all lower-order cumulant spectra that we try to reconstruct from an estimated nth-order cumulant spectrum. For our example, to obtain an asymptotically unbiased and consistent estimate of the power spectrum, we have to repeat the conventional method from the beginning for power spectrum estimation.

The Zurbenko method [1986] described in this section allows the reconstruction of asymptotically unbiased and consistent lower-order cumulant spectra from an

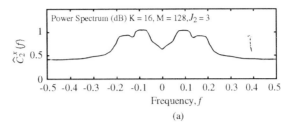

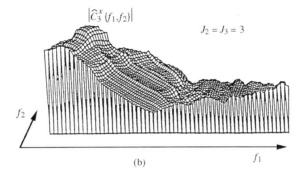

Figure 4.8 Direct conventional estimates of (a) power spectrum and (b) magnitude bispectrum for the AR(4) process described by (4.23) using $K = 16$, $M = 128$, and $J_2 = J_3 = 3$.

estimated nth-order cumulant spectrum. Let $\{X(1), X(2), \ldots, X(N)\}$ be the sample values of a stationary random process with $N = (K-1)i + M$. The Zurbenko method proceeds as follows:

1. Generate DFT coefficients

$$F_{L,M}(\lambda) = \sum_{k=L+1}^{L+M} X(k)\exp(-j\frac{2\pi}{M}k\lambda) \qquad (4.24.1)$$

and nth-order "periodograms"

$$I_{is,M}(\lambda_1, \lambda_2, \ldots, \lambda_n) = \frac{1}{M}\prod_{k=1}^{n} F_{is,M}(\lambda_k), \qquad (4.24.2)$$

for $s = 0, 1, \ldots, K-1$ and $i > 0$. Note that the actual nth-order periodograms are obtained from $I_{is,M}(\lambda_1, \ldots, \lambda_{n-1}, \lambda_n)$ for $\lambda_1 + \lambda_2 + \cdots + \lambda_n = 0$.

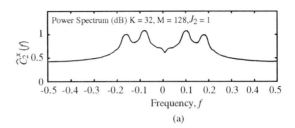

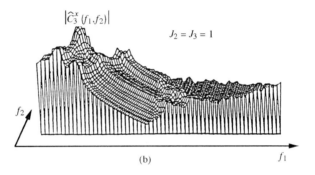

Figure 4.9 Direct conventional estimates of (a) power spectrum and (b) magnitude bispectrum for the AR(4) process described by (4.23) using $K = 32$, $M = 128$, and $J_2 = J_3 = 1$.

2. Construct nth-order moment spectra estimates

$$\hat{M}(\lambda_1, \lambda_2, \ldots, \lambda_n) = \frac{1}{K} \sum_{s=0}^{K-1} I_{is,M}(\lambda_1, \ldots, \lambda_n). \qquad (4.24.3)$$

3. Cumulant spectra estimates are constructed from

$$\hat{C}(\lambda_1, \lambda_2, \ldots, \lambda_n) = \sum (-M)^{p-1}(p-1)! \hat{M}(\underline{\lambda}_1) \cdots \hat{M}(\underline{\lambda}_p), \qquad (4.24.4)$$

where the summation extends over all partitions

$$\underline{\lambda} = \underline{\lambda}_1 + \underline{\lambda}_2 + \cdots + \underline{\lambda}_p, \quad |\underline{\lambda}_k| = 0 \quad \text{for} \quad k = 1, 2, \ldots, p, \ 1 \le p \le n.$$

Section 2.2.2 describes how to generate all possible partitions when $p = 2, 3, 4$. Zurbenko [1986] shows that $\hat{C}(\lambda_1, \ldots, \lambda_n)$ is an asymptotically unbiased and consistent estimate for any $\lambda_1 + \cdots + \lambda_n = 0$, $n > 1$ provided $1 \ll i \ll M \ll N$.

4.4 COMPLEX DEMODULATES CLASS OF CONVENTIONAL METHODS

An equivalent procedure for the direct estimation of higher-order moment spectra was suggested by Godfrey [1965], based on complex demodulates. The procedure is general in the sense that, once the complex demodulates have been computed, moment spectra of any order can be easily generated. Assuming that the DFT coefficients $\{F_x^{(i)}(\lambda)\}$, $\lambda = 0, 1, \ldots, M-1$ of the ith-record have been computed in (4.19), the complex demodulates approach proceeds as follows [Godfrey, 1965; Huber et al., 1971]:

1. Apply a narrow bandpass filter and then shift the frequencies to zero. That is

$$F_k^{(i)}(\lambda) = \begin{cases} F_x^{(i)}(\lambda + k), & |k| \leq J \\ 0, & \text{otherwise} \end{cases} \quad (4.25.1)$$

and

$$\tilde{F}_k^{(i)}(\lambda) = \begin{cases} F_x^{(i)}(\lambda + k), & |k| \leq 2J \\ 0, & \text{otherwise} \end{cases} \quad (4.25.2)$$

where $J = J_i$, $i = 1, 2, \ldots, n-1$ are the indexes defined in (4.20).

2. Generate the **complex demodulates** by transforming back to the time domain.

$$Z_s^{(i)}(\lambda) = \sum_{k=-J'/2}^{J'/2} F_k^{(i)}(\lambda) \exp\left\{j\left(\frac{2\pi sk}{J'}\right)\right\} \quad (4.26.1)$$

and

$$\tilde{Z}_s^{(i)}(\lambda) = \sum_{k=-J'/2}^{J'/2} \tilde{F}_k^{(i)} \exp\left\{j\left(\frac{2\pi sk}{J'}\right)\right\} \quad (4.26.2)$$

where $J' \geq 2J$.

3. Estimate the nth-order moment spectrum of the ith-record by

$$\hat{M}_n^{(i)}(\lambda_1, \ldots, \lambda_{n-1}) =$$

$$= \frac{1}{J'\Delta_n^{n-1}} \sum_{s=0}^{J'-1} Z_s^{(i)}(\lambda_1) \cdots Z_s^{(i)}(\lambda_{n-1}) \tilde{Z}_s^{(i)*}(\lambda_1 + \cdots + \lambda_{n-1}),$$

$$(4.27)$$

$$i = 1, 2, \ldots, K.$$

4. Average as in (4.22) to obtain the moment spectrum estimate of the given data.

It is easy to verify that (4.27) and (4.20) are equivalent. Hence, the complex demodulates approach offers an alternative way of obtaining direct conventional estimates of higher-order moment spectra.

4.5 STATISTICAL PROPERTIES OF CONVENTIONAL METHODS

The statistical properties of the indirect and direct conventional approaches for higher-order spectrum estimation have been studied by Rosenblatt and Van Ness [1965], Van Ness [1966], Brillinger and Rosenblatt [1967], Brillinger [1981], Hinich [1982], Rao and Gabr [1984], Rosenblatt [1985], and Lii and Rosenblatt [1989]. In general, the indirect and direct methods provide different estimates. They become identical, however, under certain conditions that we show in this section. It has been shown that the conventional methods (both indirect and direct) provide asymptotically unbiased and consistent estimates. Furthermore, each higher-order spectrum estimate is asymptotically complex Gaussian (e.g., for large M and N), and uncorrelated with all other higher-order spectrum estimates at different frequencies.

4.5.1 Properties of Bispectral Estimates

Let us first consider the case of the bispectrum. Assume that $M_2^x(\omega)$ and $M_3^x(\omega_1, \omega_2)$ are respectively the true power spectrum and bispectrum of a strictly stationary zero-mean random process. Let $\hat{M}_3^x(\omega_1, \omega_2)$ be a consistent bispectrum estimate computed by indirect or direct conventional methods using a single realization of the random process of length N. The key result associated with these methods is that for sufficiently large record size M and total length N, both direct and indirect approaches provide approximately unbiased estimates, viz:

$$E\left\{\hat{M}_3^x(\omega_1, \omega_2)\right\} \cong M_3^x(\omega_1, \omega_2) \qquad (4.28)$$

with asymptotic variances

$$\begin{aligned}\operatorname{var}\{Re[\hat{M}_3^x(\omega_1,\omega_2)]\} &\cong \operatorname{var}\{I_m[\hat{M}_3^x(\omega_1,\omega_2)]\} \\ &\cong \frac{1}{2}\sigma_3^2(\omega_1,\omega_2),\end{aligned} \qquad (4.29.1)$$

where

$$\sigma_3^2(\omega_1,\omega_2) = \begin{cases} \dfrac{VL_3^2}{MK} M_2^x(\omega_1) M_2^x(\omega_2) M_2^x(\omega_1+\omega_2) & \text{(indirect)} \\[2mm] \dfrac{N_3^2}{MK} M_2^x(\omega_1) M_2^x(\omega_2) M_2^x(\omega_1+\omega_2) & \text{(direct)} \end{cases} \quad (4.29.2)$$

for $0 < \omega_2 < \omega_1$, where K is the number of records, M the number of samples per record, and V is the total energy of the bispectrum window, which is unity for a rectangular window; L_3 is defined by (4.4) or (4.5) and $N_3 = M/(2J_3+1)$ where J_3 is defined by (4.20). In order for the variance to tend to zero as $M \to \infty, k \to \infty$ we should take $2J+1$ to be $O(\sqrt{K'})$. From (4.29) it is apparent that if a rectangular window is used with the indirect method and $L_3 = N_3$, the two conventional methods give approximately the same estimates. Let us note that (4.29) presents the bispectrum variance $\sigma_3^2(\omega_1,\omega_2)$ for frequencies (ω_1,ω_2) that lie inside the principal region and **not** on its boundaries. Exact expressions of the bispectrum variance on the boundaries of the principal region (e.g. $\omega_1 = \omega_2, \omega_1 = 0, \omega_2 = 0, \omega_2 = \pi - \omega_1$) can be found in Van Ness [1966], as well as in Lii and Rosenblatt [1989].

Brillinger and Rosenblatt [1967] and Rosenblatt [1985] showed that for large M and N, the error bicoherence

$$\left[\hat{M}_3^x(\omega_1,\omega_2) - M_3^x(\omega_1,\omega_2)\right] \Big/ \sigma_3(\omega_1,\omega_2) \sim N_c(0,1) \qquad (4.30)$$

is approximately complex Gaussian variant with mean zero and variance equal to one. Another equally important large sample result that follows from the asymptotic results developed by Brillinger and Rosenblatt [1967] is that these statistics can be treated as independent random variables over the grid in the principal domain if the grid width is larger than or equal to the bispectrum bandwidth; i.e., $\hat{M}_3^x(\omega_j,\omega_k)$ and $\hat{M}_3^x(\omega_r,\omega_s)$ are independent for $j \neq r$ or $k \neq s$ if $|\omega_{j+1}-\omega_j| \geq 2\pi\Delta_3$ or $|\omega_{r+1}-\omega_r| \geq 2\pi\Delta_3$, where

$$\Delta_3 \triangleq \begin{cases} \dfrac{1}{L_3} & \text{(indirect)} \\[2mm] \dfrac{1}{N_3} & \text{(direct)}. \end{cases} \qquad (4.31)$$

The asymptotic independence and Gaussianity imply that the magnitude squared bicoherence statistic [Hinich, 1982; Ashley et al., 1986]

$$ch_3(\omega_1,\omega_2) = \frac{|\hat{M}_3^x(\omega_1,\omega_2) - M_3^x(\omega_1,\omega_2)|^2}{\sigma_3^2(\omega_1,\omega_2)} \qquad (4.32)$$

is approximately a noncentral chi-square statistic with 2 degrees of freedom. Ashley et al. [1986] reported that this approximation holds for samples as small as $N = 256$ if $\Delta_n \cong \sqrt{N}$. If the process has zero bispectrum, $M_3^x(\omega_1, \omega_2) = 0$, and $ch_3(\omega_1, \omega_2)$ is central chi-square variant with 2 degrees of freedom.

Brillinger and Rosenblatt [1967] point out that the value of $L_n(N_n)$ should decrease as the order (n) of higher-order spectra increases in order to maintain reasonable stability of the estimate. In fact, they suggest that the window bandwidth in estimating the bispectrum should be the square root of the window bandwidth used in estimating the power spectrum; i.e., $L_3 = \sqrt{L_2}$ $(N_3 = \sqrt{N_2})$. Rao and Gabr [1984] report that the choice of $L_3(N_3)$ can be made on the basis of the following considerations: (i) $L_3(N_3)$ should be less than the square root of the record sample size M; i.e. $L_3 \leq \sqrt{M}$, and (ii) $L_3(N_3)$ should be less than $L_2(N_2)$, the window bandwidth used in estimating the power spectrum ($L_3 < L_2$).

Example 4.4

Consider a real Gaussian white noise process $\{G(k)\}$, $k = 1, 2, \ldots, N$ with $E\{G(k)\} = 0$ and $E\{G^2(k)\} = \gamma_2^G$. Hence, its power spectrum is given by $M_2^G(\omega) = \gamma_2^G$. Using a direct or indirect conventional method to estimate its bispectrum, $\hat{M}_3^G(\omega_1, \omega_2)$, and assuming sufficiently large values of M and N, we obtain from (4.28)–(4.30) that

$$\frac{\hat{M}_3^G(\omega_1, \omega_2)}{\sigma_3(\omega_1, \omega_2)} \sim N_c(0, 1), \quad 0 < \omega_2 < \omega_1,$$

where

$$\sigma_3^2(\omega_1, \omega_2) \cong \begin{cases} \dfrac{V \cdot L_3^2}{M \cdot K}(\gamma_2^G)^3, & \text{(indirect)} \\ \dfrac{N_3^2}{M \cdot K}(\gamma_2^G)^3, & \text{(direct)}. \end{cases}$$

In other words, a consistent bispectrum estimate of a Gaussian white noise process is approximately complex Gaussian variant with a zero mean and the variance proportional to $(\gamma_2^G)^3$. Notice that $\sigma_3^2(\omega_1, \omega_2) \to 0$ as $M \to \infty$ or $K \to \infty$. □

Example 4.5

Consider a nonminimum phase moving average (MA) system whose transfer function is

$$H(z) = (1 - 0.869z)(1 + 1.1z + 0.617z^2)(1 - 0.85z^{-1})(1 - 1.2z^{-1} + 0.45z^2). \quad (4.33)$$

The system is driven by zero-mean non-Gaussian (one-sided exponentially distributed) white noise $W(k)$, whose variance and skewness are $\gamma_3^w = 1$ and $\gamma_3^w = 2$. Consequently, the bispectrum of the resulting output sequence, $\{Y(k)\}$, is given by

$$M_3^y(\omega_1, \omega_2) = \gamma_3^w H(\omega_1) H(\omega_2) H^*(\omega_1 + \omega_2). \tag{4.34}$$

Let us assume that we are given $\{Y(k)\}$, $k = 1, 2, \ldots, N$ and that we estimate its bispectrum $\hat{M}_3^y(\omega_1, \omega_2)$ using the indirect conventional method. To study the effects of the number of records K and the type of the window, we use the sample mean-square error criterion for measuring the accuracy of $\hat{M}_3^y(\omega_1, \omega_2)$ as an estimate of $M_3^y(\omega_1, \omega_2)$. The sample mean-square error (MSE) is defined as

$$\text{MSE} \triangleq \frac{1}{L} \sum_m \sum_r \left| R_e \left[\hat{M}_3^y(\omega_m, \omega_r) \right] - R_e \left[M_3^y(\omega_m, \omega_r) \right] \right|^2 + \left| I_m \left[\hat{M}_3^y(\omega_m, \omega_r) \right] - I_m \left[M_3^y(\omega_m, \omega_r) \right] \right|^2, \tag{4.35}$$

where the summations are taken over all frequencies (ω_m, ω_r) in the principal region of the bispectrum, and L is the total number of frequency pairs.

Figure 4.10 illustrates the MSE as a function of K for $M = 128$ and $L_3 = 7$ using 2-d rectangular, Parzen, and Sasaki bispectral windows with the indirect method. The ensemble of MSE plots corresponds to different signal realizations of the output process. From Figure 4.10 it is apparent that the MSE decreases as K increases. □

4.5.2 Properties of Higher-Order Spectrum Estimation

The derivation of statistical properties of cumulant spectrum estimates of order $n \geq 4$ is more complicated because the fourth- or higher-order cumulants of a zero-mean strictly stationary random process are not just the moments but are of the form given in (2.5) or (4.3). Nevertheless, Lii and Rosenblatt [1989] showed that error cumulant spectrum estimates obtained using conventional methods are **asymptotically jointly Gaussian** with mean zero. Consequently, conventional methods provide asymptotically unbiased higher-order spectrum estimates; i.e.,

$$E\left\{\hat{C}_n^x(\omega_1, \ldots, \omega_{n-1})\right\} \cong C_n^x(\omega_1, \ldots, \omega_{n-1}). \tag{4.36}$$

The asymptotic variance of nth-order cumulant spectrum estimates can be found in [Lii and Rosenblatt, 1989].

Let us consider the asymptotic variance of indirect trispectrum estimates and

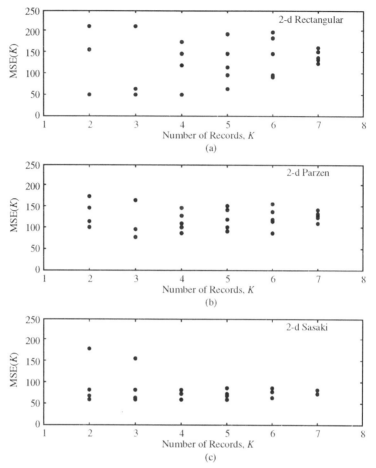

Figure 4.10 The MSE of indirect bispectrum estimates as a function of the number of records K; (a) 2-d rectangular window, (b) 2-d Parzen window, (c) 2-d Sasaki window. Results are shown for five different realizations of the output process.

compare it with that of the bispectrum. That is

$$\sigma_4^2(\omega_1,\omega_2,\omega_3) \cong \frac{L_4^3 \cdot V}{M \cdot K} C_2^x(\omega_1) C_2^x(\omega_2) C_2^x(\omega_3) C_2^x(\omega_1 + \omega_2 + \omega_3) \qquad (4.37)$$

for $0 < \omega_3 < \omega_2 < \omega_1$ where V is the total energy of a 3-d trispectrum window and $C_2^x(\omega)$ is the true power spectrum of the process. Comparing (4.37) with (4.29), we observe that the key difference between them lies in the term L_3^2 versus L_4^3. This observation suggests that the value of L_4 should be chosen to be less than L_3 in order to maintain reasonable variance for the trispectrum estimate. In fact, if we choose L_2 for power spectrum estimation, then L_n should be chosen such that

$$L_n \leq L_2^{\frac{1}{n-1}}, \qquad n = 3, 4, \ldots \qquad (4.38)$$

in order to achieve smooth estimates of nth-order cumulant spectra.

From the expressions (4.29) and (4.37), it is apparent that the higher-order spectra variance associated with conventional estimators can be reduced by: (i) increasing the number of records, (ii) reducing the size of the region of support of the window in the cumulant domain (L_n) or increasing the size of the frequency smoothing window (J_n), and (iii) increasing the record size M. However, increasing the number of records (K) is demanding on computer time and may introduce potential nonstationarities. On the other hand, frequency domain averaging over large polyhedrons of size ($2J_{n+1}$) or use of cumulant domain windows with small values of L_n reduce frequency resolution and may increase bias. In the case of "short length" data, the K could be increased by using overlapping records.

The conventional methods have the advantage of ease of implementation (FFT algorithms can be used) and provide good estimates for sufficiently long data. However, because of the "uncertainty principle" of the Fourier transform, the ability of the conventional methods to resolve harmonic components in polyspectra domains is limited. Raghuveer and Nikias [1985] point out that this could pose a problem in detecting quadratic phase coupling at closely spaced frequency pairs.

4.6 TEST FOR ALIASING WITH THE BISPECTRUM

Aliasing will not occur in discrete-time signals if the following conditions hold: (a) the equivalent continuous-time signal is band-limited at frequency f_o, and (b) the signal is sampled with sampling period T such that $\frac{1}{T} \geq 2f_o$. Aliasing is usually avoided in practice by low-pass filtering a continuous-time signal to eliminate its frequency content above frequency f_o, and then sampling the filtered signal at or above twice the cutoff frequency f_o. However, there are applications, such as economics and social sciences, where we measure directly the discrete-time signal, and therefore sampling designs of continuous-time signals are impossible. Hence, it is of value to test the discrete-time signal for the presence of aliasing.

Hinich and Wolinsky [1988] introduced a test for aliasing using bispectral analysis. Their *test is for stationary stochastic signals* that have nonzero bispectrum, and utilizes an overlooked fundamental property of the bispectrum principal domain. They point out, correctly, that there has been surprisingly persistent confusion in the literature regarding the triangular form of the principal domain of the bispectrum of a signal. Assuming that $X(t)$ is a real zero-mean stationary continuous-time stochastic signal, the principal domain of its bispectrum in the (f_1, f_2) plane is the cone $C = \{f_1, f_2 : 0 \leq f_1, 0 \leq f_2 \leq f_1\}$ illustrated in Figure 4.11. Now suppose the process is band-limited at frequency f_o. As such, the bispectrum cuts off at

$f_1 = f_o, f_2 = \pm f_o$ and $f_1 + f_2 = \pm f_o$. Consequently, the continuous-time support set is the isoceles right triangle $\{f_1, f_2 : 0 \leq f_1, \leq f_o, \ f_2 \leq f_1, \ f_1 + f_2 = f_o\}$.

The principal domain of the equivalent discrete-time sequence $X(nT)$, where $T = 1/(2f_o)$ is the triangle $\{f_1, f_2 : 0 \leq f_1 \leq (1/2T), \ f_2 \leq f_1, \ 2f_1 + f_2 = 1/T\}$ in the cone C. From Figure 4.11, we see that this principal domain consists of two triangular regions; namely, $OT = \{f_1, f_2 : f_2 \leq f_1, \ (1/2T) \leq f_1 + f_2 \leq (1/T) - f_1\}$ and $IT = \{f_1, f_2 : f_2 \leq f_1, \ f \leq f_1 + f_2 \leq (1/2T)\}$. Hinich and Wolinsky [1988] showed that *when there is no aliasing in $X(nT)$, its bispectrum is zero in the OT triangle*. This easily follows from the fact that the IT region is identical to the continuous-time support set of a signal which is band-limited at frequency f_o.

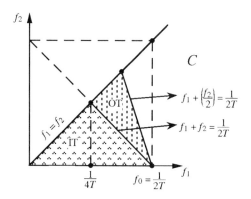

Figure 4.11 Continuous-time and discrete-time principal domains of the bispectrum

Let $\hat{M}_3^x(\omega_1, \omega_2)$ denote a consistent estimator of the bispectrum $M_3^x(\omega_1, \omega_2)$ for a grid of equally spaced discrete frequencies given a realization of size N of $\{X(nT)\}(\omega_i = 2\pi f_i)$. This estimator can be computed by either indirect or direct conventional methods. The test for aliasing [Hinich and Wolinsky, 1988] proceeds to form the magnitude squared bicoherence

$$\Lambda(\omega_1, \omega_2) = \frac{|\hat{M}_3^x(\omega_1, \omega_2)|^2}{\sigma_3^2(\omega_1, \omega_2)} \qquad (4.39)$$

where $\sigma_3^2(\omega_1, \omega_2)$ is defined by (4.29.2), and to generate the detection statistic

$$\Lambda = \sum_{(\omega_1, \omega_2)} \Lambda(\omega_1, \omega_2), \quad \text{for } (\omega_1, \omega_2) \epsilon \ OT. \qquad (4.40)$$

Under no aliasing, Λ is approximately central chi-squared with 2L degrees of freedom, where L is the number of grid points in the OT triangular region. On the other hand, under aliasing, Λ is approximately non-central chi-squared with 2L degrees of freedom and is a positive noncentrality parameter.

Hinich and Wolinsky [1988] applied the test for aliasing on stock price series corresponding to N=1,000 consecutive days. The resolution bandwidth in the principal domain was chosen to be 0.03 (1/day). There were L=24 points in the OT triangle. Their results overwhelmingly indicated that the data were aliased for sampling period T=1 day.

4.7 BISPECTRUM COMPUTATION ON POLAR RASTERS

4.7.1 Introduction

To reduce computational complexity associated with the use of higher-order spectra (bispectrum, trispectrum), considerable work has been done to develop efficient procedures for computing polyspectra with the utilization of FFT algorithms. However, the emphasis was originally placed on **rectangular grid** computations. Bessios and Nikias [1991] introduced fast computation algorithms for the bispectrum on a *polar raster*, in the triangular region that corresponds to angles $0° \leq \theta \leq 45°$. This region is enough for a complete description of the bispectrum of a real signal (see Figure 4.12).

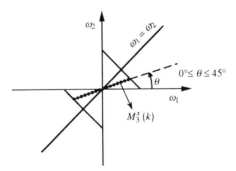

Figure 4.12   Bispectrum computation on a polar raster

These algorithms compute the bispectrum values by employing the time-domain signal directly, bypassing the computation of third-order moments. In particular, three different algorithms were introduced; namely, the direct approach (DA), the zero-padding approach (ZPA), and the super-FFT approach (SFA). These methods can be applied on both deterministic and stochastic signals, and can be extended to trispectrum computation on polar rasters. We saw in Chapter 2 that the bispectrum of a real signal has 12 symmetry regions, whereas its trispectrum has 96 symmetry regions [Pflug et al., 1992]. It is, therefore, of paramount importance to have algorithms to compute higher-order spectra in only one of their symmetry regions and directly from the time domain.

4.7.2 Problem Formulation

Let $\{x(k)\}$, $k = 0, 1, \ldots, M-1$ be a real deterministic signal with bispectrum

$$M_3^x(\omega_1, \omega_2) = X(\omega_1)X(\omega_2)X^*(\omega_1 + \omega_2), \tag{4.41}$$

where

$$X(\omega) = \sum_{k=0}^{M-1} x(k)\exp\{-j\omega k\}. \tag{4.42}$$

The problem is to compute $M_3^x(\omega_1, \omega_2)$ on a polar coordinate grid; e.g., along slices with angles θ_i, where $0° \leq \theta_i \leq 45°$, as shown in Figure 4.12.

For a slice of angle θ, the limits for the frequencies will be:

$$0 \leq \omega_1 \leq \frac{\pi}{1+\tan\theta}, \quad 0 \leq \omega_2 \leq \frac{\pi \cdot \tan\theta}{1+\tan\theta}, \quad 0 \leq \omega_1 + \omega_2 \leq \pi.$$

Note that we consider only positive frequencies. If we discretize the above frequency intervals by taking $\frac{N}{2}$ samples, then $\omega_1 = \frac{2\pi}{(1+\tan\theta)N} \cdot \lambda$, $\lambda = 0, 1, \ldots, \frac{N}{2} - 1$, and (4.43) becomes

$$\begin{aligned} X_1(\lambda) &= \sum_{k=0}^{M-1} x(k)\exp\left\{-j\frac{2\pi}{(1+\tan\theta)N}\lambda \cdot k\right\}, \\ \lambda &= 0, 1, \ldots, \frac{N}{2} - 1. \end{aligned} \tag{4.43.1}$$

Similarly, we have that

$$X_2(\lambda) = \sum_{k=0}^{M-1} x(k)\exp\left\{-j\frac{2\pi \tan\theta}{N(1+\tan\theta)} \cdot \lambda \cdot k\right\}, \tag{4.43.2}$$

$$X_3(\lambda) = \sum_{k=0}^{M-1} x(k)\exp\left\{j\frac{2\pi}{N} \cdot \lambda \cdot k\right\}. \tag{4.43.3}$$

Consequently, the bispectral points along a slice of angle θ will be given by

$$\begin{aligned} M_{3,\theta}^x(\lambda) &= X_1(\lambda)X_2(\lambda)X_3(\lambda), \\ \lambda &= 0, 1, \ldots, \frac{N}{2} - 1. \end{aligned} \tag{4.44}$$

4.7.3 The Direct Approach (DA)

The DA method proceeds by computing (4.31.1)–(4.43.3) with a discrete Fourier transform and by forming a triple product (4.44). Clearly, DA is not the most computationally efficient method when M, N are large. For computation of the bispectrum along I slices, the DA's total number of multiplications is $\sim O[(\frac{3}{2}MN + N) \cdot I]$, where M is the length of the signal and N is the number of bispectral points we want to compute along a slice of angle θ.

4.7.4 The Zero-Padding Approach (ZPA)

The basic idea behind this algorithm is to apply zero-padding on the signal $\{x(k)\}$ up to length $\left\{\text{int}\left[\frac{(1+\tan\theta)}{\tan\theta}\right] \cdot N\right\}$, and then apply a FFT algorithm, where $\text{int}[\cdot]$ denotes the integer. In particular, the following steps are taken. For a specific value of angle θ,

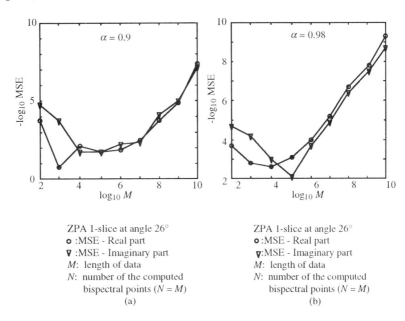

ZPA 1-slice at angle $26°$
○ :MSE - Real part
▼ :MSE - Imaginary part
M: length of data
N: number of the computed bispectral points ($N = M$)
(a)

ZPA 1-slice at angle $26°$
○ :MSE - Real part
▼ :MSE - Imaginary part
M: length of data
N: number of the computed bispectral points ($N = M$)
(b)

Figure 4.13 Mean-squared Error (MSE) between the true and computed bispectrum via ZPA of $x(k) = \alpha^k \sin(0.4\pi k)$. (a) $\alpha = 0.9$, (b) $\alpha = 0.98$. The MSE is shown as a function of $M (M = N)$.

1. Zero-pad the sequence $x(k)$ up to length $J = N \cdot \text{int}\left[\frac{(1+\tan\theta)}{\tan\theta}\right]$; e.g.,

$$y(k) = \begin{cases} x(k), & k = 0, 1, \ldots, M-1 \\ 0, & k = M, \cdots, J. \end{cases} \quad (4.45.1)$$

2. Compute a J-point FFT of $y(k)$

$$Y(\lambda) = FFT[y(\lambda)], \quad \lambda = 0, 1, \ldots, J-1. \quad (4.45.2)$$

3. Do the following approximations:

$$\begin{aligned} X_1(\lambda) &\cong Y\left(\lambda \cdot \frac{J}{N(1+\tan\theta)}\right), \\ X_2(\lambda) &\cong Y\left(\lambda \cdot \frac{J}{N(1+\tan\theta)}\right), \\ X_3(\lambda) &\cong Y^*\left(\lambda \cdot \frac{J}{N}\right), \end{aligned} \quad (4.45.3)$$

for $\lambda = 0, 1, \ldots, \frac{N}{2} - 1$.

However, for more accurate computation of the above terms, one may utilize a linear interpolation formula as follows. Define

$$d_1(\lambda) \triangleq \lambda \cdot \frac{J}{N(1+\tan\theta)}, \quad d_2(\lambda) \triangleq \lambda \cdot \frac{J\tan\theta}{N(1+\tan\theta)},$$

and

$$d_3(\lambda) \triangleq \lambda \cdot \frac{J}{N}. \quad (4.45.4)$$

Then, for $I_\lambda < d_i(\lambda) < I_\lambda + 1$, $i = 1, 2, 3$, where I_λ is an integer, generate

$$X_i(\lambda) \cong Y(I_\lambda) + (d_i(\lambda) - I_\lambda) \cdot [Y(I_\lambda + 1) - Y(I_\lambda)], \quad i = 1, 2$$

and

$$X_3(\lambda) \cong Y^*(I_\lambda) + (d_3(\lambda) - I_\lambda) \cdot [Y^*(I_\lambda + 1) - Y^*(I_\lambda)]. \quad (4.45.5)$$

4. The bispectrum along a slice of angle θ will be given by the triple product (4.44).

Let us note that there is a set of angles θ where the quantity $\frac{1}{\tan\theta}$ **is an exact integer number** which, in turn, implies that the quantity $\frac{1+\tan\theta}{\tan\theta}$ is also an exact integer. Consequently, for these angles there will be no need for

interpolation to compute the bispectrum accurately [Bessios and Nikias, 1991].

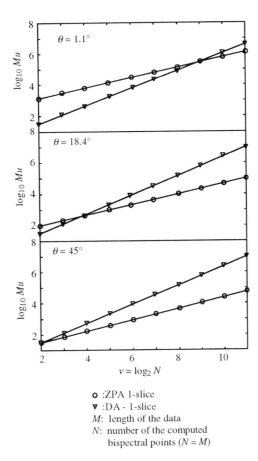

Figure 4.14 Number of multiplications (M_u) as a function of $N(N = M)$ for 1-slice computation of the bispectrum by the DA and ZPA methods

The total computational complexity of the ZPA method when the bispectrum is computed along I slices is $\sim O\left[\sum_{i=1}^{I} \log N_i + \frac{5}{2} N \cdot I\right]$, where int $[\cdot]$ denotes integer part, $N_i = \left\{\text{int}\left[\frac{1+\tan\theta_i}{\tan\theta_i}\right] \cdot N\right\}, i = 1, \ldots, I;$ and θ_i is the angle of the ith slice.

4.7.5 The "Super-FFT Approach" (SFA)

This method is an extension of the ZPA method. With the SFA, however, one has the flexibility to compute bispectral points along several slices at different angles by implementing only one FFT that corresponds to a slice that has the smallest angle. In particular, one applies zero-padding on $\{x(k)\}$ up to length $\left\{N \cdot \operatorname{int}\left[\frac{(1+\tan\theta_o)}{\tan\theta_o}\right]\right\}$, where θ_o is the smallest angle, and then uses the FFT algorithm. The following steps are taken:

1. Form the sequence

$$y(k) = \begin{cases} x(k), & k = 0, 1, \ldots, M-1 \\ 0, & k = M, \ldots, J_o \end{cases} \quad (4.46.1)$$

where

$$J_o = \left\{\operatorname{int}\left[\frac{1+\tan\theta_o}{\tan\theta_o}\right] \cdot N\right\},$$

where θ_o is the smallest angle not equal to zero, along which we want to begin the computation of bispectrum slices.

2. Compute a J_o-point FFT of $y(k)$:

$$Y(\lambda) = \operatorname{FFT}[y(k)], \quad \lambda = 0, 1, \ldots, J_o - 1. \quad (4.46.2)$$

3. The bispectral points along a slice of angle θ_i are given by (4.44) where

$$\begin{aligned} X_1(\lambda) &\cong Y\left(\lambda \cdot \frac{J_o}{(1+\tan\theta_i) \cdot N}\right) \\ X_2(\lambda) &\cong Y\left(\lambda \cdot \frac{J_o \cdot \tan\theta_i}{(1+\tan\theta_i) \cdot N}\right), \\ X_3(\lambda) &\cong Y^*\left(\lambda \cdot \frac{1+\tan\theta_i}{\tan\theta_i}\right), \\ \lambda &= 0, 1, \ldots, \frac{N}{2} - 1. \end{aligned} \quad (4.46.3)$$

For more accurate computation of the above terms one may again utilize the following linear interpolation formulas. Define

$$\begin{aligned} d_1(\lambda) &= \lambda \cdot \frac{J_o}{N \cdot (1+\tan\theta_i)}, \\ d_2(\lambda) &= \lambda \cdot \frac{J_o \tan\theta_i}{N \cdot (1+\tan\theta_i)}, \\ d_3(\lambda) &= \lambda \cdot \frac{1+\tan\theta_i}{\tan\theta_i}, \end{aligned} \quad (4.46.4)$$

and then for $I_\lambda < d_j(\lambda) < I_\lambda + 1$, $j = 1, 2, 3$, (I_λ is an integer), obtain:

$$X_j(\lambda) = Y(I_\lambda) + (d_j(\lambda) - I_\lambda) \cdot [Y(I_\lambda + 1) - Y(I_\lambda)], \quad j = 1, 2$$
$$X_3(\lambda) = Y^*(I_\lambda) + (d_3(\lambda) - I_\lambda) \cdot [Y^*(I_\lambda + 1) - Y^*(I_\lambda)]. \quad (4.46.5)$$

Step (iii) of the SFA method can be repeated for different angles θ_i.

The computational complexity of the SFA method is $\sim O[N' \log N' + \frac{5}{2} N \cdot I]$, where $N' = \left\{ \text{int} \left[\frac{1 + \tan \theta_o}{\tan \theta_o} \right] \cdot N \right\}$, and θ_o is a small angle (different from zero; chosen arbitrarily). One can choose θ_o to be the smallest among prechosen angles of I slices.

Example 4.6

Consider the energy signal

$$x(k) = \alpha^k \sin(0.4\pi k)$$

with two different values of α; namely, $\alpha = 0.9$ and $\alpha = 0.98$. Figure 4.13 illustrates the MSE between the true and the computed bispectrum using the ZPA algorithm along a slice of $\theta = 26°$. From this figure, we see that the MSE decreases as the number of bispectral points along a slice increases. This is not surprising, given the interpolation formula used by the ZPA method. □

Example 4.7

This example compares the DA, ZPA, and SFA methods in terms of the number of multiplications required to compute a bispectrum. Figure 4.14 illustrates multiplications (Mu) as a function of N ($N = M$) for the DA and ZPA methods. From this figure, it is apparent that ZPA requires fewer multiplications than DA for large values of N, and especially when the slice angle θ gets closer to 45°.

Figure 4.15 illustrates the comparative performance between the DA and SFA methods for 5-slice, 20-slice, and 40-slice computation of the bispectrum within angles $8.33° \leq \theta \leq 45°$. Clearly, SFA is faster than DA, especially as N increases. □

4.7.6 Trispectrum Computation on a Polar Raster

The problem now is to compute the trispectrum along one slice which is found in one of the 96 symmetry regions. That region is defined as follows:

$$\omega_1 \geq \omega_2 \geq \omega_3,$$
$$\omega_1 + \omega_2 + \omega_3 \leq \pi, \qquad (4.47)$$
$$\omega_1, \omega_2, \omega_3 \geq 0.$$

Let us assume that we want to compute the trispectrum along a slice of angles θ and ϕ, where θ is the angle between the projection of the slice on the (ω_1, ω_2) plane and the ω_1-axis, and ϕ is the angle between the slice and its projection on the (ω_1, ω_2) plane. The limits of the angles θ and ϕ for the region defined by (4.47) are

$$0° \leq \theta \leq 45° \quad \text{and} \quad 0 \leq \phi \leq 35.264°. \qquad (4.48)$$

Note that for the special case where $\phi = 0°$, the problem becomes bispectrum computation along a slice of angle θ, where $0° \leq \theta \leq 45°$.

It can be shown that the projections of the slice with angles θ and ϕ, on the three axes $\omega_1, \omega_2, \omega_3$, are given by [Bessios and Nikias, 1991]:

$$\begin{aligned}
0 \leq \omega_1 \leq A &= \frac{\pi}{\left(\frac{\tan\phi}{\cos\theta} + \tan\theta + 1\right)} \\
0 \leq \omega_2 \leq B &= \frac{\pi}{\left(\frac{\tan\phi}{\sin\theta} + \cot\theta + 1\right)} \\
0 \leq \omega_3 \leq C &= \frac{\pi}{[1 + \cot\phi \cdot (\sin\theta + \cos\theta)]}.
\end{aligned} \qquad (4.49)$$

Consequently, the trispectrum along a slice of angles θ and ϕ within the limits specified by (4.48) is given by

$$\begin{aligned}
M^x_{4,\theta,\phi}(\lambda) &= X_1(\lambda) \cdot X_2(\lambda) \cdot X_3(\lambda) \cdot X_4(\lambda), \\
\lambda &= 0, 1, \ldots, \frac{N}{2} - 1,
\end{aligned} \qquad (4.50)$$

where

$$X_1(\lambda) = \sum_{k=1}^{M-1} x(k) \exp\left\{-j\frac{2C}{N} \cdot \lambda \cdot k\right\}, \qquad (4.51.1)$$

$$X_2(\lambda) = \sum_{k=0}^{M-1} x(k) \exp\left\{-j\frac{2B}{N} \cdot \lambda \cdot k\right\}, \qquad (4.51.2)$$

$$X_3(\lambda) = \sum_{k=0}^{M-1} x(k) \exp\left\{-j\frac{2A}{N} \cdot \lambda \cdot k\right\}, \qquad (4.51.3)$$

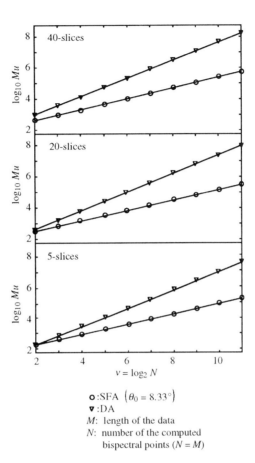

Figure 4.15 Number of multiplications (M_u) as a function of $N(N = M)$ for 5-slice, 20-slice, and 40-slice computation of the bispectrum by DA and SFA methods

$$\begin{aligned} X^*(\lambda) &= \sum_{k=0}^{M-1} x(k)\exp\left\{j\frac{2\pi}{N} \cdot \lambda \cdot k\right\}, \\ X_4(\lambda) &= X^*(\lambda). \end{aligned} \quad (4.51.4)$$

The DA method performs direct computation of (4.51.1)–(4.51.4), then generates the trispectrum slice via (4.50).

The ZPA method, on the other hand, forms the sequence

$$x'(k) = \begin{cases} x(k), & k = 0, 1, \ldots, M-1 \\ 0, & k = M, \ldots, N_2 \end{cases} \quad (4.52.1)$$

where $N_2 = N \cdot J$, $J = \text{int}[1 + \cot\phi(\sin\theta + \cos\theta)]$ and computes the N_2-point

FFT,
$$X(\lambda) = \text{FFT}[x'(k)]. \qquad (4.52.2)$$

Then, the ZPA method performs the interpolations $X_1(\lambda)$ = interpolation of $\{X(I_\lambda), X(I_\lambda + 1)\}$,
where
$$I_\lambda < \frac{J \cdot \lambda}{1 + \cot\phi(\sin\theta + \cos\theta)} < I_\lambda + 1, \qquad (4.53.1)$$

$X_2(\lambda)$ = interpolation of $\{X(I_\lambda), X(I_\lambda + 1)\}$, where
$$I_\lambda < \frac{J \cdot \lambda}{1 + \cot\theta + \frac{\tan\phi}{\sin\theta}} < I_\lambda + 1, \qquad (4.53.2)$$

and $X_3(\lambda)$ = interpolation of $\{X(I_\lambda), X(I_\lambda + 1)\}$, where
$$I_\lambda < \frac{J \cdot \lambda}{1 + \tan\theta + \frac{\tan\phi}{\cos\theta}} < I_\lambda + 1. \qquad (4.53.3)$$

These interpolations are of the same form as those used earlier in bispectrum computation. Finally, $X_4(k) \equiv X^*(k \cdot J), k = 0, 1, \ldots, \frac{N}{2} - 1$. The trispectrum follows from (4.50).

The SFA method chooses initial angles ϕ_o, θ_o which correspond to the smallest angles of the slices along which we want to compute the trispectrum. The SFA proceeds to form the sequence
$$x'(k) = \begin{cases} x(k), & k = 0, 1, \ldots, M - 1 \\ 0, & k = M, \ldots, N'_2 \end{cases} \qquad (4.45.1)$$

where $N'_2 = N \cdot J'$, $J' = \text{int}[1 + \cot\phi_o(\sin\theta_o + \cos\theta_o)]$, and to apply the N'_2-point FFT gives
$$X(\lambda) = \text{FFT}[x'(k)], \quad \lambda = 0, 1, \ldots, N'_2 - 1. \qquad (4.54.2)$$

The steps of the ZPA method are then followed by substituting J with J' and using for each slice its angle (ϕ_i, θ_i).

4.7.7 Extension to Stochastic Signals

The extension of the DA, ZPA, and SFA methods for bispectrum (trispectrum) computation of stochastic signals is somewhat straightforward. Given a sample realization of a stochastic signal, we first segment the data into K records of M samples each, then apply these methods to generate $M_{3,\theta}^{(i)}(\lambda)$ or $M_{4,\theta,\phi}(\lambda)$, $\lambda = 0, 1, \ldots, N - 1$ ($N \leq M$), for $i = 1, 2, \ldots, K$. Frequency-domain smoothing may be performed along each of the slices. The final estimate is obtained by averaging; e.g.,

$$\hat{M}_{3,\theta}^x(\lambda) = \frac{1}{K}\sum_{i=1}^{K} M_{3,\theta}^{(i)}(\lambda)$$

or

$$\hat{M}_{4,\theta,\phi}^x(\lambda) = \frac{1}{K}\sum_{i=1}^{K} M_{4,\theta,\phi}^{(i)}(\lambda).$$

However, the statistical properties of these estimates are yet to be determined.

4.8 SUMMARY

In this chapter we discussed conventional methods for the estimation of higher-order spectra of stochastic and deterministic signals. In particular, we introduced the indirect method, the direct (or FFT-based) method, and the complex demodulates approach. The statistical properties of higher-order spectrum estimates of stochastic signals were also discussed. Finally, a test for nonstationarity or aliasing using the bispectrum has been discussed and FFT-based computation of bispectrum/trispectrum on polar rasters has been presented.

Conventional methods and their statistical properties for higher-order spectrum estimation have been published extensively in the statistics and engineering literature. Important publications include those of Brillinger [1965], Brillinger and Rosenblatt [1967], Rosenblatt and Van Ness [1965], Rao and Gabr [1984], Huber et al. [1971], and Zurbenko [1986]. The asymptotic statistical properties of conventional estimates are also discussed by Rosenblatt [1985] and Lii-Rosenblatt [1989]. The test for aliasing or nonstationarity can be found in a publication by Hinich and Wolinsky [1988]. FFT-based computation of the bispectrum or trispectrum on polar rasters is given by Bessios and Nikias [1991].

REFERENCES

Ashley, R., D. Patterson, and M. Hinich, "A Diagnostic Test for Nonlinear Serial Dependence in Time Series Fitting Errors," *J. Time Series Analysis,* **7**, pp. 165–178, 1986.

Bessios, A. G. and C. L. Nikias, "FFT-Based Bispectrum Computation on Polar Rasters," *IEEE Transactions on Signal Processing*, **39**(11), pp. 2535–2539, November 1991.

Brillinger, D. R., **Time Series, Data Analysis and Theory**, New York: Holt, Rinehart and Winston, expanded ed., 1981.

Brillinger, D. R., "An Introduction to Polyspectra," *Ann. Math. Statist.*, **36**, pp. 1351–1374, 1965.

Brillinger, D. R., and M. Rosenblatt, "Asymptotic Theory of Estimates of k-th Order Spectra", in **Spectral Analysis of Time Series**, B. Harris (ed.), New York: Wiley, pp. 153–188, 1967.

Godfrey, M. D., "An Exploratory Study of the Bispectrum of Economic Time Series," *Appl. Statist.*, **14**, pp. 48–69, 1965.

Helland, K. N., E. C. Itsweire, and K. S. Lii, "A Program for Computation of Bispectra with Application to Spectral Energy Transfer in Fluid Turbulence," *Adv. Eng. Software*, **7**(1), pp. 22–27, 1985.

Hinich, M. J., and C. S. Clay, "The Application of the Discrete Fourier Transform in the Estimation of Power Spectra Coherence and Bispectra of Geophysical Data," *Rev. Geophys.*, **6**(3), pp. 347–363, 1968.

Hinich, M. J., "Testing for Gaussianity and Linearity of a Stationary Time Series," *J. Time Series Analysis*, **3**(3), pp. 169–176, 1982.

Hinich, M. J., and M. A. Wolinsky, "A Test for Aliasing Using Bispectral Analysis," *J. American Statistical Association*, **83**(402), pp. 499–502, 1988.

Huber, P. J., B. Kleiner, T. Gasser, and G. Dumermuth, "Statistical Methods for Investigating Phase Relations in Stationary Stochastic Processes," *IEEE Transactions Audio Electroacoustics*, **AV-19**, pp. 78–86, 1971.

Kay, S. M., **Modern Spectral Estimation**, Englewood Cliffs, NJ: Prentice Hall, 1988.

Lii, K. S., and K. N. Helland, "Cross-Bispectrum Computation and Variance Estimation," *ACM Trans. on Math. Soft.*, **7**(3), pp. 284–294, September 1981.

Lii, K. S., and M. Rosenblatt, "Asymptotic Normality of Cumulant Spectral Estimates," *Theoretical Probability*, 1989.

Marple, S. L., **Digital Spectral Analysis with Applications**, Englewood Cliffs, NJ: Prentice Hall, 1987.

Nikias, C. L., and M. R. Raghuveer, "Bispectrum Estimation: A Digital Signal Processing Framework," *Proceedings IEEE*, **75**(7), pp. 869-892, July 1987.

Raghuveer, M. R. and C. L. Nikias, "Bispectrum Estimation: A Parametric Approach," *IEEE Trans. Acous., Speech, and Signal Processing*, **ASSP-33**(5), pp. 1213–1230, October 1985.

Rao, T. S. and M. M. Gabr, "An Introduction to Bispectral Analysis and Bilinear Time Series Models," **Lecture Notes in Statistics**, **24**, New York: Springer-Verlag, 1984.

Rao, T. S., and M. M. Gabr, "The Estimation of the Bispectral Density Function and the Detection of Periodicities in a Signal," *J. Multivariate Analysis*, **27**(2), pp. 457–477, November, 1988.

Rosenblatt, M. and J. W. Van Ness, "Estimation of the Bispectrum," *Ann. Math. Statist.*, **36**, pp. 1120–1136, 1965.

Rosenblatt, M., **Stationary Sequences and Random Fields**, Boston: Birkhauser, 1985.

Sasaki, K., T. Sato, and Y. Yamashita, "Minimum Bias Windows for Bispectral Estimates," *J. Sound Vibr.*, **40**, pp. 139–148, 1975.

Van Ness, J. W., "Asymptotic Normality of Bispectral Estimates," *Ann. Math. Statist.*, **37**, pp. 1257–1272, 1966.

Zurbenko, I. G., **The Spectral Analysis of Time Series**, North-Holland Series in **Statistics and Probability 2** Elsevier Science Publishers, B. V., Amsterdam, 1986.

5

HIGHER-ORDER CEPSTRA (POLYCEPSTRA)

5.1 INTRODUCTION

In the earlier chapters, we have seen the motivation, definitions, properties, and conventional methods for estimation of higher-order spectra for both stochastic and deterministic signals. In this chapter we introduce the concept of the cepstrum of higher-order spectrum and show its connection with cepstrum analysis and homomorphic deconvolution. The cepstrum operation as a signal processing technique was originally introduced by Bogert, Healy, and Tukey [1963]. Oppenheim [1969] introduced homomorphic filtering as a class of nonlinear signal processing techniques which contain, as a special case, the concept of the cepstrum. Homomorphic filtering, or cepstral analysis, is a nonparametric technique and thus it is effective on a wide class of signals, including those which are nonminimum phase and characterized by poles and zeros. Although the cepstrum of a signal is a simple nonlinear transformation, it is very rich in terms of properties and applications. It has been used in speech, geophysics, sonar, biomedicine, image processing, etc. (See, for example, Oppenheim and Schafer [1989, Chapter 12] and references therein, Pitas and Venetsanopoulos [1990, Chapter 7] and Hassab [1989].)

This chapter provides a detailed exposition of the properties and computation of cepstra of higher-order spectra (i.e., polycepstra), as well as their application(s) to nonminimum phase signal reconstruction and deconvolution. Several additional concepts associated with cross-polycepstra and complex cepstrum computation from higher-order correlations are also discussed. Polycepstra are introduced for deterministic energy signals and for stochastic linear non-Gaussian signals.

5.2 THE COMPLEX CEPSTRUM OF DETERMINISTIC ENERGY SIGNALS

5.2.1 Signal Assumptions

Consider an autoregressive moving average (ARMA) energy sequence $\{h(k)\}$. Its Z transform is generally nonminimum phase and can be written as

$$H(z) = A\, z^{-r} I(z^{-1}) O(z) \qquad (5.1)$$

where A is a constant, r is an integer, and

$$I(z^{-1}) = \frac{\prod_{i=1}^{L_1}(1 - a_i z^{-1})}{\prod_{i=1}^{L_3}(1 - c_i z^{-1})}, \qquad (5.2)$$

$$O(z) = \prod_{i=1}^{L_2}(1 - b_i z) \qquad (5.3)$$

are the minimum and maximum phase components, respectively, with $|a_i| < 1$, $|b_i| < 1$ and $|c_i| < 1$. Poles $\{c_i\}$ and zeros $\{a_i\}$ are inside the unit circle whereas zeros $\{1/b_i\}$ are outside. Let us note that there is no particular reason for excluding maximum phase poles in (5.3). Cepstrum analysis would work for maximum phase poles as well. The minimum phase component of the signal is given by

$$i(k) = \frac{1}{2\pi} \int_{-\pi}^{+\pi} I(\omega) \exp\{j\omega k\} d\omega, \; k \geq 0, \text{ (causal sequence)}, \qquad (5.4.1)$$

whereas the maximum phase component is given by

$$o(k) = \frac{1}{2\pi} \int_{-\pi}^{+\pi} O(\omega) \exp\{j\omega k\} d\omega, \; k < 0, \text{ (anti-causal sequence)} \qquad (5.4.2)$$

where $I(\omega) = I(e^{-j\omega})$ and $O(\omega) = O(e^{j\omega})$. A linear convolution operation, "$*$", relates $i(k)$, $o(k)$ with $h(k)$ as follows:

$$h(k) = i(k) * o(k)$$

$$= \sum_{n=-\infty}^{+\infty} i(n) o(k-n), -\infty < k < \infty \qquad (5.4.3)$$

5.2.2 Definition

If we now compute the complex logarithm $\log[H(z)]$, eqs. (5.1)–(5.3) become

$$\log[H(z)] = \log(A) + \log(z^{-r}) + \sum_{i=1}^{L_1} \log(1 - a_i z^{-1}) + \sum_{i=1}^{L_2} \log(1 - b_i z) \quad (5.5)$$
$$- \sum_{i=1}^{L_3} \log(1 - c_i z^{-1}).$$

The term z^{-r} corresponds to a linear phase shift which is suppressed in higher-order spectrum domains of the signal. Amplification parameter A does not contribute to the shape of the signal, and therefore, will be assumed positive. As such, (5.5) becomes

$$\log[H(z)] = \log(|A|) + \sum_{i=1}^{L_1} \log(1 - a_i z^{-1}) + \sum_{i=1}^{L_2} \log(1 - b_i z) \quad (5.6)$$
$$- \sum_{i-1}^{L_3} \log(1 - c_i z^{-1}).$$

The *complex cepstrum*, $c_h(m)$, of the signal is defined as the inverse Z transform of $\log[H(z)]$ ([Oppenheim and Schafer, 1989]) and is given by

$$c_h(m) = \frac{1}{2\pi j} \oint_c \log[H(z)] z^{m-1} dz$$

$$= Z_1^{-1}[\log[H(z)]] \quad (5.7)$$

where the contour of integration C is within the region of convergence which includes the circle. The complex cepstrum can also be obtained using the inverse Fourier transform

$$c_h(m) = \frac{1}{2\pi} \int_{-\pi}^{+\pi} \log[H(\omega)] \exp\{j\omega m\} d\omega$$

$$= F_1^{-1}[\log|H(\omega)| + j\phi_h(\omega)] \quad (5.8)$$

where

$$H(\omega) = |H(\omega)| \cdot \exp\{j\phi_h(\omega)\}. \quad (5.9)$$

Note that $Z_1[\cdot], Z_1^{-1}[\cdot]$ denote the 1-d Z transform and inverse Z transform respectively. Also, $F_1[\cdot]$ and $F_1^{-1}[\cdot]$ denote the 1-d Fourier transform and inverse

Fourier transform, respectively. This notation will be used extensively in this chapter. Combining (5.6) and (5.7) we obtain the following general expression of the complex cepstrum:

$$c_h(m) = \begin{cases} \log|A|, & m = 0 \\ -\dfrac{A^{(m)}}{m}, & m > 0 \\ \dfrac{B^{(-m)}}{m}, & m < 0 \end{cases} \quad (5.10)$$

where

$$A^{(m)} = \sum_{i=1}^{L_1} a_i^m - \sum_{i=1}^{L_3} c_i^m, \quad (5.11)$$

$$B^{(m)} = \sum_{i=1}^{L_2} b_i^m$$

are cepstral *parameters* which contain the minimum and maximum phase information, respectively. Note that for MA signals, the second term of $A^{(m)}$ in (5.11) would be missing.

5.2.3 Properties

The complex cepstrum of a signal has the following properties: [Oppenheim and Schafer, 1989]

(i) The complex cepstrum, $c_h(m)$, decays at least as fast as $1/|m|$. The cepstral parameters $A^{(m)}$, $B^{(m)}$ decay exponentially with $|m|$.

(ii) $c_h(m)$ always has infinite duration.

(iii) If $h(k)$ is real, its complex cepstrum is also real.

(iv) $B^{(m)} = 0$ for all $m < 0$ if and only if $h(k)$ is minimum phase. Similarly, $A^{(m)} = 0$ for all m if and only if $h(k)$ is maximum phase.

(v) If $y(k) = h(k) * x(k)$, then $c_y(m) = c_h(m) + c_x(m)$, i.e., the linear convolution operation becomes a summation in the complex cepstrum domain. A consequence of this property and (5.4.3) is that $c_h(m) = c_i(m) + c_o(m)$ where $c_i(m)$, $c_o(m)$ are the complex cepstra of $i(k)$ and $O(k)$, respectively. In fact,

$$c_i(m) = -\frac{A^{(m)}}{m}, m > 0 \text{ and } c_o(m) = \frac{B^{(-m)}}{m}, m < 0.$$

5.2.4 Computation and Phase Unwrapping

For the complex cepstrum to exist, it is required that $\log|H(\omega)|$ is guaranteed to be continuous. However, $\phi_h(\omega)$ can be discontinuous because at each frequency ω, any integer multiple of 2π can be added, i.e.,

$$\phi_h(\omega) = \arg[H(\omega)] + 2\pi k(\omega), \tag{5.12}$$

where $-\pi \leq \arg[H(\omega)] \leq \pi$ is the principal value of the phase obtained from an arctangent subroutine and $k(\omega)$ is integer. For $\phi_h(\omega)$ to be a continuous curve, $k(\omega)$ should take appropriate integer values to "unwrap" the principal value of the phase, $\arg[H(\omega)]$.

Example 5.1

Consider a sequence $y(k)$ that has the Fourier transform

$$H(\omega) = (e^{-j\omega} - a)(e^{-j\omega} - b),$$

where $a = 0.5$ and $b = 0.25$. Clearly, this is a maximum phase FIR sequence because both its zeros lie outside the unit circle at locations $1/a = 2$ and $1/b = 4$ in the z-plane. Figure 5.1(a) illustrates the principal value, $\arg[H(\omega)]$, of the phase $\phi_h(\omega)$ obtained from an arctangent subroutine. Note the points of discontinuity at $\omega \approx \pm 1.2$. Figure 5.1(b) shows the integer values that $k(\omega)$ should take in (5.12) to unwrap the phase. The continuous $\phi_h(\omega)$ obtained from phase unwrapping is illustrated in Figure 5.1(c). □

In Example 5.1 we described a simple case where $k(\omega)$ can be easily determined and therefore phase unwrapping is somewhat straightforward. Oppenheim and Schafer [1989] point out, however, that in many practical cases the principal-value discontinuities cannot be reliably detected. As such, the phase unwrapping problem becomes a very difficult one to deal with in practice. Although the computation of the complex cepstrum appears to be simple in (5.8), the need for phase unwrapping on $\phi_h(\omega)$ before we apply $F_1^{-1}[\cdot]$, makes it a difficult one. In fact, phase unwrapping can be thought of as the "fundamental curse" of cepstrum-based signal processing algorithms. We show later in this chapter how the use of higher-order spectra results in the computation of the complex cepstrum, $c_h(m)$, *without* the need for phase unwrapping algorithms.

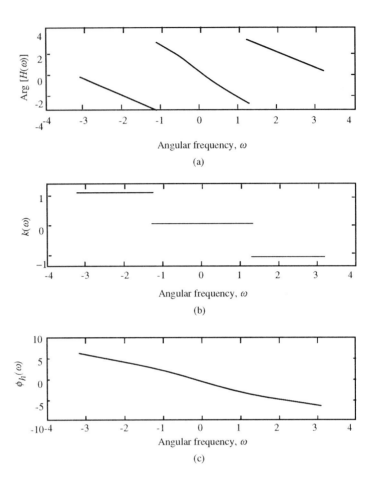

Figure 5.1 (a) Principal value of phase obtained from an arctangent subroutine, (b) Integer values that $k(\omega)$ should take to unwrap the phase, (c) Continuous phase obtained from phase unwrapping.

Tribolet [1977] introduced a phase unwrapping algorithm which assumes that we know the derivative of the phase, $\phi'_h(\omega) = d\phi_h(\omega)/d\omega$, at points $\omega_\lambda = \frac{2\pi}{N}\lambda$, $\lambda = 0, 1, ..., N-1$. The phase derivative can be computed exactly at $\omega = \omega_\lambda$ using [Oppenheim and Schafer, 1989]

$$\begin{aligned} \phi'_h(\omega_\lambda) &= I_m\left\{\frac{H'(\omega)}{H(\omega)}\right\} \\ &= \left\{\frac{F_1[-jkh(k)]}{F_1[h(k)]}\right\} \end{aligned} \quad (5.13)$$

If $\phi_h(\omega_\lambda)$ is the estimated unwrapped phase, Tribolet's algorithm is based on the recursive formula [1977]

$$\hat{\phi}_h(\omega_\lambda) = \hat{\phi}_h(\omega_{\lambda-1}) + \frac{\Delta\omega}{2}\{\phi_h'(\omega_\lambda) + \phi_h'(\omega_{\lambda-1})\} \qquad (5.14.1)$$

where initially $\Delta\omega = 2\pi/N$ and

$$\hat{\phi}_h(\omega_\lambda) = \arg[H(\omega)] + 2\pi k(\omega_\lambda). \qquad (5.14.2)$$

Tribolet points out that $\phi_h(\omega_\lambda)$ is consistent for some E_Z as the integer $k(\omega_\lambda)$ exists, such that

$$\left|\hat{\phi}_h(\omega_\lambda) - \arg[H(\omega)] - 2\pi k(\omega_\lambda)\right| < E_z < \pi. \qquad (5.14.3)$$

Consequently, if $\phi_h(\omega_\lambda)$ is not consistent, then $\Delta\omega$ is chosen to be smaller (halved) and the whole procedure is repeated until (5.14.3) is satisfied. The choice of E_z clearly determines the performance of the algorithm. Phase unwrapping algorithms have been found to be very sensitive to noise, even at very high signal-to-ratio (SNR) environments [Petropulu and Nikias, 1990].

5.3 THE DIFFERENTIAL CEPSTRUM OF DETERMINISTIC ENERGY SIGNALS

The differential cepstrum was introduced by Polydoros and Fam [1981] as a new tool for computing the cepstral parameters $A^{(m)}$, $B^{(m)}$ defined in (5.11), discarding the need for phase unwrapping algorithms. The differential cepstrum is shift invariant and its relationship to the complex cepstrum is simple.

5.3.1 Definition

The differential cepstrum of $h(k)$ (or $H(z)$) is defined by

$$d_h(m) = Z_1^{-1}\left[\frac{dH(z)/dz}{H(z)}\right] \qquad (5.15.1)$$

or equivalently in terms of Fourier transformer $H(\omega)$ by

$$d_h(m) = \frac{1}{2\pi j}\int_{-\pi}^{+\pi} \frac{H'(\omega)}{H(\omega)}\exp\{j\omega(m-1)\}d\omega, \qquad (5.15.2)$$

where

$$H'(\omega) = dH(\omega)/d\omega.$$

Polydoros and Fam [1981] showed that the differential cepstrum can be expressed directly in terms of the cepstral parameters $A^{(m)}$ and $B^{(m)}$ as follows.

$$d_h(m) = \begin{cases} A^{(m-1)}, & m \geq 2 \\ 0, & m = 1 \\ -B^{(-m+1)}, & m \leq 0. \end{cases} \tag{5.16}$$

Combining (5.10), (5.11), and (5.16), we see that the complex cepstrum is related to the differential cepstrum by

$$c_h(m) = -\frac{1}{m} d_h(m+1) \ (m \neq 0) \tag{5.17}$$

5.3.2 Properties

Properties (ii) – (iv) of the complex cepstrum also apply to the differential cepstrum due to the relationship shown in (5.17). Additional properties of the differential cepstrum include:

(i) The differential cepstrum decays exponentially with $|m|$.

(ii) If $y(k) = h(k) * y(k)$, then $d_y(m) = d_h(m) + d_y(m)$. Consequently, $d_h(m) = d_i(m) + d_o(m)$, where $d_i(m)$ and $d_o(m)$ are the differential cepstra of $i(k)$ and $o(k)$, respectively. Actually, $d_i(m) = A^{(m-1)}, m \geq 2$ and $d_o(m) = -B^{(-m+1)}, m \leq 0$. Note that $d_h(1) = 0$.

(iii) *Shift invariance*: This property is of fundamental importance and is not shared by the complex cepstrum. If we denote the differential cepstrum of $H(z)$ in (5.1) by $d_h^{(r)}(m)$, where integer r corresponds to the linear phase shift z^{-r}, then

$$d_h^{(r)}(m) = \begin{cases} d_h^{(0)}(m), & m \neq 1 \\ -r, & m = 1. \end{cases} \tag{5.18}$$

This property implies that the differential cepstrum is invariant under a time delay of the input sequence, except for the first sample, which is actually a time delay measure [Polydoros and Fam, 1981].

(iv) *Scale invariance*: From the division in (5.15), it easily follows that $d_h(m)$ is independent of the scalar value A defined in (5.1).

(v) The differential cepstra $d_i(m)$, $d_o(m)$ of the minimum and maximum phase components of the impulse response, respectively, are related to the impulse responses by the following recursive relations:

$$i(k) = -\frac{1}{k} \sum_{m=2}^{k+1} d_i(m) i(k-m+1), \text{ for } k \geq 1 \qquad (5.19.1)$$

$$o(k) = -\frac{1}{k} \sum_{m=k+1}^{0} d_o(m) o(k-m+1), \text{ (for } k \leq -1 \text{ and } i(0) = o(0) = 1.) \qquad (5.19.2)$$

Note that $i(k) \to 0$ as $k \to \infty$, and $o(k) \to 0$ as $k \to -\infty$. Combining (5.19) with (5.16) we obtain the recursive relationship between cepstral parameters and impulse response, viz:

$$i(k) = -\frac{1}{k} \sum_{m=2}^{k+1} A^{(m-1)} i(k-m+1); \; k \geq 1 \qquad (5.20.1)$$

$$o(k) = \frac{1}{k} \sum_{m=k+1}^{0} B^{(1-m)} o(k-m+1); \; k \leq -1 \qquad (5.20.2)$$

Recursive relations (5.20.1) and (5.20.2) are fundamental. Knowing either $c_h(m)$ or $d_h(m)$, i.e., knowing cepstral parameters $A^{(m)}$ and $B^{(m)}$, relations (5.20) should allow us to reconstruct $i(k)$ and $o(k)$, and then from (5.4.3), the signal $h(k)$ itself.

5.3.3 Computation

The differential cepstrum can be computed by treating (5.15.2) as a DFT, thereby leading to

$$\hat{d}_h(m) = -j \frac{1}{N} \sum_{\lambda=0}^{N-1} \frac{H'(\omega_\lambda)}{H(\omega_\lambda)} \exp j\omega_\lambda(m-1), \qquad (5.21)$$

where $\omega_\lambda = \frac{2\pi}{N} \lambda$, and

$$H(\omega_\lambda) = \sum_{k=0}^{N-1} h(k) \exp\{-j\omega_\lambda k\}, \text{ and} \qquad (5.22.1)$$

$$\frac{dH(\omega_\lambda)}{d\omega_\lambda} = H'(\omega_\lambda) = -j \sum_{k=0}^{N-1} k \, h(k) \exp\{-j\omega_\lambda k\}. \qquad (5.22.2)$$

With the use of the differential cepstrum we avoid phase unwrapping algorithms at the cost, however, of more severe aliasing since now [Oppenheim and Schafer, 1989]

$$\hat{d}_h(m) = \frac{1}{m} \sum_{r=-\infty}^{+\infty} (m+rN)\, d_h(m+rN). \qquad (5.23)$$

5.4 THE POWER CEPSTRUM OF DETERMINISTIC AND STOCHASTIC SIGNALS

The cepstrum of the power spectrum (i.e., power cepstrum) can be defined for both deterministic energy signals and stationary random processes. In the case of deterministic signals, the energy spectrum is defined as

$$M_2^h(z) = H(z)H(z^{-1}), \qquad (5.23.1)$$

whereas in the case of stochastic signals we can always find a linear model $H(z)$ that can represent the power spectrum of a stochastic process exactly, as follows:

$$C_2^h(z) = \gamma_2^w H(z)H(z^{-1}), \qquad (5.23.2)$$

In other words, the stochastic process may be seen as being generated by a white noise-driven linear model $H(z)$ where γ_2^w is the variance of the driving noise. Without loss of generality, we assume $\gamma_2^w = 1$.

5.4.1 Definition

The power cepstrum of a signal is sometimes referred to as the real cepstrum and is defined as [Bogert, et al. 1963]:

$$p_h(m) = Z_1^{-1}\left[\log\{H(z)H(z^{-1})\}\right], \qquad (5.24.1)$$

or equivalent in terms of the Fourier transform $H(\omega)$ by

$$p_h(m) = \frac{1}{2\pi} \int_{-\pi}^{+\pi} \log|H(\omega)|^2 \exp\{j\omega m\}d\omega. \qquad (5.24.2)$$

The power cepstrum can be expressed directly in terms of $A^{(m)}$ and $B^{(m)}$ (the cepstral parameters) as follows [Nikias and Liu, 1990]:

$$p_h(m) = \begin{cases} \log|A^2|, & m = 0 \\ -\frac{1}{m}\left[A^{(m)} + B^{(m)}\right], & m > 0. \\ \frac{1}{m}\left[A^{(-m)} + B^{(-m)}\right], & m < 0. \end{cases} \qquad (5.25)$$

Combining (5.25) and (5.10), we obtain [Oppenheim and Schafer, 1989]:

$$p_h(m) = c_h(m) + c_h^*(-m), \tag{5.26}$$

which implies that the power cepstrum is the conjugate-symmetric part of the complex cepstrum.

5.4.2 Properties

Properties (i) and (ii) of the complex cepstrum are also true for the power cepstrum. Additional properties of the power cepstrum are:

(i) $p_h(m)$ is a symmetric sequence, i.e., $p_h(m) = p_h(-m)$. As such, it does not discriminate minimum phase from maximum phase sequences.

(ii) $p_h(m)$ does not depend upon the phase $\phi_h(\omega)$, and thus it is much easier to compute. However, $p_h(m)$ contains only the magnitude information ($|H(\omega)|$) of the signal. *From (5.25) we conclude that the sums of $A^{(m)}$ and $B^{(m)}$ (the cepstral parameters) contain magnitude information only.*

(iii) If $y(k) = h(k) * x(k)$, then $p_y(m) = p_h(m) + p_x(m)$. However, since $h(k) = i(k) * o(k)$, we obtain $p_h(m) = p_i(m) + p_o(m)$ where $p_i(m)$ and $p_o(m)$ are the power cepstra of $i(k)$ and $o(k)$, respectively. Combining (5.10), property (v) of the complex cepstrum, and (5.25), we obtain

$$p_i(m) = \begin{cases} -\frac{1}{m}A^{(m)} &, \quad m > 0 \\ \frac{1}{m}A^{(-m)} &, \quad m < 0 \end{cases}$$

$$p_o(m) = \begin{cases} -\frac{1}{m}B^{(m)} &, \quad m > 0 \\ \frac{1}{m}B^{(-m)} &, \quad m < 0 \end{cases}$$

(iv) The sequence $h(k)$ cannot be recovered from $p_h(m)$. Since $p_h(m)$ contains only magnitude information, it is only the autocorrelation function of $h(k)$ that can be recovered. If $m_2^h(\tau)$ denotes the autocorrelation of $h(k)$; e.g., $m_2^h(\tau) = h(\tau) * h(-\tau)$, then from (5.20) and (5.25) we obtain

$$m_2^h(\tau) = -\frac{1}{\tau}\sum_{m=2}^{\tau+1}\left[A^{(m-1)} + B^{(m-1)}\right]m_2^h(\tau - m + 1), \tag{5.27}$$

where

$$m_2^h(0) = 1 \text{ and } \tau \geq 1.$$

Note that $m_2^{h*}(\tau) = m_2^h(-\tau)$. So, even when we know $A^{(m)}$ and $B^{(m)}$ (the cepstral parameters), (5.27) will still allow us to compute the autocorrelation sequence of the signal.

(v) The Eucledian distance between power cepstral coefficients is a meaningful measure of difference between minimum phase (or all pole) spectra, useful for pattern recognition and especially in speech processing.

5.4.3 Computation

The computation of the power cepstrum or real cepstrum does not require phase unwrapping algorithms. However, by its very definition, it suppresses the phase information of the signal. Combining (5.23.2) and (5.24.1), and assuming that $\gamma_2^w = 1$, and that $P_h(z)$ is the Z transform of the power cepstrum, we obtain

$$P_h(z) = \log C_2^h(z). \tag{5.28.1}$$

Differentiating with respect to z in a region of convergence of $C_2^h(z)$ (which contains the unit circle), we obtain

$$\frac{\partial P_h(z)}{\partial z} = \frac{1}{C_2^h(z)} \frac{\partial C_2^h(z)}{\partial z} \tag{5.28.2}$$

or

$$C_2^h(z) \cdot z \cdot \frac{\partial P_h(z)}{\partial z} = z \frac{\partial C_2^h(z)}{\partial z}. \tag{5.28.3}$$

From (5.28.3) it follows that the autocorrelation sequence $m_2^h(n)$ and the power cepstrum $p_h(n)$ are related through

$$m_2^h(n) * [-n p_h(n)] = -n\, m_2^h(n), \text{ or} \tag{5.29.1}$$

$$\sum_{\ell=1}^{\infty} \left[A^{(\ell)} + B^{(\ell)} \right] m_2^h(n-\ell) = -\frac{1}{2} n\, m_2^h(n). \tag{5.29.2}$$

From (5.11) it follows that in practice we can always truncate (5.29.2) to obtain

$$\sum_{\ell=1}^{p} \left[A^{(\ell)} + B^{(\ell)} \right] m_2^h(n-\ell) = -\frac{1}{2} n \cdot m_2^h(n). \tag{5.30}$$

This is a reasonable approximation because $A^{(\ell)}$ and $B^{(\ell)}$ decay exponentially. The integer p may be chosen sufficiently large so that $(A^{(\ell)} + B^{(\ell)}) \approx 0$ for $\ell > p$.

Assuming that the autocorrelation sequence $m_2^h(n)$ is available (in practice it is estimated from either deterministic or stochastic signals), from (5.30) we can form the overdetermined system of equations

$$\begin{bmatrix} m_2^h(0) & m_2^h(-1) & \cdots & m_2^h(1-p) \\ m_2^h(1) & m_2^h(0) & \cdots & m_2^h(2-p) \\ \vdots & \vdots & \cdots & \vdots \\ m_2^h(M) & m_2^h(M-1) & \cdots & m_2^h(M-p) \end{bmatrix} \begin{bmatrix} A^{(1)} + B^{(1)} \\ \vdots \\ A^{(p)} + B^{(p)} \end{bmatrix} = -\frac{1}{2} \begin{bmatrix} m_2^h(1) \\ 2m_2^h(2) \\ \vdots \\ (M+1)m_2^h(M+1) \end{bmatrix} \quad (5.31)$$

and solve it via least-squares or singular value decomposition (SVD). An alternative computation of the power cepstrum from (5.29.1) is via FFT algorithms. It follows that

$$p_h(n) = \frac{1}{n} F_1^{-1} \left\{ \frac{F_1\left[n\, m_2^h(n)\right]}{F_1\left[m_2^h(n)\right]} \right\}, \quad (5.32)$$

where $F_1[\cdot]$ and $F_1^{-1}[\cdot]$ denote the 1-d Fourier transform and inverse Fourier transform respectively.

Example 5.2

Consider a multipath distortionless channel with input $w(k)$ and output $y(k)$ [Hassab, 1989]:

$$y(k) = w(k) + \alpha w(k - D), \quad (5.33.1)$$

where $|\alpha| \neq 1$ and D is an integer time delay ($D \neq 0$). The autocorrelation of $y(k)$ and $w(k)$ are related through

$$m_2^y(\tau) = \left[1 + \alpha^2\right] m_2^w(\tau) + \alpha m_2^w(\tau - D) + \alpha m_2^w(\tau + D), \quad (5.33.2)$$

and their power spectra

$$C_2^y(\omega) = [1 + \alpha^2 + 2\alpha \cos(\omega D)] C_2^w(\omega), \quad (5.33.3)$$

where

$$|H(\omega)|^2 = 1 + \alpha^2 + 2\alpha \cos(\omega D).$$

The time delay, D, is obtained from the time difference between the two highest peaks of $m_2^y(\tau)$ provided that the time delay, D, is greater than the correlation

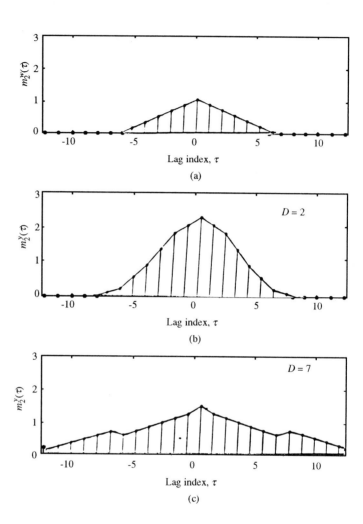

Figure 5.2 Multipath distortionless channel example with $\alpha = 0.7$. (a) Input autocorrelation, (b) output autocorrelation when time delay $D = 2$, (c) output autocorrelation when $D = 7$.

time of the signal $w(k)$. According to Hassab [1989] there are sonar applications where small time delays and small source bandwidths (thus large correlation times) exit, and the autocorrelation, $m_2^w(\tau)$, will lack the resolvability needed. Figure 5.2 illustrates $m_2^w(\tau)$ and $m_2^y(\tau)$ for time delays $D = 2$ and $D = 7$. From this figure it is apparent that for $D = 2$ $m_2^y(\tau)$ is nonresolvable, whereas for $D = 7$ the delay is obtained from the time difference between the two highest peaks.

The purpose of this example is to demonstrate that power cepstrum of $C_2^y(\omega)$ can be used to resolve the aforementioned nonresolvable case; e.g., $D = 2$.

$$\log\left[C_2^y(\omega)\right] = \log\left[(1+\alpha^2)C_2^w(\omega)\right] + \log\left[1 + A\cos(\omega D)\right], \tag{5.33.4}$$

where $A = 2\alpha/(1+\alpha^2)$. However, since $|\alpha| \neq 1$,

$$\log[1 + A\cos(\omega D)] = \sum_{\lambda=1}^{\infty} \frac{(-1)^{\lambda+1}}{\lambda}(A\cos(\omega D))^\lambda. \tag{5.33.5}$$

If we express $(\cos(\omega D))^\lambda$ in terms of $\cos(\omega \lambda D)$ and compute the inverse Fourier transform of the equations above, we find that the power cepstrum of $[1+A\cos(\omega D)]$ exhibits a *dominant* term $|A + \frac{A^3}{3} + \cdots|$, at the time delay D [Hassab, 1989]. Figure 5.3 illustrates the power cepstrum, $p_y(m)$, for $D = 2$ and $D = 7$, and clearly demonstrates that both cases can be resolved. □

5.5 THE BICEPSTRUM AND TRICEPSTRUM OF DETERMINISTIC AND STOCHASTIC SIGNALS

The purpose of this section is to introduce the cepstrum of the bispectrum (i.e., bicepstrum) and its computation for both deterministic and stochastic signals. It is established at cepstral coefficients $A^{(m)}$ and $B^{(m)}$ of $h(k)$ can be computed directly from third-order moments or cumulants without the use of a phase unwrapping algorithm. The extension to fourth-order statistics (tricepstrum) is also discussed.

Let $H(z)$ in (5.1) be the transfer function of a LTI system and the system input be a zero-mean non-Gaussian, white, i.i.d. signal with skewness γ_3^w.

Then the output bispectrum, $C_3^x(z_1, z_2)$, exits and is given by

$$C_3^x(z_1, z_2) = \gamma_3^w H(z_1)H(z_2)H\left(z_1^{-1} \cdot z_2^{-1}\right). \tag{5.34.1}$$

Let us note that the bispectrum of $\{X(k)\}$ is defined as

$$C_3^x(z_1, z_2) = \sum_{\tau_1=-\infty}^{+\infty}\sum_{\tau_2=-\infty}^{+\infty} c_3^x(\tau_1, \tau_2) z_1^{-\tau_1} z_2^{-\tau_2}, \tag{5.34.2}$$

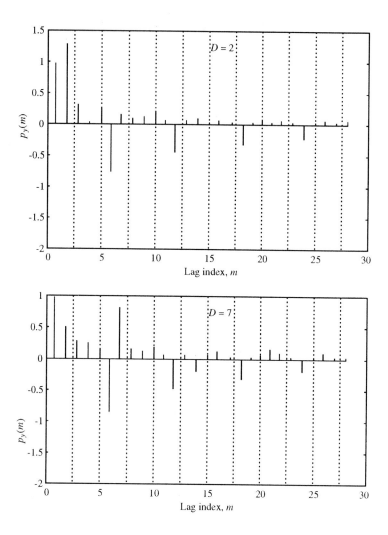

Figure 5.3 Multipath distortionless channel example with $\alpha = 0.7$. (a) Power cepstrum of output autocorrelation when $D = 2$, (b) $D = 7$.

where $c_3^x(\tau_1, \tau_2) = E\{X(k)X(k+\tau_1)X(k+\tau_2)\}$ is the third-order moment or cumulant sequence and that $\{X(k)\}$ is a linear non-Gaussian stationary random process.

On the other hand, if $\{h(k)\}$ is a deterministic energy signal, its bispectrum is given by

$$M_3^h(z_1, z_2) = H(z_1)H(z_2)H\left(z_1^{-1} \cdot z_2^{-1}\right). \tag{5.35}$$

Comparing (5.35) with (5.34) we see the duality between the stochastic and deterministic cases. Consequently, and without loss of generality, we'll employ for both deterministic and linear non-Gaussian signals

$$C_3^h(z_1, z_2) = H(z_1)H(z_2)H\left(z_1^{-1} \cdot z_2^{-1}\right). \tag{5.36}$$

as the bispectrum and $m_3^h(\tau_1, \tau_2)$ as the third-order moment sequence.

Note that for stochastic signals the skewness γ_3^w is embedded in (scalar) A of (5.1).

5.5.1 Definition

Pan and Nikias [1988] introduced the bicepstrum (and tricepstrum) as the inverse 2-d Z transform of the log bicepstrum, $C_3^h(z_1, z_2)$, viz:

$$b_h(m, n) = Z_2^{-1}\left\{\log\left[C_3^h(z_1, z_2)\right]\right\}, \tag{5.37.1}$$

where

$$\begin{aligned} B_h(z_1, z_2) &= \log\left[C_3^h(z_1, z_2)\right] \\ &= \log|A^3| + \log\left[I\left(z_1^{-1}\right)\right] + \log\left[I\left(z_2^{-1}\right)\right] \\ &+ \log\left[I(z_1 \cdot z_2)\right] + \log\left[O(z_1)\right] + \log\left[O(z_2)\right] \\ &+ \log\left[O\left(z_1^{-1} \cdot z_2^{-1}\right)\right]. \end{aligned} \tag{5.37.2}$$

They showed that

$$b_h(m, n) = \begin{cases} \log|A^3| & , \; m = 0, n = 0 \\ -\frac{1}{n}A^{(n)} & , \; m = 0, n > 0 \\ -\frac{1}{m}A^{(m)} & , \; n = 0, m > 0 \\ \frac{1}{m}B^{(-m)} & , \; n = 0, m < 0 \\ -\frac{1}{n}B^{(-n)} & , \; m = 0, n < 0 \\ -\frac{1}{n}B^{(n)} & , \; m = n > 0 \\ \frac{1}{n}A^{(-n)} & , \; m = n < 0 \\ 0 & , \; \text{otherwise} \end{cases} \tag{5.38}$$

where $A^{(m)}$ and $B^{(m)}$ are the cepstral coefficients defined by (5.11). Figure 5.4 illustrates the bicepstrum sequence $b_h(m,n)$.

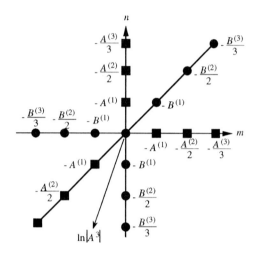

Figure 5.4 The bicepstrum of a nonminimum phase linear non-Gaussian process or a deterministic signal.

Comparing Figure 5.4 and equation (5.38) with equation (5.10), it is apparent that

$$c_h(m) = b_h(m,0) = b_h(0,m) = b_h(-m,-m), \quad m \neq 0 \tag{5.39}$$

i.e., *the bicepstrum is the complex cepstrum of $h(k)$ along three straight lines, and is zero elsewhere.*

Similarly, the differential bicepstrum, $db_h(m,n)$, is defined as [Pan and Nikias, 1988]

$$db_h(m,n) = Z_2^{-1}\left\{\frac{1}{C_3^h(z_1,z_2)}\frac{\partial C_3^h(z_1,z_2)}{\partial z_1}\right\} \tag{5.40.1}$$

and is shown to be equal to

$$db_h(m,n) = \begin{cases} A^{(m-1)} &, \ n=0, m \geq 2 \\ -B^{(1-m)} &, \ n=0, m \leq 0 \\ -A^{(-n)} &, \ n \leq -1, m = n+1 \\ B^{(n)} &, \ n \geq 1, m = n+1 \\ 0 &, \ \text{otherwise.} \end{cases} \tag{5.40.2}$$

Figure 5.5 illustrates the differential bicepstrum. Comparing (5.40.2) with (5.16) we see that the *differential bicepstrum is the differential cepstrum of the signal along two straight lines, and is zero elsewhere.*

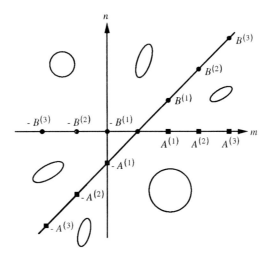

Figure 5.5 The differential bicepstrum of a nonminimum phase linear non-Gaussian process or a deterministic signal

5.5.2 Properties

Several properties are associated with the bicepstrum of nonminimum phase signals.

(i) The minimum and maximum phase components $i(k)$ and $o(k)$, respectively, can be reconstructed separately via (5.20.1) and (5.20.2). This is achieved by computing the bicepstrum from third-order statistics of the signal and then obtaining $A^{(m)}$ and $B^{(m)}$ (the cepstral coefficients) from (5.38).

(ii) The bicepstrum is flexible enough to be applied on both deterministic and stochastic signals. On the other hand, the complex cepstrum can only be applied on deterministic signals. Although the power cepstrum can also be applied on both deterministic and stochastic signals, it suppresses the phase information and thus is nonreversible.

(iii) Due to identity (5.39) the bicepstrum possesses all properties of the complex cepstrum. However, as we demonstrate next, *the bicepstrum can be computed without using phase unwrapping algorithms.*

(iv) The bicepstrum can be used for nonminimum phase system identification when the system is driven by non-Gaussian white noise (nonzero skewness) and the input is inaccessible. It will work without any a priori knowledge of the type of the system. It will not require model selection criteria because it will reconstruct the system impulse response directly, bypassing estimation of model parameters. The length of $i(k)$ and $o(k)$ may be determined by the algorithm itself; recursions with (5.20.1) and (5.20.2) could be terminated when successive values of $|i(k)|$ and $|o(k)|$ stay below a threshold value γ, where γ is very small (depending on word length).

5.5.3 Computation

Since $B_h(z_1, z_2)$ in (5.37.2) is analytic in its region of convergence (which contains the unit circle), we can perform partial differentiation with respect to z_1 or z_2 in that region, e.g.,

$$\frac{\partial B_h(z_1, z_2)}{\partial z_1} = \frac{1}{C_3^h(z_1, z_2)} \frac{\partial C_3^h(z_1, z_2)}{\partial z_1} \quad (5.41.1)$$

or

$$C_3^h(z_1, z_2) \cdot z_1 \cdot \frac{\partial B_h(z_1, z_2)}{\partial z_1} = z_1 \frac{\partial C_3^h(z_1, z_2)}{\partial z_1}. \quad (5.41.2)$$

From (5.41) it follows that third-order statistics (moments or cumulants), $m_3^h(n, \ell)$, are related to the bicepstrum, $b_h(n, \ell)$, via the linear convolution formula

$$\begin{aligned} m_3^h(n, \ell) * [-nb_h(n, \ell)] &= \sum_{\tau=-\infty}^{+\infty} \sum_{p=-\infty}^{+\infty} (-\tau) b_h(\tau, p) m_3^h(n - \tau, \ell - p) \quad (5.42) \\ &= -n \cdot m_3^h(n, \ell). \end{aligned}$$

Pan and Nikias [1988] showed that the use of differential bicepstrum results in a similar convolution relationship to (5.42). If (5.38) is substituted into (5.42), we obtain the cepstral equation [Pan and Nikias, 1988]

$$\sum_{\tau=1}^{\infty} \left\{ A^{(\tau)} \left[m_3^h(n - \tau, \ell) - m_3^h(n + \tau, \ell + \tau) \right] + B^{(\tau)} \left[m_3^h(n - \tau, \ell - \tau) - m_3^h(n + \tau, \ell) \right] \right\}$$
$$= -n m_3^h(n, \ell),$$
(5.43)

which provides a *direct relationship* between cepstral coefficients, $A^{(\tau)}$ and $B^{(\tau)}$, and third-order statistics.

From the definition of $A^{(\tau)}$ and $B^{(\tau)}$ in (5.11), and the original assumptions that: $|a_i| < 1, |b_i| < 1$, and $|c_i| < 1$ for all $\{i\}$, it follows that in practice we can always truncate (5.43) and obtain the approximate cepstral equation

$$\sum_{i=1}^{p}\left\{A^{(i)}\left[m_3^h(n-i,\ell)-m_3^h(n+i,\ell+i)\right]\right\} + \sum_{j=1}^{q}\left\{B^{(j)}\left[m_3^h(n-j,\ell-j)-m_3^h(n+j,\ell)\right]\right\}$$
$$\cong -n \cdot m_3^h(n,\ell). \tag{5.44}$$

The integers p and q may be chosen using [Pan and Nikias, 1988]

$$\begin{aligned} p &= \ln c/\ln a \\ q &= \ln c/\ln b \end{aligned} \tag{5.45.1}$$

where

$$\max[|a_i|,|c_i|] < a < 1$$
$$\max[|b_i|] < b < 1 \tag{5.45.2}$$

and c is a very small constant (say 10^{-4}). so $A^{(\tau)} = 0$ for $\tau > p$ and $B^{(\tau)} = 0$ for $\tau > q$. Clearly, this approach requires some a priori knowledge about the pole-zero location of $H(z)$; i.e., the values (a) and (b) in (5.45.2).

Assuming that the third-order statistics $m_3^h(n,\ell)$ are available (in practice their estimates), we can define $w = \max[p,q]$ and $z = \lfloor w/2 \rfloor$ (where w and z are integers) and choose $m = -w, \cdots, 0, \cdots, w$ and $\ell = -z, \cdots, 0, \cdots, z$ to form the overdetermined system of equations

$$\mathbf{Ya} = \mathbf{y} \tag{5.46.1}$$

where $\mathbf{Y}$ is $[(2w+1) \times (2z+1)] \times (p+q)$ matrix whose entries are expressions of the form $\{m_3^h(\mu,\lambda) - m_3^h(\sigma,\tau)\}$.

$\mathbf{a} = [A^{(1)}, \cdots, A^{(p)}, B^{(1)}, \cdots, B^{(q)}]^T$ is a $(p+q) \times 1$ set of unknown parameters, and $\mathbf{y}$ is a $[(2w+1) \times (2z+1)] \times 1$ vector whose entries are expressions of the form $\{-n \cdot m_3^h(n,\ell)\}$.

For example, if $p = q = 20$, then $w = 20, z = 10$, and $\mathbf{Y}$ is a 1681×40 matrix.

The least-squares solution to (5.46.1) is

$$\mathbf{a} = [\mathbf{Y}^T\mathbf{Y}]^{-1}\mathbf{Y}^T\mathbf{y}. \tag{5.46.2}$$

Note that a more stable solution of (5.46.1) can be obtained with the SVD algorithm. One of the advantages of getting parameters $\{A^{(k)}\}$, $\{B^{(k)}\}$ directly from third-order statistics (or their estimates) via (5.46.2) is that we avoid computations associated with (5.37), which requires bicepstrum estimation procedures and phase unwrapping algorithms. The limitation, however, is that if poles or zeros come close to the unit circle, the computational cost of (5.46.2) will rise because the values of p, q become large.

An alternative method for computing the cepstral coefficients in (5.38) that does not require phase unwrapping can be based on two-dimensional (2-d) fast Fourier transform (FFT) operations. From (5.41.2) we obtain

$$n \cdot b_h(n, \ell) = F_2^{-1} \left\{ \frac{F_2 \left[n \cdot m_3^h(n, \ell) \right]}{F_2 \left[m_3^h(n, \ell) \right]} \right\} \qquad (5.47)$$

where $F_2[\cdot]$ denotes a 2-d Fourier transform and $F_2^{-1}[\cdot]$ its inverse. The big computational advantage of using (5.47) instead of (5.46.2) is for signals (or systems) with pronounced resonances or antiresonances; i.e., when p and q are very large. The size of the region of support of $F_2[\cdot]$ or $F_2^{-1}[\cdot]$ should be chosen greater than or equal to $2 \cdot \max[p, q]$.

Example 5.3

Consider the nonminimum phase MA system with transfer function [Pan and Nikias 1988].

$$H(z) = (1 - 0.869z)(1 + 1.1z + 0 \cdot 617z^2)(1 - 0 \cdot 85z^{-1})(1 - 1.2z^{-1} + 0.45z^{-2}).$$

Its zero location and impulse response, $h(k)$, are shown in Figure 5.6. The true values of cepstral coefficients $\{A^{(k)}\}$ and $\{B^{(k)}\}$ of $h(k)$ are also illustrated in Figure 5.6.

The system $H(z)$ is driven by zero-mean, non-Gaussian, white noise with skewness $\gamma_3^w = 2$. Two different lengths of output data have been used in this example; namely, 8×128 (8 segments of data with 128 samples per segment) and 32×128. From the output sequence, the third-order statistics (cumulants are identical to moments in this case because the sequence is zero-mean) are estimated as described in Chapter 4. The cepstral coefficients $\{A^{(k)}\}$ and $\{B^{(k)}\}$ are computed via the least-squares (LS) approach described by (5.46.1) and (5.46.2) with $p = q = 20$. Recursive relations (5.20.1) and (5.20.2) are then utilized to compute $i(k)$ and $o(k)$ and finally $H(\omega) = I(\omega) \cdot O(\omega)$. The magnitude, $|H(\omega)|$, and phase, $\phi_h(\omega)$, at each ω have been computed using 100 different output process realizations, and their sample mean and standard deviation have been obtained. The results are illustrated in Figure 5.7 and compared to the true $|H(\omega)|$ and $\phi_h(\omega)$. This figure clearly demonstrates the effectiveness of the bicepstrum method for nonminimum phase system identification. As expected, the longer the length of the data, the less is the bias and variance of $|H(\omega)|$ and $\phi_h(\omega)$ at each frequency ω. □

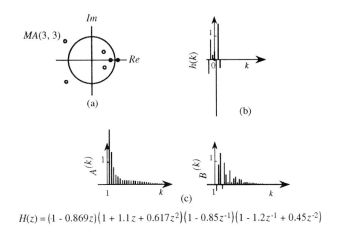

Figure 5.6 Nonminimum phase MA system with pronounced resonances: (a) zero-location, (b) impulse response, and (c) $\{A^{(k)}\}$ and $\{B^{(k)}\}$ the cepstral coefficients. ([Pan and Nikias, 1988] © 1988, IEEE)

Example 5.4

This is an example of an ARMA system with three poles and one zero inside the unit circle, and two zeros outside as shown in Figure 5.8 [Pan and Nikias, 1988]. Figure 5.9 illustrates the true and estimated (via bicepstrum least-squares approach) magnitude and phase responses for 8×128 and 32×128 lengths of output data. For this example, $p = 25$ and $q = 15$ were selected. From Figure 5.9 it is apparent that bias and variance of estimated $|H(\omega)|$ and $\phi_h(\omega)$ are small even with an 8×128 length of output data. □

5.5.4 The Tricepstrum

The equivalent to expression (5.36) in the trispectrum domain is

$$C_4^h(z_1, z_2, z_3) = H(z_1)H(z_2)H(z_3)H\left(z_1^{-1} \cdot z_2^{-1} \cdot z_3^{-1}\right) \qquad (5.48)$$

which may well represent the trispectrum of a deterministic energy signal, $h(k)$, or that of a linear non-Gaussian process with non-zero measure of kurtosis. Its tricepstrum is defined as [Pan and Nikias, 1988]

$$t_h(m, n, \ell) = Z_3^{-1}\left\{\log\left[C_4^h(z_1, z_2, z_3)\right]\right\} \qquad (5.49)$$

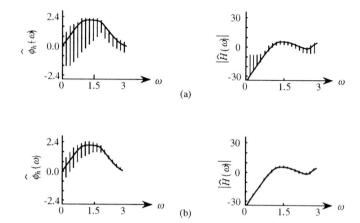

Figure 5.7 (a) Nonminimum phase MA system identified via bicepstrum using least-squares method ($p = q = 20$). Lengths of data are (a) 8×128, (b) 32×128. ([Pan and Nikias, 1988] © 1988 IEEE)

and can be shown to be

$$t_h(m, n, \ell) = \begin{cases} \ln|A| & , \quad m = n = \ell = 0 \\ -\frac{1}{m}A^{(m)} & , \quad m > 0, n = \ell = 0 \\ -\frac{1}{n}A^{(n)} & , \quad n > 0, m = \ell = 0 \\ -\frac{1}{\ell}A^{(\ell)} & , \quad \ell > 0, m = n = 0 \\ \frac{1}{m}B^{(-m)} & , \quad m < 0, n = \ell = 0 \\ \frac{1}{n}B^{(-n)} & , \quad n < 0, m = \ell = 0 \\ \frac{1}{\ell}B^{(-\ell)} & , \quad \ell < 0, m = n = 0 \\ -\frac{1}{n}B^{(n)} & , \quad m = n = \ell > 0 \\ \frac{1}{n}A^{(n)} & , \quad m = n = \ell < 0 \\ 0 & , \quad \text{otherwise.} \end{cases} \qquad (5.50)$$

From (5.10) and (5.50) it is apparent that

$$c_h(m) = t_h(m, 0, 0) = t_h(0, m, 0) = t_h(0, 0, m) = t_h(-m, -m, -m), \quad m \neq 0. \quad (5.51)$$

It is also straightforward to verify that the following identity holds between the fourth-order moments, $m_4^h(n, \ell, \tau)$, of signal $h(k)$ and its tricepstrum $t_h(n, \ell, \tau)$:

$$m_4^h(n, \ell, \tau) * [-n t_h(n, \ell, \tau)] = -n \cdot m_4^h(n, \ell, \tau) \qquad (5.52.1)$$

Chapter 5

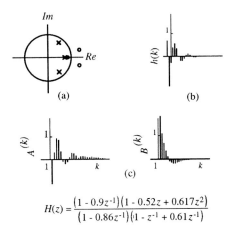

Figure 5.8 Nonminimum phase ARMA system with pronounced resonances: (a) pole-zero location, (b) impulse response, and (c) $\{A^{(k)}\}$, $\{B^{(k)}\}$ cepstral coefficients. ([Pan and Nikias, 1988] © 1988 IEEE)

or, equivalently, $\sum_{\tau=1}^{\infty} \{A(\tau) [m_4^h(n-\tau,\ell,p) - m_4^h(n+\tau,\ell+\tau,p+\tau)] +$

$B^{(\tau)} [m_4^h(n-\tau,\ell-\tau,p-\tau) - m_4^h(n+\tau,\ell,p)] \} = -n \cdot m_4^h(n,\tau,p).$ (5.52.2)

Tricepstral equation (5.52.2) is the equivalent to (5.43) for the bicepstrum. It can be truncated and be solved for $\{A^{(m)}\}$ and $\{B^{(m)}\}$ via the least-squares or SVD algorithm. The fourth-order moment $m_4^h(m,\ell,\tau)$ could be replaced by fourth-order cumulants, $c_4^x(n,\ell,\tau)$, in (5.52.1) and (5.52.2) if $X(k)$ is a stationary linear non-Gaussian process. Usually, $X(k)$ is generated by convolving a deterministic energy signal (impulse response) $h(k)$ with a non-Gaussian stationary white process $W(k)$; i.e., $X(k) = h(k) * W(k)$.

5.6   THE CEPSTRUM OF BICOHERENCY

We saw in Chapter 2 that a normalized bispectrum or bicoherency is a function that, in general, combines the bispectrum, $C_3^h(\omega_1, \omega_2)$, and the power spectrum, $C_2^h(\omega)$, as follows:

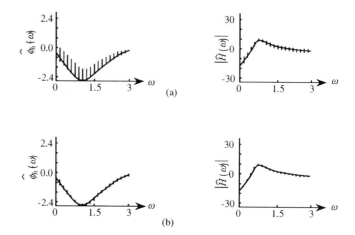

Figure 5.9 Nonminimum phase]ARMA system with pronounced resonances identified via bicepstrum using least-squares method ($p = 25, q = 15$). Lengths of data are (a) 3×128, (b) 32×128. ([Pan and Nikias, 1988] © 1988 IEEE)

$$P_3^h(\omega_1, \omega_2) = \frac{C_3^h(\omega_1, \omega_2)}{\sqrt{C_2^h(\omega_1)C_2^h(\omega_2)C_2^h(\omega_1 + \omega_2)}} \qquad (5.53)$$

However, for deterministic energy signals or linear non-Gaussian processes the power spectrum has the form of (5.23.2) and the bispectrum is given by (5.36). Consequently, the bicoherency takes the form

$$P_3^h(\omega_1, \omega_2) = \frac{A_1 C_3^h(\omega_1, \omega_2)}{|C_3^h(\omega_1, \omega_2)|}, \qquad (5.54)$$

where

$$C_3^h(\omega_1, \omega_2) = |C_3^h(\omega_1, \omega_2)| \exp\{j\Psi_3^h(\omega_1, \omega_2)\} \qquad (5.55)$$

with the magnitude and phase bispectra taking the form

$$|C_3^h(\omega_1, \omega_2)| = |H(\omega_1)||H(\omega_2)||H(\omega_1 + \omega_2)| \qquad (5.56.1)$$

$$\Psi_3^h(\omega_1, \omega_2) = \phi_h(\omega_1) + \phi_h(\omega_2) - \phi_h(\omega_1 + \omega_2). \qquad (5.56.2)$$

Without loss of generality γ_2^w and γ_3^w are assumed to be equal to one.

5.6.1 Definition

Combining (5.54) – (5.56), we obtain

$$P_3^h(\omega_1, \omega_2) = \exp\{j\Psi_3^h(\omega_1, \omega_2)\} \tag{5.57}$$

and thus the cepstrum of bicoherency

$$\begin{aligned} bic_h(m, n) &= F_2^{-1}\left[\ln\left[P_3^h(\omega_1, \omega_2)\right]\right] \\ &= F_2^{-1}\left[\Psi_3^h(\omega_1, \omega_2)\right] \end{aligned} \tag{5.58}$$

is the inverse 2-d Fourier transform of the phase of the bispectrum. Nikias and Liu [1990] showed that

$$bic_h(m,n) = \frac{1}{2}\begin{cases} -\frac{1}{m}\left[A^{(m)} - B^{(m)}\right] &, m > 0, n = 0 \\ -\frac{1}{m}\left[A^{(-m)} - B^{(-m)}\right] &, m < 0, n = 0 \\ -\frac{1}{n}\left[A^{(n)} - B^{(n)}\right] &, n > 0, m = 0 \\ -\frac{1}{n}\left[A^{(-n)} - B^{(-n)}\right] &, n < 0, m = 0 \\ \frac{1}{m}\left[A^{(m)} - B^{(m)}\right] &, m = n > 0 \\ \frac{1}{m}\left[A^{(-m)} - B^{(-m)}\right] &, m = n < 0 \\ \ln|A_1| &, m = n = 0 \\ 0 &, \text{otherwise} \end{cases} \tag{5.59.1}$$

which, in turn, implies that

$$bic_h(m, 0) = -\frac{1}{2m}\left[A^{(m)} - B^{(m)}\right], m > 0. \tag{5.59.2}$$

From (5.25) and (5.59.1), it is obvious that *the differences of* cepstral parameters $A^{(m)}$ and $B^{(m)}$ contain the phase information while their sums contain the magnitude information. The extension of this definition to the cepstrum of tricoherency is straightforward.

5.6.2 Properties

The cepstrum of bicoherency decays at least as fast as $1/|m|$ and has an infinite duration. Additional distinct properties are:

(i) $bic_h(m,n)$ is an antisymmetric sequence ; i.e., $bic_h(m,n) = -bic_h(-m,-n)$. Consequently, it does discriminate minimum phase from maximum phase sequences.

(ii) Combining (5.56.2), (5.58), and (5.59.2), we see that the phase $\phi_h(\omega)$ is entirely based on the sequence $-\frac{1}{m}\left[A^{(m)} - B^{(m)}\right]$, $m > 0$. As such, the cepstrum of bicoherency isolates the phase information of the signal the same way that the cepstrum of the power spectrum isolates the magnitude information.

(iii) Defining

$$d(m) = \begin{cases} j\frac{1}{2m}\left[A^{(|m|)} - B^{(|m|)}\right] & m \neq 0 \\ 0 & m = 0, \end{cases} \quad (5.60)$$

Petropulu and Nikias [1990] showed that the phase of the signal is given by

$$\phi_h(\omega) = F_1[d(m)] \quad (5.61)$$

where $F_1[\cdot]$ denotes a 1-d Fourier transform.

5.6.3 Computation

There are two equivalent ways of computing $bic_h(m,\ell)$ from higher-order statistics of the signal. The first one is based on the 2-d convolution formula [Nikias and Liu, 1990]

$$m_3^h(n,\ell) * m_3^p(n,\ell) * [-nbic_h(n,\ell)] =$$

$$m_3^p(n,\ell) * [-nm_3^h(n,\ell)] - \frac{1}{2}m_3^h(n,\ell) * [-nm_3^p(n,\ell)] \quad (5.62.1)$$

where $m_3^h(n,\ell)$ are the third-order moments of $h(k)$ and $m_3^p(n,\ell)$ are the "third-order moments of the autocorrelation sequence" defined by

$$m_3^p(n,\ell) = F_2^{-1}\left[C_2^h(\omega_1)C_2^h(\omega_2)C_2^h(\omega_1+\omega_2)\right]. \quad (5.62.2)$$

The $bic_h(m,0)$ can be computed from (5.62.1) by forming a linear system of equations and solving it via least-squares or SVD algorithms.

The second approach is based on the formula

$$bic_h(n,\ell) = \frac{1}{n}F_2^{-1}\left[\frac{F_2\{n \cdot m_3^h(n,\ell)\}}{F_2\{m_3^h(n,\ell)\}} - \frac{1}{2}\frac{F_2\{n \cdot m_3^p(n,\ell)\}}{F_2\{m_3^p(n,\ell)\}}\right] \quad (5.63)$$

which can be computed with a 2-d FFT algorithm.

5.6.4 Computation of $A^{(m)}$, $B^{(m)}$ from Second- and Third-Order Statistics

Cepstral coefficients $A^{(m)}$, $B^{(m)}$ and thus the complex cepstrum, $c_h(m)$, of $h(k)$ can be computed from the cepstrum of the power spectrum, $p_h(m)$, and the cepstrum of bicoherency, $bic_h(m, n)$. The procedure is as follows:

1. Using second-order statistics, $m_2^h(\tau)$, and (5.31) or (5.32) compute $p_h(m)$. Hence,
$$[A^{(m)} + B^{(m)}] = -m \cdot p_h(m) \qquad (5.64.1)$$

2. Using third-order statistics, $m_3^h(\tau, \rho)$, and $m_3^p(\tau, \rho)$, and (5.62) or (5.63) compute $bic_h(m, 0)$. It follows that
$$[A^{(m)} + B^{(m)}] = -2 \cdot m \cdot bic_h(m, 0) \qquad (5.64.2)$$

3. Combining (5.64.1) and (5.64.2), we obtain the solution
$$A^{(m)} = -\frac{m}{2}[p_h(m) + 2 \cdot bic_h(m, 0)]$$
$$B^{(m)} = \frac{m}{2}[2 \cdot bic_h(m, 0) - p_h(m)] \qquad (5.65)$$

Table 5.1 summarizes the notation of polycepstra, as well as their definitions, expressions in terms of $A^{(m)}$ and $B^{(m)}$, and computation. It should be emphasized that all these results hold only for deterministic energy signals or stationary linear non-Gaussian processes.

5.7 HIGHER-ORDER SPECTRUM FACTORIZATION AND INDEX FOR LINEARITY

The power spectrum factorization problem has been an important problem in signal processing applications. It states that any given power spectral density function, $C_2^t(\omega)$, can be decomposed exactly as a product of minimum phase and maximum phase components as follows [Orfanidis, 1985]
$$C_2^t(\omega) = T(\omega)T^*(\omega). \qquad (5.66)$$

TABLE 5.1 POLYCEPSTRA: NOTATIONS AND THE EQUATIONS TO FIND THE APPROPRIATE DEFINITION, EXPRESSIONS IN TERMS OF $A^{(m)}$ AND $B^{(m)}$, COMPUTATION

	Notation	Definition	In terms of $A^{(m)}, B^{(m)}$	Computation
Complex cepstrum	$c_h(m)$	(5.7)	(5.10)	(5.8), (5.12)
Differential cepstrum	$d_h(m)$	(5.15)	(5.16)	(5.21), (5.22)
Power cepstrum	$p_h(m)$	(5.24)	(5.25)	(5.31), (5.32)
Bicepstrum	$b_h(m,n)$	(5.37.1)	(5.38)	(5.44) and (5.38) or (5.47)
Tricepstrum	$t_h(m,n,\ell)$	(5.49)	(5.50)	(5.52.2) and (5.50)
Cepstrum of bicoherency	$bic_h(m,n)$	(5.58)	(5.59)	(5.62), (5.63)

Nikias and Raghuveer [1987] defined the bispectrum factorization as one of finding a LTI filter with frequency response function $H(\omega)$ to match a given bispectrum, $C_3^h(\omega_1, \omega_2)$, as follows:

$$C_3^h(\omega_1, \omega_2) = H(\omega_1)H(\omega_2)H^*(\omega_1 + \omega_2). \tag{5.67}$$

However, Tekalp and Erdem [1989] pointed out that a given bispectrum can not always be decomposed exactly to the form of (5.67). According to Tekalp and Erdem [1989], a necessary and sufficient condition for a bispectrum, $C_3^h(\omega_1, \omega_2)$, to be factorized in the form of (5.67) is that its bicepstrum

$$b_h(m,n) = F_2^{-1}\left[\ln\left[C_3^h(\omega_1,\omega_2)\right]\right] \tag{5.68}$$

satisfies (5.38). In other words, the bispectrum $C_3^h(\omega_1, \omega_2)$ is factorizable if and only if its bicepstrum, $b_h(m,n)$, takes the form

$$b_h(m,n) = \begin{cases} c_h(m), & n=0, m=n \\ c_h(n), & m=0 \\ c_h(-n), & m=n \\ 0, & \text{otherwise} \end{cases} \tag{5.69}$$

where $c_h(m)$ is a complex cepstrum sequence.

Based on the aforementioned bispectrum factorizability result, Erdem and Tekalp [1990] proposed the following index to quantify factorization:

$$\rho = \left\{ \frac{|b_h(0,0) - \ell n(\beta)|^2 + 3\sum_{m \neq 0} |b_h(m,0)|^2}{|b_h(0,0) - \ell n(\beta)|^2 + \sum_{(m,n) \neq (0,0)} |b_h(m,n)|^2} \right\}^{\frac{1}{2}}, \quad (5.70)$$

where

$$\beta = 1/a_o^3 \text{ and}$$

$$a_m = F_1^{-1}[\exp\{F_1[-b_h(m,0)]\}].$$

We note that $0 < \rho \leq 1$ (with $\rho = 1$ for a factorizable bispectrum). This index measures the fraction of energy that may be attributed to linear mechanisms. It is important to note that the bispectrum of linear non-Gaussian processes can always be factorized into the form of (5.67). As such, index ρ in (5.70) may also be used to distinguish between a linear non-Gaussian process ($\rho = 1$) and a nonlinear process ($\rho < 1$). This test for linearity is equivalent to that described in section 2.5.4, which is based on the bicoherence index.

The bispectrum factorization conditions and index ρ can be generalized to higher-order spectra directly.

5.8 INVERSE FILTER RECONSTRUCTION

One of the big advantages of polycepstra in practice is their ability to reconstruct the inverse filter for deconvolution in a very simple manner. Let $h(k)$ be the impulse response of the filter and $H(\omega)$ its Fourier transform. Assuming $H(\omega) \neq 0$, the inverse filter is defined

$$F(\omega) = 1/H(\omega) \quad (5.71)$$

with impulse response $f(k)$. If $A^{(m)}$ and $B^{(m)}$ are the cepstral parameters of $H(\omega)$, then $-A^{(m)}, -B^{(m)}$ are the cepstral parameters of $F(\omega)$. Consequently, using $A^{(m)}$ and $B^{(m)}$ of $H(\omega)$ we can reconstruct $f(k)$ from (5.20) as follows:

$$f_i(k) = -\sum_{m=2}^{k+1} [-A^{(m-1)}] f_i(k-m+1); \quad k \geq 1 \quad (5.72.1)$$

$$f_o(k) = \frac{1}{k} \sum_{m=k+1}^{0} [-B^{(1-m)}] f_o(k-m+1); \quad k \leq -1 \quad (5.72.2)$$

$$f_i(0) = f_o(0) = 1$$

and
$$f(k) = f_i(k) * f_o(k) \qquad (5.72.3)$$

This fundamental property of polycepstra is utilized in blind deconvolution (or equalization) where the system transfer function is identified first from output data and then its inverse filter is reconstructed [Hatzinakos and Nikias, 1991]. The inverse filter reconstruction is easily achieved by changing the sign of estimated cepstral parameters and using (5.72.1), (5.72.2), and (5.72.2).

Example 5.5

This is a digital communications example that illustrates the effectiveness of the tricepstrum for channel estimation and deconvolution (i.e., equalization). The channel is modeled as a FIR filter with coefficients and zero locations shown in Table 5.2. The transmitted sequence $X(k)$ (i.e., driving process) is a white binary sequence taking the values $(-1, +1)$ with equal probability. The additive Gaussian noise $W(k)$ is assumed to be zero-mean white or colored with autocorrelation function $c_2^w(\tau) = 0.3^{|\tau|}, \tau = 0, \pm 1, \pm 2, \ldots$. Figure 5.10 illustrates the channel model and the equalization filter. From this figure it is apparent that

$$Y(k) = \sum_{i=0}^{L} h(i) X(k-i) + W(k), \qquad (5.73)$$

where $Y(k)$ is the received sequence. The problem is to estimate $h(k)$ coefficients from $Y(k)$, and then to utilize $h(k)$ and $Y(k)$ to recover the transmitted sequence $X(k)$.

TABLE 5.2 NONMINIMUM PHASE IMPULSE RESPONSE COEFFICIENTS AND TRANSFER FUNCTION ZERO LOCATION FOR EXAMPLE 5.5

Impulse Response	Zero Location
$h(0) = 0.2197$	$z_1 = 1.8 + j0.5$
$h(1) = -0.747$	$z_2 = 1.8 - j0.5$
$h(2) = 0.6085$	$z_3 = -0.2$
$h(3) = 0.1533$	

The performance of the tricepstrum approach is demonstrated in terms of probability of error of the reconstructed input, assuming different SNR's and lengths of

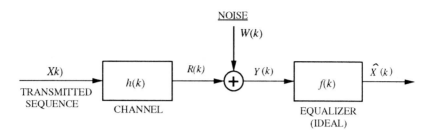

Figure 5.10 The communication channel (low-pass equivalent), the noise source and the equalizer.

data [Hatzinakos and Nikias, 1989]. By rewriting (5.73) we obtain

$$Y(k) = R(k) + W(k) \quad \text{(time domain)} \tag{5.74.1}$$

or, equivalently,

$$C_2^y(\omega) = C_2^r(\omega) + C_2^w(\omega) \quad \text{(power spectrum domain)}. \tag{5.74.2}$$

Since $W(k)$ is assumed to be independent of $R(k)$, the trispectrum domain relationship is given by

$$C_4^y(\omega_1,\omega_2,\omega_3) = C_4^r(\omega_1,\omega_2,\omega_3) + C_4^w(\omega_1,\omega_2,\omega_3) \quad \text{(trispectrum domain)} \tag{5.74.3}$$

where

$$C_4^r(\omega_1,\omega_2,\omega_3) = C_4^h(\omega_1,\omega_2,\omega_3)C_4^x(\omega_1,\omega_2,\omega_3).$$

Note that $C_4^w(\omega_1,\omega_2,\omega_3) = 0$ because $W(k)$ is Gaussian. In addition $C_4^x(\omega_1,\omega_2,\omega_3) = \gamma_4^x$ because $X(k)$ is white. Consequently,

$$C_4^y(\omega_1,\omega_2,\omega_3) = \gamma_4^w C_4^h(\omega_1,\omega_2,\omega_3); \quad \text{(trispectrum domain)} \tag{5.74.4}$$

i.e., the trispectrum of the received noisy sequence $Y(k)$ equals the trispectrum of the channel impulse response times a constant γ_4^x, which is the kurtosis of the transmitted signal.

Combining (5.49) and (5.74.4) we obtain

$$\begin{aligned} t_y(m,n,\ell) &= t_h(m,n,\ell) + C \cdot \delta(m,n,\ell) \quad \text{(tricepstrum domain)} \\ &=_h (m,n,\ell), \ (m,n,\ell) \neq (0,0,0); \end{aligned} \tag{5.74.5}$$

i.e., the tricepstrum of the received sequence is identical to the tricepstrum of the channel impulse response except for $(m, n, \ell) = (0, 0, 0)$.

Let us assume that we are given N samples of the zero-mean sequence $\{Y(k)\}$. Then the following steps are undertaken:

1. Estimate fourth-order cumulants $\hat{c}_4^y(\tau_1, \tau_2, \tau_3)$ utilizing

$$\begin{aligned}\hat{c}_4^y(\tau_1, \tau_2, \tau_3) &= \hat{m}_4^y(\tau_1, \tau_2, \tau_3) - \hat{m}_2^y(\tau_1)\hat{m}_2^y(\tau_2 - \tau_3) \\ &\quad - \hat{m}_2^y(\tau_2)\hat{m}_2^y(\tau_1 - \tau_3) - \hat{m}_2^y(\tau_3)\hat{m}_2^y(\tau_1 - \tau_2)\end{aligned} \quad (5.75)$$

and the estimation procedures described in Chapter 4.

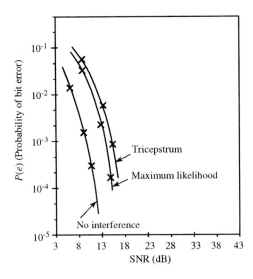

Figure 5.11 Probability of error performance for a binary signal and the channel described in Table 5.2. The noise is additive *white* Gaussian. ([Hatzinakos and Nikias, 1989] © 1989 IEEE)

2. Compute cepstral parameters $\hat{A}^{(\tau)}$ and $\hat{B}^{(\tau)}$ by solving a linear system of equations generated from (5.52.2), or by utilizing a 3-d FFT in the formula

$$mm_y^t(m,n,\ell) = F_3^{-1}\left\{\frac{F_3[m\cdot \hat{c}_4^y(m,n,\ell)]}{F_3[\hat{c}_4^y(m,n,\ell)]}\right\} \quad (5.76)$$

and then combine (5.76) and (5.50). $F_3[\cdot]$ and $F_3^{-1}[\cdot]$ denote forward and inverse 3-d Fourier transforms, respectively.

3. The estimated impulse response of the channel, $h(k)$, is computed from

$$\hat{i}(k) = -\frac{1}{k}\sum_{\tau=2}^{k+1}\hat{A}^{(\tau-1)}\hat{i}(k-\tau+1); \quad k \geq 1$$

$$\hat{o}(k) = \frac{1}{k}\sum_{m=k+1}^{0}\hat{B}^{(1-\tau)}\hat{o}(k-\tau+1); \quad k \leq -1$$

$$\hat{h}(k) = \hat{i}(k) * \hat{o}(k)$$

whereas the impulse response of the equalizer, $\hat{f}(k)$, follows from (5.72.1) – (5.72.3).

4. The reconstructed input sequence, $\hat{X}(k)$, is obtained from

$$\hat{X}(k) = \hat{f}(k) * Y(k)$$

and by passing $\hat{X}(k)$ through a detection device. The detection device assumes zero threshold; i.e., if $\hat{X}(k) > 0$ then $\hat{X}(k) = +1$ and if $\hat{X}(k) < 0$, $\hat{X}(k) = -1$.

Table 5.3 illustrates the estimated channel impulse response $h(k)$, for different lengths of received data and different lengths of cepstral parameters. As expected, the reconstruction accuracy improves with the increase of both lengths. Figure 5.11 illustrates the probability of error ($P(e)$) performance of the tricepstrum method and the maximum likelihood (ML) method (which assumes *exactly* known impulse response $f(k)$ for the equalizer) for white Gaussian noise. From Figure 5.11 it appears that the $P(e)$ performance achieved by the tricepstrum method is close to that of the ML. Figure 5.12 illustrates $P(e)$ performance of these methods for colored Gaussian noise. However, in this case, the ML method assumes that the noise is white (mismatch condition). We see that the performance of the tricepstrum method approaches that of the ML-mismatch for this example [Hatzinakos and Nikias, 1989]. □

TABLE 5.3 COMPUTED IMPULSE RESPONSE COEFFICIENTS BY THE TRICEPSTRUM METHOD SNR = ∞

Length of tricepstrum parameters	$p = q = 4$			$p = q = 7$			$p = q = 15$		
Length of time data	128	256	512	128	256	512	128	256	512
$\hat{h}(0)$	0.2	0.181	0.257	0.146	0.185	0.23	0.244	0.19	0.22
$\hat{h}(1)$	−0.747	−0.747	−0.747	−0.747	−0.747	−0.747	−0.747	−0.747	−0.747
$\hat{h}(2)$	0.834	0.621	0.58	0.86	0.625	0.63	0.75	0.633	0.606
$\hat{h}(3)$	−0.016	0.104	0.174	−0.048	0.115	0.149	0.045	0.121	0.152

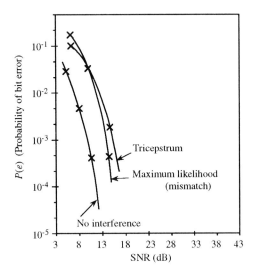

Figure 5.12 The probability of error performance for a binary signal and the channel of Table 5.2. The noise is additive *colored* Gaussian. The maximum likelihood (ML) method assumes white Gaussian noise (mismatch situation). ([Hatzinakos and Nikias, 1989] © 1989 IEEE)

Example 5.6

Consider an underwater multipath channel response of the general form

$$h(k) = \sum_{i=1}^{M} A_i \delta(k - D_i)$$

where $\delta(k)$ is the Kronecker delta function. Let the received signal at a hydrophone sensor be

$$Y(k) = h(k) * x(k) + W(k) = \sum_{i=0}^{M} A_i x(k - D_i) + W(k) \quad \text{(time domain)}$$

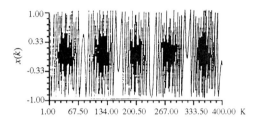

Figure 5.13 The FM signal

where $x(k)$ is a FM signal and $W(k)$ is colored Gaussian noise with autocorrelation $c_2^w(\tau)$. Assuming that $x(k)$ is known, the problem is to estimate the scaling amplitudes $\{A_i\}$ and time delays $\{D_i\}$ from a finite length record of $\{Y(k)\}$. In the bispectrum domain it holds that

$$M_3^y(\omega_1, \omega_2) = M_3^h(\omega_1, \omega_2) \cdot M_3^x(\omega_1, \omega_2). \tag{5.77}$$

Note that the bispectrum of the colored Gaussian noise is identically zero. Since the FM signal $x(k)$ i=has pronounced resonances, the third-order cumulant sequence of $Y(k)$ should be prefiltered by [Petropulu and Nikias, 1988]

$$m_3^u(\tau, \rho) = m_3^e(\tau, \rho) * m_3^y(\tau, \rho) \tag{5.78.1}$$

or

$$M_3^u(\omega_1, \omega_2) = M_3^e(\omega_1, \omega_2) M_3^y(\omega_1, \omega_2), \tag{5.78.2}$$

where

$$M_3^e(\omega_1, \omega_2) = \frac{[M_3^x(\omega_1, \omega_2) * M_3^z(\omega_1, \omega_2)]}{M_3^x(\omega_1, \omega_2)}; \tag{5.79}$$

to obtain a new third-order cumulant sequence $c_3^u(\tau, \rho)$. Note that $M_3^z(\omega_1, \omega_2)$ is the bispectrum of a "window" signal $z(k) = a^k, 0 < a < 1$. In other words, $m_3^u(\tau, \rho)$ is the third-order moment sequence of the signal

$$\begin{aligned} u(k) &= [z(k)x(k)] * h(k) \\ &= f(k) * h(k). \end{aligned} \tag{5.80}$$

In summary, a bicepstrum-based high-resolution time delay estimation method proceeds as follows [Petropulu and Nikias, 1988]:

1. Since $x(k)$ is known, generate $f(k) = a^k x(k)$ and compute its third-order moment sequence $m_3^f(\tau, \rho)$; using $m_3^f(\tau, \rho)$ in (5.47) compute its bicepstrum $b_f(m, n)$.

2. From the observed signal $Y(k)$, estimate its third-order cumulant sequence $\hat{m}_3^y(\tau, \rho)$ and then compute $\hat{m}_3^u(\tau, \rho)$ utilizing (5.78.1).

3. Using $\hat{c}_3^u(\tau, \rho)$ in (5.47) compute the bicepstrum $\hat{b}_u(m, n)$. Clearly, the bicepstrum of the underwater multipath channel is given by $\hat{b}_h(m, n) = \hat{b}_u(m, n) - \hat{b}_f(m, n)$.

4. Reconstruct $h(k)$ from $\hat{b}_h(m, n)$ utilizing (5.38), (5.20.1), and (5.20.2). The channel amplitudes and time delays follow from $h(k)$. Note that no data segmentation should be performed here for the computation of third-order statistics because $x(k)$ is a deterministic signal.

Simulation examples are shown for the case of $h(k) = \delta(k) + 0.58(k-3)$, $x(k) = \cos[2\pi(6.28 \cdot 10^{-3})k^2]$ (see Figure 5.13) and $c_2^w(\tau) = (0.8)^\tau \cdot (0.56\tau)$. Note that in this case $D_3 = 3$ which is much less than the length of the FM signal's autocorrelation.

Table 5.4 illustrates Monte-Carlo simulation results (50 runs) when SNR=10 dB, $z(k) = \exp\{-0.15k\}$, and the length of $Y(k)$ is 4096. In particular, the bias and variance of the estimated channel amplitudes A_0, A_1, A_2, and A_3 have been computed utilizing the bicepstrum method.

5.9 THE CROSS-BICEPSTRUM OF DETERMINISTIC AND STOCHASTIC SIGNALS

5.9.1 Preliminaries

Just as we can generalize from the power spectrum to second-order cross-spectra calculated between two distinct (in general) signals, we can similarly generalize from the usual third- (or higher-) order "auto"-spectra to third- (or higher-) order cross-spectra, where now the cross-spectrum is taken among three (or more) generally

TABLE 5.4 THE BICEPSTRUM METHOD FOR HIGH-RESOLUTION TIME DELAY ESTIMATION. BIAS AND VARIANCE OF ESTIMATED AMPLITUDES OF MULTIPATH CHANNEL ARE SHOWN. THE ADDITIVE NOISE IS COLORED.

	Bicepstrum	
	bias	variance
A_i	$\times 10^{-3}$	$\times 10^{-3}$
A_0	2.03	0.215
A_1	6.14	1.60
A_2	5.57	1.41
A_3	0.90	0.04

distinct signals. Third-order cross-spectra (called "cross-bispectra") have already been seen through limited applications, primarily with problems of time-delay and bearing estimation [Nikias and Raghuveer, 1987].

The cross-bispectrum of three signals is defined as the triple product

$$C_{x_1x_2x_3}(\omega_1,\omega_2) = \gamma_3^w H_1(\omega_1)H_2(\omega_2)H_3^*(\omega_1+\omega_2) \qquad (5.81)$$

where $H_i(\omega)$ is the Fourier transform of the deterministic signal $h_i(k)$, $i = 1, 2, 3$. This definition holds for linear non-Gaussian random processes, $X_i(k)$, which are generated by three systems represented by transfer functions $H_i(\omega)$ (or impulse responses $h_i(k), i = 1, 2, 3$) [Brooks and Nikias, 1991]. The three systems are driven by the same zero-mean, non-Gaussian i.i.d. noise process $W(k)$ with skewness γ_3^w; i.e.,

$$X_i(k) = h_i(k) * W(k), i = 1, 2, 3. \qquad (5.82)$$

The cross-bispectrum $C_{x_1x_2x_3}(\omega_1,\omega_2)$ is the 2-d Fourier transform of the cross-third order cumulant sequence $c_{x_1x_2x_3}(\tau,\rho)$, where

$$c_{x_1x_2x_3}(\tau,\rho) = E[X_1(k+\tau)X_2(k+\rho)X_3(k)]. \qquad (5.83)$$

On the other hand, the cross-bispectrum of three deterministic energy signals is defined by

$$M_{h_1h_2h_3}(\omega_1,\omega_2) = H_1(\omega_1)H_2(\omega_2)H_3^*(\omega_1+\omega_2) \qquad (5.84)$$

which is the 2-d Fourier transform of the cross-third-order moment sequence $m_{h_1h_1h_3}(\tau,\rho)$ where

$$m_{h_1h_1h_3}(\tau,\rho) = \sum_{k=-\infty}^{+\infty} h_1(k+\tau)h_2(k+\rho)h_3(k). \tag{5.85}$$

Comparing (5.84) with (5.81), we observe that $C_{x_1x_2x_3}(\omega_1,\omega_2) = M_{h_1h_2h_3}(\omega_1,\omega_2)$ if $\gamma_3^w = 1$. Clearly, the *form* of the cross-bispectrum is the same in both cases within a magnitude factor, and this establishes the duality between the problem of reconstructing three deterministic signals from their cross-bispectrum and that of identifying three independent nonminimum phase systems simultaneously from the cross-bispectrum of their outputs. Brooks and Nikias [1992] point out that we can, in fact, achieve the same result in a more general problem with different inputs $w_i(k)$ to each system, if the inputs have a nonzero cross-third-order moment only at the zeroth lag, i.e.,

$$E[W_1(k+\tau)W_2(k+\rho)W_3(k)] = \gamma_3^w \delta(\tau,\rho).$$

Cross-third-order moments do not preserve all of the symmetry properties of the third-order moment sequence, and consequently the cross-bispectrum does not preserve all the symmetries of the bispectrum. However, the following symmetric relationships are not hard to establish [Brooks and Nikias, 1992]:

$$\begin{aligned} c_{x_1x_2x_3}(\tau,\rho) &= c_{x_2x_1x_3}(\rho,\tau) \\ &= c_{x_3x_2x_1}(-\tau,\rho-\tau) = c_{x_2x_3x_1}(\rho-\tau,-\tau) \\ &= c_{x_1x_3x_2}(\tau-\rho,-\rho) = c_{x_3x_1x_2}(-\rho,\tau-\rho) \end{aligned} \tag{5.86}$$

and

$$\begin{aligned} C_{x_1x_2x_3}(z_1,z_2) &= C_{x_2x_1x_3}(z_2,z_1) \\ &= C_{x_3x_2x_1}\left(z_1^{-1}\cdot z_2^{-1}, z_2\right) = C_{x_2x_3x_1}\left(z_2, z_1^{-1}\cdot z_2^{-1}\right) \\ &= C_{x_3x_1x_2}\left(z_1^{-1}\cdot z_2^{-1}, z_1\right) = C_{x_1x_3x_2}\left(z_1, z_1^{-1}\cdot z_2^{-1}\right). \end{aligned} \tag{5.87}$$

Also, for real signals

$$C_{x_1x_2x_3}(z_1,z_2) = C^*_{x_1x_2x_3}\left(z_1^{-1}, z_2^{-1}\right). \tag{5.88}$$

5.9.2 Signal Assumptions

We assume that each signal has no zeros on the unit circle and that its z-transform can be written in terms of its poles and zeros as

$$H_i(z) = A_i \cdot z^{-r_i} \cdot I_i(z^{-1}) O_i(z), \tag{5.89}$$

where the A_i are gain constants, the r_i are integer linear factors,

$$I_i(z^{-1}) = \frac{\prod_{j=1}^{L_i 1}(1 - a_{ij}z^{-1})}{\prod_{j=1}^{L_i 3}(1 - c_{ij}z^{-1})} \tag{5.90}$$

is the minimum phase component, and

$$O_i(z) = \Pi_{j=1}^{L_i 2}(1 - b_{ij}z) \tag{5.91}$$

is the maximum phase component, with zeros a_{ij} and poles c_{ij} inside the unit circle and zeros b_{ij} outside the unit circle (i.e., $|a_{ij}| < 1, |b_{ij}| < 1$, and $|c_{ij}| < 1$). Maximum phase poles could be included as well if desired and are omitted here only for simplicity. Thus we make no inherent assumptions that the signal is causal or of finite extent.

5.9.3 Definition

Substituting (5.89) into (5.81) we get

$$\begin{aligned} C_{x_1 x_2 x_3}(z_1, z_2) &= \left(\gamma_3^w \prod_{i=1}^{3} A_i z^{-r_i}\right) I_1(z_1^{-1}) O_1(z_1) I_2(z_2^{-1}) O_2(z_2) \\ &\cdot I_3(z_1 \cdot z_2) O_3(z_1^{-1} \cdot z_2^{-1}). \end{aligned} \tag{5.92}$$

Since we are unable to separate the contributions of the individual signals to the magnitude and linear phase factors in the first parentheses on the right hand side of (5.92), we will assume without further loss of generality that $\gamma_3^w = 1$, $A_i > 0$ and $r_i = 0, i = 1, 2, 3$. Now with the conditions above, we can write

$$CB_{123}(z_1, z_2) = \ln\{C_{x_1 x_2 x_3}(z_1, z_2)\} \tag{5.93}$$

and define the cross-bicepstrum as

$$cb_{123}(m, n) = Z_2^{-1}[CB_{123}(z_1, z_2)]. \tag{5.94.1}$$

Substituting (5.92) and (5.93) into (5.94) we get [Brooks and Nikias 1992]

$$cb_{123}(m,n) = \begin{cases} \ln A, & m = n = 0 \\ -\frac{1}{m}A_1^{(m)}, & m > 0, n = 0 \\ \frac{1}{m}B_1^{(-m)}, & m < 0, n = 0 \\ -\frac{1}{n}A_2^{(n)}, & m = 0, n > 0 \\ \frac{1}{n}B_2^{(-n)}, & m = 0, n < 0 \\ \frac{1}{n}A_3^{(-n)}, & m = n < 0 \\ -\frac{1}{n}B_3^{(n)}, & m = n > 0 \\ 0, & \text{otherwise} \end{cases} \qquad (5.94.2)$$

with

$$A_i^{(\tau)} = \sum_{j=1}^{L_{i1}}(a_{ij})^\tau - \sum_{j=1}^{L_{i3}}(c_{ij})^\tau$$

$$B_i^{(\tau)} = \sum_{j=1}^{L_{i2}}(b_{ij})^\tau. \qquad (5.94.3)$$

We can see that the result in (5.94.2) and (5.94.3) means that the cross-bicepstrum is non-zero on three lines only in the (m,n) plane and that on each of these lines we find the complex cepstrum of *one* of the three original signals. On the m-axis we find the complex cepstrum of $h_1(k)$, on the n-axis that of $h_2(k)$, and on the diagonal $(m = n)$ that of $h_3(k)$, reversed with respect to the cepstrum domain variable n.

5.9.4 Properties

The bicepstrum in section 5.5 can be as a special case of the cross-bicepstrum when the three signals are the same. Thus, the cross-bicepstrum inherits many of the properties of the bicepstrum. Brooks and Nikias [1992] point to the following additional properties:

1. If $c_{h_1}(m)$ is the complex cepstrum of $h_i(k)$, then

$$c_{h_1}(m) = cb_{123}(m, 0)$$
$$c_{h_2}(m) = cb_{123}(0, m)$$
$$c_{h_3}(m) = cb_{123}(-m, -m).$$

2. If $d_{h_i}(m)$ is differential cepstrum of $h_i(k)$, then

$$d_{h_1}(m) = -(m-1)cb_{123}(m-1, 0), \ m \neq 1$$
$$d_{h_2}(m) = -(m-1)cb_{123}(0, m-1), \ m \neq 1$$
$$d_{h_3}(m) = -(m-1)cb_{123}(1-m, 1-m), \ m \neq 1.$$

3. We can view the calculation of the cross-bicepstrum as a means of calculating the complex cepstrum of three signals *simultaneously*. We show in the next section that this goal can be achieved *without* resorting to phase unwrapping or other assumptions.

4. The blindness of the cross-bicepstrum to linear phase differences between the three signals could be an important advantage in detection or feature extraction applications where unknown linear phase shifts can present difficulties.

The extension of the method to the cross-tricepstrum is straightforward.

5.9.5 Computation

Taking derivatives of (5.94.1) separately with respect to z_1 and to z_2, after some algebra we obtain

$$C_{x_1 x_2 x_3}(z_1, z_2) \cdot z_i \cdot \frac{\partial C B_{123}(z_1, z_2)}{\partial z_i} = z_i \frac{\partial C_{x_1 x_2 x_3}(z_1, z_2)}{\partial z_i} \qquad (5.95)$$

for $i = 1, 2$. Taking inverse 2-d z-transforms, we arrive at a pair of 2-d convolution equations

$$C_{x_1 x_2 x_3}(m, n) * [-m \cdot cb_{123}(m, n)] = -m \cdot C_{x_1 x_2 x_3}(m, n)$$

$$C_{x_1 x_2 x_3}(m, n) * [-n \cdot cb_{123}(m, n)] = -n \cdot C_{x_1 x_2 x_3}(m, n)$$

or, equivalently,

$$\sum_{\tau=-\infty}^{+\infty} \sum_{\rho=-\infty}^{+\infty} (-\tau) \cdot cb_{123}(\tau,\rho) c_{x_1 x_2 x_3}(m-\tau, n-\rho) = -m \cdot c_{x_1 x_2 x_3}(m,n)$$

$$\sum_{\tau=-\infty}^{+\infty} \sum_{\rho=-\infty}^{+\infty} (-\rho) \cdot cb_{123}(\tau,\rho) c_{x_1 x_2 x_3}(m-\tau, n-\rho) = -n \cdot c_{x_1 x_2 x_3}(m,n) \quad (5.96)$$

Substituting (5.94.2) into (5.96) we obtain a direct relationship between the cross-third-order cumulants and the cepstral parameters $A_i^{(m)}$ and $B_i^{(m)}$, $i=1,2,3$, viz:

$$\sum_{\tau=1}^{\infty} \left\{ A_1^{(\tau)} c_{x_1 x_2 x_3}(m-\tau, n) - B_1^{(\tau)} c_{x_1 x_2 x_3}(m+\tau, n) - A_3^{(\tau)} c_{x_1 x_2 x_3}(m+\tau, n+\tau) \right.$$

$$\left. + B_3^{(\tau)} c_{x_1 x_2 x_3}(m-\tau, n-\tau) \right\} = -m \cdot c_{x_1 x_2 x_3}(m,n).$$

$$\sum_{\tau=1}^{\infty} \left\{ A_2^{(\tau)} c_{x_1 x_2 x_3}(m, n-\tau) - B_2^{(\tau)} c_{x_1 x_2 x_3}(m, n+\tau) - A_3^{(\tau)} c_{x_1 x_2 x_3}(m+\tau, n+\tau) \right.$$

$$\left. + B_3^{(\tau)} c_{x_1 x_2 x_3}(m-\tau, n-\tau) \right\} = -n \cdot c_{x_1 x_2 x_3}(m,n). \quad (5.97)$$

From (5.94.3) and the fact that $|a_{ij}| < 1, |b_{ij}| < 1$ and $|c_{ij}| < 1$, the sums in (5.97) can be truncated as closely as desired. Assume that we truncate $A_i^{(\tau)}$ to p_i terms and $B_i^{(\tau)}$ to q_i terms, where p_i and q_i are chosen as in (5.45.1) according to the proximity of the poles and zeroes of $h_i(k)$ to the unit circle. Choose points (m_1, n_1) and (m_2, n_2) in the cross-third-order moment plane. Then we can rewrite (5.97) as the pair of vector dot products [Brooks and Nikias, 1992].

$$\mathbf{R}_i^T \mathbf{c}_i = \mathbf{r}_i, \quad (5.98)$$

$i = 1, 2$, where $\mathbf{R}_i$ is a $(w_i \times 1)$ vector of cross-third-order moments shifted from the reference (m_i, n_i), $r_1 = -m \cdot c_{x_1 x_2 x_3}(m_1, n_1)$, $r_2 = -n \cdot c_{x_1 x_2 x_3}(m_2, n_2)$. $\mathbf{c}$ is a $(w_i \times 1)$ vector of unknown cepstral parameters corresponding to the ith and the 3rd signals, and $w_i = p_i + q_i + p_3 + q_3$. Writing $M_i > w_i$ such equations at appropriately chosen points (m_i, n_i) in the third-order moment plane, we can find least-squares or SVD estimates of the $\mathbf{c}_i$. *Note that in contrast to the bicepstrum method in (5.46.2), here the overdetermined matrices consist not of differences of third-order moments but of cross-third-order moments themselves.*

Returning to (5.96) and taking Fourier transforms we get [Brooks and Nikias, 1991]

$$m \cdot cb_{123}(m,n) = F_2^{-1} \left\{ \frac{F_2[m \cdot c_{x_1 x_2 x_3}(m,n)]}{C_{x_1 x_2 x_3}(\omega_1, \omega_2)} \right\}$$

and

$$n \cdot cb_{123}(m,n) = F_2^{-1}\left\{\frac{F_2\left[n \cdot c_{x_1 x_2 x_3}(m,n)\right]}{C_{x_1 x_2 x_3}(\omega_1,\omega_2)}\right\} \qquad (5.99)$$

which provides an alternative way of computing the cross-bicepstrum $cb_{123}(m,n)$. The advantage of this method is reduced computational complexity due to the growing size of the system of equations necessary as w_i gets large (i.e., many poles and zeros close to the unit circle).

Once we have computed the A's and B's for each signal we can separately reconstruct the minimum and maximum phase factors of each signal via (5.20.1) and (5.20.2).

In practice the cross-bicepstrum method is subject to two types of errors: statistical error resulting from the substitution of third-order cumulant estimates computed from (possibly noisy) finite length data records for the true third-order cumulants in (5.98) or (5.99), and approximation error resulting from truncation of the cepstral parameters (and error of the least-squares (or SVD) solution which results from this truncation).

Example 5.7

Here a system identification problem is considered where the observed signals $X_i(k)$ are the outputs of three different LTI systems $h_i(k)$ excited by the same non-Gaussian white noise input. The input noise process was generated to be an exponentially distributed, zero-mean white process, with skewness $\gamma_3^w = 1$. It is assumed that we have data of length $N = K \times L$ where K and L are integers. We will use the notation $K \times L$ to describe the partition of a particular data set into K segments of length L each.

The cross-bicepstrum method proceeds as follows:

1. Segment the data into K blocks of length L each. For each block, estimate the cross-third-order cumulants of the three outputs, then average over the K blocks.

2. Form (5.98) and solve it via least-squares (or SVD) to obtain cepstral parameters $A_i^{(m)}$ and $B_i^{(m)}, i = 1, 2, 3$.

3. Generate $h_i(k)$ utilizing (5.20.1), (5.20.2), and (5.4.3).

4. From the estimated $h_i(k)$ calculate the magnitude and phase of the responses $H_i(\omega)$.

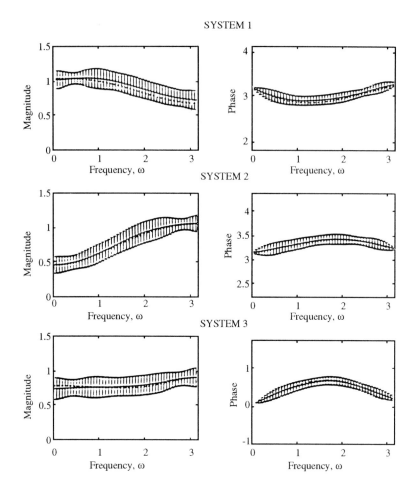

Figure 5.14 Nonminimum phase ARMA systems simultaneously identified via the cross-bicepstrum method. The length of output data is 16 × 32..

In this example, 100 Monte-Carlo realizations were performed and the mean and standard deviation of the resulting estimates of the magnitude and phase of each frequency response $H_i(\omega)$ were computed after normalizing magnitudes and correcting for any linear phase factors. This example uses the following three mixed-phase ARMA (1,3) systems [Brooks and Nikias, 1990]:

$$H_1(z) = \frac{(z+5)(z-4)(z+0.1)}{(z-0.2)}$$

$$H_2(z) = \frac{(z+5)(z-4)(z-0.2)}{(z+0.1)}$$

$$H_3(z) = \frac{(z+2(1+3\sqrt{3}))(z+2(1-\sqrt{3}))(z-0.1)}{z+0.2}$$

The values $p_i = q_i = 4$ were chosen for this example. Figures 5.14 and 5.15 show results for 16×32 and 128×32 data respectively. We see good estimates, with improved variance as the data length increased. □

5.10 ANALYTIC PERFORMANCE EVALUATION OF THE COMPLEX CEPSTRUM AND BICEPSTRUM

5.10.1 Preliminaries and Problem Definition

Let signal $z(k)$ be given by

$$Z(k) = \alpha(k) + W(k) \text{ (time domain)},$$

$$k = 0, 1, \cdots, N, \tag{5.100}$$

where $\alpha(k)$ is a deterministic energy signal and $W(k)$ is a stationary Gaussian white noise process with mean zero and variance σ^2. In the frequency domain (5.100) becomes

$$Z(\omega) = A(\omega) + W(\omega), \text{ (Fourier domain)} \tag{5.101}$$

where $Z(\omega)$, $A(\omega)$, and $W(\omega)$ are the Fourier transforms of $z(k)$, $\alpha(k)$, and $W(k)$, respectively, taken over finite length of size N. The Fourier transform is defined as where $\omega = \frac{2\pi}{N}\lambda, \lambda = 0, 1, \cdots, N-1$ are discrete frequencies. *In all equations that follow, frequency ω is assumed discrete.*

The *complex cepstrum* of $Z(k), c_z^{(n)}$ is defined by

$$Z(\omega) = \sum_{k=0}^{N-1} Z(k) \exp\{-j\omega k\} \tag{5.102}$$

$$\begin{aligned} c_z(n) &= F_1^{-1}\{\log(Z(\omega))\} \tag{5.103} \\ &= F_1^{-1}\{\log(A(\omega))\} + F_1^{-1}\left\{\log\left(1 + \frac{W(\omega)}{A(\omega)}\right)\right\} \\ &= c_\alpha(n) + c_w(n) \text{ (cepstrum domain)}, \end{aligned}$$

where $c_\alpha(n)$ is the complex cepstrum of the deterministic signal $\alpha(n)$ and $c_w(n)$ is the *error contribution* due to the presence of the additive noise $W(k)$. On the other

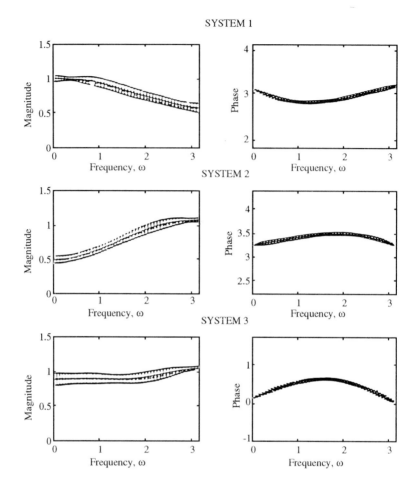

Figure 5.15 Nonminimum phase ARMA systems simultaneously identified via the cross-bicepstrum method. The length of output data is 128 × 32.

hand, the *bicepstrum* of $Z(k)$, $b_z(m,n)$, is given by

$$b_z(m,n) = F_2^{-1}\{\log(C_3^z(\omega_1,\omega_2))\} \tag{5.104}$$

$$= F_2^{-1}\{\log(C_3^\alpha(\omega_1,\omega_2))\} + F_2^{-1}\left(1 + \frac{C_3^e(\omega_1,\omega_2)}{C_3^\alpha(\omega_1,\omega_2)}\right)$$

$$= b_\alpha(m,n) + b_e(m,n) \quad \text{(bicepstrum domain)},$$

where $C_3^z(\omega_1,\omega_2)$ and $C_3^\alpha(\omega_1,\omega_2)$ are the bispectra of $z(k)$ and $\alpha(k)$ respectively, and $C_3^e(\omega_1,\omega_2)$ is defined by

$$C_3^e(\omega_1,\omega_2) = C_{\alpha w}(\omega_1,\omega_2) + C_2^w(\omega_1,\omega_2). \tag{5.105.1}$$

Thus $C_3^e(\omega_1,\omega_2)$ consists of the noise bispectrum $C_3^w(\omega_1,\omega_2)$ and $C_{\alpha w}(\omega_1,\omega_2)$; i.e., all cross-bispectral terms between signal and noise. The $C_{\alpha w}(\omega_1,\omega_2)$ is defined as follows:

$$\begin{aligned} C_{\alpha w}(\omega_1,\omega_2) &= A(\omega_1)A(\omega_2)E\{W^*(\omega_1+\omega_2)\} + A(\omega_1)A^*(\omega_1+\omega_2)E\{W(\omega_2)\} \\ &+ A(\omega_2)A^*(\omega_1+\omega_2)E\{W(\omega_1)\} + A(\omega_1)E\{W^*(\omega_1+\omega_2)W(\omega_2)\} \\ &+ A(\omega_2)E\{W^*(\omega_1+\omega_2)W(\omega_1)\} + A^*(\omega_1+\omega_2)E\{W(\omega_1)W(\omega_2)\} \end{aligned}$$
(5.105.2)

where $E\{\cdot\}$ denotes expectation operation.

Since $c_\alpha(n) = b_\alpha(m,0)$, the *error contribution* in the bicepstrum due to additive noise $W(k)$ is given by

$$c_e(n) = b_e(n,0), \tag{5.105.3}$$

which is different from $c_w(n)$. Petropulu and Nikias [1990] derived approximate analytic expressions for the mean and variance of errors $c_w(n)$ and $c_e(n)$. These expressions are presented in the two sections that follow.

5.10.2 The Complex Cepstrum Evaluation

From (5.103), the error due to additive white Gaussian noise in the frequency domain is given by

$$\begin{aligned} CW(\omega) &= F_1\{c_w(m)\} \\ &= \frac{1}{2}\ln\left\{\frac{(A_R(\omega)+W_R(\omega))^2 + (A_I(\omega)+W_I(\omega))^2}{|A(\omega)|^2}\right\} \\ &+ j\left\{\tan^{-1}\left(\frac{A_I(\omega)+W_I(\omega)}{A_R(\omega)+W_R(\omega)}\right) - \tan^{-1}\left(\frac{A_I(\omega)}{A_R(\omega)}\right)\right\} \\ &= CW_R(\omega) + jCW_I(\omega), \end{aligned}$$
(5.106)

where $A(\omega) = A_R(\omega) + jA_I(\omega)$. Note that the subscripts R, I stand for real and imaginary parts, respectively.

If we define the first term of $CW_I(\omega)$ in (5.106) as

$$Q(\omega) = \tan^{-1}\left[\frac{A_I(\omega)+W_I(\omega)}{A_R(\omega)+W_R(\omega)}\right], \tag{5.107}$$

then

$$E\{CW(\omega)\} = E\{CW_R(\omega)\} + jE\{Q(\omega)\} - j\tan^{-1}\left[\frac{A_I(\omega)}{A_R(\omega)}\right] \tag{5.108.1}$$

and
$$E\{c_w(m)\} = F_I^{-1}[E\{CW(\omega)\}]. \tag{5.108.2}$$

It can be shown that

$$E\{CW_R(\omega)\} = \frac{1}{2}E\left\{\log\left[1 + \frac{|W(\omega)|^2}{|A(\omega)|^2}\right]\right\}$$

$$-\sum_{k=1}^{\infty}\frac{2^{2k-1}}{k}E\left\{\left(\frac{A_R(\omega)W_R(\omega) + A_I(\omega)W_I(\omega)}{|A(\omega)|^2 + |W(\omega)|^2}\right)^{2k}\right\}, \tag{5.109.1}$$

where

$$E\left\{\log\left[1 + \frac{|W(\omega)|^2}{|A(\omega)|^2}\right]\right\} = -e^{\beta^2}Ei(\beta^2) \tag{5.109.2}$$

and

$$\beta^2 = \frac{|A(\omega)|^2}{\sigma^2 N}. \tag{5.109.3}$$

$Ei(\cdot)$ is the exponential integral function [Gradsteyn and Ryzhik, 1980; p.926]. Although β is a function of ω, it will be referred to as β for simplicity. Also,

$$E\left\{\left(\frac{A_R(\omega)W_R(\omega) + A_I(\omega)W_I(\omega)}{|A(\omega)|^2 + |W(\omega)|^2}\right)^{2k}\right\}$$

$$= \frac{2c}{\pi}I(k, 2k)\sum_{i=0,\text{even}}^{2k}\binom{2k}{i}A_R^i A_I^{2k-i}B\left(\frac{2k-i+1}{2}, \frac{i+1}{2}\right), \tag{5.109.4}$$

where [Gradsteyn and Ryzhik, 1980; p. 311]

$$I(1,0) = 1$$

$$I(0,1) = -e^{\beta^2}Ei(\beta^2)$$

$$I(0,2) = ce^{\beta^2}Ei(-\beta^2) + \frac{1}{|A(\omega)|^2}$$

$$I(0,n) = \frac{(-c)^{n-1}}{(n-1)!}\sum_{k=1}^{n-1}(k-1)!(-\beta^2)^{-k} - \frac{(-c)^{n-1}}{(n-1)!}e^{\beta^2}Ei(-\beta^2), n > 2 \tag{5.109.5}$$

$$I(i,1) = (-1)^{i-1}|A(\omega)|^{2i}e^{\beta^2}Ei(-\beta^2) + \sum_{m=1}^{i}(m-1)!(-|A(\omega)|^2)^{i-m}c^{-m}$$

and for $k \neq 1$

$$I(i,k) = \frac{1}{(k-1)}I(i-1, k-1) - \frac{c}{k-1}I(i, k-1) \tag{5.109.6}$$

$$c = \frac{1}{\sigma^2 N}, \tag{5.109.7}$$

and $B(m,n)$ is the beta function [Gradsteyn and Ryzhik, 1980; p. 948.].
Finally,

$$E\{Q(\omega)\} = -2\Psi \frac{\beta}{\sqrt{\pi}} \int_0^1 e^{-\beta^2 x^2} \Phi\left(b\sqrt{1-x^2}\right) dx$$

$$+ \sqrt{\pi}\beta u_1 \int_0^{\sin(|\Psi|)} e^{-\beta^2 x^2} \Phi\left(\beta\sqrt{1-x^2}\right) dx \qquad (5.110.1)$$

where

$$u_1 = \begin{cases} 1, & \text{if } \frac{\pi}{2} \geq \Psi \geq 0 \\ -1, & \text{if } 0 > \Psi \geq -\frac{\pi}{2} \end{cases}$$

$$\Phi(\lambda) = \frac{2}{\sqrt{\pi}} \int_0^{\lambda} e^{-t^2} dt, \text{ and} \qquad (5.110.2)$$

$$\Psi = \tan^{-1} \frac{A_R(\omega)}{A_I(\omega)}. \qquad (5.110.3)$$

To calculate the *second-order moment* of $CW(\omega)$ we need to evaluate the term $E\{CW(\omega_1)\,CW^*(\omega_2)\}$ in the following cases:

(I) for $\omega_1 \neq \omega_2$

$$E\{CW(\omega_1)CW^*(\omega_2)\} = E\{CW(\omega_1)\}E\{CW^*(\omega_2)\} \qquad (5.111)$$

where both product terms have been calculated in (5.108).

(II) for $\omega_1 = \omega_2 = \omega$

$$E\{CW(\omega_1)CW^*(\omega_2)\} = E\{|CW(\omega)|^2\} = E\{CW_R^2(\omega)\} + E\{CW_I^2(\omega)\}. \qquad (5.112)$$

However,

$$E\{CW_R^2(\omega)\} \cong \frac{1}{8}\beta^2 e^{\beta^2} \left[\frac{\partial^2}{\partial \nu^2}\{\beta^{-2\nu}\Gamma(\nu,\beta^2)\}|_{\nu=1}\right]$$

$$-\frac{\beta^2}{2}\left[e^{\beta^2}(1+\beta^2)Ei(-\beta^2)+1\right] \qquad (5.112.2)$$

where $\Gamma(\nu,\mu)$ is the incomplete gamma function [Gradsteyn and Ryzhik, p 940]. On the other hand,

$$E\{CW_I^2(\omega)\} = \frac{\beta}{\sqrt{\pi}} \int_0^{\sin(|\Psi|)} \left(\sin^{-1}\left(\sqrt{1-x^2}\right)\right)^2 e^{-\beta^2 x^2} \Phi\left(\beta\sqrt{1-x^2}\right) dx$$

$$-2\pi \int_0^{\sin(|\Psi|)} \sin^{-1}\left(\sqrt{1-x^2}\right) e^{-\beta^2 x^2} \Phi\left(\beta\sqrt{1-x^2}\right) dx$$

$$+ \left(\pi^2 + 2\Psi^2 - 2\pi u_1 \Psi\right) \int_0^{\sin(|\Psi|)} e^{-\beta x^2} \Phi\left(\beta\sqrt{1-x^2}\right) dx$$

$$+ \frac{\pi^2}{12} e^{-\beta^2} + \left[\left(u_1 \frac{\pi}{2}\right) - \Psi\right]^2 - 2\left[\left(u_1 \frac{\pi}{2}\right) - \Psi\right] E(Q(\omega)) \qquad (5.112.3)$$

where $E\{Z(\omega)\}$ and u_1 are given by (5.107) and (5.110.1), respectively.

(III) for $\omega_1 = -\omega_2$

$$E\{CW(\omega_1)CW^*(\omega_2)\} = E\{CW^2(\omega)\}$$

$$= E\left\{CW_R^2(\omega)\right\} - E\left\{CW_I^2(\omega)\right\} + 2jE\{CW_R(\omega)CW_I(\omega)\}. \qquad (5.113.1)$$

The first two terms have already been determined, while the third is given by

$$E\{CW_R(\omega)CW_I(\omega)\} = -\tan^{-1}\left(\frac{A_I(\omega)}{A_R(\omega)}\right) E\{CW_R(\omega)\}$$

$$+ \frac{c}{\pi} \int_0^\infty \int_0^{2\pi} \ln\left(1 + \frac{r^2}{|A(\omega)|^2}\right) \tan^{-1}\left(\frac{A_I(\omega) + r\sin\phi}{A_R(\omega) + r\cos\phi}\right) re^{-cr^2} dr d\phi$$

$$+ \frac{c}{\pi} \int_0^\infty \int_0^{2\pi} \frac{A_R(\omega)r\cos\phi + A_I(\omega)r\sin\phi}{|A(\omega)|^2 + r^2} \tan^{-1}\left(\frac{A_I(\omega) + r\sin\phi}{A_R(\omega) + r\cos\phi}\right) re^{-cr^2} dr d\phi$$
$$(5.113.2)$$

where c is given by (5.109.7). Taking into account (5.111) through (5.113), if

$$c'(m,n) = F_2^{-1}[E\{CW(\omega_1)CW^*(\omega_2)\}]$$

then

$$E\left\{c_w(m)c_w^*(m)\right\} = c'(m,-m) \qquad (5.114)$$

which gives the second-order moment of the error contribution in the complex cepstrum due to the presence of additive white Gaussian noise.

5.10.3 The Bicepstrum Evaluation

Under the assumption that $C_{aw}(\omega_1,\omega_2)$ of (5.105.1) can be ignored when it is compared with $C_3^w(\omega_1,\omega_2)$, we have that $C_3^e(\omega_1,\omega_2) \approx C_3^w(\omega_1,\omega_2)$. Consequently, the error bicepstrum is given by

$$BE(\omega_1,\omega_2) \approx \ln\left(1 + \frac{C_3^w(\omega_1,\omega_2)}{C_3^a(\omega_1,\omega_2)}\right). \qquad (5.115)$$

Van Ness [1966] and Lii and Rosenblatt [1990] have shown that $C_3^w(\omega_1,\omega_2)$ converges in distribution to a complex Gaussian random variable $[R+jI]$ where R and

I are independent over the bicepstrum domain grid if the grid width is larger than or equal to the bandwidth Δ_0, with mean zero and variances:

$$\begin{aligned}
\sigma_R^2(\omega_1,\omega_2) &= \sigma_I^2(\omega_1,\omega_2) = \tfrac{1}{2}\sigma_B^2(\omega_1,\omega_2) \\
&= \frac{M^2\sigma^6}{2N}[1 + \delta(\omega_1 - \omega_2) + \delta(2\omega_1 + \omega_2) + \delta(\omega_1 + 2\omega_2) \\
&+ \delta(\omega_1) + \delta(\omega_2) + \delta(\omega_1 + \omega_2) \\
&+ \delta(\omega_1)\delta(\omega_2)]
\end{aligned} \quad (5.116)$$

with

$$\delta(x) = \begin{cases} 1, & \text{if } x = 0 \\ 0, & \text{otherwise.} \end{cases}$$

Note that if we substitute $A(\omega)$ and $W(\omega)$ in (5.103) for $C_3^a(\omega_1,\omega_2)$ and $C_3^w(\omega_1,\omega_2)$, respectively, where (ω_1,ω_2) corresponds to a grid of values $(\Delta_0 m, \Delta_0 n)$, then (5.103) becomes the error bicepstrum expression (5.115). Therefore, the complex cepstrum analysis also applies here after some modifications that will take into account the variance of $C_3^w(\omega_1,\omega_2)$ given by (5.116). The quantities that will be affected are:

(I) $\beta(\omega)$ of (5.109.3) becomes

$$\beta^2(\omega_1,\omega_2) = \frac{|C_3^a(\omega_1,\omega_2)|^2}{2\sigma_B^2(\omega_1,\omega_2)} \quad (5.117.1)$$

and

(II) $c(\omega)$ in (5.109.7) becomes

$$c(\omega_1,\omega_2) = \frac{1}{2\sigma_B^2(\omega_1,\omega_2)}. \quad (5.117.2)$$

Also, all other defined quantities become two-dimensional, including in the Fourier transforms.

The derivation of all these formulas, as well as their proofs and Monte-Carlo verification experiments, are given by Petropulu and Nikias [1990].

5.11 SUMMARY

In this chapter we have introduced the definition, properties, computation, and interrelationships of the cepstra of higher-order spectra (or polycepstra) as well as their application to nonminimum phase signal reconstruction and deconvolution. Polycepstra have been introduced for both deterministic signals and linear non-Gaussian random processes.

The definition and properties of the real cepstrum or power cepstrum of a signal can be found in the paper by Bogert, Healy, and Tukey [1963]. Oppenheim [1968,1969], Oppenheim and Schafer [1989], and Tribolet [1977] provide a detailed exposition of the complex cepstrum and its applications to signal analysis problems. The differential cepstrum and its properties can be found in the paper by Polydoros and Fam [1981]. The cepstrum of higher-order spectra and its computation is discussed by Pan and Nikias [1988]. The definition and properties of the cepstrum of the bicoherency is given by Nikias and Liu [1990].

The cross-bicepstrum and cross-tricepstrum for the analysis of multichannel signals is discussed by Brooks and Nikias [1992]. The analytic performance evaluation of the bicepstrum is given by Petropulu and Nikias [1990]. Finally, the discussion on the higher-order factorization problem using polycepstra can be found in Tekalp and Erdem [1989].

REFERENCES

Bogert, B. P., M. J. R. Healy, and J. W. Tukey, "The Quefrequency Analysis of Time Series for Echoes: Cepstrum, Pseudo-Auto Covariances, Cross-Cepstrum, and Saphe Cracking," *Proc. Symposium Time Series Analysis*, M. Rosenblatt (ed.), pp. 209–243, New York: J. Wiley, NY, 1963.

Brooks, D. H., and C. L. Nikias, "The Cross-Bicepstrum: Properties and Applications for Signal Reconstruction and System Identification," *Proc. ICASSP'91*, pp. 3433-3436, Toronto, Canada, May, 1991.

Erdem, A. T., and A. M. Tekalp, "On the Measure of the Set of Factorizable Polynomial Bispectra," *IEEE Transactions Acoust., Speech, and Signal Processing,* **38**(9), pp. 1637–1639, September, 1990.

Gradsteyn, I. S., and I. B. Ryzhik, **Table of Integrals, Series and Products**, New York: Academic Press, 1980.

Hassab, J. C., **Underwater Signal and Data Processing**, Boca Raton, FL: CRC Press, 1989.

Hatzinakos, D. and C. L. Nikias, "Estimation of Multipath Channel Response in Frequency Selective Channels," *IEEE Journal Selected Areas in Communications*, 7(1), pp. 12–19, January, 1989.

Hatzinakos, D., and C. L. Nikias, "Blind Equalization Using a Tricepstrum Based Algorithm," *IEEE Trans. Communications*, **39**, pp. 669–682, May, 1991.

Lii, K. S., and M. Rosenblatt, "Asymptotic Normality of Cumulant Spectral Estimates," *Theoretical Probability*, 1990.

Nikias, C. L. and M. R. Raghuveer, "Bispectrum Estimation: A Digital Signal Processing Framework," *Proceedings IEEE*, **75**(7), pp. 869–891, July, 1987.

Nikias, C. L. and F. Liu, "Bicepstrum on Computation Based on Second and Third-Order Statistics with Applications," *Proceedings ICASSP 90*, pp. 2381–2384, Albuquerque, NM, April, 1990.

Oppenheim, A. V., "Generalized Superposition," *Information Control*, **11**(5-6), pp. 528-536, Nov.– Dec., 1968.

Oppenheim, A. V., "Generalized Linear Filtering," Chap. 8 in **Digital Processing of Signals**, P. Gold and C. M. Rader (eds.), New York: McGraw-Hill, 1969.

Oppenheim, A. V., and R.W. Schafer, **Discrete-Time Signal Processing**, Englewood Cliffs, NJ: Prentice Hall, 1989.

Orfanidis, S. F., **Optimum Signal Processing**, Chapter 3, New York: Macmillan, 1985.

Pan, R., and C. L. Nikias, "The Complex Cepstrum of Higher-Order Cumulants and Nonminimum Phase System Identification," *IEEE Transactions Acoust., Speech, and Signal Processing*, **ASSP-36**(2), pp. 186–205, February, 1988.

Petropulu A. P. and C. L. Nikias, "Cumulant Cepstrum of FM Signals and High-Resolution Time Delay Estimation," *Proceedings ICASSP '88*, pp. 2642–2645, New York, NY, April, 1988.

Petropulu, A. P., and C. L. Nikias, "The Complex Cepstrum and Bicepstrum: Analytic Performance Evaluation in the Presence of Gaussian Noise," *IEEE Trans. Acoust., Speech, and Signal Processing*, **38**(7), pp. 2346–1256, July, 1990.

Petropulu, A. P., and C. L. Nikias, "Signal Reconstruction from the Phase of the Bicepstrum," *IEEE Transactions on Acoust., Speech, and Signal Processing*, **40**(3), pp. 601–610, March, 1992.

Pitas, I., and A. N. Venetsanopoulos, **Nonlinear Digital Filters: Principles and Applications**, Boston: Kluwer Academia Publishers, 1990.

Polydoros, A., and A. T. Fam, "The Differential Cepstrum Definition and Properties," *Proc. IEEE Int. Symposium on Circuits and Systems*, pp. 77–80, April 1981.

Tekalp, A. M., and A. T. Erdem, "Higher-Order Spectrum Factorization in One and Two Dimensions with Applications in Signal Modeling and Nonminimum Phase Systems Identification," *IEEE Trans. Acoust., Speech, and Signal Processing*, **37**, pp. 1537–1549, October, 1989.

Tribolet, J. M., "A New Phase Unwrapping Algorithm," *IEEE Trans. Acoustics, Speech, and Signal Processing*, **ASSP-25**(2), pp. 170–177, April, 1977.

Van Ness, J. W., "Asymptotic Normality of Bispectral Estimates," *Ann Math. Statis.*, **37**, pp. 2257–2272, 1966.

6

NONPARAMETRIC METHODS FOR SIGNAL RECOVERY FROM HIGHER-ORDER SPECTRA

6.1 INTRODUCTION

Nonparametric methods for signal reconstruction from higher-order spectra are algorithms which recover the Fourier magnitude and the Fourier phase of a signal from its bispectrum or trispectrum. The methods are nonparametric in the sense that they do not fit a parametric model (MA, AR, or ARMA) to the data in order to solve the problem. The nonparametric methods require a priori knowledge of the bispectrum or trispectrum of a signal.

In this chapter, we study several different families of nonparametric methods for signal recovery:

1. Phase recovery algorithms, which can be recursive or nonrecursive in nature.

2. Phase and magnitude recovery algorithms based on the properties of polycepstra.

3. Signal recovery methods from only the phase of the bispectrum.

4. Signal reconstruction from the bispectrum based on the method of Projections Onto Convex Sets (POCS).

5. Signal recovery in a two-channel blind deconvolution scenario.

We begin our study of nonparametric methods by discussing the phase recovery problem and its practical importance.

6.2 MAGNITUDE AND PHASE ESTIMATION FROM HIGHER-ORDER SPECTRA

For a *linear non-Guassian* signal, knowledge of its higher-order spectrum (spectrum of order greater than 2) allows one to recover both its Fourier-magnitude (within a scale factor) and its Fourier-phase (apart from a linear phase shift). In this section, we present Fourier-magnitude and Fourier-phase retrieval algorithms assuming knowledge of the bispectrum magnitude and phase, respectively. The extension of these algorithms to higher-order spectra such as the trispectrum is straightforward.

Let the signal $X(k)$ have Fourier transform $X(\omega)$. Then, the following expression relates the bispectrum phase, $\Psi_3^x(\omega_1, \omega_2)$, to the Fourier-phase, $\phi_x(\omega)$,

$$\Psi_3^x(\omega_1, \omega_2) = \phi_x(\omega) + \phi_x(\omega_2) - \phi_x(\omega_1 + \omega_2), \tag{6.1}$$

while the bispectral magnitude is related to the Fourier-magnitude through

$$|C_3^x(\omega_1, \omega_2)| = |X(\omega_1)| \cdot |X(\omega_2)| \cdot |X(\omega_1 + \omega_2)|. \tag{6.2}$$

The problem is how to recover $\phi_x(\omega)(|X(\omega)|)$ from $\Psi_3^x(\omega_1,\omega_2)(|C_3^x(\omega_1,\omega_2)|)$. Taking the logarithm of both sides of (6.2), we obtain

$$\ln |C_3^x(\omega_1, \omega_2)| = \ln |X(\omega_1)| + \ln |X(\omega_2)| + \ln |X(\omega_1 + \omega_2)|. \tag{6.3}$$

Comparing (6.1) and (6.3), we easily conclude that phase recovery algorithms can also be used as magnitude recovery algorithms, and vice versa, provided that $|X(\omega)| \neq 0$.

6.3 PHASE RECOVERY ALGORITHMS

One of the most appealing features of the bispectrum is that it preserves the Fourier phase of a signal. In this section we describe several different phase retrieval algorithms, and we comment on their advantages and limitations. Most of these algorithms are based on (6.1) and solve for the Fourier phase, $\phi_x(\omega)$. In a practical situation, however, the bispectral phase, $\Psi_3^x(\omega_1, \omega_2)$, can only be computed from:

$$\hat{\Psi}_3^x(\omega_1, \omega_2) = \arctan \frac{Im[C_3^x(\omega_1, \omega_2)]}{Re[C_3^x(\omega_1, \omega_2)]} \tag{6.4}$$

where $C_3^x(\omega_1, \omega_2)$ is the bispectrum and $Im[\cdot]$, $Re[\cdot]$ denote imaginary and real parts, respectively. Even if we ignore the estimation errors that result from the utilization of bispectrum estimation methods, the computed phase, $\hat{\Psi}_3^x(\omega_1,\omega_2)$, differs from the true phase, $\Psi_3^x(\omega_1,\omega_2)$ by $2\pi k(\omega_1,\omega_2)$, where $k(\omega_1,\omega_2)$ can only take integer values. This phase difference is of no importance to the bispectrum as it does not affect its value. However, depending on the phase retrieval algorithm, it may lead to an erroneous Fourier phase, $\hat{\phi}_x(\omega)$. For some of the algorithms that will be discussed in subsequent sections, the calculated Fourier phase, $\hat{\phi}_x(\omega)$, derived from the computed bispectrum phase ($\hat{\Psi}_3^x(\omega_1,\omega_2)$, differs from the true Fourier phase ($\hat{\phi}_x(\omega)$) by only a linear phase term, i.e., $\hat{\phi}_x(\omega) = \phi_x(\omega) + 2\pi k(\omega)$, where $k(\omega)$ is an integer function [Lii and Rosenblatt, 1982; Bartelt and Lohman, et al., 1984]. This linear phase difference is usually of no importance since it corresponds only to a time shift of the signal. If this is not the case, then a two-dimensional phase unwrapping method such as the one described by Dudgeon and Mersereau [1984] for the estimated bispectral phase is required before the application of a phase retrieval algorithm.

6.3.1 The Brillinger Algorithm [Brillinger, 1977]

The phase retrieval algorithm, as proposed by Brillinger, is a recursive scheme that utilizes all the bispectral values. Assigning to ω_1 the discrete values $0, 1, ..., k$ and to ω_2, the values $k, k-1, ..., 0$, Equation (6.1) yields $(k+1)$ equations. Summing these equations, we obtain

$$\sum_{i=0}^{k}\Psi_3^x(i,k-i) = 2\sum_{i=0}^{k}\phi_x(i) - (k+1)\phi_x(k) = 2\sum_{i=0}^{k-1}\phi_x(i) - (k-1)\phi_x(k) \quad (6.5)$$

which can be solved for $\phi_x(k)$ as follows:

$$\begin{aligned}\phi_x(k) &= \frac{2}{k-1}\sum_{i=0}^{k-1}\phi_x(i) - \frac{1}{k-1}\sum_{i=0}^{k}\Psi_3^x(i,k-i) \\ &= \frac{1}{k-1}\left[2\sum_{i=0}^{k-1}\phi_x(i) - S(k)\right], \quad k=2,3,\cdots,N\end{aligned} \quad (6.6)$$

where

$$S(k) = \sum_{i=0}^{k}\Psi_3^x(i,k-i).$$

Note that $k = N$ corresponds to $\omega = \pi$. To implement the recursion in (6.6) we need two initial values, i.e., $\phi_x(0)$ and $\phi_x(1)$. To determine these two values, we first assume that $\phi_x(0) = 0$, and $\phi_x(N) = 0$. Actually, $\phi_x(N) = k\pi, k = 0, \pm 1, \pm 2, \ldots$; hence, the assumption $\phi_x(N) = 0$ will imply that the signal will be time shifted.

Using this assumption on $\phi_x(N)$, $\phi(1)$ can be computed as follows [Matsuoka and Ulrych, 1984]:

$$\phi_x(1) = \sum_{i=2}^{N}\left\{\frac{S(i)-S(i-1)}{i(i-1)}\right\} + \frac{\phi_x(N)}{N}. \qquad (6.7)$$

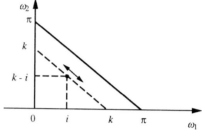

Figure 6.1 The bispectral phase lines used in the Brillinger algorithm

From (6.6) we see that the Fourier phase, $\phi_x(\omega)$, is computed by summing along the diagonal bispectral phase lines as illustrated in Figure 6.1. It can easily be shown that (6.6) and (6.7) would lead to an erroneous Fourier phase, $\hat{\phi}_x(\omega)$, if instead of the true bispectral phase, the modulo 2π phase, $\hat{\Psi}_3^x(\omega_1,\omega_2)$, was used (i.e., $\hat{\phi}_x(\omega) - \phi_x(\omega) \neq 2\pi k(\omega)$).

The Brillinger algorithm is recursive, and as such it is sensitive to errors, especially in the estimation of the initial phase value $\phi_x(1)$.

6.3.2 The Lii-Rosenblatt Algorithm [Lii and Rosenblatt, 1982]

The Lii-Rosenblatt algorithm is also recursive, but it utilizes only one line of bispectral values. Summing bispectral phases that correspond to $\omega_1 = 0, 1, \ldots, k-1$ and $\omega_2 = 1$ we obtain

$$\phi_x(k) = -\sum_{i=1}^{k}\Psi_3^x(i,1) - k\phi_x(1) + \phi_x(0), \quad k = 2, 3, \ldots, N. \qquad (6.8)$$

Again, $\phi_x(0) = 0$ and $\phi_x(1)$ is set arbitrarily, which establishes the position of the time signal. With these initial conditions, (6.8) would still give a correct Fourier phase, $\hat{\phi}_x(k)$, even if the modulo 2π bispectral phase, $\hat{\Psi}_3^x(i,1)$, was used, instead of $\Psi_3^x(i,1)$. Setting $\phi_x(1) = 0$, (6.8) leads to

$$\hat{\phi}_x(k) - \phi_x(k) = -\sum_{i=0}^{k-1} 2\pi \cdot n(i,1), \qquad (6.9)$$

where $n(\cdot)$ is an integer function. In other words, $\hat{\phi}_x(k)$ differs from $\phi_x(k)$ by an integer multiple of 2π.

This algorithm is essentially limited to the use of the bispectral phase values along the line $\Psi_3^x(\omega, i), i = 0, 1, \ldots, N$. Consequently, not all the available information is taken into account, which may cause serious problems in the case of band-limited signals. For example, let us consider a band-limited process, with ω_L and ω_H denoting the low and high cut off frequencies, respectively (Figure 6.2). If $\omega_L > 1$, the Lii-Rosenblatt algorithm will fail to compute the Fourier-phase, since the utilized bispectral values are all zero.

The Lii-Rosenblatt algorithm is also sensitive to errors due to its recursive nature.

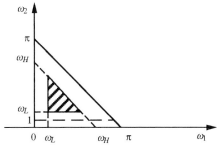

Figure 6.2 Bispectral values available for phase computation of band-limited signals

6.3.3 The Bartelt-Lohman-Wirnitzer Algorithm [Bartelt and Lohman, et. al, 1984]

The Bartelt-Lohman-Wirnitzer algorithm is basically a generalization of the Lii-Rosenblatt algorithm. From (6.1) we can write

$$\phi_x(i) = \phi_x(j) + \phi_x(i-j) - \Psi_3^x(i-j, j). \tag{6.10}$$

So, for $j = 1$ we have

$$\phi_x(i) = \phi_x(1) + \phi_x(i-1) - \Psi_3^x(i-1, 1) \tag{6.11}$$

with the initial condition $\phi_x(0) = 0$, and $\phi_x(1)$ being arbitrary. The phase can be recovered from (6.11) as follows:

$$\phi_x(2) = \phi_x(1) + \phi_x(1) - \Psi_3^x(1, 1) \tag{6.12.1}$$

$$\phi_x(3) = \phi_x(1) + \phi_x(2) - \Psi_3^x(2, 1) \tag{6.12.2}$$
$$\vdots$$

$$\phi_x(N) = \phi_x(1) + \phi_x(N-1) - \Psi_3^x(N-1,1) \qquad (6.12.3)$$

where the value of $\phi_x(1)$ is substituted into (6.12.1) which then gives $\phi_x(2)$. Similarly, substitution of $\phi_x(2)$ into (6.12.2) yields $\phi_x(3)$, and this procedure is repeated until all phase values are determined. The same procedure in (6.10) is repeated for $j = 2, 3, \ldots N$. Hence, each phase $\phi_x(p)$ has $(p-1)/2$ independent representations if p is odd and $p/2$ representations if p is even.

To improve the signal-to-noise ratio (SNR) of the reconstruction we can average the different representations. Note that the exponential factors, $\exp\{j\phi_x(p)\}$, should be averaged, instead of summing up phases directly, since each representation of $\phi_x(p)$ can be computed only up to modulo 2π. Hence,

$$\exp\{j\phi_x(p)\} = (\text{const.}) \cdot \sum_q \exp\{j(\phi_x(q) + \phi_x(p-q) - \Psi_3^x(p-q,q))\}. \qquad (6.13)$$

The Bartelt-Lohman-Wirnitzer algorithm, although recursive in nature, does not require phase unwrapping methods and utilizes all bispectral values. More robust initialization procedures for all three recursive algorithms, namely the Brillinger, Lii-Rosenblatt, and Bartelt-Lohman-Wirnitzer have been proposed by Pan and Nikias [1987].

6.3.4 The Matsuoka-Ulrych Algorithm [Matsuoka and Ulrych, 1984]

All the phase estimation algorithms discussed so far are recursive. A nonrecursive approach that utilizes all bispectral values was proposed by Matsuoka and Ulrych [1984]. Their phase recovery method proceeds as follows: Form all possible equations similar to (6.1) for $\omega_1 = 1, 2, \ldots, N/2$ and $\omega_2 = \omega_1, \omega_1 + 1, \ldots, N - \omega_1$, i.e.,

$$\begin{aligned}
\Psi_3^x(1,1) &= 2\phi_x(1) - \phi_x(2) \\
\Psi_3^x(1,2) &= \phi_x(1) + \phi_x(2) - \phi_x(3) \\
&\vdots \\
\Psi_3^x(1,N-1) &= \phi_x(1) + \phi_x(N-1) - \phi_x(N) \\
\Psi_3^x(2,2) &= 2\phi_x(2) - \phi_x(4) \\
\Psi_3^x(2,3) &= \phi_x(2) + \phi_x(3) - \phi_x(5) \\
&\vdots \\
\Psi_3^x(2,N-2) &= \phi_x(2) + \phi_x(N-2) - \phi_x(N) \\
&\vdots \\
\Psi_3^x(N/2,N/2) &= 2\phi_x(N/2) - \phi_x(N), \text{ if } N \text{ even.}
\end{aligned} \qquad (6.14)$$

By taking $\phi_x(N) = 0$, the equations in (6.14) can be combined into the matrix form

$$\mathbf{A}\boldsymbol{\Phi}_x = \boldsymbol{\Psi}_3^x \qquad (6.15)$$

where
$$\Phi_x = [\phi_x(1), \ldots, \phi_x(N-1)],$$
$$\Psi_3^x = \left[\Psi_3^x(1,1), \Psi_3^x(1,2), \ldots, \Psi_3^x(1, N-1), \Psi_3^x(2,2), \ldots, \Psi_3^x(\frac{N}{2}, \frac{N}{2})\right]^T$$

and

$$\mathbf{A} = \begin{bmatrix} 2 & -1 & 0 & 0 & 0 & \cdots & 0 \\ 1 & 1 & -1 & 0 & 0 & \cdots & 0 \\ \cdot & \cdot & \cdot & \cdot & \cdot & \cdots & \cdot \\ \cdot & \cdot & \cdot & \cdot & \cdot & \cdots & \cdot \\ 1 & 0 & 0 & 0 & 0 & \cdots & 1 \\ 0 & 2 & 0 & -1 & 0 & \cdots & 0 \\ 0 & 1 & 1 & 0 & -1 & \cdots & 0 \\ \cdot & \cdot & \cdot & \cdot & \cdot & \cdots & \cdot \\ \cdot & \cdot & \cdot & \cdot & \cdot & \cdots & \cdot \\ \cdot & \cdot & \cdot & \cdot & \cdot & \cdots & 0 \end{bmatrix}, \begin{array}{l} ((\frac{N}{2})^2 \times (N-1)) \text{ if } N \text{ is even,} \\[1em] (\frac{(N-1)(N+1)}{4} \times (N-1)) \text{ if } N \text{ is odd.} \end{array}$$

It can be shown that $\mathbf{A}$ is a full rank matrix [Matsuoka and Ulrych, 1984]. From (6.15), Φ_x is obtained as the least-squares solution:

$$\Phi_x = (\mathbf{A}^T \mathbf{A})^{-1} \mathbf{A}^T \Psi_3^x. \tag{6.16}$$

This phase recovery algorithm utilizes all available bispectral phase values and it is non-recursive. Hence it does not suffer from accumulated errors. Unfortunately, (6.16) fails to provide a correct solution for $\phi_x(i)$ when a modulo 2π bispectral phase is used instead of the true value.

In the sequel, two methods will be presented which deal with the bispectral phase unwrapping problem.

METHOD I [Marron, et al., 1990]

Let $\hat{\Psi}_3^x(\omega_1, \omega_2)$ denote the modulo 2π bispectral phase, i.e.,

$$\Psi_3^x(\omega_1, \omega_2) = \hat{\Psi}_3^x(\omega_1, \omega_2) + 2\pi \cdot k(\omega_1, \omega_2) \tag{6.17}$$

where $\hat{\Psi}_3^x(\omega_1, \omega_2)$ lies in the region $[0, 2\pi]$ and $k(\omega_1, \omega_2)$ is an integer function. The phase unwrapping problem is basically the estimation of the integers $k(\omega_1, \omega_2)$.

The matrix equation of (6.15) can be written as

$$\mathbf{A}\Phi_x = \Psi_3^x + 2\pi \mathbf{K} \tag{6.18}$$

where
$$\mathbf{K} = [k(1,1), k(1,2), \ldots, k(1, N-1), k(2,2), k(2,3), \ldots, k(2, N-2), k(N/2, N/2)]^T,$$
and $\mathbf{A}$ was defined in (6.15).

Let $\mathbf{C}$ be a matrix such that
$$\mathbf{C} \cdot \mathbf{A} = \mathbf{0}. \tag{6.19}$$

Then from (6.18) and (6.19), we obtain
$$\mathbf{C} \cdot \mathbf{K} = -\frac{1}{2\pi} \cdot \mathbf{C} \cdot \hat{\boldsymbol{\Psi}}_3^x. \tag{6.20}$$

Hence, given the modulo 2π phase vector $\hat{\boldsymbol{\Psi}}_3^x$, we can solve (6.20) for $\mathbf{K}$, and thus unwrap the bispectral phase. In their paper, Marron among others [1990] showed that when $\mathbf{C}$ is specified to have the maximum possible rank ($\frac{(N-2)^2}{4}$), all vectors $\mathbf{K}$ that satisfy (6.20) result in equivalent Fourier phase vectors in the sense that the corresponding time signals are shifted versions of the true signal. To facilitate the determination of $\mathbf{K}$ from (6.20) it would be better to choose $\mathbf{C}$ to be a triangular matrix. Equation (6.19) implies that the elements of each row of $\mathbf{C}$ are the coefficients of linear combinations of bispectral phase equations that sum to zero. Note that such a combination is the following:

$$\Psi_3^x(i,j) - \Psi_3^x(i-1, j+1) - \Psi_3^x(1,j) + \Psi_3^x(1, i-1) = 0. \tag{6.21}$$

Hence, each row of $\mathbf{C}$ needs to contain only four nonzero elements. Using (6.21), we construct the matrix $\mathbf{C}$. For example, when $N = 6$ and

$$\mathbf{A} = \begin{bmatrix} 2 & -1 & 0 & 0 & 0 \\ 1 & 1 & -1 & 0 & 0 \\ 1 & 0 & 1 & -1 & 0 \\ 1 & 0 & 0 & 1 & -1 \\ 1 & 0 & 0 & 0 & 1 \\ 0 & 2 & 0 & -1 & 0 \\ 0 & 1 & 1 & 0 & -1 \\ 0 & 1 & 0 & 1 & 0 \\ 0 & 0 & 2 & 0 & 0 \end{bmatrix},$$

the four rows of $\mathbf{C}$ are formed by the coefficients of each element of the vector $\boldsymbol{\Psi}_3^x$ in (6.21) for $(i=3, j=3)$, $(i=2, j=4)$, $(i=2, j=3)$ and $(i=2, j=2)$, i.e.,

$$\mathbf{C} = \begin{bmatrix} 0 & 1 & -1 & 0 & 0 & 0 & 0 & -1 & 1 \\ 1 & 0 & 0 & -1 & -1 & 0 & 0 & 1 & 0 \\ 1 & 0 & -1 & -1 & 0 & 0 & 1 & 0 & 0 \\ 1 & -1 & -1 & 0 & 0 & 1 & 0 & 0 & 0 \end{bmatrix}.$$

For this particular ordering of rows, **C** is upper triangular. After **C** is constructed, **K** can be computed from (6.20). Since the system of equations of (6.20) is underdetermined, the first $N-1$ elements of **K** can be set to zero, while the remaining elements can be computed by back substitution.

The vector on the right side of (6.20) contains only integer entries, hence the elements of **K** computed from (6.20) should be integers. However, in the presence of noise, the right hand side of (6.20) may contain noninteger entries. Provided that the maximum error due to noise in the bispectrum phase estimation is less than $\pi/4$, the computed elements of **K** can be approximated by their nearest integer values.

METHOD II [Rangoussi and Giannakis, 1991]

An alternative approach for phase unwrapping, which yields identical results as those from Method I, was proposed by Rangoussi and Giannakis [1991].

For $\hat{\Psi}_3^x(\omega_1, \omega_2)$ that satisfies (6.17), the following holds:

$$\hat{\Psi}_3^x(\omega_1, \omega_2) = [\hat{\phi}_x(\omega_1) + \hat{\phi}_x(\omega_2) - \hat{\phi}_x(\omega_1, \omega_2)] \text{ modulo } 2\pi, \ 0 \leq \omega_1, \omega_2 \leq N-1 \quad (6.22)$$

where $\hat{\phi}_x(\cdot)$ denotes the modulo 2π Fourier phase. In (6.22) the last modulo 2π operation is necessary because we add two principal arguments and subtract one. In fact if $\hat{\phi}_x(\omega)$ lies in $[0, 2\pi]$, then the right side of (6.22) lies in $[-2\pi, 4\pi]$, while $\hat{\Psi}_3^x(\omega_1, \omega_2)$ lies in $[0, 2\pi]$. Hence,

$$\hat{\Psi}_3^x(\omega_1, \omega_1) + 2\pi k(\omega_1, \omega_2) = \hat{\phi}_x(\omega_1) + \hat{\phi}_x(\omega_2) - \hat{\phi}_x(\omega_1, \omega_2), \quad (6.23)$$

$0 \leq \omega_1, \omega_2 \leq N-1$, where $k(\omega_1, \omega_2)$ takes values $\{-1, 0, 1\}$. In a matrix form, (6.23) can be written as

$$\hat{\boldsymbol{\Psi}}_3^x + 2\pi \cdot \mathbf{K} = \mathbf{A} \cdot \hat{\boldsymbol{\Phi}}_x \quad (6.24)$$

where $\hat{\boldsymbol{\Psi}}_3^x, \hat{\boldsymbol{\phi}}_x$ were defined as $\boldsymbol{\Psi}_3^x$ and $\boldsymbol{\phi}_x$ in (6.15) and **K** as in (6.18). From (6.24) we obtain the solution

$$\hat{\boldsymbol{\Phi}}_x = (\mathbf{A}^T \mathbf{A})^{-1} \mathbf{A}^T [\hat{\boldsymbol{\Psi}}_3^x + 2\pi \mathbf{K}] = \hat{\boldsymbol{\Phi}}_x^{LS} + 2\pi (\mathbf{A}^T \mathbf{A})^{-1} \mathbf{A}^T \mathbf{K}, \quad (6.25)$$

with $\hat{\boldsymbol{\Phi}}_x^{LS}$ being the least-squares solution of (6.16), if $\hat{\boldsymbol{\Psi}}_3^x$ is replaced by its modulo 2π estimate, $\hat{\boldsymbol{\Psi}}_3^x$. From equation (6.25), $\hat{\boldsymbol{\Phi}}_x^{LS}$ would be an acceptable solution if $\det[A^T A] = 1$. In this case, since both **A** and **K** have integer entries, $\hat{\boldsymbol{\Phi}}_x^{LS}$ would only differ from $\hat{\boldsymbol{\Phi}}_x$ by an integer multiple of 2π. The reader may easily verify that the Lii-Rosenblatt and Bartelt-Lohman-Wirnitzer algorithms need no phase unwrapping because when they are written in the form of (6.25), $\det[\mathbf{A}^T \mathbf{A}] = 1$.

However, in this case (Method II) det$[\mathbf{A}^T \mathbf{A}] \neq 1$, and thus phase unwrapping is necessary. The phase unwrapping approach proposed by Rangoussi and Giannakis [1991] is the following:

Step 1: Utilize the Bartelt-Lohman-Wirnitzer algorithm to obtain the values of the principal argument, $\hat{\boldsymbol{\Phi}}$

Step 2: Having available values of $\hat{\boldsymbol{\Psi}}_3^x$, apply (6.24) to compute the vector $\mathbf{K}$. Because sample estimates, $\hat{\boldsymbol{\Psi}}_3^x$, are used, the resulting estimate of $\mathbf{K}$ may not have integer entries. In this case, round $\mathbf{K}$ to the nearest integer vector.

Step 3: Use rounded $\hat{\mathbf{K}}$ in (6.25), to compute $\hat{\boldsymbol{\phi}}_x$.

Applying the Matsuoka-Ulrych method in practice, either phase unwrapping algorithm just described could be used to derive meaningful phase recovery results.

6.3.5 Phase Reconstruction from a 45° Slice of the Bispectral Phase

The reconstruction of the Fourier phase of a two-dimensional deterministic signal from its bispectral phase was considered by Dianat and Raghuveer [1990]. The application of their method to 1-d deterministic signals or linear non-Gaussian processes is straightforward.

Let $X(\mathbf{k}), \mathbf{k} = [k_1, k_2]^T$ be the discrete 2-d signal, and $X(\boldsymbol{\omega}), \boldsymbol{\omega} = [\omega_1, \omega_2]^T$ is its Fourier transform. Its bispectrum is given by

$$C_3^x(\boldsymbol{\omega}_1, \boldsymbol{\omega}_2) = X(\boldsymbol{\omega}_1) X(\boldsymbol{\omega}_2) X^*(\boldsymbol{\omega}_2 + \boldsymbol{\omega}_2) \tag{6.26}$$

where $\boldsymbol{\omega}_i = [\omega_{1i}, \omega_{2i}]^T, i = 1, 2$.

Let
$$C_3^x(\boldsymbol{\omega}_1, \boldsymbol{\omega}_2) = |C_3^x(\boldsymbol{\omega}_1, \boldsymbol{\omega}_2)| \cdot \exp\{j \Psi_3^x(\boldsymbol{\omega}_1, \boldsymbol{\omega}_2)\}$$
and
$$X(\boldsymbol{\omega}) = |X(\boldsymbol{\omega})| \exp\{j \phi_x(\boldsymbol{\omega})\}.$$
Then, if $\boldsymbol{\omega}_1 = \boldsymbol{\omega}_2 = \boldsymbol{\omega}$, it holds that

$$\Psi_3^x(\boldsymbol{\omega}) \equiv \Psi_3^x(\boldsymbol{\omega}, \boldsymbol{\omega}) = 2\phi_x(\boldsymbol{\omega}) - \phi_x(2\boldsymbol{\omega}). \tag{6.27}$$

If we are given $N \times N$ samples of $C_3^x(\boldsymbol{\omega}, \boldsymbol{\omega})$, and define

$$\Phi(\boldsymbol{\lambda}) \stackrel{\Delta}{=} DFT_2\{\phi_x(\boldsymbol{\omega})\} \tag{6.28.1}$$

$$\Psi(\lambda) \stackrel{\Delta}{=} DFT_2\{\Psi_3^x(\omega)\} \qquad (6.28.2)$$

where $DFT_2[\cdot]$ denotes the 2-d discrete Fourier transform, then from (6.27) we obtain

$$\sum_{\lambda} \Psi(\lambda)\exp(j\omega^T\lambda) = 2\sum_{k} \Phi(\lambda)\exp(j\omega^T\lambda) - \sum_{\lambda} \Phi(\lambda)\exp\{j2\omega^T\lambda\} \qquad (6.29)$$

where $\omega = [\frac{2\pi}{N}i, \frac{2\pi}{N}j]^T$, and $\lambda = [\lambda_1, \lambda_2]$. To recover $\Phi(k)$ from $\Psi(\lambda)$, the coefficients of similar terms on both sides of (6.29) must be equated. If either λ_1 or λ_2 is odd, it is easy to show that

$$\Phi(\lambda) = \frac{1}{2}\Psi(\lambda). \qquad (6.30)$$

The case where both λ_1 and λ_2 are even is more complicated. Suppose that $N = 2^M$. If both λ_1 and λ_2 are even, from (6.29) we obtain

$$\Psi(\lambda_1,\lambda_2) = 2\Phi(\lambda_1,\lambda_2) - \Phi(\frac{\lambda_1}{2},\frac{\lambda_2}{2}) - \Phi(\frac{\lambda_1}{2},\frac{\lambda_2}{2}+\frac{N}{2}) \begin{array}{l} -\Phi(\frac{N}{2}+\frac{\lambda_1}{2},\frac{N}{2}+\frac{\lambda_2}{2}) \\ -\Phi(\frac{N}{2}+\frac{\lambda_1}{2},\frac{\lambda_2}{2}) \end{array} \qquad (6.31)$$

where the notation of $\Phi(\lambda_1,\lambda_2)$ and $\Psi(\lambda_1,\lambda_2)$ was used instead of $\Phi(\lambda)$ and $\Psi(\lambda)$. Based on (6.30) and (6.31), Dianat and Raghuveer [1990] proposed the following recursive scheme:

Step 1: If λ_1 or λ_2 is odd, then $\Phi(\lambda)$ is given by (6.30).

Step 2: If either $\frac{\lambda_1}{2}$ or $\frac{\lambda_2}{2}$ is odd then

$$\begin{aligned}\Phi(\lambda_1,\lambda_2) &= \frac{1}{2}[\Psi(\lambda_1,\lambda_2) + \Phi(\frac{\lambda_1}{2},\frac{\lambda_2}{2}) + \Phi(\frac{\lambda_1}{2},\frac{N}{2}+\frac{\lambda_2}{2}) \\ &+ \Phi(\frac{N}{2}+\frac{\lambda_1}{2},\frac{N}{2}+\frac{\lambda_2}{2}) + \Phi(\frac{N}{2}+\frac{\lambda_1}{2},\frac{\lambda_2}{2})].\end{aligned} \qquad (6.32)$$

Step 3: Repeat Step 2 successively for cases where either $(\lambda_1/2)\cdot \ell$ or $(\lambda_2/2)\cdot \ell$ is odd, ℓ taking the values $2, 3, \ldots, \log_2(N-1)$.

Step 4:
$$\Phi(0,0) = \Psi(0,0) + \Phi(0,\frac{N}{2}) + \Phi(\frac{N}{2},0) + \Phi(\frac{N}{2},\frac{N}{2}). \qquad (6.33)$$

Step 5: Reconstruct the phase, $\phi_x(\omega)$, as

$$\phi_x(\omega) = IDFT_2\{\Phi(\lambda)\} \qquad (6.34)$$

where $IDFT_2\{\cdot\}$ denotes the inverse 2-d DFT.

The method is computationally attractive since it can be implemented by a 2-d fast Fourier transform. However, due to the use of the inverse 2-d DFT in (6.34) and the multiplication by $\frac{1}{2}$ of $\Psi(\lambda)$ in (6.30) and (6.32), the method will yield an erroneous Fourier phase, $\phi_x(\omega)$, if a modulo 2π bispectral phase estimate is used instead of $\Psi_3^x(\omega)$. Note that this method does not utilize all bispectral values for phase recovery.

Several 2-d phase estimation algorithms are also discussed in [Swami, Giannakis, and Mendel, 1989].

6.4 PHASE AND MAGNITUDE ALGORITHMS BASED ON POLYSPECTRA/POLYCEPSTRA PROPERTIES

The purpose of this section is to describe nonparametric phase and magnitude recovery algorithms that are based on properties of polyspectra and polycepstra.

6.4.1 Preliminaries

Consider the ARMA energy sequence $\{X(k)\}$. If the Z transform of $X(k)$, i.e., $X(z)$, has no zeros or poles on the unit circle, the bicepstrum of $X(k)$, i.e., $b_x(m,n)$, was found in Chapter 5 to be zero everywhere except on the axes $m = 0, n = 0$ and the diagonal $m = n$. Furthermore, the values of the bicepstrum on these lines are simple relations of the cepstral coefficients of $X(k)$, as described by (5.38).

The bicoherency or normalized bispectrum of $X(k)$, i.e., $P_3^x(\omega_1,\omega_2)$, was defined in (5.53). Considering the complex cepstrum of the bicoherency, we obtain

$$bic_x(m,n) = F_2^{-1}\{\ln[P_3^x(\omega_1,\omega_2)]\} = F_2^{-1}\left\{\ln[\frac{C_3^x(\omega_1,\omega_2)}{|C_3^x(\omega_1,\omega_2)|}]\right\} = j \cdot F_2^{-1}\{\Psi_3^x(\omega_1,\omega_2)\},$$
(6.35)

where $F_2^{-1}\{\cdot\}$ denotes inverse 2-d Fourier transform. Note that (6.35) is valid for the case of deterministic energy signals and linear non-Gaussian processes. What (6.35) implies is that the *cepstrum of the bicoherency is the inverse 2-d Fourier transform of the bispectral phase multiplied by j*. In Chapter 5, it was also shown that the cepstrum of the bicoherency is everywhere zero except on the lines $m = 0$, $n = 0$, and $m = n$. On these lines $bic_x(m,n)$ is a function of the differences of the cepstral coefficients, as shown by (5.59.1).

6.4.2 Bispectrum Signal Reconstruction (BSR)

By definition, the complex cepstrum of $X(k)$ is given by

$$c_x(m) = F_1^{-1}\{\ln X(\omega)\} = F_1^{-1}\{\ln |X(\omega)|\} + jF_1^{-1}\{\phi_x(\omega)\}$$

$$c_x(m) = \begin{cases} -\dfrac{A^{(m)}}{m}, & m > 0 \\ \dfrac{B^{(-m)}}{m}, & m < 0 \end{cases} \quad (6.36)$$

where $\phi_x(\omega) = \arg[X(\omega)]$ is an odd and continuous function of ω [Oppenheim and Schafer, 1989]. Substituting (5.10) and (5.25) into (6.36), we obtain the following representation of $\phi_x(\omega)$ in terms of the signal's cepstrum coefficients [Petropulu and Nikias, 1992]:

$$\phi_x(\omega) = F_1\{d(m)\}, 0 \leq \omega \leq N - 1 \quad (6.37.1)$$

where

$$d(m) = \begin{cases} \dfrac{j}{2m}[A^{(|m|)} - B^{(|m|)}], & m \neq 0 \\ 0, & m = 0. \end{cases} \quad (6.37.2)$$

Considering (6.35) and the fact that

$$bic_x(m, 0) = -\dfrac{1}{2m}[A^{(m)} - B^{(m)}] \text{ when } m > 0, \quad (6.38)$$

another expression for $\phi_x(\omega)$ is the following

$$\phi_x(\omega) = F_1\{d'(m, 0)\}, \ 0 \leq \omega \leq N - 1 \quad (6.39.1)$$

$$d'(m, n) = \begin{cases} F_2^{-1}\{\Psi_3^x(\omega_1, \omega_2)\}, & m \neq 0 \\ 0 & m = 0. \end{cases} \quad (6.39.2)$$

After some algebraic manipulation, (6.39.1) leads to

$$\phi_x(\omega) = \dfrac{1}{2N - 1}\sum_{i=-N+1}^{N-1} \Psi_3^x(\omega, i), \ \omega = 0, \ldots, N - 1. \quad (6.40)$$

Note that (6.40) requires no initial conditions. Since the bispectrum suppresses all linear phase information, the reconstructed phase corresponds to a linear shifted version of the true phase of $X(k)$. It is important to note that since the differences of the cepstral coefficients contain the bispectral phase information [Compare (6.35) and (6.38)], we do not have to estimate directly the bispectral phase in (6.37). Therefore, utilizing (6.37) we do not have to worry about phase unwrapping problems; instead, first we compute the bicepstrum of $X(k)$ as described in section 5.5.3, and then we calculate the cepstrum coefficients utilizing (5.38) and substitute them

TABLE 6.1 PHASE RECOVERY ALGORITHMS

	Recursive	Phase Unwrapping Required	All Bispectrum Values Utilized
Brillinger	Yes	Yes	Yes
Lii-Rosenblatt	Yes	No	No
Bartelt-Lohman Wirnitzer	Yes	No	Yes
Matsuoka-Ulrych	No	Yes	Yes
Dianat-Raghuveer	No	Yes	No
Nonparametric BSR	No	Yes	Yes
Polycepstra-Based BSR	No	No	Yes

into (6.37). Note that considering only the differences of the cepstrum coefficients is equivalent to retaining only phase information and discarding all magnitude information. In scenarios where the Fourier magnitude is corrupted the error will appear only in the sum of the cepstrum coefficients.

The phase estimation formula (6.37) fails when the Z transform of the signal has zeros or poles on the unit circle. As the zeros of $X(z)$ approach the unit circle, the cepstrum coefficients $\{A^{(m)}, m = 1, 2, \ldots, p\}$ and $\{B^{(m)}, m = 1, 2, \ldots, q\}$ decay slower with increasing m. Therefore, larger size 2-d Fourier transforms have to be considered for bicepstrum estimation if (5.47) is used, or a larger **Y** matrix has to be constructed if (5.46.1) is employed. If the sizes of the 2-d Fourier transforms used in (5.47) are not large enough, aliasing will occur in the estimated cepstrum coefficients. To test whether or not the utilized Fourier transform size, i.e., $L \times L$, is large enough, check if $A^{(m)}$ and $B^{(m)}$ have both decayed to zero within the interval $(1, L/2)$. If not, larger size Fourier transforms have to be considered. A way to avoid using large 2-d Fourier transforms will be discussed in section 6.4.5, where a third-order moment window is introduced to force all zeros and poles away from the unit circle.

The phase recovery formula (6.40) can be applied on signals with zeros on the unit circle. However, since it involves the bispectral phase directly, it requires the utilization of phase unwrapping algorithms. In addition to the existing 2-d phase unwrapping scheme introduced by Dudgeon and Mersereau [1984], another scheme, based on the properties of the bispectrum, is described in the next section. The properties of the phase recovery algorithms described so far are summarized in Table 6.1.

6.4.3 Two-Dimensional Phase Unwrapping

Let
$$b_x(z_1, z_2) = \ln\{C_3^x(z_1, z_2)\} = \ln|C_3^x(z_1, z_2)| + j\Psi_3^x(\omega_1, \omega_2). \tag{6.41}$$

then
$$\frac{\partial b_x(z_1, z_2)}{\partial z_1} = \frac{1}{C_3^x(z_1, z_2)} \cdot \frac{\partial C_3^x(z_1, z_2)}{\partial z_1} \tag{6.42}$$

and after multiplying both sides of (6.42) by z_1, we obtain

$$C_3^x(z_1, z_2) \cdot Z_2\{-m \cdot b_x(m,n)\} = Z_2\{-m \cdot m_3^x(m,n)\} \tag{6.43}$$

where $b_x(m,n) = Z_2^{-1}\{b_x(z_1, z_2)\}$ is the bicepstrum of $X(k)$ and $m_3^x(m,n)$ is the third-order moment sequence of $X(k)$. Note that $Z_2^{-1}\{\cdot\}$ denotes inverse 2-d Z transform. Hence, from (6.43) we get

$$b_x(m,n) = F_2^{-1}\{\ln[C_3^x(\omega_1, \omega_2)]\} = \frac{1}{m} F_2^{-1}\left\{\frac{F_2\{m \cdot m_3^x(m,n)\}}{C_3^x(\omega_1, \omega_2)}\right\}, \quad m \neq 0 \tag{6.44}$$

where $b_x(0,0) = 0$, and $b_x(0,n) = b_x(n,0)$. The unwrapped bispectral phase is then computed from (6.41) and (6.44) as follows:

$$\Psi_3^x(\omega_1, \omega_2) = -j(F_2\{b_x(m,n)\} - \ln|C_3^x(\omega_1, \omega_2)|). \tag{6.45}$$

Example 6.1

The performance of the BSR method for phase recovery is tested for a specific example that includes additive noise. The model

$$Z(k) = h(k) + W(k)$$

is considered, where $W(k)$ is zero-mean Gaussian white noise of variance γ_2^w and $h(k) = \{0.7, 1, 0.8, -0.7, 0.6, -0.5, 0.4, -0.3, 0.2, -0.1\}$.

It is assumed that $L = 50$ realizations of $Z(k)$ are available with length 128 samples in each realization, i.e., $Z_i(k) = h(k) + W_i(k), i = 1, 2, \ldots, L$. The bicepstrum is estimated based on (5.47) where the third-order moment sequence is taken to be the ensemble average of the third-order moment sequences of each $Z_i(k)$. The averaging is performed to reduce the variance in bicepstrum estimates. From the estimated bicepstrum, and applying (5.38), the cepstral coefficients of $h(k)$ are computed, and applying (5.38), the cepstral coefficients of $h(k)$ are computed. The Fourier-phase of $h(k)$ is computed based on (6.37) and the results are shown in Figure 6.3 for different noise levels. The signal-to-noise ratio is defined by

$$\text{SNR} \triangleq \frac{1}{N\gamma_2^w} \cdot \sum_{k=1}^{N} h^2(k)$$

where $N = 10$ is the length of the FIR sequence $h(k)$.

The illustrated true phase in Figure 6.3 is obtained using (6.37), assuming that no noise is present. Apparently the phase reconstruction is very good even for low SNR's. □

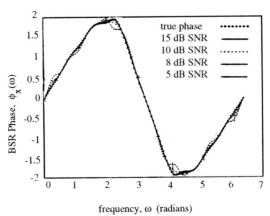

Figure 6.3 Phase Reconstruction Using BSR, as described in Example 6.1

6.4.4 Duality Between Phase and Magnitude Recovery

Let us rewrite (6.3) in the following way:

$$\ln |C_3^x(\omega_1, \omega_2)|^2 - \ln |X(\omega_1 + \omega_2)|^4 = \ln |X(\omega_1)|^2 + \ln |X(\omega_2)|^2 - \ln |X(\omega_1 + \omega_2)|^2. \tag{6.46}$$

Comparing (6.1) and (6.46), we observe that there exists a duality between the following quantities:

$$\Psi_3^x(\omega_1, \omega_2) \leftrightarrow \ln |C_3^x(\omega_1, \omega_2)|^2 - \ln |X(\omega_1 + \omega_2)|^4 \tag{6.47.1}$$

and

$$\phi_x(\omega) \leftrightarrow \ln |X(\omega)|^2. \tag{6.47.2}$$

Consequently, we may apply (6.40) to compute $\ln |X(\omega)|^2$ as follows:

$$\ln |X(\omega)|^2 = \frac{1}{2N-1} \sum_{i=-N+1}^{N-1} \{\ln |C_3^x(\omega, i)|^2 - \ln |X(\omega + i)|^4\}, \quad \omega = 0, \ldots, N-1. \tag{6.48}$$

It can easily be shown that

$$\frac{1}{2N-1} \sum_{i=-N+1}^{N-1} \ln|X(\omega+i)|^4 = 2p_x(0) \qquad (6.49)$$

where $p_x(0)$ is the 0-th lag of the power cepstrum of $X(\omega)$. Hence, from (6.48) and (6.49) we obtain

$$\ln|X(\omega)|^2 = \frac{1}{2N-1} \sum_{i=-N+1}^{N-1} \ln|C_3^x(\omega,i)|^2 - 2 \cdot p_x(0). \qquad (6.50)$$

Although $p_x(0)$ is not known, $|X(\omega)|$ can be computed from (6.50) within a constant, $\exp\{2 \cdot p_x(0)\}$.

Two versions of the Bispectrum Signal Recovery (BSR) procedure for phase and magnitude reconstruction are summarized in Table 6.2.

TABLE 6.2   PHASE AND MAGNITUDE RECOVERY BASED ON
POLYSPECTRA/POLYCEPSTRA PROPERTIES

	Phase Estimation	Magnitude Estimation
Nonparametric BSR	$\phi_x(\omega) = \frac{1}{2N+1} \sum_{i=-N}^{N} \Psi_3^x(\omega,i)$ $\omega = 0, \ldots, N-1$	$\ln(\alpha|X(\omega)|^2) = \frac{1}{2N+1} \sum_{i=-N}^{N} \ln|C_3^x(\omega,i)|^2$ $\omega = 0, \ldots, N-1$ α: positive constant
Polycepstra-Based SBR	$\phi_x(\omega) = F_1\{d(m)\}, \omega = 0, \ldots, N-1$ $d(m) = \begin{cases} \frac{j}{2m}[A(|m|) - B(|m|)], & m \neq 0 \\ 0, & m = 0 \end{cases}$	$\ln|X(\omega)|^2 = F_1\{p_x(m)\}, \omega = 0, \ldots, N-1$ $p_x(m) = \begin{cases} -\frac{1}{m}[A(m) + B(m)], & m > 0 \\ \frac{1}{m}[A(-m) + B(-m)], & m < 0 \end{cases}$

6.4.5 Treating Singularities in the Vicinity of the Unit Circle

All polycepstra-based schemes cannot accommodate zeros or poles on the unit circle. In the noise free case, if the Z-transform of the sequence $X(k)$ has singularities on the unit circle, by multiplying $X(k)$ by α^k, we force the singularities to move inside ($|\alpha| < 1$), or outside ($|\alpha| > 1$) of the unit circle.

Consider now the case where $X(k)$ is corrupted by additive stationary, zero-mean Gaussian noise, $W(k)$, i.e.,

$$Y(k) = X(k) + W(k) \qquad (6.51)$$

and that we have access only to the observed sequence $Y(k)$. Multiplication of $Y(k)$ by α^k will have the undesired effect of turning the noise into a nonstationary process. On the other hand, in the third-order moment domain, the noise, with the aforementioned properties, is suppressed, i.e.,

$$m_3^y(\tau,\rho) = m_3^x(\tau,\rho). \tag{6.52}$$

Let $m_3^z(\tau,\rho)$ be a windowed version of $m_3^y(\tau,\rho)$:

$$m_3^z(\tau,\rho) = w(\tau,\rho)m_3^y(\tau,\rho), \tag{6.53}$$

where
$$w(\tau,\rho) = \alpha^{\tau+\rho}. \tag{6.54}$$

From the definition of the bispectrum and (6.52) we have

$$C_3^y(\alpha^{-1}z_1, \alpha^{-1}z_2) = C_3^x(\alpha^{-1}z_1, \alpha^{-1}z_2) = X(\alpha^{-1}z_1)X(\alpha^{-1}z_2)X(\alpha^2 z_1^{-1} z_2^{-1}). \tag{6.55}$$

Let $Z(k)$ be the sequence that corresponds to $m_3^z(\tau,\rho)$. Then, from (6.53)–(6.55) and the Z-transform properties we obtain

$$Z(k) = c \cdot \alpha^k \cdot X(k - k_o) \tag{6.56}$$

where k_o is an integer denoting the time shift suppressed in the bispectrum domain, and c is a real constant denoting the scale that was also lost in the bispectrum domain. Hence, the role of the third-order moment window in (6.53) is to push the zeros or the poles of $X(z)$ away from the unit circle. (Of course, there may be cases where a zero or pole that does *not* reside close to the unit circle, will come closer to the unit circle after the windowing, creating a problem.) Therefore, a suitable value for the window constant has to be selected. One way to check if a particular value for α is appropriate is to apply (6.53) and (6.54), and then from the bispectrum of the windowed moments find the corresponding cepstrum coefficients. If there is still a singularity close to the unit circle, the computed cepstral coefficients will not decay fast enough within their chosen length.

Third-Order Moment Window Properties

Let $A_x^{(m)}$ and $B_x^{(m)}$ be the cepstral coefficients of $X(k)$. Then it holds that the cepstral coefficients of $Z(k)$ are

$$A_z^{(m)} = A_x^{(m)}\alpha^m \tag{6.57}$$

$$B_z^{(m)} = B_x^{(m)} \alpha^{-m}. \tag{6.58}$$

The above equations can easily be verified using (5.11), which shows the relationship between the signal's cepstral parameters with its poles and zeros.

Suppose we observe the signal $Y(k)$, which is

$$Y(k) = X(k) * h(k) + W(k) \tag{6.59}$$

where $W(k)$ is the noise, as defined in (6.51). The application of the window on the third-order moment sequence of $Y(k)$ is equivalent to applying the same window on the third-order moment sequences of $X(k)$ and $h(k)$ separately. This can simply be drawn by transforming (6.59) into its cepstral coefficients domain where

$$A_y^{(m)} = A_x^{(m)} + A_h^{(m)} \tag{6.60.1}$$

$$B_y^{(m)} = B_x^{(m)} + B_h^{(m)}. \tag{6.60.2}$$

The windowing of $m_3^y(\tau,\rho)$ by $w(\tau,\rho) = \alpha^{\tau+\rho}$ leads to the multiplication of $\{A_y^{(m)}, B_y^{(m)}\}$ by α^m and α^{-m}, respectively or from (6.60), to the multiplication of both $\{A_x^{(m)}, B_x^{(m)}\}$ and $\{A_h^{(m)}, B_h^{(m)}\}$ by α^m and α^{-m}, respectively. The latter is the same as if $m_3^x(\tau,\rho)$ and $m_3^h(\tau,\rho)$ had been windowed by $w(\tau,\rho)$ separately.

6.5 SIGNAL RECONSTRUCTION FROM ONLY THE PHASE OF THE BISPECTRUM

It has been well established in the signal processing literature that under certain conditions, a signal may be reconstructed using only partial Fourier domain information. Several schemes have been proposed to reconstruct a signal from some samples of its Fourier phase or magnitude, or both [Oppenheim and Lim, 1981; Stark, 1987].

In this section, we first present a brief literature review of the existing signal reconstruction schemes from partial Fourier-domain or time-domain information and their practical applications. The rest of this section is devoted to incomplete higher-order spectra (HOS) information signal reconstruction from incomplete higher-order spectra (HOS) information. Since one of the key properties of HOS is to preserve phase information, the emphasis in our discussion is placed on signal reconstruction algorithms from only the phase of the bispectrum.

6.5.1 The Importance of Signal Recovery from Partial Information

In signal reconstruction from partial information, there are two chief problem definitions, namely, the phase retrieval from magnitude information and the reverse. In the phase retrieval problem, the Fourier magnitude of the signal is known and the Fourier phase is to be calculated. This case arises in many applications in x-ray crystallography where the intensity of the reflected x-ray is proportional to the magnitude of a quantity called the structure factor [Ramachandran and Srinivasan, 1970]. It has been shown that the structure factor is related to the electron density distribution through a Fourier transform. The structure factor is a complex number and only its magnitude can be computed. Most of the theoretical approaches in crystallography are focused toward obtaining the structure factor's phase.

Imaging through atmospheric turbulence is another area where phase retrieval is important. When the resolution of an object falls below the diffraction limit of the telescope, by employing interferometric techniques it is possible to obtain some Fourier magnitude information about the object [Fienup, 1979]. If the recovery of the phase from the estimated magnitude is possible, then one could obtain diffraction limited images through a turbulent atmosphere.

There are other applications where the magnitude retrieval is important. For example, consider the case of a signal which is transmitted through an unknown channel. If the channel is known to be zero (or linear) phase then the Fourier phase of the observed signal equals the phase of the transmitted signal (within a linear phase shift). Such situations occur in long-term exposure to atmospheric turbulence or when an image is blurred by a defocused lens with circular aperture stops [Andrews, 1977].

Since the early 1970s, several schemes have been proposed for the reconstruction of a signal from partial information. First, Gerchberg and Saxton [1972] introduced an iterative procedure, called the G-S algorithm, for the restoration of a signal from the magnitude of its Fourier transform. Later, Papoulis [1975] employed the G-S algorithm to recover a sequence based on the knowledge of a part of its time or frequency representation. Fienup [1978] used similar ideas to develop a procedure for the reconstruction of a sequence from only its magnitude. The importance of the phase in signal recovery has been pointed out several times, and iterative schemes for signal reconstruction from Fourier phase only have been presented, as well as the conditions under which such a reconstruction is possible. [Hayes, Lim, and Oppenheim, 1980; Oppenheim and Lim, 1981; Tom, Quatieri, et al., 1981; Oppenheim, Hayes, et al., 1980; Quatieri and Oppenheim, 1981]. Youla and Webb [1982] and Sezan and Stark [1982] employed the method of projections on convex sets to recover a signal from partial information.

6.5.2 Conditions for Signal Reconstruction from the Phase of the Bispectrum

For a minimum phase sequence, the log magnitude of its Fourier transform and its Fourier phase form a Hilbert transform pair [Oppenheim and Schafer, 1989]. Hence, the signal's Fourier magnitude can be computed from its Fourier phase and vice versa. Consequently, the knowledge of only the Fourier phase or magnitude of a minimum phase signal can lead to the unique reconstruction of the signal.

The conditions under which a general FIR sequence can be reconstructed from its Fourier phase only were stated by Hayes, Lim, and Oppenheim [1980] in the form of the following theorem:

Theorem 6.1

Let $x(k)$ and $y(k)$ be two FIR sequences which are zero outside the interval $[0, N-1]$, and their Z transforms have no zeros on the unit circle, nor in reciprocal pairs. Let $\theta_x(\omega)$ and $\theta_y(\omega)$ be the Fourier phases of $x(k)$ and $y(k)$, respectively. If $\theta_x(\omega) = \theta_y(\omega)$ at $N-1$ distinct frequencies in the interval $0 < \omega < \pi$, then $x(k) = \alpha y(k)$, for some positive constant α.

The proof is given in [Hayes, et al., 1980]. The conditions under which a FIR sequence can be reconstructed from its bispectral phase only are stated in the following lemma [Petropulu and Nikias, 1992]:

Lemma 6.1

Let $x(k)$ and $y(k)$ be FIR sequences of length N that have Z transforms with no zeros on the unit circle nor in reciprocal pairs. Let $\Psi_3^x(\omega_1,\omega_2)$ and $\Psi_3^y(\omega_1,\omega_2)$ be their bispectral phases. Suppose we sample the bispectral phases at $L = 2^v > 2N-1$ equispaced frequency points (v is any integer). If $\Psi_3^x(\omega_1,\omega_2) = \Psi_3^x(\omega_1,\omega_2)$ at discrete frequency pairs within the nonredundant bispectrum region $\{0 \leq \omega_1+\omega_2 \leq \pi, \omega_2 \leq \omega_1, \omega_1 \geq 0\}$, then $x(k) = \alpha y(k - k_0)$ for some positive constant α and some integer k_0.

Proof

From the bispectrum phase we can reconstruct the Fourier phase within a linear phase term. Let $\phi_x(\omega)$ and $\phi_y(\omega)$ be the phases reconstructed from $\Psi_3^x(\omega_1,\omega_2)$ and $\Psi_3^y(\omega_1,\omega_2)$, respectively. Then $\phi_x(\omega) = \phi_y(\omega) + 2\pi \cdot k_1(\omega)$ and $\phi_y(\omega) = \phi_y(\omega) + 2\pi \cdot k_2(\omega)$, where $\phi_x(\omega)$ and $\phi_y(\omega)$ are the Fourier phase of $x(k)$ and $y(k)$, respectively. Since $\phi_x(\omega)$ and $\phi_y(\omega)$ correspond to $x(k - k_1)$ and $y(k - k_2)$, respectively (k_1 and k_2 are integers), and $\phi_x(\omega) = \phi_y(\omega)$, based on theorem 6.1 we have that $x(k) = \alpha \cdot y(k - k_0)$ where α is a positive constant and $k_0 = k_2 - k_1$ is some integer.

Petropulu and Nikias [1992] proposed the following algorithm to reconstruct a FIR sequence that meets the conditions of Lemma 6.1 from its bispectral phase only.

6.5.3 The Bispectrum Iterative Reconstruction Algorithm (BIRA)

Let $x(k)$ be a FIR sequence of length N with a Z transform which has no zeros on the unit circle nor in reciprocal pairs. As in Lemma 6.1, we assume that the bispectrum of $x(k)$ is sampled at $L = 2^v$ (v is any integer), where $L > 2 \cdot N - 1$ equispaced points. The BIRA algorithm recovers a scaled and shifted version of $x(k)$ based only on the signal's bispectral phase. As shown in section 5.5.1, this bispectral phase information is contained in the differences of the cepstrum coefficients of $x(k)$ [see (5.58) and (5.59.2)]. Hence, the algorithm basically reconstructs a scaled and shifted version of $x(k)$ from the differences of its cepstral coefficients only. The BIRA algorithm proceeds as follows:

Initialization:

Estimate the bicepstrum of $x(k)$ (see section 5.5.3), and from that compute the values of $A^{(m)} - B^{(m)}$, i.e.,

$$D(m) = A^{(m)} - B^{(m)}, \quad m = 1, 2, \ldots, r \tag{6.61.1}$$

where $r = \max(p, q)$ (p and q are the lengths of $A^{(m)}$ and $B^{(m)}$ respectively). When the Fourier magnitude is corrupted, only the differences of the computed cepstral coefficients contain undistorted information. Note that

$$D(m) = 2m \cdot bic_x(m, 0). \tag{6.61.2}$$

Set each sum of the cepstral coefficients to some arbitrary initial value, i.e.,

$$A^{(m;0)} + B^{(m;0)} = 0, \quad m = 1, 2, \ldots, r \tag{6.62.1}$$

where $A^{(m;0)}$ and $B^{(m;0)}$ denote the values of the cepstral coefficients at iteration $i = 0$. Note that according to (5.25), equation (6.62.1) is equivalent to

$$-m \cdot p_x(m;0) = 0, \quad m = 1,\ldots,r \qquad (6.62.2)$$

where $p_x(m;0)$ is the power cepstrum of $x(k)$ at iteration $i = 0$.

Step 1: Solve (6.61) and (6.62) for cepstral coefficients:

$$A^{(m;i)} = \frac{D(m) - m \cdot p_x(m;i)}{2}$$

$$B^{(m;i)} = \frac{-D(m) - m \cdot p_x(m;i)}{2}, \qquad (6.63)$$

$$m = 1, 2, \ldots, r$$

where $\{i\}$ is the iteration index, taking values $i = 1, 2, 3, \ldots$

Step 2: Compute $x(k;i)$ from the equations

$$x(k;i) = F_1^{-1}\{e^{F_1\{c_x(m;i)\}}\}, k = 0,\ldots, M-1 \qquad (6.64.1)$$

$$c_x(m;i) = \begin{cases} -\frac{1}{m}A^{(m;i)}, & m > 0 \\ 0, & m = 0 \\ \frac{1}{m}B^{(-m;i)}, & m < 0 \end{cases} \qquad (6.64.2)$$

where $x(k;i)$ is the computed sequence at iteration (i) and $M(M > 2r)$ is the length of the Fourier transform used in (6.64.1).

Step 3: Generate the sequence $Y(k;i)$ as follows:

$$y(k;i) = x(k;i)R_N(k), \quad k = 0,\ldots, M-1 \qquad (6.65.1)$$

where

$$R_N(k) = \begin{cases} 0, & N - k_0 \le k \le M - k_0 - 1 \\ 1, & \text{otherwise} \end{cases} \qquad (6.65.2)$$

where k_0 is the time shift introduced to the signal due to its reconstruction from its cepstrum coefficients. Due to this shift, $x(k;i)$ will appear in the interval $[-k_0, N - k_0 - 1]$. When $x(k;i)$ is computed based on (6.64.1), the sample $x(-k;i)$ will appear at position $M - k$, as long as $M > N$. The value of k_0 is not known, but for the moment let us assume that we can compute it.

Step 4: Calculate the power cepstrum of $y(k;i)$, i.e.,

$$p_y(m;i) = F_1^{-1}\{\ln|Y(\omega;i)|^2\} \qquad (6.66.1)$$

and set
$$p_x(m;i+1) = p_y(m;i). \qquad (6.67)$$

Repeat Step 1–Step 4 until the reconstructed sequence $x(k;i)$ remains unchanged as $\{i\}$ increases. In other words, if we define

$$E_i = \sum_{k=0}^{M-1}[x(k;i) - x(k;i-1)]^2, \qquad (6.68)$$

the algorithm stops at iteration $i = I$ if $E_I < \delta$ where δ is very small.

It can be easily shown that the BIRA algorithm consists of successive applications of the phase substitution and time-limiting operations [Petropulu and Nikias, 1992]. Hence, based on the theory of nonexpansive maps, it is guaranteed to converge [Tom, et al., 1981].

At the point of convergence, due to Lemma 6.1, we have that

$$x(k;I) = \alpha x(k - k_0). \qquad (6.69)$$

Steps 1–4 correspond to the BIRA(k_0) algorithm. The time shift k_0 required in Step 3 can be computed as follows. Initially guess some value for k_0 within $[0, N-1]$. Apply Steps 1–4 checking E_i at successive iterations. The BIRA(k_0) will not reach convergence if k_0 is incorrect. A simple test for convergence at iteration I is to compute the cepstral coefficients of $y(k;I)$ and then take their differences, i.e., $\hat{D}(m), m = 1, \ldots, r$. If $\hat{D}(m) \neq D(m)$ (in the mean-square error sense) then the algorithm has not converged and we have to guess another value for k_0 within $[0, N-1]$. After a finite number of trials the algorithm will reach a convergent solution.

The computation of k_0 can be implemented as an external loop connected to the BIRA(k_0) algorithm. The whole structure then corresponds to the BIRA algorithm, and is illustrated in Table 6.3.

The BIRA algorithm was described so far for deterministic signals $x(k)$. However, it also applies to stochastic non-Gaussian processes that are generated by exciting FIR system with impulse response $C_{x(k)}$, by a non-Gaussian white input process. In this case, it is absolutely necessary to use the bicepstrum for the computation of the cepstrum coefficients of the random process, while for the deterministic case the complex cepstrum could also be used.

Chapter 6   243

Convergence Improvement of BIRA

By setting an upper bound to the energy of the cepstrum or equivalently to the energy of $\frac{A^{(m;i)}}{m}$ and $\frac{B^{(m;i)}}{m}$ of the signal $x(k;i)$ at each iteration $\{i\}$, the convergence rate of the algorithm can be substantially improved. At each iteration $\{i\}$, we set

$$E(i) = \sum_{m=1}^{r} \left\{ \left(\frac{A^{(m;i)}}{m}\right)^2 + \left(\frac{B^{(m;i)}}{m}\right)^2 \right\} \leq \mathcal{E}. \tag{6.70}$$

To guarantee faster convergence of the algorithm, the following step (Step 1a) should be included immediately after Step 1 in the BIRA(k_0):

Step 1(a):

$$A^{(m,i)} = \begin{cases} A^{(m;i)}, & E(i) \leq \mathcal{E} \\ \sqrt{\frac{\mathcal{E}}{E(i)}} \cdot A^{(m;i)}, & E(i) > \mathcal{E} \end{cases} \tag{6.71.1}$$

$$B^{(m,i)} = \begin{cases} B^{(m;i)}, & E(i) \leq \mathcal{E} \\ \sqrt{\frac{\mathcal{E}}{E(i)}} \cdot B^{(m;i)}, & E(i) > \mathcal{E}. \end{cases} \tag{6.71.2}$$

For more discussion on the bounds of (6.71) see Sezan and Stark [1982] and Petropulu [1990]. Even further reduction in the number of iterations can be achieved if separate bounds on the energies of the cepstral coefficients are available, viz:

$$E_a(i) = \sum_{m=1}^{r} \left(\frac{A^{(m;i)}}{m}\right)^2 \leq \mathcal{E}_a, \tag{6.72.1}$$

$$E_b(i) = \sum_{m=1}^{r} \left(\frac{B^{(m;i)}}{m}\right)^2 \leq \mathcal{E}_b. \tag{6.72.2}$$

To incorporate these bounds in each iteration of the algorithm, we should include the following step, Step 1(b), instead of Step 1(a), immediately after Step 1:

TABLE 6.3 (a) THE BIRA (K_0) ALGORITHM AND (b) THE BIRA ALGORITHM

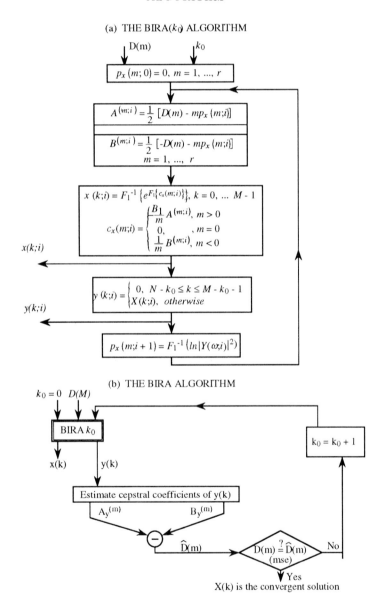

Step 1(b):

$$A^{(m,i)} = \begin{cases} A^{(m;i)}, & E_a(i) \leq \mathcal{E}_a \\ \sqrt{\frac{\mathcal{E}_a}{E_a(i)}} \cdot A^{(m;i)}, & E_a(i) > \mathcal{E}_a \end{cases} \qquad (6.73.1)$$

$$B^{(m,i)} = \begin{cases} B^{(m;i)}, E_b(i) \leq \mathcal{E}_b \\ \sqrt{\frac{\mathcal{E}_b}{E_b(i)}} \cdot B^{(m;i)}, & E_b(i) > \mathcal{E}_b. \end{cases} \qquad (6.73.2)$$

Of course Step 1(a) or Step 1(b) can be included only if we are given some a priori information about the total energies of the cepstral coefficients, $\frac{A^{(m)}}{m}$ and $\frac{B^{(m)}}{m}$ (combined or separately) of the signal $x(k)$.

Example 6.2

The BIRA algorithm is applied to recover an image that has been distorted magnitude-wise. The recovery is based entirely on the phase of the bispectrum of the observed image. It was assumed that the image was transmitted line-by-line through the linear phase channel $h(n) = \{0.8, 1., 0.8\}$, and as a result, its magnitude was distorted. In Figure 6.4, the originally transmitted image (14 × 14) is shown, as well as the received image (14 × 16).

A typical Fourier-magnitude plot for a horizontal line off the "NU" image is shown in Figure 6.5. The deep nulls in this plot correspond to zeros in the vicinity of the unit circle. Therefore, in order to apply cepstra-based operations, a windowing on the third-order moment sequence of every line of the received image is necessary (for window description see section 6.4.5). For this particular example, the window was taken to be

$$w(m, n) = (0.9)^{m+n}.$$

The BIRA algorithm was applied on the exponentially weighted cepstral coefficients, calculated from windowed third-order moment sequences. The reconstructed sequence was multiplied by the inverse exponential window. The reconstructed images obtained after 1, 10, and 30 iterations, respectively, are shown in Figure 6.4 (b,c, and d, respectively). □

Example 6.3

In this example, the BIRA is applied for the recovery of an image that has been

original image

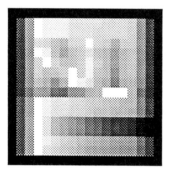

Image convolved with
$h(n) = (0.8, 1, 0.8)$
(a)

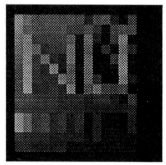

Recovery after 1 iteration
(b)

Recovery after 10 iterations
(c)

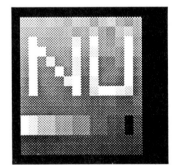

Recovery after 30 iterations
(d)

Figure 6.4 Recovery of an image convolved with a linear phase channel, with BIRA ([Petropulu and Nikias, 1992] © 1992, IEEE)

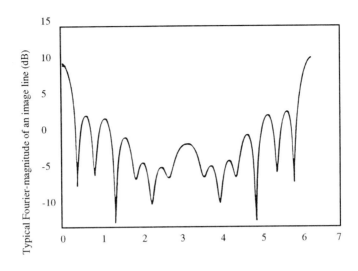

Figure 6.5 Typical Fourier-magnitude plot of an image line

corrupted by additive zero-mean, white Gaussian noise. The reconstruction is based entirely on the bispectral phase of the observed image.

The originally transmitted image (60 × 60) is shown in Figure 6.6. The processing was made on a line-by-line basis. It was assumed that the same image was received by an array of 10 different receivers. The third-order moment sequence was taken to be the ensemble average of the third-order moments from the 10 receivers. A third-order moment window ($\alpha = 0.82$) was applied on the third-order moment sequence of each receiver. The recovered images for 30 and 25 dB SNR are shown in Figure 6.6 against the observed images at one receiver under the same SNR. For this image recovery example the BIRA algorithm executed 50 iterations. □

Example 6.4

In this example, the BIRA algorithm with energy constraints is compared to the iterative scheme proposed by Oppenheim, et al. [1980], which is based on the adaptive relaxation technique.

Let the FIR sequence be

$$x(k) = (0.3, 1, -0.4, 0.7, -0.1).$$

The BIRA algorithm, with and without energy constraints, is applied to re-

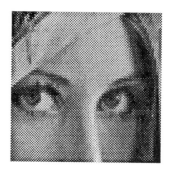

original image

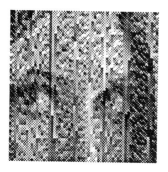

noisy image at 25 dB

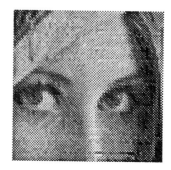

recovery at 25 dB

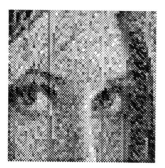

noisy image at 30 dB

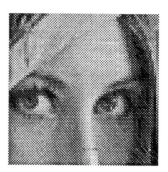

recovery at 30 dB

Figure 6.6 Recovery of an image corrupted by additive white Gaussian noise, with BIRA ([Petropulu and Nikias, 1992] © 1992 IEEE)

construct $x(k)$ from only the phase of its bispectrum, or equivalently, from the differences of its cepstral coefficients. The adaptive relaxation technique is also employed for the reconstruction of $x(k)$ from its Fourier phase, where the Fourier phase has been computed from the bispectrum phase using (6.37).

Figure 6.7 illustrates the mean-squared error between the true sequence and the reconstructed sequence versus the number of iterations for the following methods: BIRA, the adaptive relaxation technique, and BIRA with energy constraints. In BIRA with energy constraints, the two numbers $(\mathcal{E}_a, \mathcal{E}_b)$ denote the bands on the energy of the cepstral coefficients.

The goal of this comparison is not to show that one scheme is better than the other, since the adaptive relaxation scheme utilizes no a priori information about the sequence. Figure 6.7 demonstrates how much improvement can be achieved in the BIRA convergence rate, when a priori information about $x(k)$ is available.

In Figure 6.8, the experiments are repeated for the case of additive white Gaussian noise at 25 dB SNR. The rate of convergence of BIRA with energy constraints is very high in the first 5 iterations, but then slows down. On the other hand, the adaptive-relaxation scheme has a uniform rate of convergence, except at some isolated points. Note that those isolated points are due to the fact that the relaxation parameter is not restricted to lie within the interval (0,2). This result indicates that we can run BIRA with energy constraints for a few iterations and then switch to the adaptive relaxation algorithm. The mean-squared error versus the number of iterations for the "hybrid" scheme is also shown in Figure 6.8. Clearly, the convergence rate is greatly improved. □

6.6 SIGNAL RECONSTRUCTION FROM THE BISPECTRUM BASED ON THE METHOD OF POCS

An iterative method for the reconstruction of the impulse response of a LTI system from knowledge of some bispectral samples only was proposed by Cetin [1989; 1991]. Cetin's algorithm is based on the method of Projections Onto Convex Sets (POCS) [Bregman, 1965; Youla and Webb, 1982; Stark 1987].

Let $G(k)$ be a random process generated by an LTI system with impulse response $h(k)$, driven by white non-Gaussian noise. Cetin's algorithm reconstructs $h(k)$ from the bispectral samples $\{C_3^g(\omega_{1i}, \omega_{2j}), i = 1, 2, \ldots, L; j = 1, 2, \ldots, L\}$ in an iterative fashion. If $c_h(m, \ell)$ denotes the complex cepstrum of $h(k)$ at the ℓ-th iteration, then the algorithm consists of cycles of the form:

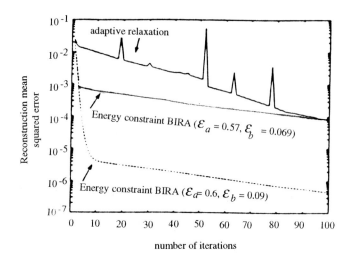

Figure 6.7 The mean-squared error between the original sequence and the reconstructed one versus the number of iterations, for BIRA, the adaptive relaxation technique, and BIRA with energy constraints, for different energy bounds

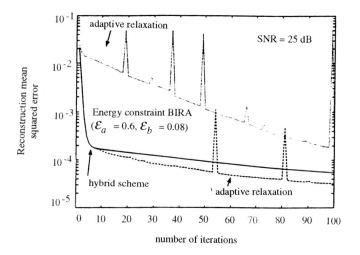

Figure 6.8 Mean-squared error versus the number of iterations for the adaptive relaxation technique, and BIRA with energy constraints and the "hybrid" scheme for SNR = 25 dB ([Petropulu and Nikias, 1992] © 1992 IEEE)

$$i = 1, 2, \ldots, L$$

$$j = 1, 2, \ldots, L$$

$$c_h(m;\ell) = c_h(m;\ell-1) + \lambda_m[\cos(\omega_{1i}m) + \cos(\omega_{2j}m) + \cos([\omega_{1i}+\omega_{2j}]m)]$$
$$m = -q \ldots \rho, \ldots, p$$

(6.74.1)

$$\lambda_m = \frac{\ln|c_3^g(\omega_{1i},\omega_{2j})| - Re\{c_h(\omega_{1i};\ell-1) + c_h(\omega_{2j};\ell-1) + c_h(\omega_{1i}+\omega_{2j};\ell-1)\}}{\sum_{r=-q}^{q}[\cos(\omega_{1i}r) + \cos(\omega_{2j}r) + \cos((\omega_{1i}+\omega_{2j})r)]^2}$$

$$\ell = \ell + 1$$

(6.74.2)

$$c_h(m;\ell) = c_h(m;\ell-1) + \lambda_p[\sin(\omega_{1i}m) + \sin(\omega_{2j}m) - \sin([\omega_{1i}+\omega_{2j}]m)]$$
$$m = -q, \ldots, 0, \ldots, p$$

(6.74.3)

$$\lambda_p = \frac{\arg[c_3^q(\omega_{1i},\omega_{2j})] - Im\{c_h(\omega_{1i};\ell-1) + c_h(\omega_{2j};\ell-1) - c_h(\omega_{1i}+\omega_{2j};\ell-1)\}}{\sum_{r=-q}^{q}[\sin(\omega_{1i}r) + \sin(\omega_{2j}r) - \sin((\omega_{1i}+\omega_{2j})r)]^2}$$

for $\omega_{1i} \neq 0$ and $\omega_{2j} \neq 0$, $\ell = \ell + 1$ (6.74.4)

where $c_h(m;0) = \delta(m), m = -q, \ldots, 0, \ldots, p, c_h(w;\ell)$ denotes the Fourier transform of the complex cepstrum of $h(k)$ at iteration ℓ, and p,q are integers chosen so that $c_h(m;\ell)$ is zero for $m > p, m < -q$. Equations (6.74.1) through (6.74.4) describe one cycle. Initially, $\ell = 1$. However, cycles are repeated until a satisfactory error difference in successive cycles is obtained. The final result of the iteration is the cepstrum of the sequence $\hat{h}(k)$, that is the closest sequence to $h(k)$ with respect to the distance

$$\|\hat{h} - h\|_*^2 = \frac{1}{2\pi}\int_{-\pi}^{\pi}|\log H(\omega) - \log \hat{H}(\omega)|^2 d\omega \quad (6.75)$$

where $H(\omega), \hat{H}(\omega)$ are the Fourier transforms of $h(k)$ and $\hat{h}(k)$, respectively.

The derivation of the algorithm can be found in [Cetin, 1991]. If bounds of the energy of the cepstrum $c_h(m)$ are given a priori, i.e.,

$$\sum_{m=-q}^{p} c_h^2(m) \leq \mathcal{E} \quad (6.76)$$

then the following steps can be included in the algorithm right after (6.74.4)

$$c_h(m;\ell) = \begin{cases} c_h(m;\ell-1), & \text{if } E(\ell-1) \le \mathcal{E}_0 \\ \left(\dfrac{\mathcal{E}_0}{\sum_r c_h^2(r;k-1)}\right)^{1/2}, & \text{if } E(\ell-1) > \mathcal{E}_0 \end{cases} \quad (6.77.1)$$

$$\ell = \ell + 1 \quad (6.77.2)$$

where

$$E(\ell) = \sum_{r=-q}^{p} c_h^2(r;\ell). \quad (6.77.3)$$

Faster convergence can be achieved with the utility of energy bounds on the cepstrum coefficients.

Example 6.5 [Cetin, 1989]

The iterative reconstruction algorithm based on the POCS method is tested for the reconstruction of a specific sequence. Let $\mathbf{h} = \{\ldots, 0, 0, 1, 0.75, h(0) = 1, -0.5, 0.5, 1, 0, 0, \ldots\}$ be the impulse response of a LTI system. It is assumed that the bispectrum of $\mathbf{h}$, i.e., $C_3^h(\omega_1, \omega_2)$, was corrupted by noise. The signal-to-noise ratios of the bispectrum magnitude and phase are respectively defined by

$$\text{SNRM} = \frac{\sum_{i,\ell} |C_3^h(\omega_{1i}, \omega_2)|^2 j}{\sum_{i,\ell} |N_m(\omega_{1i}, \omega_{2j})|^2} = 15.56$$

$$\text{SNRM} = \frac{\sum_{i,\ell} \arg\left[C_3^h(\omega_{1i}, \omega_2 j)\right]^2}{\sum_{i,\ell} |N_p(\omega_{1i}, \omega_2)|^2 j} = 16.13$$

where $N_m(\omega_{1i}, \omega_{2\ell})$ and $N_p(\omega_{1i}, \omega_{2\ell})$ are the additive noise components that corrupt the magnitude and phase bispectrum, respectively.

The reconstruction is based on the bispectral samples $c_3^h(\omega_{1i}, \omega_{2j})$, where $\omega_{1i} = \omega_{2i} = \frac{2\pi i}{64}$, $i = 0, 1, 2, \ldots, 32$ and $\omega_{1i} = \frac{2\pi i}{64}, \omega_{2i} = 0, i = 0, 1, \ldots, 32$. The absolute minimum error achieved after 15 iterations is $\| e_{15} \|_1 = \| \mathbf{h} - \mathbf{h}_{15} \|_1 = 0.28$. If we were given that $\| h \| = 1.647$, then the resultant minimum error would have been 0.16. Obviously, the energy information improves the reconstruction. □

6.7 BLIND DECONVOLUTION USING BIRA

A situation encountered often in applications like communications, seismology, ocean acoustics, sonar, radar etc., is when an unknown signal propagates through a multipath environment of unknown transfer function. Blind deconvolution is the reconstruction of the original signal from a reverberated version of it. Obviously the problem, as stated, contains too many unknowns to be solved uniquely, unless something is known about the distinguishable characteristics of the signal or the reverberation channel. Most of the algorithms proposed in the past deal with specific applications where some properties of the two signals are known. Stockham, et al. [1975] developed a scheme that recovers reverberated sound and applied it to restore old recordings. The proposed technique was not able to estimate phase distortions associated with old recordings; however, since the human ear is relatively insensitive to phase this was not a critical problem. The deconvolution of cyclostationary signals with known statistics has been addressed by Cerrato and Eisenstein [1977]. The blind deconvolution has been approached as an optimization problem whose cost function is related to how well the reverberated signal constraints are satisfied [Smith, et al., 1983]. The reduction of the cost function was achieved by simulated annealing, a process analogous to the cooling of a melt to produce an ordered crystal. However, the solution depends upon the chosen cost function. For the case that nonnegativity and extend constraints are known for both signals, an iteratively blind deconvolution method has been suggested by Ayers and Dainty [1988]. Such constraints may be available in image reconstruction problems, but generally this information is not available. In addition, the convergence properties of the iterative scheme were uncertain and the uniqueness of the solution was not guaranteed. A more general approach was followed by Li and Kurkjian [1983] who suggested to consider two different measurements of the reverberated signal, each of which is the convolution of the same signal with two different channels. For the noise-free case, and for deterministic signals, Li and Kurkjian suggested that the phase of the cross-spectrum of the two measurements equals the phase of a sequence which is a function of the two channels only. This sequence could be reconstructed from phase only information [Hayes, et al., 1980] under the assumption that the arrival sequence at each receiver had no zero-phase convolutional component. Then the two channels could be reconstructed from this sequence based on an exhaustive search. In addition to the computational load that the exhaustive search introduced, especially when the lengths of the channels are long, this algorithm failed to work in the case where additive observation noise is present in each measurement. Petropulu and Nikias [1991] proposed a solution to the blind deconvolution problem by means of higher-order cepstra. The proposed algorithm utilizes information collected by two sensors, in recursive, and applies for both determinist and stochastic signals, in the noise free or additive Gaussian noise cases. The rest of this section is devoted to the description of this particular algorithm.

6.7.1 The Blind Deconvolution Scheme

Let $s(k)$ be a deterministic signal that propagates through a nondispersive, multipath environment where additive zero-mean Gaussian noise is present. Suppose we set two receivers, well separated from each other, to record the propagating signal. The two recorded sequences are:

$$y_i(k) = s(k) * h_i(k), \ i = 1, 2 \quad (6.78)$$

where $h_i(k)$ and $h_2(k)$ are the two channels seen by the receivers. The blind deconvolution problem is to estimate $s(k), h_1(k)$, and $h_2(k)$ by observing only the sequences $y_1(k)$ and $y_2(k)$.

Let us assume that (1) $h_1(n)$ and $h_2(n)$ are FIR sequences, (2) $s(n), h_1(n)$, and $h_2(n)$ have no zeros on the unit circle, (3) there are no zero-pole cancellations between $s(n), h_1(n)$ and $s(n), h_2(n)$, and (4) $h_1(n)$ and $h_2(n)$ have no common zeros.

From (6.78) and (5.38) we can write the following cepstral coefficient equations:

$$A_{y_i}^{(m)} = A_s^{(m)} + A_{h_i}^{(m)}, i = 1, 2; m = 1, \ldots, p \quad (6.79.1)$$

$$B_{y_i}^{(m)} = B_s^{(m)} + B_{h_i}^{(m)}, i = 1, 2; m = 1, \ldots, q. \quad (6.79.2)$$

Subtraction of the two equations of (6.79.1) for $i = 1$ and $i = 2$ yields

$$A_{y_1}^{(m)} - A_{y_2}^{(m)} = A_{h_1}^{(m)} - A_{h_2}^{(m)}. \quad (6.80.1)$$

Similarly, from (6.79.2), we obtain

$$B_{y_1}^{(m)} - B_{y_2}^{(m)} = B_{h_1}^{(m)} - B_{h_2}^{(m)}. \quad (6.80.2)$$

Consider the sequences

$$h_{\min}(k) = Z_1^{-1}\{I_i(z^{-1}) \cdot (I_2(z^{-1}))^*\}, \quad k = 0, \ldots, L_1, -1 \qquad (6.81.1)$$

$$h_{\max}(k) = Z_1^{-1}\{O_1(z) \cdot [O_2(z)]^*\}, \quad k = 0, \ldots, L_2, -1 \qquad (6.81.2)$$

where $I_i(z^{-1})$ and $O_i(z)$ are the minimum and maximum phase components of the channel $h_i(k)$, respectively [see (5.1)]. From (6.81) it is easy to show that the minimum and maximum cepstral coefficients of $h_{\min}(k)$ are $A_{h_1}^{(m)}$ and $A_{h_2}^{(m)}$ respectively, while the corresponding cepstral coefficients of $h_{\max}(k)$ are $B_{h_2}^{(m)}$ and $B_{h_1}^{(m)}$.

Suppose that the lengths of $h_{\min}(k)$ and $h_{\max}(k)$ are L_1 and L_2, respectively. It can be shown that based on assumptions 1r, 2r, and 4r, the sequences $h_{\min}(k)$ and $h_{\max}(k)$ can be reconstructed from the differences of their cepstral coefficients, i.e., $A_{h_1}^{(m)} - A_{h_2}^{(m)}$ and $B_{h_1}^{(m)} - B_{h_2}^{(m)}$, respectively [Petropulu and Nikias, 1992]. The reconstruction can be achieved using the BIRA algorithm described in Section 6.5.3. Once $h_{\min}(k)$ and $h_{\max}(k)$ are reconstructed, the coefficients $A_{h_1}^{(m)}, A_{h_2}^{(m)}, B_{h_1}^{(m)}$, and $B_{h_2}^{(m)}$ can be computed. The cepstral coefficients of the signal $s(k)$ result from (6.79) for either $i = 1$ or for $i = 2$. Since in practice these two sets of coefficients will be slightly different, we can average them to produce the cepstral coefficients of $s(k)$. Finally the signal $s(k)$ can be reconstructed from $A_s^{(m)}, B_s^{(m)}$ using

$$s(k) = F_1^{-1}\{e^{F_1\{c_s(m)\}}\} \qquad (6.82.1)$$

where

$$c_s(m) = \begin{cases} -\frac{1}{m} A_s^{(m)}, & m > 0 \\ 0, & m = 0 \\ \frac{1}{m} B_s^{(-m)}, & m < 0. \end{cases} \qquad (6.82.2)$$

As already mentioned, besides the cepstral coefficient differences, the reconstruction of $h_{\min}(k)$ and $h_{\max}(k)$ requires knowledge of the length L_1 and L_2. If there is no such information available, we proceed as follows. Initially, we start with $L_1 = 2$ and apply the BIRA on the differences $A_{h_1}^{(m)} - A_{h_2}^{(m)}$. If the BIRA does not converge to a solution it means that there is no FIR sequence of length $L_1 = 2$ that corresponds to the given differences of cepstra coefficients. In this case we increase L_1 by one and repeat the same step until a convergent solution is reached. Petropulu and Nikias [1992b] we showed that the first convergent solution reached by BIRA corresponds to the correct length L_1. The length of L_2 is computed in an analogous way.

6.7.2 Computation of the Differences of Cepstral Coefficients

Depending on the nature of the signal to be recorded (deterministic or stochastic) and the presence of additive noise, the differences of the cepstral coefficients required in (6.80) can be computed as follows.

Noise-free case, $s(k)$ deterministic

In this case the observed sequences $y_i(k)$ are given by (6.78). Computation of the complex cepstra $c_{y_i}(m)$ leads to the cepstral coefficients $A_{y_i}^{(m)}$ and $B_{y_i}^{(m)}$, hence to their differences, through (6.80.2). For the special case that $h_i(k)$ are both either minimum phase or maximum phase, the power cepstrum could be used for the computation of $A_{h_1}^{(m)} - A_{h_2}^{(m)}$ or $B_{h_1}^{(m)} - B_{h_2}^{(m)}$, respectively, as follows. From (5.25) we get

$$-m \cdot p_{y_1}(m) + m \cdot p_{y_2}(m) = A_{y_1}^{(m)} + B_{y_1}^{(m)} - A_{y_2}^{(m)} - B_{y_2}^{(m)} \\ = A_{h_1}^{(m)} - A_{h_2}^{(m)} \quad (6.83)$$

where it was assumed that both channels were minimum phase. If both channels were maximum phase, then (6.83) would have led to the differences $B_{h_1}^{(m)} - B_{h_2}^{(m)}$.

Noise-free case, $s(k)$ stochastic

Let us assume that $s(k)$ is the output of a LTI system with repulse response $g(k)$, driven by non-Gaussian white noise $n(k)$, i.e.,

$$s(k) = g(k) * n(k). \quad (6.84)$$

Then, for the general case of mixed phase $h_i(k)$, the differences $A_{h_1}^{(m)} - A_{h_2}^{(m)}$ and $B_{h_1}^{(m)} = B_{h_2}^{(m)}$ can only be computed from the bicepstra of the observation sequences; i.e.,

$$mb_{y_1}(m,0) + mb_{y_2}(m,0) = A_{h_1}^{(m)} - A_{h_2}^{(m)} \quad (6.85.1)$$

$$-mb_{y_1}(m,m) + mb_{y_2}(m,m) = B_{h_1}^{(m)} - B_{h_2}^{(m)}. \quad (6.85.2)$$

For the special case that both $h_1(k)$ and $h_2(k)$ are minimum or maximum phase,

TABLE 6.4 METHODS FOR THE COMPUTATION OF THE CEPSTRAL COEFFICIENT DIFFERENCES REQUIRED BY THE BLIND DECONVOLUTION SCHEME BASED ON BIRA

	$h_i(k)$, noise free		$h_i(k)$, Gaussian noise	
$s(k)$	minimum or maximum phase	nonminimum phase	minimum or maximum phase	nonminimum phase
	cepstrum	cepstrum	bicepstrum	bicepstrum
	power cepstrum	bicepstrum		
deterministic	bicepstrum			
colored non-Gaussian	power cepstrum bicepstrum	bicepstrum	bicepstrum	bicepstrum

the power cepstrum could also lead to the correct differences by means of (6.83).

Additive Gaussian noise, $s(k)$ deterministic or stochastic

Consider the case where the observed sequences at the two sensors are corrupted by additive zero-mean Gaussian noise; i.e.,

$$x_i(k) = y_i(k) + W_i(k) \qquad (6.86)$$

where $y_i(k)$ is given by (6.78) and $W_1(k), W_2(k)$ are Gaussian processes. Then only the bicepstra of $x_i(k)$ can lead to the correct cepstral coefficient differences, i.e.,

$$-mb_{x_1}(m,0) + mb_{x_2}(m,0) = A_{h_1}^{(m)} - A_{h_2}^{(m)} \qquad (6.87.1)$$

$$-mb_{x_1}(m,m) + mb_{x_2}(m,m) = B_{h_1}^{(m)} - B_{h_2}^{(m)}. \qquad (6.87.2)$$

The different procedures for the computation of the cepstral coefficient differences are summarized in Table 6.4.

Example 6.6

The blind deconvolution algorithm is applied to the reconstruction of an image that has gone through reverberation.

Suppose that the "NU" image (14 × 14) shown in Figure 6.9 was transmitted through a multipath and non dispersive environment. In order to recover the original image, it is assumed that the outputs of two receivers are available. The receivers record two distorted versions of the transmitted image. Let the two blurring channels be $h_1(n) = (0.3, -0.6, 1, -0.2, 0.5)$ and $h_2(n) = (-0.5, 0.2, 1, -0.6, 0.3, -0.1)$. Assuming that the transmission of the original image is made line-by-line, the recorded images at the two receivers are shown in Figure 6.9(a) and (b), respectively.

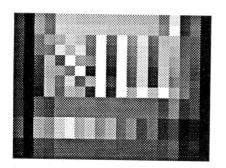
(a) Image recorded at sensor 1.

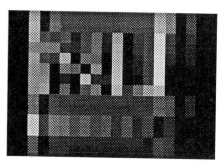
(b) Image recorded at sensor 2.

(c) Original image.

(d) Recovered image.

Figure 6.9 Image recovery based on the blind deconvolution scheme (noise-free case) ([Petropulu and Nikias, 1993] © 1993 IEEE)

Let $y_1(k)$ and $y_2(k)$ be two lines observed at the two receivers. The third-order moments of these sequences are computed as

$$m_3^{y_i}(m,n) = \frac{1}{N_i} \sum_k y_i(k) y_i(k+m) y_i(k+n), \quad i=1,2,$$

for $-20 \leq m, n \leq 20$, with $N_1 = 18$ and $N_2 = 19$. The bicepstra of $Y_1(k)$ and $Y_2(k)$ are computed from

$$b_{y_i}(m,n) = \frac{1}{m} F_2^{-1} \left\{ \frac{F_2\{m\{m_3^{y_i}(m,n)\}\}}{F_2\{m_3^{y_1}(m,n)\}} \right\}$$

where the 2-d Fourier transforms were implemented with FFT's of length $L = 128$. Using (5.38) the cepstral coefficients of $y_i(k)$, and hence their differences, were computed.

The reconstructed image from the blind deconvolution scheme is also shown in Figure 6.9 against the original image. It is important to note that in order to apply cepstra operations on each line, one has to perform a windowing on the third-order moment sequence of each line, i.e., the moments were multiplied by $w(m, n) = 0.98^{(m+n)}$. The effect of this window is to move the zeros that are close, or on the unit circle, away from the unit circle (see also section 6.4.5). This windowing was necessary because almost every line of the image has zeros that reside in the vicinity of the unit circle. □

Example 6.7

The blind deconvolution algorithm is tested for the reconstruction of a transient signal that propagates in the presence of reverberation and additive noise. Consider the model of (6.87) with $h_1(k) = (0.4, 1.0, 0.7, -0.2), h_2(k) = (0.2, 1.0, 0.6, -0.1)$ and $s(n)$ the transient illustrated in Figure 6.10. The additive noise processes $W_1(k)$ and $W_2(k)$ is considered to be zero-mean Gaussian and white.

The third-order moments of $y_1(k)$ and $y_2(k)$ are estimated by segmenting the data into N segments of M samples each, and averaging out the third-order moment of each segment. The bicepstra of $y_1(k)$ and $y_2(k)$ are computed using (5.47), and from there the differences of the cepstral coefficients of the two channels are computed using (5.38) and (6.90). The BIRA algorithm is applied on these differences and the values of L_1 and L_2 that correspond to the first convergent solution are found to be $L_1 = 5$ and $L_2 = 4$. The reconstructed signal $\hat{s}(n)$ is also illustrated in Figure 6.10 for different noise levels, SNR = 10 dB and 5 dB. The values of N and M are indicated on Figure 6.10. Clearly, good reconstructions can be obtained by the blind deconvolution algorithm based on BIRA. □

6.8 SUMMARY

In this chapter we have described nonparametric methods for the recovery of signals from their higher-order spectra (e.g., bispectrum or trispectrum). These methods are nonparametric in the sense that they recover the Fourier phase (magnitude) of a

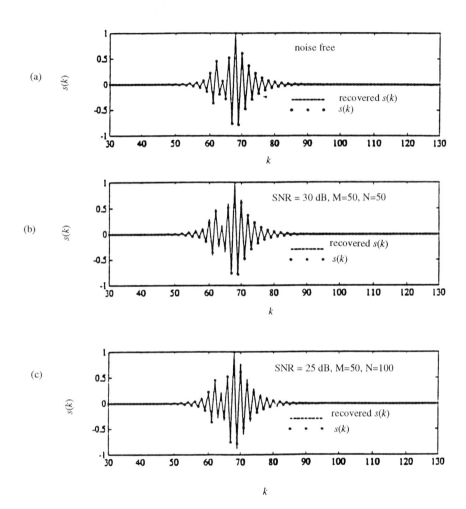

Figure 6.10 Transient signal reconstruction using blind deconvolution scheme. (a) The true, (b) the reconstructed sequences for SNR = 30 dB, $M = 50$, $N = 50$, and (c) the reconstructed sequence for SNR = 25 dB, $M = 50$, $N = 100$. ([Petropulu and Nikias, 1993] © 1993 IEEE)

signal from its higher-order spectrum phase (magnitude) directly, either recursively or nonrecursively, without the need of parametric models. In particular, we have presented signal recovery algorithms based on the properties of polyspectra and polycepstra, as well as on the method of projections onto convex sets (POCS). We have described signal recovery methods from only the phase of the bispectrum. Finally, we have introduced a two-channel blind deconvolution algorithm based on phase recovery algorithms and the properties of polycepstra.

Recursive phase recovery algorithms have been introduced by Brillinger [1977], Lii and Rosenblatt [1982], and Bartelt, et al. [1984]. A nonrecursive phase recovery algorithm has been published by Matsuoka and Ulrych [1984]; and phase unwrapping algorithms for their method have been developed by Marron, et al. [1990] and Rangoussi and Giannakis [1991]. Dianat and Raghuveer [1990] have treated the phase recovery problem along the diagonal slice of the bispectral phase of a 2-d signal. Phase and magnitude recovery algorithms based on the properties of polycepstra, as well as a two-channel blind deconvolution technique, have been introduced by Petropulu and Nikias [1990; 1991]. Cetin [1989] introduced a signal reconstruction method from the bispectrum using the method of POCS. Finally, nonparametric methods for signal recovery from higher-order spectra have recently been published by Sadler and Giannakis [1992].

REFERENCES

Andrews, H. C., and B. R. Hunt, **Digital Image Restoration**, Englewood Cliffs, NJ: Prentice Hall, 1977.

Ayers, G. R., and J. C. Dainty, "Iterative Blind Deconvolution Method and Its Applications," *Optics Letters*, **13**(7), pp. 547–549, July, 1988.

Bartelt, H., A. W. Lohman, and B. Wirnitzer, "Phase and Amplitude Recovery from Bispectra," *Applied Optics*, **23**, pp. 3121–3129, 1984.

Bregman, L. M., "The Method of Successive Projections for Finding a Common Point of Convex Sets," *Dokl. Akad. Nauk. USSR*, **162**(3), pp. 487–490, 1965.

Brillinger, D. R., "The Identification of a Particular Nonlinear Time Series System," *Biometrika*, **64**, pp. 509–515, 1977.

Cerrato, L. R., and B. A. Eisenstein, "Deconvolution of Cyclostationary Signals," *IEEE Trans. Acoustics, Speech and Signal Processing*, **ASSP-25**(6), pp. 466–476, December, 1977.

Cetin, A. E., "An Algorithm for Signal Reconstruction from Bispectrum," *Proc. ICASSP'89*, pp. 1330–1332, April, 1989.

Cetin, A. E., "An Iterative Algorithm for Signal Reconstruction from Bispectrum," *IEEE Trans. on Signal Processing*, **39**(12), pp. 2621-2628, December, 1991.

Dianat, S. A., and M. R. Raghuveer, "Fast Algorithms for Bispectral Reconstruction of Two-Dimensional Signals," *Proc. ICASSP'90*, pp. 2377–2379, April, 1990.

Dudgeon, D. E., and R. M. Mersereau, **Multidimensional Signal Processing**, Englewood Cliffs, NJ: Prentice Hall, 1984.

Fienup, J. R., "Reconstruction of an Object from the Modulus of its Fourier Transform," *Opt. Lett.*, **3**, p. 27, July, 1978.

Fienup, J. R., "Space Object Imaging Through the Turbulent Atmosphere," *Opt. Eng.*, pp. 529–534, 1979.

Gerchberg, R. W., and W. O. Saxton, "A Practical Algorithm for the Determination of Phase from Image and Diffraction Plane Picture," *Optik*, **35**, pp. 237–246, 1972.

Gerchberg, R. W., and W. O. Saxton, "Super Resolution Through Error Average Reduction," *Opt. Acta*, **21**, pp. 709-720, 1974.

Hayes, M. H., J. S. Lim, and A. V. Oppenheim, "Signal Reconstruction from the Phase or Magnitude," *IEEE Transactions Acoustics, Speech and Signal Processing*, **28**, pp. 672–680, December, 1980.

Li, Y. T., and A. L. Kurkjian, "Arrival Time Determination Using Iterative Signal Reconstruction from the Phase of the Cross-Spectrum," *IEEE Trans. on Acoustics, Speech and Signal Processing*, **ASSP-31**(2), pp. 502–504, April, 1983.

Lii, K. S., and M. Rosenblatt, "Deconvolution and Estimation of Transfer Function Phase and Coefficients for Non-Gaussian Linear Processes," *Ann. Statist.*, **20**, pp. 1195–1208, 1982.

Marron, J. C., P. P. Sanchez, and R. C. Sullivan, "Unwrapping Algorithm for Least-Squares Phase Recovery from the Modulo 2π Bispectrum Phase," *J. Opt. Soc. Am.* **A7**(1), pp. 14–20, January, 1990.

Matsuoka, T., and T. J. Ulrych, "Phase Estimation Using the Bispectrum," *Proceedings of IEEE*, **72**, pp. 1403–1411, October, 1984.

Oppenheim, A. V., M. H. Hayes, and J. S. Lim, "Iterative Procedures for Signal Reconstruction from Phase," *Proc. SPIE Conference*, **SPIE-231**, pp. 121–129, 1980.

Oppenheim, A. V., and J. S. Lim, "The Importance of Phase in Signals," *Proc. IEEE*, **69**, pp. 529–541, May, 1981.

Oppenheim, A. V., and R. W. Schafer, **Discrete-Time Signal Processing**, Englewood Cliffs, NJ: Prentice Hall, 1989.

Pan, R., and C. L. Nikias, "Phase Reconstruction in the Trispectrum Domain", *IEEE Transactions on Acoustics, Speech and Signal Processing*, **ASSP-35**, pp. 895–897, June, 1987.

Papoulis, A., "A New Algorithm in Spectral Analysis and Bandlimited Extrapolation," *IEEE Trans. on Circuits & Systems*, **CAS-22**(9), pp. 735–742, September, 1975.

Petropulu, A., "Signal/Image Reconstruction from the Phase of the Bispectrum," Ph.D. Thesis, Northeastern University, Boston, December, 1990.

Petropulu, A. P., and C. L. Nikias, "Blind Deconvolution Using Reconstruction from Partial Information, Based on Higher-Order Spectra," *Proc. ICASSP'91*, pp. 1757–1760, Toronto, Canada, May, 1991.

Petropulu, A. P., and C. L. Nikias, "Signal Reconstruction from the Phase of the Bispectrum," *IEEE Transactions on Signal Processing*, **ASSP40**(3), pp. 601–610, March, 1992.

Quatieri, T. H., and A. V. Oppenheim, "Iterative Techniques for Minimum Phase Signal Reconstruction from Phase or Magnitude," *IEEE Transactions Acoustics, Speech and Signal Processing*, **ASSP-29**(6), pp. 1187–1193, December, 1981.

Ramachandran, G. N., and R. Srinivasan, **Fourier Methods in Crystallography**, New York: Wiley Interscience, 1970.

Rangoussi, M., and G. B. Giannakis, "FIR Modeling Using Log-Bispectra: Weighted Least-Squares Algorithms and Performance Analysis," *IEEE Trans. on Circuits and Systems*, **38**(3), pp. 281–296, March, 1991.

Sadler, B., and G. B. Giannakis, "Shift and Rotation Invariant Object Reconstruction Using the Bispectrum", *J. Optical Society Amer.*, **9**, pp. 57–69, January, 1992.

Sezan, M. I., and H. Stark, "Image Restoration by the Method of Convex Projections: Part 2 – Applications and Numerical Results," *IEEE Trans. on Medical Imaging*, **MI-1**(2), pp. 95–101, October, 1982.

Smith, W. E., W. E. Barrett and R. G. Paxman, "Reconstruction of Objects from Codes Images by Simulated Annealing," *Optics Letters*, **8**(4), pp. 199–201, April, 1983.

Stark, Henry, **Image Recovery, Theory and Applications**, New York: Academic Press, 1987.

Stockham, T. G., T. M. Cannon and R. B. Ingebretsen, "Blind Deconvolution Through Digital Signal Processing," *Proceedings of the IEEE*, **63**(4), pp. 678–692, April, 1975.

Swami, A., G. B. Giannakis, and J. M. Mendel, "A Unified Approach to Modeling Multichannel ARMA Processes," *Proc. ICASSP-89*, pp. 2182–2185, Glasgow, Scotland, 1989.

Tom, V. T., T. H. Quatieri, M. H. Hayes, and J. H. McClellan, "Convergence of Iterative Nonexpansive Signal Reconstruction Algorithms," *IEEE Trans. on Acoust., Speech and Signal Processing*, **ASSP-29**(5), pp. 1052–1058, October, 1981.

Youla, D. C., and H. Webb, "Image Restoration by the Method of Convex Projections: Part I – Theory," *IEEE Trans. on Medical Imaging*, **MI-I**(2), pp. 81–94, October, 1982.

7

PARAMETRIC METHODS FOR THE ESTIMATION OF HIGHER-ORDER SPECTRA

7.1 INTRODUCTION

One of the widely used approaches in system identification and power spectrum estimation has been the construction of a white noise driven, linear time-invariant model from a given realization of a random signal. If the driving noise process is assumed to be Gaussian, a nonminimum phase system will be identified as being a minimum phase system when the input to the system is inaccessible. Hence, the main motivation behind the use of non-Gaussian, white noise driven parametric models for higher-order spectrum estimation is to recover both the magnitude and phase response of the system.

Consider the real and stable ARMA process $\{X(k)\}$ given by

$$\sum_{i=0}^{p} a(i)X(k-i) = \sum_{j=0}^{q} b(j)W(k-j) \qquad (7.1.1)$$

where $\{W(k)\}$ are independent and identically distributed (i.i.d.) random variables, independent from $\{X(m)\}$ for $m < k$, with zero-mean and

$$\text{cum}\{W(k), W(k+\tau_1), \cdots, W(k+\tau_{n-1})\} = \gamma_n^w \delta(\tau_1, \ldots, \tau_{n-1}) \qquad (7.1.2)$$

i.e., $\{W(k)\}$ is n^{th}-order white. Note that both $\{W(k)\}$ and $\{X(k)\}$ are non-Gaussian. The higher-order spectra of the system input and output are related by

$$C_n^x(\omega_1, \ldots, \omega_{n-1}) = \gamma_n^w H(\omega_1) \cdots H(\omega_{n-1}) H^*(\omega_1 + \omega_2 + \cdots + \omega_{n-1}) \qquad (7.2)$$

where $H(\omega)$ is the frequency response function of the ARMA system given by

$$H(\omega) = \frac{\sum_{k=0}^{q} b(k)e^{-j\omega k}}{\sum_{k=0}^{p} a(k)e^{-j\omega k}}; \qquad a(0) = 1. \tag{7.3}$$

For classical system identification problems, least-squares estimation is almost exclusively used because it yields maximum likelihood estimates of the parameters of Gaussian white-noise driven systems and because the equations obtained are in a linear form involving autocorrelation lags. However, least-squares methods cannot correctly identify a nonminimum phase system from its output only since they are based on autocorrelation measurements of the output.

The identifiability of the magnitude and phase of $H(\omega)$ from output observations only, depending on whether $W(k)$ is Gaussian or not, is determined as follows:

1. If $\{W(k)\}$ is Gaussian and $H(\omega)$ is minimum phase, the autocorrelation-based methods can identify both the phase and the magnitude of $H(\omega)$.

2. If $\{W(k)\}$ is Gaussian and $H(\omega)$ is nonminimum phase, no method can correctly recover the phase of $H(\omega)$.

3. If $\{W(k)\}$ is non-Gaussian and $H(\omega)$ is nonminimum phase, autocorrelation-based methods can only correctly identify the magnitude of $H(\omega)$.

4. If $\{W(k)\}$ is non-Gaussian and $H(\omega)$ is nonminimum phase, higher-order spectra-based methods can estimate both the phase and the magnitude of $H(\omega)$, without any knowledge of the actual distribution of $\{W(k)\}$.

In this chapter, we address the nonminimum phase system identification problem using higher-order statistics when the system is MA, noncausal AR, or ARMA. A thorough review of the existing parametric methods based on higher-order statistics is given by Mendel [1991].

7.2 MA METHODS

The equation that describes a $MA(q)$ (moving average of order q) process $\{X(k)\}$ can be obtained from (7.1.1) by setting $p = 0$, i.e.,

$$X(k) = \sum_{i=0}^{q} b(i)W(k-i) \tag{7.4}$$

where q is the order of the MA model. The identification problem requires the estimation of the parameters $\{b(i), i = 0, 1, \ldots, q\}$ from the third-order (or higher-order) cumulants of $X(k)$, or equivalently, the system's impulse response, $h(k) = b(k)$, for $k = 0, 1, \ldots, q$ and $h(k) = 0$ for $k > q$. For convenience we set $b(0) = 1$.

7.2.1 Closed Form Solutions

Consider the model
$$Y(k) = X(k) + n(k) \tag{7.5}$$
where $X(k)$ is given by (7.4), and $\{n(k)\}$ is Gaussian noise, independent of $W(k)$. Under the assumption that $W(k)$ is third-order white, the third-order cumulant of $Y(k)$ is given by
$$\begin{aligned} c_3^y(\tau, \rho) &= E\{Y(k)Y(k+\tau)Y(k+\rho)\} \\ &= \gamma_3^w \sum_{i=0}^{\infty} b(i)b(i+\tau)b(i+\rho). \end{aligned} \tag{7.6}$$

Evaluating (7.6) at $\tau = q$, $\rho = k$ yields
$$c_3^y(q, k) = \gamma_3^w b(k) b(q). \tag{7.7}$$
For $k = 0$, (7.7) becomes
$$c_3^y(q, 0) = \gamma_3^w b(q). \tag{7.8}$$
Combining the last two equations and assuming that $\gamma_3^w \neq 0$ (i.e., $W(k)$ is not symmetrically distributed), we obtain
$$b(k) = \frac{c_3^y(q, k)}{c_3^y(q, 0)}, \qquad k = 0, 1, \ldots, q. \tag{7.9}$$

Equation (7.9) relates the impulse response of the MA model to the third-order cumulants of the system output. It is known as the $c(q, k)$ formula, and it was first introduced by Giannakis [1987]. The $c(q, k)$ formula provides a closed form solution to the MA parameter estimation problem; however, it requires exact knowledge of the MA order, and since it uses only two cumulant lags for the estimation of each $b(k)$, it does not smooth out the effect of additive noise on the estimated third-order cumulants.

Next we are going to present two methods that compute the MA parameters recursively, based on third- and second-order cumulant estimates of the output data. The first method was proposed by Swami and Mendel [1989] and proceeds as follows.

For the model of (7.5), where $W(k)$ is zero-mean, i.i.d., non-Gaussian, and $n(k)$ is zero-mean, i.i.d. noise with symmetric pdf (Gaussian or not), independent of $X(k)$, the third-order cumulants of $Y(k)$ are given by (7.6) while the autocorrelation of $Y(k)$ is given by

$$c_2^y(\tau) = \gamma_2^w \sum_{k=0}^{q} b(k)b(k+\tau) + \gamma_2^n \delta(\tau). \tag{7.10}$$

Evaluating (7.6) and (7.10) at $\tau = \pm q$ we obtain

$$b(q) = \frac{c_3^y(q,q)}{c_3^y(-q,-q)}; \quad \gamma_2^w = \frac{c_2^y(q)c_3^y(-q,-q)}{c_3^y(q,q)}; \quad \gamma_3^w = \frac{[c_3^y(-q,-q)]^2}{c_3^y(q,q)}. \tag{7.11.1}$$

Based on these values, the MA parameters can be computed recursively as:

$$b(q-m) = \frac{f_1(m)}{2} + \frac{b(q)f_2(m) - f_3(m)}{2[b(q) - f_1(m)]}, \quad m = 1, \ldots, \lfloor \frac{q}{2} \rfloor \tag{7.11.2}$$

$$b(m) = \frac{f_1(m) - b(q-m)}{b(q)}, \quad m = 1, \ldots, \lfloor \frac{q}{2} \rfloor \tag{7.11.3}$$

where

$$f_1(m) \triangleq \frac{c_2^y(q-m)}{\gamma_2^w} - \sum_{k=1}^{m-1} b(k)b(k+q-m)$$
$$= b(q-m) + b(m)b(q)$$

$$f_2(m) \triangleq \frac{c_3^y(m-q,m-q)}{\gamma_3^w} - \sum_{k=1}^{m-1} b^2(k)b(k+q-m)$$
$$= b(q-m) + b^2(m)b(q)$$

$$f_3(m) \triangleq \frac{c_3^y(q-m,q-m)}{\gamma_3^w} - \sum_{k=1}^{m-1} b(k)b^2(k+q-m)$$
$$= b^2(q-m) + b(m)b^2(q).$$

The noise variance γ_2^n does not have to be known, and it can actually be computed once the MA parameters are computed, by evaluating (7.10) at $\tau = 0$. The algorithm assumes knowledge of the MA order and furthermore requires that $b(q-k) \neq b(q)[1-b(k)]$, for $k = 0, 1, \ldots, \frac{q}{2}$. The latter assumption is required in order to avoid division of zero-by-zero in (7.11.2); however, it renders the algorithm rather impractical. Also, the method cannot be extended to the fourth-order cumulant domain.

The second recursive MA estimation method was proposed by Tugnait [1990] as an alternative to the $c(q,k)$ algorithm.

Consider the model of (7.5), where $W(k)$ is i.i.d., zero-mean, non-Gaussian and the noise process $n(k)$ is zero-mean, i.i.d., independent of $W(k)$. Evaluating the third-order cumulant sequence of $X(k)$ at lags τ and $\tau + m$, we obtain

$$\begin{aligned}c_3^x(\tau, \tau + m) &= E\{X(k)X(k+\tau)X(k+\tau+m)\} \\ &= \gamma_3^w \sum_{k=0}^{q} b(k)b(k+\tau)b(k+\tau+m) \\ &= \gamma_3^w \sum_{k=0}^{q} b(k)g(k+\tau; m),\end{aligned} \quad (7.12.1)$$

where

$$g(k; m) = b(k)b(k+m), \quad (7.12.2)$$

and m is a constant. Taking the Z transform of (7.12.1) we obtain

$$C_3^x(z; m) = \gamma_3^w B(z^{-1}) G(z; m), \quad (7.13.1)$$

where $B(z)$ is the Z transform of $b(k)$ and $G(z; m)$ is the Z transform of $g(k; m)$, or equivalently

$$G(z; m) = B(z) * [z^m B(z)]. \quad (7.13.2)$$

The autocorrelation of $X(k)$ is given by

$$c_2^x(\tau) = E\{X(k)X(k+\tau)\} = \gamma_2^w \sum_{k=0}^{q} b(k)b(k+\tau) \quad (7.14)$$

and its Z transform by

$$C_2^x(z) = \gamma_2^w B(z) B(z^{-1}). \quad (7.15)$$

Combining (7.13.1) and (7.15) we obtain

$$G(z; m) C_2^x(z) = \varepsilon B(z) C_3^x(z; m) \quad (7.16)$$

where $\varepsilon = \gamma_2^w / \gamma_3^w$. The time-domain equivalent of (7.16) is

$$\sum_{k=0}^{q} b(k)b(k+m) c_2^x(\tau - m) = \varepsilon \sum_{k=0}^{q} b(k) c_3^x(\tau - k, \tau - k + m). \quad (7.17)$$

Since the observed sequence is $Y(k)$, (7.17) has to be expressed in terms of the cumulants of $Y(k)$. Due to the noise assumptions, we have that $c_2^x(\tau) = c_2^y(\tau) \neq 0$, and $c_3^x(\tau_1, \tau_2) = c_3^y(\tau_1, \tau_2)$ if $\tau_1 \neq 0$ or $\tau_2 \neq 0$.

Setting $m = q$ in (7.17) and replacing the cumulant of $X(k)$, with the cumulants of $Y(k)$, we get

$$\sum_{k=1}^{q} b(k) c_3^y(k - \tau, q) - \varepsilon'[c_2^y(\tau) - \gamma_2^n \delta(\tau)] = -c_3^y(-\tau, q) \quad (7.18)$$

where $\varepsilon' = \frac{b(q)}{\varepsilon}$.

Setting $\tau = -q, -q+1, \cdots, -1$, we form q equations from (7.18), and based on them, a recursive solution can be obtained as follows:

For $\tau = -q$

$$-\varepsilon' c_2^y(-q) = -c_3^y(q,q), \text{ hence } \varepsilon' = \frac{c_3^y(q,q)}{c_2^y(q)}. \quad (7.19.1)$$

For $\tau = -q+i$, $i = 1, \cdots, q$, (7.18) and (7.19.1) yield

$$b(i) = \frac{\varepsilon' c_2^y(q-i) - \sum_{k=0}^{i-1} b(k) c_3^y(k+q-i,q)}{c_3^y(q,q)}, \quad 1 < i < q \quad (7.19.2)$$

$$b(q) = \varepsilon' c_2^y(q)/c_3^y(0,q). \quad (7.19.3)$$

The extension of this recursive algorithm to fourth-order statistics was also presented by Tugnait [1990]. This algorithm assumes knowledge of the model order q.

Using (7.17) with $m = 0$, a recursive algorithm was also proposed by Giannakis and Mendel [1989].

7.2.2 Optimization Methods

Consider the diagonal third-order cumulant slice of process $Y(k)$ described in (7.5), i.e.,

$$\begin{aligned} c_3^y(\tau,\tau) &= E\{Y(k)Y^2(k+\tau)\} \\ &= \gamma_3^w \sum_{i=o}^{q} b(i)b^2(i+\tau), \quad \tau = -q, \cdots, 0, \cdots, q. \end{aligned} \quad (7.20)$$

Lii and Rosenblatt [1982] suggested two methods for the estimation of the MA coefficients based on (7.20): (i) a nonlinear least-squares approach, and (ii) a linear programming approach. Both methods estimate the third-order cumulants $\hat{c}_3^y(\tau,\tau)$, as described in Chapter 4, and then proceed as follows.

Nonlinear Least Squares Approach:

This method estimates the parameters $b(i)$ by minimizing the performance criterion

$$\sum_{\tau=-q}^{q} \left[\hat{c}_3^y(\tau,\tau) - \gamma_3^w \sum_{i=0}^{q} b(i)b^2(i+\tau) \right]^2 \quad (7.21)$$

with respect to the $q+2$ unknowns $b(i), i = 1, \cdots, q$ and γ_3^w where $\{b(0) = 1\}$.

Linear Programming Approach:

In this approach, one employs first an autocorrelation-based method to estimate the MA coefficients that accurately reflect the autocorrelation structure of the data, i.e., $\{b^{(2)}(k)\}$. The resulting transfer function of the MA model will then be

$$B(z) = \sum_{k=0}^{q} b^{(2)}(k) z^{-k} \qquad (7.22)$$

with roots $r_j, j = 1, \ldots, q$ such that $\mid r_j \mid < 1$ (minimum phase).

An accurate estimate of the distribution of the roots can be obtained by taking the conjugated inverse of an appropriate number of r_j's. The linear programming method seeks to find the set of roots and thus the coefficients $\{b(k)\}$ that minimize (7.21).

A suboptimal implementation of Lii and Rosenblatt's [1982] nonlinear approach was proposed by Mendel and Wang [1991]. They built a structure that resembles a neural network structure where each value of the third-order cumulant $c_3^y(m,n)$ in the nonredundant region is to be learned, and the neuron weights are the MA coefficients. The training sequences are estimates of the second and third-order cumulants. The convergence properties of this method have not been studied.

7.2.3 Parameter Estimation Based on ARMA and AR Approximations

Let us consider the nonminimum phase process

$$X(k) = \sum_{i=-L_1}^{L_2} b(i) W(k+i) \qquad (7.22)$$

with system transfer function

$$B(z) = \sum_{i=-L_1}^{L_2} b(i) z^{-i} = I(z^{-1}) O(z), \qquad (7.23)$$

where $I(z^{-1})$ and $O(z)$ are polynomials with all their zeros inside and outside the unit circle, respectively. Then, from (7.22) and (7.23) we obtain

$$X(z) = I(z^{-1}) O(z) W(z). \qquad (7.24)$$

Causal (forward) ARMA Model:

From (7.24) we can write

$$\frac{X(z)}{I(z^{-1})} = O(z)W(z) \qquad (7.25)$$

which may be approximated arbitrarily close by a stable ARMA (p, L_2) model of the form

$$\sum_{i=0}^{p} c(i)X(k-i) = \sum_{i=0}^{L_2} t(i)W(k+i); \quad c(0) = t(0) = 1, \qquad (7.26.1)$$

where

$$\sum_{i=0}^{p} c(i)z^{-i} \simeq \frac{1}{I(z^{-1})}. \qquad (7.26.2)$$

The AR order p in (7.26) should be sufficiently large for an accurate ARMA approximation, and its value depends upon the location of the zeros of $I(z^{-1})$. The $AR(p)$ part of (7.26), i.e., the coefficients $c(i)$, contain the minimum phase component of the system transfer function $B(z)$.

Assuming that $\{X(k)\}$ is stationary, multiplying both sides of (7.26.1) by $X(k-\tau)X(k-\rho)$ and taking expectations, we obtain

$$\sum_{i=0}^{p} c(i)c_3^x(-\tau+i, -\rho+i) = \sum_{i=0}^{L_2} t(i)c_3^{wxx}(-\tau-i, -\rho-i) \qquad (7.27.1)$$

where $c_3^x(\tau, P)$ is the third-order cumulant of $X(k)$ and

$$c_3^{wxx}(-\tau-i, -\rho-i) \stackrel{\Delta}{=} E\{W(k+i)X(k-\tau)X(k-\rho)\} \qquad (7.27.2)$$

is the cross-cumulant sequence of $W(k)$ and $X(k)$. Combining (7.22) and (7.27.2), $c_3^{wxx}(-\tau-i, -\rho-i)$ can be written as

$$c_3^{wxx}(-\tau-i, -\rho-i) = \gamma_3^w \sum_{m=-L_1}^{L_2} \sum_{n=-L_1}^{L_2} b(m)b(n)\delta(-\tau-i+m, -\rho-i+n). \qquad (7.28)$$

Substitution of (7.28) into (7.27.1) yields

$$\sum_{i=0}^{p} c(i)c_3^x(-\tau+i, -\rho+i) = \sum_{i=0}^{L_2} \sum_{m=-L_1}^{L_2} \sum_{n=-L_1}^{L_2} t(i)b(m)b(n)\delta(-\tau-i+m, -\rho-i+n)$$
$$= RHS(\tau, \rho).$$
$$(7.29)$$

Note that $RHS(\tau, \rho)$ is zero if τ or ρ lies outside the interval $[-L_1 - L_2, L_2]$.

From (7.29), the following system of equations is formed.
$$\mathbf{c}_i^x \cdot \mathbf{c} = \mathbf{b}_i, \quad i = 1, 2, \ldots, L_1 + L_2, \tag{7.30}$$
where
$$\mathbf{c} = [1, c(1), c(2), \ldots, c(p)]^T, \quad ((p+1) \times 1),$$
$$\mathbf{b}_i = [RHS(L_2, L_2 + 1 - i), 0, \ldots, 0]^T, \quad ((p+1) \times 1),$$
and $\mathbf{c}_i^x$ is the nonsymmetric Toeplitz matrix
$$\mathbf{c}_i^x = \begin{bmatrix} c_3^x(-L_2, -L_2 - 1 + i) & \mathbf{m}_i^T, \\ \mathbf{r}_i & \mathbf{M}_i \end{bmatrix}, \quad ((p+1) \times (p+1))$$
where
$$\mathbf{m}_i = [c_3^x(-L_2+1, -L_2+i), c_3^x(-L_2+2, -L_2+1+i), \ldots, c_3^x(-L_2+p, -L_2+p-1+i)]^T, \quad (p \times 1)$$
$$\mathbf{r}_i = [c_3^x(-L_2-1, -L_2-2+i), c_3^x(-L_2-2, -L_2-3+i), \ldots, c_3^x(-L_2-p, -L_2-p-1+i)]^T, \quad (p \times 1)$$

$$\mathbf{M}_i = \begin{bmatrix} c_3^x(-L_2, -L_2-1+i) & \cdots & c_3^x(-L_2+p-1, -L_2+p-2+i) \\ \vdots & & \vdots \\ c_3^x(-L_2-p+1, -L_2-p+i) & \cdots & c_3^x(-L_2, -L_2-1+i). \end{bmatrix}, \quad (p \times p).$$

From (7.21) we can form the overdetermined system of equations
$$\begin{bmatrix} \mathbf{M}_1 \\ \mathbf{M}_2 \\ \vdots \\ \mathbf{M}_{L_1+L_2} \end{bmatrix} \begin{bmatrix} c(1) \\ \vdots \\ c(p) \end{bmatrix} = - \begin{bmatrix} r_1 \\ r_2 \\ \vdots \\ r_{L_1+L_2} \end{bmatrix} \tag{7.31}$$
or
$$\mathbf{M}_c \tilde{\mathbf{c}} = -\mathbf{r}_c \tag{7.32}$$
and obtain the least-squares [or singular value decomposition (SVD)] solution
$$\tilde{\mathbf{c}} = -[\mathbf{M}_c^T \mathbf{M}_c]^{-1} \mathbf{M}_c^T \mathbf{r}_c. \tag{7.33}$$

The minimum phase components of the system's magnitude and phase response are given by
$$|B_I(\omega)| = \left|\frac{1}{C(\omega)}\right| \tag{7.34.1}$$
$$\phi_I(\omega) = \text{Arg}\left[\frac{1}{C(\omega)}\right], \tag{7.34.2}$$
where
$$C(\omega) = 1 + \sum_{i=1}^{p} c(i) e^{-j\omega i} \tag{7.34.3}$$

Anticausal (backward) ARMA Model:

Equation (7.24) can also be written as

$$\frac{X(z)}{O(z)} = W(z)I(z^{-1}), \qquad (7.35)$$

and similarly approximated arbitrarily close by a backward stable $ARMA(q, L_1)$ model of the form

$$\sum_{i=0}^{q} d(i)X(k+i) = \sum_{i=0}^{L_1} u(i)W(k-i); \quad d(0) = u(0) = 1, \qquad (7.36.1)$$

$$\sum_{i=0}^{q} d(i)z^i \simeq \frac{1}{O(z)}. \qquad (7.36.2)$$

The maximum phase information of $B(z)$ is contained in the $AR(q)$ part of (7.36), i.e., the coefficients $d(i)$. The third-order recursion of (7.36) is given by

$$\sum_{i=0}^{q} d(i) c_3^x(\tau - i, \rho - i) = \gamma_3^w \sum_{i=0}^{L_1} \sum_{m=-L_1}^{L_2} \sum_{n=-L_1}^{L_2} u(i)b(m)b(n)\delta(\tau + i + m, \rho + i + n)$$

$$= RHS(z, \rho). \qquad (7.37)$$

Note that $RHS(\tau, \rho)$ in the above equation is zero if τ or ρ lies outside the internal $[-L_1 - L_2, L_1]$.

Along the lines of the derivation of (7.30) we can derive the overdetermined system of equations

$$\mathbf{M}_a \, \tilde{\mathbf{d}} = -r_a \qquad (7.38)$$

where

$$\mathbf{r}_i = [c_3^x(L_1+1, L_1+2-i), c_3^x(L_1+2, L_1+3-i), \cdots, c_3^x(L_1+q, L_1+q+1-i)]^T$$

$$\mathbf{r}_a = [r_1, r_2, \cdots, r_{L_1+L_2}]^T, \tilde{\mathbf{d}} = [d(1), d(2), \cdots, d(q)]$$

$$\mathbf{M}_a = [\mathbf{M}_1, \mathbf{M}_2, \cdots, \mathbf{M}_{L_1+L_2}],$$

$$\mathbf{M}_i = \begin{bmatrix} c_3^x(L_1, L_1+1+i) & \cdots & c_3^x(L_1-q+1, L_1+2-q-i) \\ \vdots & & \vdots \\ c_3^x(L_1+q, L_1+q+1-i) & \cdots & c_3^x(L_1, L_1+1-i) \end{bmatrix}$$

which has the least-squares solution

$$\tilde{\mathbf{d}} = -[\mathbf{M}_a^T \mathbf{M}_a]^{-1} \mathbf{M}_a^T \mathbf{r}_a. \qquad (7.39)$$

The maximum phase magnitude and phase components of the system transfer function are given by

$$|B_0(\omega)| = |\frac{1}{D(\omega)}| \qquad (7.40.1)$$

$$\phi_0(\omega) = Arg[\frac{1}{D(\omega)}] \qquad (7.40.2)$$

where

$$D(\omega) = 1 + \sum_{i=0}^{q} d(i)e^{j\omega i}. \qquad (7.40.3)$$

Combining (7.34) and (7.40), the magnitude and phase response of the system of (7.23) is

$$|B(\omega)| = |B_I(\omega)| \cdot |B_0(\omega)| \qquad (7.41.1)$$

$$\phi(\omega) = \phi_I(\omega) + \phi_0(\omega). \qquad (7.41.2)$$

The performance of this identification method depends on the MA model order determination scheme, i.e., on the choice of L_1 and L_2. Note that if the MA order L_2 is found to be zero, a minimum phase system is implied, whereas if $L_1 = 0$ the system is maximum phase. The AR order of p and q of the forward and backward ARMA models, respectively, can be chosen arbitrarily large because in theory $p = q = \infty$. Therefore, their choice does not present a real challenge in practical applications. Model order selection criteria are discussed in section 7.5 of this chapter.

Extension of the parameter estimation method based on ARMA approximations to the fourth-order cumulant domain, and its performance evaluation, under different symmetric probability density functions for the driving non-Gaussian white noise, was derived by Nikias and Pan [1988].

Nikias and Chiang [1988] also suggested the use of a noncausal AR approximation to the MA system identification problem by rewriting (7.24) as

$$\frac{X(z)}{I(z^{-1})O(z)} = W(z), \qquad (7.42)$$

or in terms of third-order cumulants, as

$$\sum_{i=-\infty}^{+\infty} f(i)c_3^x(\tau - i, p - i) = 0, \quad \text{for } \tau > L_1 \text{ or } \tau < -L_2 \qquad (7.43)$$
$$\text{or } p > L_1 \text{ or } p < -L_2.$$

Methods for estimation of a finite number of $\{f(i)\}$ coefficients from cumulants were provided by Nikias and Chiang [1988]. The estimated transfer function is obtained from

$$\hat{B}(\omega) = \frac{1}{\sum_{i=-M}^{M} f(i)\exp(-j\omega i)} \qquad (7.44)$$

where M is chosen arbitrarily large.

7.2.4 The GM Method

A novel MA identification approach based on both second- and third-order statistics was proposed by Giannakis and Mendel [1989]. Their method is known as the GM algorithm.

Consider the model of (7.5), where $W(k)$ is i.i.d., zero-mean non-Gaussian and $n(k)$ is zero-mean i.i.d. with symmetric pdf and is independent of $X(k)$. Combining a diagonal slice of the third-order cumulant sequence of $X(k)$, i.e.,

$$\begin{aligned} c_3^x(\tau,\tau) &= E\{X(k)X^2(k+\tau)\} \\ &= \gamma_2^w \sum b(i)b^2(i+\tau) \end{aligned} \qquad (7.45)$$

and the autocorrelation of $X(k)$ given by (7.14), leads to the equation

$$c_2^x(\tau) + \sum_{i=1}^{q} b^2(i) c_2^x(\tau - i) = \varepsilon [c_3^x(\tau,\tau) + \sum_{i=1}^{q} b(i) c_3^x(\tau - i, \tau - i)] \qquad (7.46)$$

with $\varepsilon = \gamma_2^w / \gamma_3^w$. The derivation of (7.46) is the same as the derivation of (7.17) with $m = 0$.

Taking $\tau = -q, \cdots, 0, \cdots, 2q$, the equations resulting from (7.46) can be written in a matrix form as

$$\mathbf{M}\,\boldsymbol{\theta} = \mathbf{r} \qquad (7.47)$$

where

$$M = \begin{bmatrix} c_3^x(-q,-q) & 0 & \cdots & 0 & 0 & \cdots & 0 \\ c_3^x(-q+1,-q+1) & c_3^x(-q,-q) & \cdots & 0 & -c_2^x(q) & \cdots & 0 \\ \vdots & \vdots & & \vdots & \vdots & & \vdots \\ c_3^x(q,q) & c_3^x(q-1,q-1) & \cdots & c_3^x(0,0) & -c_2^x(q-1) & \cdots & -c_2^x(0) \\ 0 & c_3^x(q,q) & \cdots & c^x(1,1) & -c_2^x(q) & \cdots & -c_2^x(1) \\ \vdots & \vdots & & \vdots & \vdots & & \vdots \\ 0 & 0 & \cdots & c_3^x(q,q) & 0 & \cdots & -c_2^x(q) \end{bmatrix}, ((3q+1) \times (2q+1))$$

$$\boldsymbol{\theta} = [\varepsilon b(1), \cdots, \varepsilon b(q), b^2(1), \cdots, b^2(q)]^T, \ (2q \times 1)$$

$$r = [c_2^x(q), c_2^x(q-1), \cdots, c_2^x(q), 0, \cdots, 0]^T, \ (2q \times 1)$$

Chapter 7

The least-squares solution of the overdetermined system of equations is

$$\hat{\boldsymbol{\theta}}_{LS} = (\mathbf{M}^T\mathbf{M})^{-1}\mathbf{M}^T\mathbf{r} \qquad (7.48)$$

From the structure of vector $\boldsymbol{\theta}$ it can be seen that both estimates, $\hat{b}(k)$, and $\widehat{b^2(k)}$, will be computed from (7.48). Giannakis and Mendel [1989] proposed to keep the sign of $\hat{b}(k)$ and take $|\hat{b}(k)| = \{\frac{1}{2}[\hat{b}^2(k) + \widehat{b^2(k)}]\}^{1/2}$ as the amplitude of $b(k)$.

Since the observed data sequence is $Y(k)$, the matrix $\mathbf{M}$ and vector $\mathbf{r}$ have to be written in terms of the cumulants of $Y(k)$. For these particular noise assumptions, it holds that $c_2^y(\tau) = c_2^x(\tau)$, $\tau \neq 0$ and $c_3^y(\tau, \rho) = c_3^x(\tau, \rho)$. Hence, only the lag $c_2^y(0)$ is affected by the additive noise. Therefore, the solution of (7.48) can remain unaffected by removing from the matrix $\mathbf{M}$ the row which includes $c_2^y(0)$.

The GM algorithm assumes that the MA order is given.

An adaptive way of solving (7.48) has been proposed by Friedlander and Porat [1989] and is discussed in Chapter 9.

Example 7.1 [Giannakis and Mendel, 1989]

The GM method is applied to a MA parameter estimation problem from noisy output data. The true model is

$$X(k) = W(k) + 0.9\ W(k-1) + 0.385\ W(k-2) - 0.771\ W(k-3)$$

with zeros at 0.6 and $-0.75 \pm j0.85$. The input sequence $\{W(k)\}$ was taken to be an exponentially distributed sequence with $\gamma_2^w = 1$ and $\gamma_3^w = 2$, and it was generated using the GGEXN subroutine of the IMSL library. To compute the estimates $\hat{c}_2^x(\tau)$ and $\hat{c}_3^x(\tau, \tau)$, N output samples were used, which were computed by convolving the random input with the MA model. The $N = 2048$ output samples were divided into $M = 16$ records of 128 samples each. The estimates, $\hat{c}_2^x(\tau)$ and $\hat{c}_3^x(\tau, \tau)$, were computed for each record and then averaged over all records to produce the final estimate. The parameter estimation was performed for the noise-free case and for signal-to-noise ratio *(SNR)* levels of 20 dB and 10 dB. The SNR is defined as $SNR \triangleq \frac{E\{X^2(k)\}}{E\{n^2(k)\}}$.

To reduce the realization dependency, the parameter estimates were averaged over 30 Monte-Carlo runs. The mean $\pm$ standard deviation of the estimated parameters against the true ones are shown in Table 7.1. To demonstrate the robustness of the GM algorithm in model order mismatch, the estimation was also performed for orders $q = 4$ and $q = 5$. The performance of the algorithm is illustrated in the last two columns of Table 7.1 for $SNR = 20$ dB. □

TABLE 7.1 TRUE AND ESTIMATED PARAMETERS OF A MA(3) MODEL USING THE GM ALGORITHM WHEN THE TRUE ORDER IS KNOWN AND WHEN AN OVERFIT IS CONSIDERED
(30 MONTE-CARLO RUNS) [GIANNAKIS AND MENDEL, 1989]

	MA (3), $N = 2048$		
True MA parameters	True order $SNR = 10$ dB	Overfit: 4 $SNR = 20$ dB	Overfit: 5 $SNR = 20$ dB
$b(1) = 0.900$	0.9220 ± 0.0039	0.9215 ± 0.1277	0.5975 ± 0.069
$b(2) = 0.385$	0.3840 ± 0.0031	0.4689 ± 0.1029	0.1847 ± 0.1749
$b(3) = -0.771$	-0.7701 ± 0.0015	-0.7968 ± 0.0077	-0.8003 ± 0.0085
$b(4) = 0.000$	-	0.0152 ± 0.1626	0.3652 ± 0.1012
$b(5) = 0.000$	-	-	-0.1756 ± 0.1061
$\gamma_2^w = 1.$	0.9805 ± 0.0011	0.7328 ± 0.2915	0.7640 ± 0.5098
$\gamma_3^w = 2.$	1.9318 ± 0.0376	1.3878 ± 0.9312	1.4697 ± 1.6159

7.3 NONCAUSAL AR METHODS

Several methods have been suggested for determining the coefficients of the AR model

$$\sum_{i=0}^{p} a(i)X(k-i) = W(k); \qquad a(0) = 1 \qquad (7.49)$$

where $\{W(k)\}$ is zero-mean, non-Gaussian with $\gamma_2^w \neq 0$ and $\gamma_3^w \neq 0$. If the AR model is causal and stable (e.g., minimum phase), then any AR power spectrum estimation technique can be used to compute the AR coefficients [Marple, 1987]. In this section we consider only noncausal AR models. Causal AR models and third-order statistics are considered in Chapter 10 within the context of quadratic phase coupling.

7.3.1 The Huzii Method

Huzii [1981] was the first to introduce a method for estimating the parameters of a noncausal AR model from cumulants. From (7.49), we can write

$$X(z)A_1(z^{-1})A_2(z) = W(z) \qquad (7.50)$$

where $A_1(z^{-1})$ and $A_2(z)$ are polynomials with p_1 roots inside and p_2 roots outside the unit circle, respectively ($p_1 + p_2 = p$). Note that in (7.50) linear phase components and gain are omitted. Huzii's method [1981] fits a spectrally equivalent stable causal $AR(p)$ model to $X(k)$ by using second-order statistics of the data. Let $\overline{A}_2(z^{-1})$ be the equivalent minimum phase polynomial of $A_2(z)$, i.e., $|\overline{A}_2(z^{-1})| = |A_2(z)|$.

The AR coefficients of $\overline{A}(z^{-1})$ given by

$$\overline{A}(z^{-1}) = A_1(z^{-1})\overline{A}_2(z^{-1}) \tag{7.51}$$

are computed using the Yule-Walker method [Marple, 1987]. By reflecting one or more poles to their reciprocal positions outside the unit circle, we obtain 2^p spectrally equivalent models to $\overline{A}(z^{-1})$, i.e., $|A_i(z)| = |\overline{A}(z^{-1})|$ for $i = 1, 2, \ldots 2^p$. To find which $A_i(z)$ is identical to $A_1(z^{-1})A_2(z)$, the data $X(k)$ are processed through each of the inverse filters $A_i(z)$ and the resulting processes are tested for higher-order whiteness, i.e.,

$$\begin{aligned} X_i(z) &= X(z)A_i(z) \\ &= \frac{W(z)}{A_1(z^{-1})A_2(z)} \cdot A_1(z^{-1})A_2^{(i)}(z) \\ &= W(z)\frac{A_2^{(i)}(z)}{A_2(z)}. \end{aligned} \tag{7.52}$$

If $A_2^{(i)}(z) = A_2(z)$, then $X_i(k)$ is non-Gaussian white with variance and skewness identical to those of $W(k)$.

7.3.2 Cumulant-Based Solutions [Tugnait, 1987 a,b]

Let $X(k)$ be a process described by (7.49) which is observed in the presence of additive noise, i.e.,

$$Y(k) = X(k) + n(k) \tag{7.53}$$

where $n(k)$ is zero-mean, i.i.d., independent of $X(k)$. Let $\boldsymbol{\theta} = [a(1), a(2), \cdots, a(p), \gamma_2^w, \gamma_2^n]^T$ and $\boldsymbol{\psi} = [\boldsymbol{\theta}^T, \gamma_3^w, \gamma_3^n]^T$ be vectors of dimensions $((p+2) \times 1)$ and $((p+4) \times 1)$ respectively, that contain the unknown parameters. The problem of estimating $\boldsymbol{\theta}$ given a set of noisy data $Y(k), k = 1, \cdots, N$ was addressed by Tugnait [1987a], who proposed two identification methods. The first method is an extension of the Lii-Rosenblatt [1982] MA method to the noncausal AR case. In this approach, a causal $AR(p)$ model is fit to the noisy observations of (7.53). Using the Yule-Walker method and autocorrelations, the 2^p spectrally equivalent transfer functions $H_i(z) = \frac{1}{A_i(z)}$ are generated as in Huzii's method. Given the observed data $Y(k), k = 1, \cdots, N$ and the finite set of 2^p vectors $\boldsymbol{\theta}$ corresponding to each $A_i(z)$, $\boldsymbol{\psi}$ is selected to minimize the cost function

$$J_1(\boldsymbol{\psi}) = 0.5 \sum_{\tau=-L_1}^{0} \sum_{\rho=\tau}^{0} | c_3^y(\tau,\rho;\boldsymbol{\psi}) - \hat{c}_3^y(\tau,\rho) |^2, \qquad (7.54)$$

where

$$c_3^y(\tau,\rho;\boldsymbol{\psi}) = c_3^x(\tau,\rho;\boldsymbol{\psi}) + \gamma_3^n(\boldsymbol{\psi})\delta(\tau,\rho)$$

$$c_3^x(\tau,\rho;\boldsymbol{\psi}) = \gamma_3^w \sum_{i=-\infty}^{\infty} h(i)h(i+\tau)h(i+\rho)$$

are the third-order cumulant functions calculated as if $\boldsymbol{\psi}$ was the true vector, $h(i)$ is the impulse response of (7.49),

$$\hat{c}_3^y(\tau,\rho) = \frac{1}{N} \sum_{k=\max(0,-\tau,-\rho)}^{\min(N,N+\tau,N+P)} Y(k)Y(k+\tau)Y(k+\rho)$$

is the estimated third-order cumulant function, and L_1 is some large positive integer.

To estimate γ_3^w and γ_3^n, the derivative (7.54) with respect to these quantities may be set to zero while $\boldsymbol{\theta}$ is taken to have some fixed value. Hence,

$$\hat{\gamma}_3^w(\boldsymbol{\theta}) = \frac{[\sum_{\tau=-L}^{0} \sum_{\rho=\tau}^{0} \hat{c}_3^y(\tau,\rho)\overline{c}_3^x(\tau,\rho;\boldsymbol{\theta})] - \hat{c}_3^y(0,0)\overline{c}_3^x(0,0;\boldsymbol{\theta})}{[\sum_{\tau=-L}^{0} \sum_{\rho=\tau}^{0} (\overline{c}_3^x(\tau,\rho;\boldsymbol{\theta}))^2] - (\overline{c}_3^x(0,0;\boldsymbol{\theta}))^2}, \qquad (7.55)$$

and

$$\hat{\gamma}_3^n(\boldsymbol{\theta}) = \hat{c}_3^y(0,0) - \hat{\gamma}_3^w(\boldsymbol{\theta})\overline{c}_3^x(0,0;\boldsymbol{\theta}), \qquad (7.56)$$

where

$$\overline{c}_3^x(\tau,\rho;\boldsymbol{\theta}) = c_3^x(\tau,\rho;\boldsymbol{\theta})/\gamma_3^w.$$

Substituting (7.55) and (7.56) into (7.54), the vector $\boldsymbol{\theta}$ is chosen to be the one that results in the smallest J_1.

Another way of choosing $\boldsymbol{\psi}$ is to minimize a cost function that combines second- and third-order statistics, i.e., [Tugnait, 1987a]

$$J(\boldsymbol{\psi}) = J_2(\boldsymbol{\psi}) + \lambda J_1(\boldsymbol{\psi}), \qquad (7.57)$$

where $1 < \lambda < 0$ is a scalar, $J_1(\psi)$ is defined as in (7.54), and

$$J_2(\psi) = 0.5 \sum_{\tau=-L_2}^{0} [c_2^y(\tau;\theta) - \hat{c}_2^y(\tau;\theta)]^2 \qquad (7.58)$$

where

$$\hat{c}_2^y(\tau) = \frac{1}{N} \sum_{k=-\tau}^{N} Y(k)Y(k+\tau), \qquad \tau \leq 0$$

and $c_2^y(\tau;\theta)$ is the autocorrelation of $Y(k)$ calculated as if ψ was the true vector. The positive scalar λ determines the amount of the contribution of the second- or third- order cumulants. A choice for λ that equally weights errors in matching the estimated correlation and third-order cumulant function is the following:

$$\lambda = \lambda_0 \frac{\sum_{\tau=-L_1}^{0} |\hat{c}_2^y(\tau)|^2}{\sum_{\tau=-L_2}^{0} \sum_{\rho=\tau}^{0} |\hat{c}_3^y(\tau,\rho)|^2}. \qquad (7.59)$$

To minimize (7.57) with respect to ψ, several gradient-type methods can be used. The computation of $J_2(\psi)$ and its gradient was discussed by Tugnait [1987b].

The estimation algorithm resulting from the minimization of criterion (7.54) is computationally less expensive than the algorithm which minimizes criterion (7.57). However, the latter algorithm does not require polynomial factorization, whereas the former does. The minimization of criterion (7.57) becomes a nonlinear optimization problem with a risk that a global minimum may not be reached.

The Modified Akaike Criterion (MAIC)

The *MAIC* discussed by Kashyap [1980] was employed by Tugnait [1987a] to select the order of the AR model. The function of the criterion is given by

$$MAIC(p) = 2F(\theta) + (p+2)\ln(N) + N\ln(2\pi), \qquad (7.60)$$

where

$$F(\theta) = 0.5 \sum_{i=1}^{N} [\overline{Y}^2(i;\theta) / P(i;\theta) + \ln P(i;\theta)],$$

$$\overline{Y}(i;\theta) = Y(i;\theta) - \hat{Y}(i;\theta),$$

and $\hat{Y}(i;\theta)$ is the linear least-squares one-step prediction of $Y(i)$ given the samples $\{Y(k), 1 \leq k \leq i-1\}$. Assuming θ to be the true parameter vector, $P(i;\theta)$

is the prediction error variance

$$P(i;\boldsymbol{\theta}) = E\{[Y(i) - \hat{Y}(i;\boldsymbol{\theta})]^2\}$$

and N is the length of the data $Y(k)$. The parameter $\hat{\boldsymbol{\theta}}$ is the estimate of the vector $\boldsymbol{\theta}$ and $p+2$ represents the number of unknown parameters which affect the second-order statistics of the observed process. For the order selection an upper bound $\bar{p}$ on the model order p is chosen and $MAIC(p)$ is computed for $1 \leq p \leq \bar{p}$. The value of p is the one that minimizes $MAIC(p)$.

7.4 ARMA METHODS

Consider the ARMA model described by (7.1.1), which is assumed to be causal, stable, and generally nonminimum phase. As such, some of the zeros of the system transfer function $H(\omega)$ given by (7.3) may lie outside the unit circle. Next we discuss a number of different methods for the ARMA parameter estimation based on higher-order statistics.

7.4.1 Estimation of the ARMA Parameters from the System Impulse Response

For a given set of data $\{X(k)\}$, their power spectrum, $C_2^x(\omega)$, is estimated first. Then the phase of the transfer function $H(\omega)$, $\phi(\omega)$ is reconstructed from a conventional bispectrum estimate of the data by using one of the nonparametric methods described in Chapter 6. Third, the ARMA (p,q) impulse response is generated using an inverse Fourier transform on $H(\omega)$. Finally, the model parameters are obtained from the impulse response via Padé rational approximation [Lii, 1982]. This method combines conventional power spectrum and bispectrum estimation techniques.

The transfer function $H(\omega)$ is given by [Lii, 1982]:

$$H(\omega) = \left[\frac{C_2^x(\omega)}{\gamma_2^w}\right]^{1/2} \exp\{j\phi(\omega)\}. \tag{7.61}$$

A large body of literature is devoted to estimating $C_2^x(\omega)$ and γ_2^w from the available data [e.g. Marple, 1987]. In addition, several methods have been proposed for the estimation of the phase $\phi(\omega)$ from the bispectrum $C_3^x(\omega_1, \omega_2)$. These methods were discussed in Chapter 6. Thus, having computed $C_2^x(\omega)$, γ_2^w and $\phi(\omega)$ of (7.61), the impulse response of the stable and causal ARMA (p,q) model of (7.1), i.e.,

$h(k)$, is found as the inverse Fourier transform of (7.61). The ARMA parameters are computed from the identity

$$\frac{B(z)}{A(z)} = \frac{\sum_{k=0}^{q} b(k) z^{-k}}{\sum_{k=0}^{P} a(k) z^{-k}} = H(z) = \sum_{k=0}^{\infty} h(k) z^{-k} \qquad (7.62)$$

assuming that $B(z)$ and $A(z)$ have no common factors. Equation (7.62) can be rewritten in the form

$$\frac{\sum_{k=0}^{q} \hat{b}(k) z^{-k}}{\sum_{k=0}^{p} \hat{a}(k) z^{-k}} = \sum_{k=0}^{L} \hat{h}(k) z^{-k}; \qquad L > p + q. \qquad (7.63)$$

Assuming that $p > q$, (7.63) leads to

$$\{\cdots, 0, \hat{b}(0), \hat{b}(1), \cdots, \hat{b}(q), 0, \cdots\} = \{\cdots, 0, \hat{a}(0), \hat{a}(1), \cdots, \hat{a}(p), 0, \cdots\} \\ * \{\cdots, 0, \hat{h}(0), \hat{h}(1), \cdots, \hat{h}(L), 0, \cdots\}, \qquad (7.64)$$

where "*" denotes linear convolution. Since $\hat{b}(k) = 0$ for $k \geq q+1$ and $\hat{a}(0) = 1$, it follows that for any $r > p - q$,

$$\begin{bmatrix} \hat{h}(q+r) & \hat{h}(q+r-1) & \cdots & \hat{h}(q+r-p+1) \\ \hat{h}(q+r+1) & \hat{h}(q+r) & \cdots & \hat{h}(q+r-p+2) \\ \vdots & \vdots & & \vdots \\ \hat{h}(q+r+p-1) & \hat{h}(q+r+p-2) & \cdots & \hat{h}(q+r) \end{bmatrix} \begin{bmatrix} \hat{a}(1) \\ \hat{a}(2) \\ \vdots \\ \hat{a}(p) \end{bmatrix} = \begin{bmatrix} \hat{h}(q+r+1) \\ \hat{h}(q+r+2) \\ \vdots \\ \hat{h}(q+r+p) \end{bmatrix}. \qquad (7.65)$$

Also, for $k = 0, 1, 2, \cdots, q$, we have

$$\sum_{i=0}^{p} \hat{a}(i) \hat{h}(k-i) = \hat{b}(k), \qquad (7.66)$$

or in a matrix form

$$\begin{bmatrix} \hat{b}(0) \\ \hat{b}(1) \\ \vdots \\ \hat{b}(q) \end{bmatrix} = \begin{bmatrix} \hat{h}(0) & 0 & \cdots & 0 \\ \hat{h}(1) & \hat{h}(0) & \cdots & 0 \\ \vdots & \vdots & & \vdots \\ \hat{h}(q) & \hat{h}(q-1) & \cdots & \hat{h}(0) \end{bmatrix} \begin{bmatrix} 1 \\ \hat{a}(1) \\ \vdots \\ \hat{a}(p) \end{bmatrix} \qquad (7.67)$$

The orders p and q can be estimated by constructing the estimates

$$\hat{h}(t,s) = (-1)\frac{s(s-1)}{2} \det[\hat{h}(t+i-j); \quad i,j = 1,2,\cdots s] \qquad (7.68)$$

which satisfy the conditions (i) $\hat{h}(q,p) \neq 0$, and (ii) $\hat{h}(q+1,p+1) = 0$ ("breaking point"). Note that det[·] denotes the determinant of a given matrix. In practice, of course, $\hat{h}(q+1,p+1) \cong 0$.

Extension of this method using the trispectrum was published by Lii and Rosenblatt [1984, 1985].

7.4.2 Residual Time Series

Giannakis and Mendel [1989] developed a method that estimates the ARMA parameters using autocorrelations and third- or fourth-order cumulants. The basic idea of their method is to estimate the AR coefficients first, and then form a residual MA time series from which the MA parameters can be estimated by any of the methods described in section 7.2.

Utilizing autocorrelations and the Yule-Walker equations, the AR parameters can be computed from the normal equation

$$c_2^x(\tau) + \sum_{i=1}^{p} a(i)c_2^x(\tau - i) = 0, \qquad \text{for } r > q. \qquad (7.69)$$

Assuming that the ARMA does not contain all-pass factors and is free of polezero cancellations, the parameters $\{a(i)\}$ can be computed by solving the overdetermined system of equations resulting from (7.69).

An analogous equation to (7.69) holds for a slice of the third-order cumulants [Swami and Mendel, 1990], i.e.,

$$\sum_{i=0}^{p} a(i)c_3^x(\tau - i, t) = 0, \qquad x \text{ for } \tau > q. \qquad (7.70)$$

Concatenating (7.70) for $\tau = q+1, \cdots, q+p+M, M \geq 0$, an overdetermined system of equations can be obtained, i.e., $\mathbf{C}(t)\mathbf{a} = 0$ where $\mathbf{a} = [1, a(1), \cdots, a(p)]$, and the matrix $\mathbf{C}$ can be deduced from (7.70). Consistent estimates of the AR parameters can be obtained only if $\text{rank}[\mathbf{C}(t)] = p$.

For some time, it was assumed that consistent AR estimates can be obtained by using a diagonal cumulant slice in (7.70). However, by a counter-example, Tugnait [1987c] and Swami and Mendel [1989, 1990] proved that this assumption is false, and they suggested that consistent estimates of the AR parameters can be obtained from $(p+1)$ 1-d slices of the third-order cumulants, i.e.,

$$\sum_{i=0}^{p} a(i) c_3^x(\tau - i, t) = 0, \qquad \tau > q \tag{7.71}$$

$\tau = q+1, \cdots, q+p+M$, $M \geq 0$ and $t = q-p, \cdots, q$, or in a matrix form

$$\mathbf{C\,a = c}, \tag{7.72}$$

where

$$\mathbf{C} = \begin{bmatrix} c_3^x(q+1-p, q-p) & \cdots & c_3^x(q, q-p) \\ \vdots & \vdots & \vdots \\ c_3^x(q+1-p, q) & \cdots & c_3^x(q, q) \\ \vdots & \vdots & \vdots \\ c_3^x(q+M, q-p) & \cdots & c_3^x(q+p+M-1, q-p) \\ \vdots & \vdots & \vdots \\ c_3^x(q, q) & \cdots & c_3^x(q+p-1, q) \end{bmatrix}, ((p+M)(p+1) \times p),$$

$\mathbf{a} = [a(p), a(p-1), \cdots, a(1)]^T, (p \times 1)$

$\mathbf{c} = -[c_3^x(q+1, q-p), \cdots c_3^x(q+1, q), \cdots, c_3^x(q+p+M, q-p), \cdots, c_3^x(q+p+M, q)]^T,$
$((p+M)(p+1) \times 1)$.

The system of (7.72) can be solved for the AR parameters. Matrix $\mathbf{C}$ is guaranteed to have rank p [Swami, 1988; Swami and Mendel, 1989b; Giannakis and Mendel, 1990; Swami and Mendel, 1992].

In practice, if cumulant estimates are used in (7.72) the singular value decomposition can be applied on $\mathbf{C}$. The number of significant singular values of $\mathbf{C}$ is an estimate of the AR order p; the AR estimates are then obtained by using a low-rank approximation of $\mathbf{C}$ [Giannakis and Mendel, 1990].

To estimate the MA parameters of the ARMA model, the so-called residual time series method is applied. The process $X(k)$ is applied as input to the p-th order filter with transfer function $\hat{A}(z) = 1 + \sum_{i=1}^{p} \hat{a}(i) z^{-1}$, where $\{\hat{a}(i)\}$ are the AR coefficients previously computed. The residual time series is given by

$$\tilde{X}(k) = X(k) + \sum_{i=1}^{p} \hat{a}(i) X(k-i). \tag{7.73}$$

Rewriting (7.1.1) with $\hat{a}(i)$, instead of $a(i)$, and substituting it into (7.73), we obtain

$$\tilde{X}(k) = \sum_{i=0}^{q} b(i) W(k-i); \tag{7.74}$$

in other words, the residual time series is a MA process of order q. To estimate the MA coefficients of $\tilde{X}(k)$, any method described in section 7.2 can be applied.

Example 7.2 [Giannakis and Mendel, 1989]

Consider the model

$$X(k) \; -1.435843X(k-1) + 0.88666X(k-2) - 0.127898X(k-3)$$
$$= W(k) - 1.3W(k-1) - 1.4W(k-2)$$

with poles at 0.2 and $0.6179 \pm j0.5976$ and zeros at -0.7 and 2.

The data, $X(k)$, were generated as described in Example 7.1. The AR parameters are obtained by solving, with singular value decomposition, the system of equations

$$\hat{c}_3^x(-\tau,-\tau) + \sum_{k=1}^{p} a(k)\hat{c}_3^x(-\tau+k,-\tau+k) = 0; \quad \tau = 1,\cdots,12$$

where $\hat{c}_3^x(-\tau,-\tau)$ are third-order cumulant estimates obtained as described in Example 7.1, considering $N = 2048$ output samples. The MA parameters were estimated using the GM algorithm (section 7.2.4). The results are summarized in Table 7.2 for different SNR levels, and averaged over 30 Monte Carlo runs.

The impulse response of the ARMA(3,2) model is plotted in Figure 7.1, together with the estimated impulse response when $SNR = 20$ dB. For comparison purposes, the estimated minimum phase impulse response is plotted on the same graph. □

7.4.3 The Double $c(q,k)$ Algorithm

Giannakis and Swami [1990] suggested computing the ARMA parameters by applying a MA method twice. Combination of (7.1.1) and (7.2) for $n=3$ (bispectrum domain) yields

$$\begin{aligned} C_3^x(z_1,z_2) &= \gamma_3^w \frac{B(z_1)B(z_2)B(z_1^{-1}z_2^{-1})}{A(z_1)A(z_2)A(z_1^{-1}z_2^{-1})} \\ &= \gamma_3^w \frac{B_3(z_1,z_2)}{A_3(z_1,z_2)}, \end{aligned} \quad (7.75)$$

or equivalently

$$C_3^x(z_1,z_2)A_3(z_1,z_2) = \gamma_3^w B_3(z_1,z_2) \quad (7.76.1)$$

where

$$A_3(z_1, z_2) \equiv A(z_1)A(z_2)A(z_1^{-1}z_2^{-1}) = \sum_{i,j=-p}^{p} a_3(i,j)z_1^{-i}z_2^{-j} \qquad (7.76.2)$$

TABLE 7.2 TRUE AND ESTIMATED PARAMETERS (MEAN ± STANDARD DEVIATION) OF AN ARMA (3,2) MODEL USING THIRD-ORDER STATISTICS (30 MONTE-CARLO RUNS) [GIANNAKIS AND MENDEL, 1989]

True ARMA parameters		(3,2) ARMA model N=2048 noise-free			SNR = 20 dB			SNR = 10 dB		
$a(1)$	= −1.435843	−1.4269	±	0.00589	−1.3924	±	0.01519	−1.1685	±	0.2522
$a(2)$	= 0.88666	0.8852	±	0.01932	0.8328	±	0.0283	0.3971	±	0.2215
$a(3)$	= −0.127898	−0.12827	±	0.00616	−0.1174	±	0.00356	0.096	±	0.2037
$b(1)$	= −1.3	−1.3249	±	0.0066	−1.4253	±	0.0173	−1.774	±	0.1099
$b(2)$	= −1.4	−1.4558	±	0.0481	−1.30205	±	0.2173	−0.9453	±	1.7541
γ_3^w	= 1.0	1.0120	±	0.1396	1.0423	±	0.2892	1.6013	±	0.868
γ_3^w	= 2.0	2.0610	±	0.1116	2.0420	±	0.2546	2.4687	±	0.9715

$$B_3(z_1, z_2) = B(z_1)B(z_2)B(z_1^{-1}z_2^{-1}) = \sum_{i,j=-q}^{q} b_3(i,j)z_1^{-i}z_2^{-j}. \qquad (7.76.3)$$

The $a_3(i,j)$ and $b_3(i,j)$ are related to the AR and MA parameters as follows:

$$a_3(i,j) = \sum_{k=0}^{p} a(k)a(k+i)a(k+j) \qquad (7.77.1)$$

$$b_3(i,j) = \sum_{k=0}^{q} b(k)b(k+i)b(k+j). \qquad (7.77.2)$$

In the third-order cumulant domain, (7.76.1) becomes

$$\sum_{i=p}^{p}\sum_{j=-p}^{p} c_3^x(m-i, n-j)a_3(i,j)z_1^{-i}z_2^{-j} = \begin{cases} 0 & (m,n) \notin S(q) \\ \gamma_3^w b_3(m,n), & (m,n) \in S(q) \end{cases} \qquad \begin{array}{l}(7.78.1)\\ \\ (7.78.2)\end{array}$$

where $S(q) = \{0 \le m \le q, n \le m\}$ is the nonredundant region of support of $b_3(m,n)$.

Using the third-order cumulants of $X(k)$ and the least-squares approach [Giannakis and Swami, 1990; Tugnait, 1992; Giannakis and Swami, 1992a,b], the $a_3(m,n)$ can be determined from (7.78.1). Then, using the computed $a_3(i,j)$ and the third-order cumulants of $X(k)$ from (7.78.2), we obtain $b_3(m,n)$ from (7.78.2).

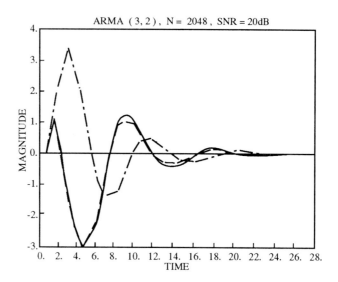

Figure 7.1 True (solid) versus estimated (dashed) versus minimum phase (dot-dashed) impulse responses of a nonminimum ARMA (3,2) model, using third-order statistics ([Giannakis and Mendel, 1989] © 1989 IEEE)

Let $h(k)$ be the transfer function of the ARMA model. Then

$$c_3^x(\tau, \rho) = \gamma_3^w \sum_{i=-\infty}^{\infty} h(i)h(i+\tau)h(i+\rho). \tag{7.79}$$

Comparing (7.79), (7.77.1), and (7.77.2) we see that $a_3(i,j)$ and $b_3(i,j)$ are the third-order cumulants of the *MA(p)* and *MA(q)* parts, respectively. Hence (7.9) can be used to obtain

$$a(i) = \frac{a_3(p,i)}{a_3(p,0)} \tag{7.80.1}$$

$$b(i) = \frac{b_3(q,i)}{b_3(q,0)}. \tag{7.80.2}$$

When p and q are unknown, better estimates of $a(i)$ and $b(i)$ are obtained if all third-order cumulant slices within $S(p)$ and $S(q)$ regions are exploited, i.e.,

$$a(i) = \frac{2}{p(p+1)} \sum_{m=0}^{p} \sum_{n=0}^{m} \frac{a_3^{1/3}(m,n) a_3(p,i)}{\left[\sum_{k=0}^{p-m} a_3(p,k) a_3(p,k+m) a_3(p,k+n)\right]^{1/3}} \tag{7.81.1}$$

$$b(i) = \frac{2}{q(q+1)} \sum_{m=0}^{q} \sum_{n=0}^{m} \frac{b_3^{1/3}(m,n) b_3(q,i)}{\left[\sum_{k=0}^{q-m} b_3(q,k) b_3(q,k+m) b_3(q,k+n)\right]^{1/3}}. \tag{7.81.2}$$

Analogous equations to (7.78) and (7.80) can be derived using fourth-order cumulants [Giannakis and Swami, 1990].

Using the symmetry properties of $a_3(m,n)$ (a cumulant sequence) and (7.80.1), equation (7.78) can be modified as follows:

$$\sum_{j=1}^{p-1}\{a(j)\ [c_3^x(m-p,n-j)+c_3^x(m-j,n-p)+c_3^x(m+p,n+p-j)$$

$$+c_3^x(m+p-j,n+p)+c_3^x(m+j,n-p+j)+c_3^x(m-p+j,n+j)]\}$$

$$+a(p)[c_3^x(m-p,n-p)+c_3^x(m+p,n)+c_3^x(m,n+p)]+\bar{a}_3(0,0)c_3^x(m,n)$$

$$+\sum_{i=2}^{p-1}\sum_{j=1}^{i-1}\{\bar{a}_3(i,j)[c_3^x(m-i,n-j)+c_3^x(m-j,n-i)+c_3^x(m+i,n+i-j)$$

$$+c_3^x(m+i-j,n+i)+c_3^x(m+j,n-i+j)+c_3^x(m-i+j,n+j)]\}$$

$$+\sum_{i=1}^{p-1}\{\bar{a}_3(i,i)\left[c_3^x(m-i,n-i)+c_3^x(m+i,n)+c_3^x(m,n+i)\right]$$

$$+\sum_{i=1}^{p-1}\bar{a}_3(i,0)\left[c_3^x(m-i,n)+c_3^x(m,n-i)+c_3^x(m+i,n+i)\right]$$

$$= -[c_3^x(m-p,n)+c_3^x(m,n-p)+c_3^x(m+p,n+p)],\quad (m,n)\notin S(q)$$
(7.82.1)

$$= -[c_3^x(m-p,n)+c_3^x(m,n-p)+c_3^x(m+p,n+p)]+\bar{B}_3(m,n),\quad (m,n)\in S(q)$$
(7.82.2)

where $\bar{a}_3(i,j) = \dfrac{a_3(i,j)}{a_3(p,0)}$ and $\bar{\beta}_3(i,j) = \gamma_3^w \beta_3(i,j)/a_3(p,0)$. Taking (7.82.1) for $m = Q+1,\cdots,Q+2p, n = Q,\cdots,m$ with $Q = \max(p,q)$, the following system of equations can be formed:

$$\mathbf{C}_a \boldsymbol{\theta}_a = \mathbf{c}_a \tag{7.83}$$

where $\mathbf{C}_a$, and $\mathbf{c}_a$ can be inferred from (7.82.1) and

$$\boldsymbol{\theta}_a = [a(1),\cdots,a(p),\bar{a}_3(0,0),\bar{a}_3(2,1),\cdots,\bar{a}_3(p-1,p-2),\bar{a}_3(1,1),\cdots,$$

$$\bar{a}_3(p-1,p-1),\bar{a}_3(1,0),\cdots,\bar{a}_3(p-1,0)]^T.$$

The solution of (7.83) leads to the AR parameters. A more robust way to estimate the AR parameters would be to use in (7.81.1) the $a_3(i,j)$ values computed as elements of $\boldsymbol{\theta}$. After $a_3(i,j)$'s have been computed, $\bar{\beta}(i,j)\ i,j = -q,\cdots,q$ can be computed from (7.82.2). Then the MA parameters are obtained from (7.80.2) or from (7.81.2).

The skewness of the input process $W(k)$ can be obtained from (7.82.2) by setting $m = n = 0$.

By simulating different ARMA models, Giannakis and Swami [1990] noted that the double $c(q,k)$ algorithm can lead to reliable results when the correct model orders are known, low-variance AR parameters are estimated, and long data records at high SNR's are available.

Example 7.3 [Giannakis and Swami, 1990]

Consider the nonminimum phase ARMA model

$$H(z) = \frac{1 - 0.5z^{-1}}{(1 - 0.8z)(1 + 0.75z^{-1})}.$$

Independent exponentially distributed random processes $W(k)$ were generated and convolved with the impulse response of the model to produce $X(k)$. Zero-mean, Gaussian, i.i.d. noise $n(k)$ was added at $SNR = 20$ dB ($SNR = E\{X^2(k)\}/E\{n^2(k)\}$).

To obtain the third-order cumulant estimates of the noisy process, the data of length $N = 1024$ were divided into $M = 4$ records of 256 samples each. The third-order cumulants of each record were computed and then averaged to produce the third-order cumulant sequence estimate of the noisy data. The resulting sequence was averaged over 100 Monte-Carlo runs.

With the AR order known to be $p = 2$, the AR parameters were obtained by solving, via least squares, the system of equations (7.83).

For MA order $q = 1$, the MA parameters were computed using (7.81.2). The estimation results are shown in Table 7.3. Figure 7.2 illustrates the estimated impulse response of the ARMA model, against the true impulse response. Although the parameter estimators exhibit high variance, the estimated impulse response seems to follow the true impulse response. □

TABLE 7.3 TRUE AND THIRD-ORDER CUMULANT-BASED PARAMETER ESTIMATES (MEAN ± STANDARD DEVIATION) OF AN ARMA (2,1) MODEL WITH KNOWN ORDERS [GIANNAKIS AND SWAMI, 1990]

ARMA(2,1) N=1024 (4×256), SNR = 20 dB, 100 Monte-Carlo Runs	
True AR coeff: 1.000-0.500-0.9375	Poles: 1.25,-0.75
True MA coeff.: 1.0000,-0.5	Zero: 0.5
Est. AR coeff.: 1.000,-0.428,-1.008 ± 0.164 ±0.215	Poles: 1.241, -0.812
Normalized $\{\hat{\beta}_3(1,k)\}_{k=2}^{3}$: $-0.029 \pm 0.502, 0.045 \pm 0.292$	
Est. MA coeff.: 1.000, -0.626 ± 0.727	Zero: 0.626

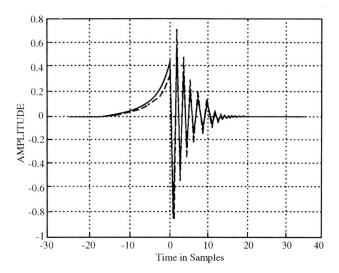

Figure 7.2 True (solid) versus estimated impulse response of an ARMA (2,1) model using the double $c(q,k)$ algorithm © IEEE, 1990

7.4.4 The Q-slice Algorithm [Swami and Mendel, 1990]

The impulse response of the model in Eq. (7.1) satisfies the equation

$$\sum_{i=0}^{p} a(i)h(k-i) = \sum_{i=0}^{q} b(i)\delta(k-i) = b(k). \tag{7.84}$$

Since the system is causal, the third-order cumulants of the output $X(k)$ are related to the system impulse response through

$$c_3^x(\tau, p) = \gamma_3^w \sum_{i=0}^{\infty} h(i)h(i+\tau)h(i+p). \tag{7.85}$$

Combining (7.84) and (7.85), we obtain

$$\sum_{i=0}^{p} a(i)c_3^x(\tau-i, k) = \gamma_3^w \sum_{i=0}^{q} b(i)h(i-\tau+k)h(i-\tau). \tag{7.86}$$

This equation serves as the basis for the estimation of the MA parameters. Since the model is causal and $b(i)$ is nonzero only for $0 \leq i \leq q$, for $\tau = q$ (7.86) yields

$$\sum_{i=0}^{p} a(i)c_3^x(q-i, k) = \gamma_3^w h(k)b(q). \tag{7.87}$$

Note that in (7.87), it was assumed that $h(0) = 1$ which comes from the assumption that $a(0) = b(0) = 1$. From (7.87), the impulse response $h(k)$ can be computed as

$$h(k) = \frac{\sum_{i=0}^{p} a(i) c_3^x(q-i, k)}{\sum_{i=0}^{p} a(i) c_3^x(q-i, 0)}. \tag{7.88}$$

Assuming that the AR parameters are estimated [i.e., solving (7.72)], the MA parameters can be computed from (7.84) as

$$b(k) = \sum_{i=0}^{k} a(i) h(k-i), \quad k = 1, \cdots, q. \tag{7.89}$$

Since q slices are required for the estimation of the q coefficients $b(k)$, the algorithm is called the "q-slice."

For a MA model, $p = 0$ and (7.88) reduces to the $c(q, k)$ formula of (7.9). Thus, the $c(q, k)$ procedure for a MA system identification can be viewed as a special case of the q-slice algorithm.

The AR parameters together with the impulse response $h(k)$ can be computed in a single step from the equation:

$$\sum_{i=1}^{p} a(i) c_3^x(q-i, k) - \gamma_3^w b(q) h(k) = -c_3^x(q, k). \tag{7.90}$$

Evaluating (7.90) for $k = 0, \cdots, Q (Q \geq q)$, we form the system of equations

$$\begin{bmatrix} c_3^x(q-1, 0) & \cdots & c_3^x(q-p, 0) \\ c_3^x(q-1, 1) & \cdots & c_3^x(q-p, 1) \\ \vdots & \ddots & \vdots \\ c_3^x(q-1, Q) & \cdots & c_3^x(q-p, Q) \end{bmatrix} \begin{bmatrix} a(1) \\ a(2) \\ \vdots \\ a(p) \end{bmatrix} + \begin{bmatrix} -\gamma_3^w b(q) \\ -\gamma_3^w b(q) h(1) \\ \vdots \\ -\gamma_3^w b(q) h(Q) \end{bmatrix} = - \begin{bmatrix} c_3^x(q, 0) \\ c_3^x(q, 1) \\ \vdots \\ c_3^x(q, Q) \end{bmatrix} \tag{7.91.1}$$

or in a compact form

$$\mathbf{C}_1 \mathbf{a} - \gamma_3^w b(q) \mathbf{h} = -\mathbf{c}_1. \tag{7.91.2}$$

The equation above can be combined with (7.72) to lead to

$$\begin{bmatrix} \mathbf{C} & \mathbf{0} \\ \mathbf{c}_1 & \mathbf{I} \end{bmatrix} \begin{bmatrix} \mathbf{a} \\ -\gamma_3^w b(q) \mathbf{h} \end{bmatrix} = - \begin{bmatrix} \mathbf{c} \\ \mathbf{c}_1 \end{bmatrix} \tag{7.92}$$

which can be solved for both the AR parameters and the impulse response. Note that from the structure of (7.92), the impulse response $h(k)$ depends on the AR estimates, but not vice versa.

The q-slice algorithm assumes knowledge of the AR and MA orders.

Example 7.4 [Swami and Mendel, 1990]

The q-slice algorithm is applied to estimate the AR and MA parameters of an ARMA model from noisy observations. The true model is

$$X(k) \begin{array}{l} -1.3X(k-1) + 1.05X(k-2) - 0.325X(k-3) \\ = W(k) - 2.95W(k-1) + 1.90W(k-z) \end{array}$$

with poles $0.4 \pm j0.7$ and 0.5 and zeros 2.0 and 0.95. The observation sequence is $Y(k) = X(k) + n(k)$. The input $W(k)$ is an i.i.d., zero-mean, one-sided exponentially distributed random process with $\gamma_2^w = 1$ and $\gamma_3^w = 2$. The process $X(k)$ is corrupted by additive noise $n(k)$, which is zero-mean Gaussian with variance such that $SNR = 20$ dB ($SNR = 10 \log \frac{\gamma_2^x}{\gamma_2^n}$). Thirty independent realizations of the noisy data, each of 2048 points, were generated. Each realization was segmented into 16 records of 128 samples each. The third-order cumulant of the noisy process was taken to be the average of the cumulants of each record.

Parameter estimates were obtained via the following methods:

M-1. The AR parameters were estimated using the SVD method and the MA parameters were estimated via the q-slice algorithm.

M-2. The AR parameters and the impulse response were obtained using (7.92).

Parameter estimates were obtained by assuming the model orders were known exactly, i.e., $p = 3, q = 2$. The results are shown in Table 7.4. The estimates obtained with the M-2 method are better in terms of bias and variance than those obtained by the M-1. □

7.4.5 Identification of ARMA Systems Via Second- and Fourth-Order Cumulant Matching

Consider the process

$$Y(k) = X(k) + n(k) \tag{7.93}$$

where $X(k)$ is described by (7.1.1) with $q = p - 1, n(k)$ is a zero-mean Gaussian process, $X(k)$ has a nonzero fourth-order cumulant, and the roots of the polynomial $\{z^p[1 + \sum_{i=1}^{p} a(i)z^{-i}]\}$ lie inside the unit circle. The latter condition guarantees that the transfer function of the ARMA model of (7.3) is exponentially stable. It is

TABLE 7.4 (a) AR PARAMETER ESTIMATES [SWAMI AND MENDEL, 1990]

$N = 2048$, $SNR = 20$ dB

Method	(p,q)	$a(1)$	$a(2)$	$a(3)$
True	(3,2)	-1.3	1.05	-0.325
M-1	(3,2)	-1.2907 ± 0.1380	1.0395 ± 0.1366	-0.3182 ± 0.1147
M-2	(3,2)	-0.7946 ± 0.1527	0.6358 ± 0.1255	0.0437 ± 0.1130

(b) MA PARAMETER ESTIMATES

$N = 2048$, $SNR = 20$ dB

Method	(p,q)	$b(1)$	$b(2)$
True	(3,2)	-2.95	1.90
M-1	(3,2)	-3.0493 ± 2.7199	2.2277 ± 2.6669
M-2	(3,2)	-0.9452 ± 0.8032	0.1735 ± 0.6240

also assumed that there are no zero-pole cancellations in the system's characteristic function.

Let $\boldsymbol{\theta} = [a(1), a(2), \cdots, a(p), b(1), b(2), \cdots, b(p-1), \gamma_2^w, \gamma_2^n]^T$ and $\boldsymbol{\psi} = [\boldsymbol{\theta}^T, \gamma_4^w]^T$ be vectors that contain the unknown parameters. In order to estimate the vector $\boldsymbol{\theta}$ from the noisy observations $Y(k), k = 1, \cdots, N$, Tugnait [1987b] proposed a criterion based on second- and fourth-order cumulants of $Y(k)$. To indicate the dependence of the system parameters on $\boldsymbol{\theta}$, (7.1.1) can be written as

$$X(k) = -\sum_{i=1}^{p} a(i;\boldsymbol{\theta})X(k-i) + \sum_{i=0}^{p-1} b(i;\boldsymbol{\theta})W(k-i). \quad (7.94)$$

Note that the variance of $W(k)$ and $n(k)$ are also functions of $\boldsymbol{\theta}$. The vector $\boldsymbol{\theta}$ is selected so that the criterion

$$J(\boldsymbol{\psi}) = J_2(\boldsymbol{\psi}) + \lambda J_1(\boldsymbol{\psi}) \quad (7.95.1)$$

is minimized, where

$$J_2(\boldsymbol{\psi}) = 0.5 \sum_{\tau=-L_1}^{0} [c_2^y(\tau;\boldsymbol{\theta}) - \hat{c}_2^x(\tau)]^2 \quad (7.95.2)$$

$$J_1(\boldsymbol{\psi}) = 0.5 \sum_{\tau_1=-L_2}^{0} \sum_{\tau_2=\tau_1}^{0} \sum_{\tau_3=\tau_2}^{0} [c_4^y(\tau_1, \tau_2, \tau_3; \boldsymbol{\psi}) - \hat{c}_4^y(\tau_2, \tau_2, \tau_3)]^2 \quad (7.95.3)$$

and $\lambda > 0$. The sequences $c_2^y(\tau; \boldsymbol{\theta})$ and $c_4^y(\tau_1, \tau_2, \tau_3; \boldsymbol{\theta})$ are the second- and fourth-order cumulants computed assuming that $\boldsymbol{\psi}$ is the true parameter vector, while

Chapter 7   295

$\hat{c}_2^y(\tau)$ and $\hat{c}_4^y(\tau_1, \tau_2, \tau_3)$ are the estimated second- and fourth-order cumulants, respectively. The scalar λ determines the contribution of the fourth-order cumulants to the criterion. Tugnait [1987b] suggested λ to be

$$\lambda = \lambda_0 \cdot \frac{\sum_{\tau=-L_1}^{0} [\hat{c}_2^y(\tau)]^2}{\sum_{\tau_1=-L_2}^{0} \sum_{\tau_2=\tau_1}^{0} \sum_{\tau_3=\tau_2}^{0} [\hat{c}_4^y(\tau_1, \tau_2, \tau_3)]^2} \qquad (7.96)$$

where $\lambda_0 > 0$ determines the relative weighting.

The computational aspects of the cumulants and their gradients are discussed in [Tugnait, 1987], where it is also shown that for $L_1 \geq 2p, L_2 \geq 2p$ and $\lambda > 0$, the minimization of (7.95) yields strongly consistent parameter estimates.

7.5 MODEL ORDER DETERMINATION

In the majority of the parameter estimation algorithms discussed in previous sections, the orders of the parametric models were assumed to be known. However, the estimation of the model order is a difficult problem. Since the input noise process is assumed to be non-Gaussian, autocorrelation-based criteria for order determination cannot estimate all-pass (i.e., phase only) factors and assume either Gaussian processes or minimum phase models. In the following we discuss model order selection criteria for non-Gaussian and nonminimum phase MA, AR, and ARMA models.

7.5.1   MA Models

Model Order Selection Based on Rank Determination

Assuming that $X(k)$ is the output of a LTI system, with input $W(k)$ being an i.i.d., non-Gaussian noise process, the following model can be adopted:

$$X(k) = \sum_{i=-L_1}^{L_2} b(i) W(k+i). \qquad (7.97)$$

Nikias [1986, 1988] proposed an algorithm to estimate the impulse response of this system based on parametric modeling of the third-order cumulants of $X(k)$, and causal and anticausal ARMA approximation models. As it was shown in section

7.2.3, the magnitude and phase response were expressed in terms of the AR parameters of two of the ARMA models. The AR part of the causal ARMA model contained the minimum phase component of the system, whereas the AR part of the anticausal ARMA model contained the maximum phase component. Assuming p and q are chosen sufficiently large, the criterion is applied to the causal ARMA(p, L_2) and anticausal ARMA(q, L_1) models to determine L_1, and L_2, respectively. For the causal case, (7.27.1) yields

$$c_3^x(-\tau,-\rho) = \sum_{i=1}^{p} c(i) c_3^x(-t+i,-\rho+i), \quad \text{for } \tau \text{ or } \rho > L_2 \quad (7.98.1)$$

and

$$c_3^x(-\tau,-\rho) \neq -\sum_{i=1}^{p} c(i) c_3^x(-\tau,-\rho+i), \quad \text{for } \tau \leq L_2 \text{ and } \rho \leq L_2. \quad (7.98.2)$$

From (7.98), the following super matrix can be formed:

$$\mathbf{A}_c = \begin{bmatrix} \mathbf{A}_c^{(0)} \\ \mathbf{A}_c^{(1)} \\ \vdots \\ \mathbf{A}_c^{(J-1)} \end{bmatrix} (J \cdot S) \times (p + S) \quad (7.99)$$

where $\mathbf{A}_c^{(i)}$ is an $S \times (p+S)$ matrix defined by

$$\mathbf{A}_c^{(i)} = \begin{bmatrix} c_3^x(-\overline{L}_2-1+p,-\overline{L}_2+i-1+p) & c_3^x(-\overline{L}_2-S,-\overline{L}_2-S+i) \\ c_3^x(-\overline{L}_2+p,-\overline{L}_2+i+p) & c_3^x(-\overline{L}_2-S+1,-\overline{L}_2-S+i+1) \\ \vdots & \vdots \\ c_3^x(-\overline{L}_2+S-2,-\overline{L}_2+S+i-2+p) & c_3^x(-\overline{L}_2-1,-\overline{L}_2+i-1) \end{bmatrix}$$

$i = 0, 1, \cdots, J - 1$ and $\overline{L}_2$ is greater than the largest possible MA order L_2. We can always take $\overline{L}_2 \geq L_1 + L_2 + 1$. The index S is selected so that $(J \cdot S) > (S + p)$.

It turns out that the rank of $\mathbf{A}_c$ is

$$\text{rank}(\mathbf{A}_c) = -\overline{L}_2 + p + S + L_2 - 1 = I_c \quad (7.100)$$

Thus, any method that determines the rank of $\mathbf{A}_c$ can be used to estimate L_2 from (7.100).

A similar procedure can be followed to determine L_1 using the anticausal ARMA(q, L_1) model. Hence, the rank can be determined from:

$$\text{rank}(\mathbf{A}_a) = -\overline{L}_1 + q + H + L_1 - 1 = I_a \quad (7.101)$$

where

$$\mathbf{A}_a = \begin{bmatrix} \mathbf{A}_a^{(0)} \\ \mathbf{A}_a^{(1)} \\ \vdots \\ \mathbf{A}_a^{(J-1)} \end{bmatrix} \quad H \times (q+H)$$

$$\mathbf{A}_a^{(i)} = \begin{bmatrix} c_3^x(\overline{L}_1 - q + 1, \overline{L}_1 - q + 1 - i), & \cdots & c_3^x(\overline{L}_1 + H, \overline{L}_1 + H - i) \\ c_3^x(\overline{L}_1 - q, \overline{L}_1 - q - i), & \cdots & c_3^x(\overline{L}_1 + H - i, \overline{L}_1 + h - i - i) \\ \vdots & \vdots & \vdots \\ c_3^x(\overline{L}_1 - q - H + 2, \overline{L}_1 - q - i - H + 2), & \cdots & c_3^x(\overline{L}_1 + 1, \overline{L}_1 - i + 1) \end{bmatrix}$$

$i = 0, 1, \cdots, J - 1$ and $\overline{L}_1$ is greater than the largest MA order L_1. We can take $\overline{L}_1 \geq L_1 + L_2 + 1$. The index H is selected so that $(JH) \geq q + H$.

Example 7.5

The effectiveness of the model order selection criterion described by (7.99 – 7.101) is demonstrated in a specific example. The rank of the matrix is determined by the SVD method. The rank I_c is estimated through:

$$e_c(i = I_c + 1) < t_r$$

$$e_c(i) = 1 - \frac{\sum_{j=1}^{i} \sigma_j^{(c)}}{\sigma_1^{(c)} + \cdots + \sigma_{S+p}^{(c)}}$$

where $\{\sigma_i^{(c)}\}, i = 1, 2, \cdots, (p + S)$ are the singular values of $\mathbf{A}_c$, and t_r is a small threshold.

The rank I_a is estimated in a similar way. Having estimated the length of the autocorrelation of the output sequence, $\hat{L}$, the threshold value, t_r, is chosen by trial and error so that $\hat{L} = \hat{L}_1 + \hat{L}_2 + 1$. Figure 7.3 illustrates the true phase and magnitude response of a nonminimum phase FIR system with parameters $\{L_1 = 3, L_2 = 2, b(-3) = -0.0125, b(-2) = 0.3, b(-1) = -1.0119, b(0) = 1.71, b(1) = -0.9275, b(2) = 0.15\}$. The minimum and maximum components of the system transfer function are illustrated in the same figure. The data were divided into M records of K samples each.

The estimated autocorrelation sequence of the output sequence is shown in Figure 7.4 for three different (M, K), i.e., (16,128), (64,128), and (128,128). From this figure it is apparent that although the autocorrelation length is $L = L_1 + L_2 + 1 = 6$, it may be approximated by $\hat{L} = 4$ or 5.

TABLE 7.5 ESTIMATED L_1, L_2 ORDERS OF MA(L_1, L_2) PROCESS FOR DIFFERENT THRESHOLD VALUES (t_r), AR ORDERS (p, q), AND LENGTHS OF DATA.

Record	\multicolumn{6}{c}{$p = q = 8, 64 \times 128$}	\multicolumn{6}{c}{$p = q = 8, 128 \times 128$}
t_r	0.04		0.03		0.02		0.027		0.025		0.02	
Trials	L_1	L_2	L_1	L_2	L_1	L_2	L_1	L_2	L_1	L_2	L_1	L_2
1	1	1	2	2	3	3	2	1	2	2	2	2
2	1	1	2	2	3	3	2	2	2	2	2	2
3	1	1	2	1	2	2	1	1	1	2	2	2
4	1	0	2	1	3	2	1	1	1	1	2	2
5	1	1	1	1	2	2	1	1	2	1	2	2
6	1	0	2	2	3	3	2	2	2	2	3	2
7	1	1	2	1	3	2	1	2	2	2	3	2
8	1	0	2	1	2	2	1	1	2	1	2	2
9	1	0	1	1	2	2	2	2	2	2	2	3
10	1	1	2	2	3	2	2	2	2	2	2	2
11	1	1	2	2	3	3	1	1	1	2	2	2
12	2	1	3	2	4	3	1	1	2	1	2	2
13	2	1	2	2	3	3	2	1	2	2	2	2
14	1	1	1	2	2	3	2	2	2	2	2	2
15	1	0	2	1	2	2	1	1	1	1	2	2
16	1	1	2	2	3	3	2	1	2	2	2	2
17	1	1	1	2	2	3	2	1	2	2	3	2
18	1	1	2	2	3	4	2	1	2	1	2	2
19	1	1	2	2	2	3	1	1	2	1	2	2
20	1	1	2	2	3	3	2	2	2	2	2	2

TABLE 7.5 CONTINUED

Record	\multicolumn{6}{c}{$p = q = 16, 64 \times 128$}					
t_r	0.0175		0.007		0.006	
Trials	L_1	L_2	L_1	L_2	L_1	L_2
1	3	2	1	2	1	2
2	3	3	2	1	2	2
3	2	2	2	1	2	2
4	2	2	1	2	2	2
5	2	2	1	2	1	2
6	3	3	1	1	1	1
7	3	2	1	1	2	1
8	2	2	1	1	1	1
9	3	1	1	1	1	1
10	3	3	2	2	2	2
11	2	2	1	3	2	3
12	2	2	1	1	2	1
13	3	2	0	1	0	1
14	3	3	2	1	2	2
15	2	2	1	2	2	2
16	3	2	2	1	2	2
17	3	3	1	1	2	2
18	2	2	2	2	2	2
19	2	2	1	1	1	1
20	3	3	2	2	3	2

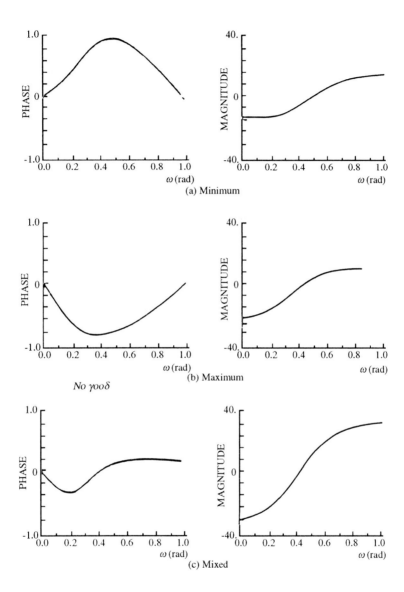

Figure 7.3 Nonminimum phase MA (3,2) process. True magnitude and phase responses: (a) minimum phase component, (b) maximum phase component, and (c) true nonminimum phase component

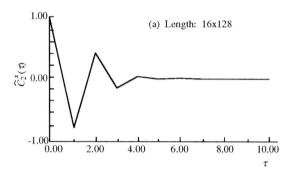

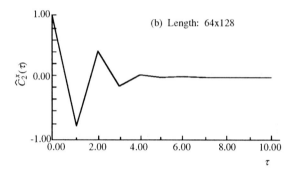

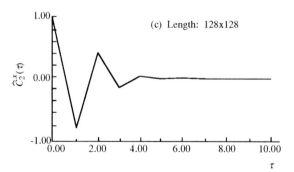

Figure 7.4 Estimated autocorrelation sequence of the MA (3,2) process for three different lengths of data. (a) 16 segments with 128 samples per segment, (b) 64 × 128, and (c) 128 × 128.

Chapter 7   301

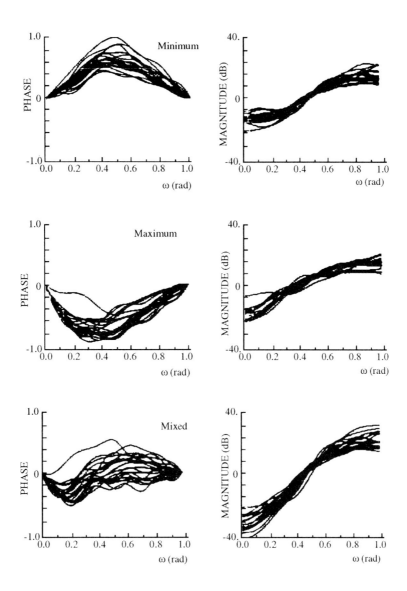

Figure 7.5 (a) Estimated phase and magnitude response of MA (3,2) process via ARMA approximation methods when $p = q = 8, M = 64, K = 128$ and $t_r = 0.03$.

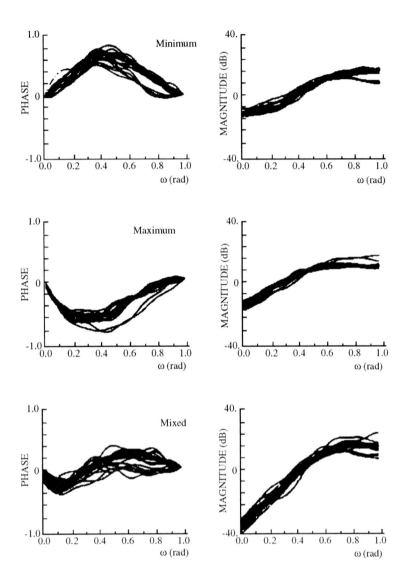

Figure 7.5 (Continued) (b) Estimated phase and magnitude response of MA (3,2) process via ARMA approximation methods when $p = q = 8, M = 64, K = 128$ and $t_r = 0.04$.

Table 7.5 shows the estimated L_1, L_2 orders for different threshold values (t_r), AR orders (p, q), lengths of data, and driving noise seeds (trials). Figure 7.5(a) and (b) illustrate the phase and magnitude response estimates using the method of section 7.2.3 for $p = q = 8, M = 64, K = 128$, and (a) $t_r = 0.03$, (b) $t_r = 0.04$. These threshold values are natural choices to guarantee $\hat{L}_1 + \hat{L}_2 = 3$ or 4. Apparently, comparing Figure 7.3 and Figure 7.5, the estimates follow closely to the true values.
□

Model Order Selection Based on a Statistical Test

For a MA(q) process $X(k)$, it holds that

$$c_3^x(q,0) \neq 0 \text{ and } c_3^x(q+1,0) = 0 \qquad (7.102)$$

which indicates that q is the lag corresponding to the last nonzero cumulant sample.

For sufficiently long data, the second-order cumulant estimate $\hat{c}_2^x(\tau)$ is approximately Gaussian with mean $c_2^x(\tau)$ and variance

$$\text{var}\{\hat{c}_2^x(\tau)\} \simeq \frac{1}{N} \sum_{j=-(N-m)}^{N-m} (1 - \frac{|j|}{N}) \left[(c_2^x(j))^2 + c_2^x(\tau+j)c_2^x(\tau-j) + c_4^x(\tau,j,0) \right], \tau \geq 0 \qquad (7.103)$$

where N is the data length [Priestley, 1981]. Based on the variance of $\hat{c}_2^x(\tau)$, Chow [1972] developed a statistical test for MA order determination. The test is valid only in the noise-free case. Based on the asymptotic Gaussianity of the third-order cumulant estimates of MA models [Lii and Rosenblatt, 1982], Giannakis and Mendel [1990] presented an analogous statistical test using third-order cumulants. The variance of $\hat{c}_3^x(\tau, e)$ is [Giannakis and Mendel, 1990]:

$$\text{var}\{\hat{c}_3^x(\tau,e)\} \simeq \frac{1}{N} \sum_{i=1}^{N} \sum_{j=-q+e}^{q+\tau} (1 - \frac{|j|}{N})[X(i)X(i+\tau)X(i+e) - \hat{c}_3^x(\tau,e)]$$

$$[X(i+j)X(i+j+\tau)X(i+j+e) - \hat{c}_3^x(\tau,e)] \qquad (7.104)$$

while its mean is $c_3^x(\tau, e)$.

Let the null hypothesis be that the MA order is q when $|\hat{c}_3^x(q+1,0)| \leq t_c$. Then the confidence $1 - \epsilon$ will be given by

$$P_r\{|\hat{c}_3^x(q+1,0)| \leq t_c\} = \int_{-t_c}^{t_c} \frac{1}{2\pi\hat{\sigma}^2} e^{-c^2/2\hat{\sigma}^2} dc = 1 - \epsilon \qquad (7.105)$$

where $\hat{\sigma}^2 = \text{var}\{\hat{c}_3^x(q+1,0)\}$. Hence, given the probability of error ϵ, the threshold t_c can be computed from tables of Gaussian distribution. If $|\hat{c}_3^x(q+1,0)| > t_c$, then the MA order is not q with probability of error ϵ, and the test has to be repeated for order $q+1$.

Example 7.6 [Giannakis and Mendel, 1990]

The statistical test for MA order determination is performed using the model

$$X(k) = W(k) + 0.9W(k-1) + 0.385W(k-2) - 0.771W(k-3).$$

The input $W(k)$ is exponentially distributed with $\gamma_2^w = \gamma_3^w = 0.25$ and $\gamma_4^w = 0.375$. The output $X(k)$ is corrupted by additive Gaussian noise $\eta(k)$. To estimate the sample cumulants, the noisy data were divided into 8 records of 128 samples each, and the sample cumulants of each record were averaged. The results were averaged over 100 Monte-Carlo runs.

The true statistics of the model are shown in the first part of Table 7.6. In the second part of Table 7.6, the sample cumulants of the data corrupted by i.i.d. Gaussian noise at $SNR = 0$ dB are shown ($SNR = E\{X^2(k)\}/E\{n^2(k)\}$). By inspection one could decide that the MA order is $q = 3$. In the third part of Table 7.6, the sample cumulants of the data corrupted by colored Gaussian noise at $SNR = 0$ dB are shown. The colored noise was generated by passing a white Gaussian process through the MA(5) model with coefficients, $\{1, -2.33, 0.75, 0.5, 0.3, -1.4\}$. The true estimated cumulants, averaged over 100 Monte-Carlo runs, are plotted in Figure 7.6, where the autocorrelation suggests an incorrect MA order $q = 4$ or 5, while the third- and fourth-order cumulants suggest $q = 3$.

To illustrate the usefulness of the statistical test, the same signal was computed with noise at $SNR=10$ dB through 20 dB. For each SNR, 100 Monte-Carlo runs were performed and for each run the variance of sample autocorrelation and the third-order cumulant log was computed via (7.103) and (7.104) respectively. Setting $1-\epsilon$ to 95%, the variance of each statistic was used in (7.105) to derive thresholds. For each run the hypothesis test was performed to decide the MA order using the derived thresholds. The number of successful selections of the MA order as a function of the SNR is plotted in Figure 7.7. Note that as the SNR becomes low, the autocorrelation degrades and finally fails. Both statistics, however, were successful at high and moderate SNR's. □

7.5.2 AR Models

In section 7.4.2 it was shown that the third-order cumulants of an MA(p,q) process

TABLE 7.6 ORDER DETERMINATION OF A MA(3) MODEL, USING SECOND- AND THIRD-ORDER CUMULANTS.

$MA(3), N = 1024(8 \times 128)$, 100 Monte Carlo Runs, $\overline{q} = 6, \gamma_2^w = \gamma_3^w = 0.25$							
True statistics							
$\{c_2^x(\tau)\}_{\tau=0}^6$	0.638	0.237	−0.077	−0.192	0.000	0.000	0.000
$\{c_3^x(\tau,0)\}_{\tau=0}^6$	0.332	0.274	−0.060	−0.192	0.000	0.000	0.000
Estimated statistics (mean ± st. dev.) w/i.i.d. Gaussian noise, SNR = 0 dB							
$\{\hat{c}_2^x(\tau)\}_{\tau=0}^6$	1.256	0.228	−0.087	−0.203	−0.008	−0.012	−0.019
	±0.115	±0.045	±0.039	±0.053	±0.041	±0.043	±0.044
$\{\hat{c}_3^x(\tau,0)\}_{\tau=0}^6$	0.321	0.257	−0.065	−0.188	−0.003	−0.007	−0.003
	±0.169	±0.128	±0.072	±0.085	±0.072	±0.069	±0.067
Estimated statistics (mean ± st. dev) w/colored Gaussian noise, SNR = 0 dB							
$\{\hat{c}_2^x(\tau)\}_{\tau=0}^6$	1.259	0.036	−0.141	−0.286	−0.232	−0.093	−0.006
	±0.105	±0.026	±0.045	±0.054	±0.049	±0.047	±0.041
$\{\hat{c}_3^x(\tau,0)\}_{\tau=0}^6$	0.298	0.255	−0.056	−0.180	−0.020	−0.009	−0.007
	±0.169	±0.109	±0.062	±0.078	±0.071	±0.072	±0.059

satisfy:

$$\sum_{i=0}^{p} a(i) c_3^x(\tau - i, t) = -c_3^x(\tau, t), \tau > q. \quad (7.106)$$

Concentrating (7.106) for $\tau = q+1, \cdots, q+p, t = q-p, \cdots, q$ leads to the system of equations

$$\mathbf{C}\,\mathbf{a} = \mathbf{c} \quad (7.107)$$

[see (7.72)] where the matrix $\mathbf{C}$, dim = $(p(p+1) \times p)$, has full rank p [Giannakis and Mendel, 1990].

Since p and q are unknown, the extended matrix $\mathbf{C}_e$ can be formed

$$\mathbf{C}_e = \begin{bmatrix} c_3^x(M_1, N_1) & \cdots & c_3^x(M_1 + N_2 - 1, N_1) \\ \vdots & \vdots & \vdots \\ c_3^x(M_1, N_2) & \cdots & c_3^x(M_1 + M_2 - 1, N_2) \\ c_3^x(M_1 + M_2 - 1, N_1) & \vdots & c_3^x(M_1 + 2M_1 - 2, N_1) \\ \vdots & \vdots & \vdots \\ c_3^x(M_1 + M_2 - 1, N_2) & \cdots & c_3^x(M_1 + 2N_2 - 2, N_2) \end{bmatrix}, (M_2(N_2 - N_1 + 1) \times M_2) \quad (7.108)$$

with $M_1 \geq q+1-p$, $M_2 \geq p$, $N_1 \leq q-p$ and $N_2 \geq q$. Swami and Mendel [1989a, b], Giannakis and Mendel [1990] showed that $\mathbf{C}_e$ has rank p if and only

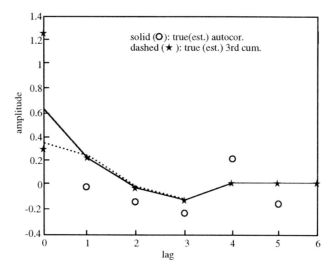

Figure 7.6 True and estimated autocorrelations and third-order cumulants of a MA (3) process corrupted by additive colored Gaussian noise. © IEEE, 1990

if the $ARMA(p,q)$ model is free of pole-zero cancellations. Hence the AR order determination reduces to a rank estimation problem.

Although the ARMA orders are unknown, we may have some knowledge about their upper bounds, i.e., $\bar{p} \geq p$ and $\bar{q} \leq q$. By taking $M_1 = \bar{q}+1, M_2 = \bar{p}, N_1 = -\bar{p}$ and $N_2 = \bar{q}$, the rank of the $\mathbf{C}_e$ matrix is p. If Singular Value Decomposition is applied for rank determination, then p equals the maximum number of nonzero singular values. In practice, since $\mathbf{C}_e$ contains third-order cumulant estimates, all singular values will be nonzero. The rule to decide on the AR order is to find the largest drop among two successive singular values of $\hat{\mathbf{C}}_e$.

Example 7.7 [Giannakis and Mendel, 1990]

The estimation of the AR order via rank determination is performed for the model

$$X(k) - 1.4X(k-1) + 0.65X(k-2) = W(k).$$

The input $W(k)$ was generated as explained in Example 7.6. Colored Gaussian noise was added to the data $X(k)$. Dividing the noisy data into 8 records of 128 samples each, the third-order cumulants were estimated by averaging the cumulants of each of the segments. The final estimates were computed by averaging the results over 100 Monte-Carlo runs. Based on these estimates, the (30×5) matrix $\hat{\mathbf{C}}_e$ was formed ($\bar{p} = 5, \bar{q} = 0$). Figure 7.8 illustrates the singular values of $\hat{\mathbf{C}}_e$ for $SNR =$

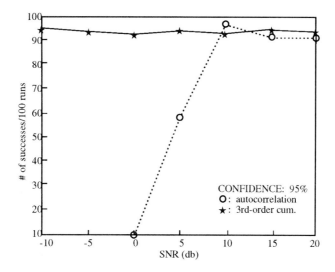

Figure 7.7 Percentage of correct MA order determination using statistical thresholds derived based on autocorrelation and third-order statistics. © IEEE, 1990

5 dB. Apparently the AR order is correctly chosen to be $p = 2$. □

7.5.3 ARMA Models

For an $ARMA(p, q)$ model, the AR order p can be determined as described in Section 7.5.2, i.e., the rank of matrix $\mathbf{C}_e$ in (7.108) equals p. Once we estimate the AR parameters, we can form the residual time series (see also section 7.4.2)

$$\tilde{X}(k) = X(k) + \sum_{i=1}^{p} \hat{a}(i) X(k-i) \qquad (7.109)$$

where $\hat{a}(i)$ are the AR estimates.

Note that $\tilde{X}(k)$ is a pure MA(q) process. Hence any method described in section 7.5.1 for the MA order determination can be applied to determine MA order q.

7.6 SUMMARY

In this chapter we introduced parametric methods for the estimation of higher-order spectra based on MA, noncausal AR, and ARMA models. These methods can also be used for nonminimum phase system identification problems from output measurements only when the system (MA, AR, or ARMA) is driven by a non-Gaussian white noise process. MA methods have been treated by Giannakis [1987], Tugnait [1990], Giannakis and Mendel [1989], Lii and Rosenblatt [1982], Nikias [1988], and

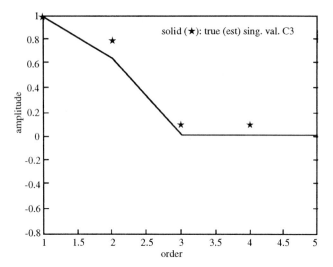

Figure 7.8 True and estimated singular values of $\mathbf{C}_e$ of an AR(2) process observed with additive colored Gaussian noise. © IEEE, 1990

Nikias and Chiang [1988]. Huzii [1981] and Tugnait [1987a, b] have treated higher-order spectral estimation methods based on noncausal AR models. Methods based on ARMA models have been published by Lii [1982], Swami and Mendel [1989, 1990], Giannakis and Mendel [1989], and Giannakis and Swami [1990].

Finally, model order selection criteria for parametric models based on higher-order statistics have been developed by Nikias [1986, 1988], Giannakis and Mendel [1990], and Tugnait [1987a, b; 1990]. A detailed description of MA, AR, and ARMA methods based on higher-order statistics is the subject of a tutorial paper by Mendel [1991].

REFERENCES

Chow, J. C., "On Estimating the Orders of an ARMA Process with Uncertain Observations," *IEEE Trans. Automat. Contr.*, **V. 17**, pp. 707–709, 1972.

Friedlander, B., and B. Porat, "Adaptive IIR Algorithms Based on Higher-Order Statistics," *IEEE Trans. Acoust., Speech and Signal Processing*, **V. 37**, pp. 485–495, 1989.

Giannakis, G. B., "Cumulants: A Powerful Tool in Signal Processing," *Proc. IEEE*, **V. 75**, pp. 1333–1334, 1987.

Giannakis, G. B., and J. M. Mendel, "Identification of Nonminimum Phase Systems Using Higher Order Statistics," *IEEE Trans. Acoust., Speech, and Signal Processing*, **V. 37**, pp. 360–377, March, 1989.

Giannakis, G. B., and J. M. Mendel, "Cumulant Based Order Determination of Non-Gaussian ARMA Models," *IEEE Trans. Acoust., Speech and Signal Processing*, **V. 38 (8)**, pp. 1411–1423, August, 1990.

Giannakis, G. B., and A. Swami, "On Estimating Non-Causal Nonminimum Phase ARMA Models of Non-Gaussian Processes," *IEEE Trans. Acoust. Speech and Signal Processing*, **V. 38**, pp. 478–495, March, 1990.

Giannakis, G. B., and A. Swami, "Author's reply to Counter-examples to: on estimating non-causal nonminimum phase ARMA models of non-Gaussian processes," *IEEE Trans. on Signal Processing*, **40**, pp. 1013–1015, April, 1992a.

Giannakis, G. B., and A. Swami, "Identifiability of General ARMA Processes Using Linear Cumulant-Based Estimators," *Automatica*, **28**(4), pp. 771–779, July, 1992b.

Huzii, M., "Estimation of Coefficient of an Autoregressive Process by Using Higher Order Moments," *J. Time Series Analysis*, **V. 2**, pp. 87–93, 1981.

Kashyap, R. L., "Inconsistency of the AIC Rule for Estimating the Order of Autoregressive Models," *IEEE Trans. Automat. Contr.*, **V. AC-25**, pp. 996–998, Oct., 1980.

Lii, K. S., "Non-Gaussian ARMA Model Identification and Estimation," *Proc. Bus. and Econ. Statistics (ASA)*, pp. 135–141, 1982.

Lii, K. S., and M. Rosenblatt, "Deconvolution and Estimation of Transfer Function Phase and Coefficients for Non-Gaussian Linear Processes," *Ann. Statist.*, **V. 10**, pp. 1195–1208, 1982.

Lii, K. S., and M. Rosenblatt, "Non-Gaussian Linear Processes, Phase and Deconvolution," *Statistical Signal Processing*, E. J. Wegnian and J. G. Smith, (eds.), pp. 51–58, 1984.

Lii, K. S., and M. Rosenblatt, "A Fourth-Order Deconvolution Technique for Non-Gaussian Linear Processes," **Multivariate Analysis VI**, P. R. Krishnaiah, (ed.), Amsterdam, The Netherlands: Elsevier, pp. 395–410, 1985.

Marple, S. L., Jr., **Digital Spectral Analysis with Applications**, Englewood Cliffs, NJ: Prentice Hall, 1987.

Mendel, J. M., "Tutorial on Higher-Order Statistics (Spectra) in Signal Processing and System Theory: Theoretical Results and Some Applications," *Proceeding of the IEEE*, **V. 79(3)**, March, 1991.

Mendel, J. M., and L. Wang, "Cumulant-Based Parameter Estimation Using Structured Networks," *IEEE Trans. Neural Networks*, **V. 2**, pp. 73–83, Jan., 1991.

Nikias, C. L., "Parametric Trispectrum Estimation," *Proceedings Third ASSP Workshop on Spectrum Estimation and Modeling*, Boston: pp. 17–20, November, 1986.

Nikias, C. L., "ARMA Bispectrum Approach to Nonminimum Phase System Identification," *IEEE Trans. on Acoust., Speech and Signal Processing*, **V. 36(4)**, pp. 513–524, April, 1988.

Nikias, C. L., and H. H. Chiang, "Higher-Order Spectrum Estimation Via Non-Causal Autoregressive Modeling and Deconvolution," *IEEE Trans. Acoustics, Speech and Signal Processing*, **V. ASSP-36(12)**, December, 1988.

Nikias, C. L., and R. Pan, "ARMA Modeling of Fourth-Order Cumulants and Phase Estimation," *Circuits, Systems, and Signal Processing*, **V. 7(13)**, pp. 291–325, 1988.

Priestley, M. B., **Spectral Analysis and Time Series**, London: Academic, 1981.

Rosenblatt, M., **Stationary Sequences and Random Fields**, Boston: Birkhauser, 1985.

Swami, A., "System Identification Using Cumulants," Ph.D. Dissertation USC SIPI Report 140, University of Southern California, Department of Electrical Engineering Systems, LA, 1988.

Swami, A., and J. M. Mendel, "Closed-Form Recursive Estimation of MA Coefficients Using Autocorrelations and Third-Order Cumulants," *IEEE Trans. Acoust., Speech and Signal Processing*, **V. 37**, pp. 1794–1795, Nov., 1989a.

Swami, A. and J. M. Mendel, "AR Identifiability Using Cumulants," in *Proc. Workshop on Higher-Order Spectral Analysis*, Vail, CO, pp. 1–6, 1989b.

Swami, A., and J. M. Mendel, "ARMA Parameter Estimation Using Only Output Cumulants", *IEEE Trans. Acoust., Speech and Signal Processing*, **V. 38**, pp. 1257–1265, July, 1990.

Swami, A., and J. M. Mendel, "Identifiability of the Parameters of an ARMA Process Using Cumulants," *IEEE Trans. on Automatic Control*, **37**, pp. 268–273, February, 1992.

Tugnait, J., "Fitting Non-Causal AR Signal Plus Noise Models to Noisy Non-Gaussian Linear Processes," *IEEE Trans. Automat. Contr.*, **V. 32**, pp. 547–552, 1987a.

Tugnait, J., "Identification of Linear Stochastic Systems Via Second-and Fourth-Order Cumulant Matching," *IEEE Trans. Inform. Theory*, **V. 33**, pp. 393–407, 1987b.

Tugnait, J., "Realization and Reduction of SISO Nonminimum-Phase Parametric Models of Non-Gaussian Linear Processes: Modeling, Parameter Estimation and System Identification," (presented only), MTNS Conference, Phoenix, AZ, 1987c.

Tugnait, J., "Approaches to FIR System Identification with Noisy Data Using Higher-Order Statistics," *IEEE Trans. Acoust., Speech and Signal Processing*, **V. 38**, pp. 1309–1317, July, 1990.

Tugnait, J., "Counterexamples to: an Estimating Non-Causal Non-minimum Phase ARMA Models of Non-Gaussian Processes," *IEEE Trans. on Signal Processing*, **40**, pp. 1011-1013, April, 1992.

8

DIRECTION OF ARRIVAL ESTIMATION AND ANALYSIS OF TRANSIENT SIGNALS

.1 INTRODUCTION

In this chapter, we study the higher-order statistics based solution of three important parameter estimation problems that arise in practical situations; namely, time delay estimation (TDE) from two sensor measurements, bearing estimation of a source relative to a line array configuration, and simultaneous parameter estimation of damped exponentials. Later in the chapter, we also discuss a number of detection schemes based on higher-order statistics.

We begin the study by describing the time delay estimation problem and its importance in source bearing and range calculation. We then go on to summarize the time delay estimation methods based on second-order statistics (cross-correlation operations), as well as their advantages and limitations. The TDE methods (nonparametric and parametric) based on higher-order statistics are then described, and their performance compared to generalized cross-correlation methods.

In the second part of the chapter, we discuss the bearing estimation problem from line array data using third- or fourth-order statistics. In particular, we show the extension of beamforming, signal subspace, and noise subspace methods to higher-order statistics domains. We also discuss bearing estimation methods which are based on the asymptotic statistics of cross-higher-order spectra and employ maximum likelihood theory.

In the third part of the chapter, we present methods for simultaneous parameter estimation of damped exponentials embedded in white or colored Gaussian noise. These methods may be seen as an extension of Prony-type techniques to third- or fourth-order statistics domains. However, these extensions are not straightforward as far as uniqueness of solutions are concerned.

In the final part of the chapter, we describe schemes based on higher-order statistics for the detection of either stochastic or deterministic signals in Gaussian noise. Applications of these detection methods to practical situations are also discussed.

8.2 TIME DELAY ESTIMATION

8.2.1 Problem Definition

Let us assume that $\{X(k)\}$ and $\{Y(k)\}$ (k : discrete time) are two spatially separated sensor measurements that satisfy the equations:

$$X(k) = S(k) + W_1(k) \tag{8.1}$$

$$Y(k) = S(k - D) + W_2(k) \tag{8.2}$$

where $\{S(k)\}$ is an unknown signal, $\{S(k-D)\}$ is a shifted and probably scaled version of $\{S(k)\}$, and $\{W_1(k)\}$ and $\{W_2(k)\}$ are unknown noise sources. The problem is to estimate the time delay D from finite length measurements of $X(k)$ and $Y(k)$. This situation arises in such application areas as sonar, radar, biomedicine, geophysics, etc. The basic approach to solve the time delay estimation problem is to shift the measurement sequence $\{X(k)\}$ with respect to $\{Y(k)\}$, and look for similarities between them. The best match will occur at a shift equal to D. In signal processing, "look for similarities" is translated into "taking the cross-correlation" between $\{X(k)\}$ and $\{Y(k)\}$. That is

$$\begin{aligned} c_{xy}(\tau) &= E\{Y(k)Y(k+\tau)\} \\ &= c_2^s(\tau - D), \quad -\infty < \tau < \infty \end{aligned} \tag{8.3}$$

provided that $\{W_1(k)\}$ and $\{W_2(k)\}$ are zero-mean stationary signals, independent with each other and with $\{S(k)\}$. Note that

$$c_2^s(\tau) = E\{S(k)S(k+\tau)\} \tag{8.4}$$

is the covariance sequence of $\{S(k)\}$ and $E\{\cdot\}$ denotes the expectation operation (for ergodic processes, the ensemble average can be replaced by time average).

The $c_{xy}(\tau)$ in (8.3) peaks at $\tau = D$. However, in practical situations, due to finite length data records and noise sources that are not exactly independent, the $c_{xy}(\tau)$ does not necessarily show a peak at the time delay D. Various window functions have been suggested to smooth the cross-correlation function in order to improve the quality of time delay estimates. These are ROTH, SCOT, PHAT, Eckart, and Hannan-Thompson maximum likelihood (or maximum likelihood) just to name a few [Carter, 1987]. The description of these windows is given later in this chapter.

8.2.2 Bearing and Range Estimation

One important application of time delay estimation methods is for source bearing and range calculation. In other words, the purpose here is to locate a source that is emitting energy. The source could be an underwater acoustic source, a radio transmitter, an earthquake epicenter, a star, etc. The approach is to place three or more receivers in a known configuration and to estimate the time delays between the receiver output signals. In this section we describe how to calculate the location of the source utilizing the estimated time delays.

Consider the geometry shown in Figure 8.1 where three receivers, A, B, and C are placed at distances d_{AB}, d_{AC}, and d_{BC} between them; r_B is the range between the source S and receiver B, v is the signal speed of propagation, which is assumed to be known, and D_{AB} and D_{BC} are the time delays to be estimated with signal processing methods.

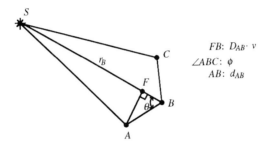

Figure 8.1 Geometry configuration of source (S) and array receivers A, B, and C

Bearing Calculation

Assuming that $r_B \gg d_{AB}$, it follows from Figure 8.1 that

$$d_{AB} \cdot \cos \theta \equiv D_{AB} \cdot v. \tag{8.5}$$

Consequently,

$$\theta \approx \cos^{-1}\left\{\frac{D_{AB} \cdot v}{d_{AB}}\right\}. \tag{8.6}$$

The angle θ is the bearing of the source S relative to receivers A and B. Hence, (8.6) is called the bearing formula.

Range Calculation

It can be shown that the range r_B of the source S is approximately [Chan, 1985]

$$r_B \approx \frac{d_{AB}\left[1 - \frac{\Delta_{AB}^2}{d_{AB}^2}\right] - d_{BC}\left[1 - \frac{\Delta_{BC}^2}{d_{BC}^2}\right]\cos\phi}{2\left\{\frac{\Delta_{AB}}{d_{AB}} + \frac{\Delta_{BC}}{d_{BC}}\cos\phi + \sin\phi\left[1 - \frac{\Delta_{BC}^2}{d_{BC}^2}\right]\right\}} \tag{8.7}$$

where $\Delta_{AB} = D_{AB} \cdot v, \Delta_{BC} = D_{BC} \cdot v$ and ϕ is $\angle ABC$. For the special case where $\phi = 180°$ (linear array), (8.7) simplifies to

$$r_B \approx \frac{d_{AB}\left[1 - \frac{\Delta_{AB}^2}{d_{AB}^2}\right] + d_{BC}\left[1 - \frac{\Delta_{BC}^2}{d_{BC}^2}\right]}{2\left[\frac{\Delta_{AB}}{d_{AB}} - \frac{\Delta_{BC}}{d_{BC}}\right]}. \tag{8.8}$$

From (8.6) it is apparent that one time delay, D_{AB}, is sufficient for bearing calculation. However, two time delays, D_{AB} and D_{BC}, are required in (8.7) or (8.8) to calculate the range.

Localization Accuracy

The errors in the estimation of time delays between the receiver signals will propagate and affect the calculation of bearing and range, and thus, the source localization accuracy.

If we denote $\sigma^2(\hat{D})$ as the variance of the estimate time delay between two receivers that are placed a distance d apart, the variance of the calculated bearing is given by

$$\sigma^2(\theta) = \sigma^2(\hat{D})\left(\frac{v}{d\sin\theta}\right)^2. \tag{8.9}$$

For a linear array of three equally spaced receivers A, B, and C, placed at distance d from each other, the variance of the calculated bearing is given by

$$\sigma^2(\theta) = \sigma^2(\hat{D}_{AC})\frac{2v^2}{3(2d\sin\theta)^2}, \tag{8.10}$$

where $\hat{D}_{AC}$ is the estimated time delay between receivers A and C. On the other hand, the variance of the calculated range for the same receiver configuration is

$$\sigma^2(r) = \sigma^2(\hat{D}_{AC})\frac{2v^2 r^4}{(d\sin\theta)^4}. \tag{8.11}$$

Closer examination of (8.10) and (8.11) reveals that $\sigma^2(\theta)$ and $\sigma^2(r)$ increase in value as the separation interval d between receivers decreases. In other words, the bearing and range calculations become more reliable as the separation distance becomes larger. However, from

(8.11) it is apparent that the longer the range of the source is, the less reliable is the range calculation. Best bearing and range calculations are achieved when the source is at the center of the array (i.e., $\theta = 90°$). *The critical issue here is to obtain good estimates of time delays between receivers, using appropriate signal processing techniques, so we can achieve as small $\sigma^2(D)$ as possible.*

8.2.3 Generalized Cross-Correlation (GCC) Method

As we have already discussed in section 8.2.1, in practice the cross-correlation function, $c_{xy}(\tau)$, can only be estimated from finite length observations of $\{X(k)\}$ and $\{Y(k)\}$ signals. However, to achieve good time delay estimates, we need to smooth the estimated cross-correlation, $\hat{c}_{xy}(\tau)$. The resulting smoothed cross-correlation, $c^w_{xy}(\tau)$, is referred to as the generalized cross-correlation (GCC), and is given by [Knapp and Carter, 1976]:

$$c^w_{xy}(\tau) = F_1^{-1}\{\hat{C}_{xy}(\omega) W(\omega)\} \tag{8.12}$$

$$= \hat{c}_{xy}(\tau) * w(\tau) \tag{8.13}$$

where

$$\hat{C}_{xy}(\omega) = F_1\{\hat{c}_{xy}(\tau)\} \tag{8.14}$$

is the estimated cross-spectrum between $\{X(k)\}$ and $\{Y(k)\}$, "$*$" stands for the linear convolution operation, and $W(\omega)$ is the Fourier transform of the smoothing window $\{w(\tau)\}$. $F_1\{\cdot\}$ and $F_1^{-1}\{\cdot\}$ denote the 1-d forward and inverse Fourier transform operations, respectively. Figure 8.2 illustrates the block diagram implementation structure of (8.13), as a time delay estimation method. An equivalent smoothing procedure is also illustrated in Figure 8.3, where $\{X(k)\}$ and $\{Y(k)\}$ are prefiltered separately before the computation of their cross-correlation function. Comparing figures 8.2 with 8.3, we observe that

$$\begin{aligned} W(\omega) &= H_1(\omega) H_2^*(\omega) \\ &= |H(\omega)|^2 \ (\text{if } H(\omega) = H_1(\omega) = H_2(\omega)) \end{aligned} \tag{8.15}$$

where $H_1(\omega)$ and $H_2(\omega)$ are the frequency transfer functions of the prefilters. Hence, to implement the procedure shown in Figure 8.3, we have to choose $H_1(\omega)$ and $H_2(\omega)$ such that (8.15) is satisfied. The estimated time delay, $\hat{D}$, corresponds to the location of the maximum of $c^w_{xy}(\tau)$. To enhance its accuracy in practical situations, several smoothing windows, $W(\omega)$, have been proposed and tested [Knapp and Carter, 1976; Carter, 1987; Hassab, 1989]. We continue the discussion by describing some of the well-known smoothing windows utilized by GCC methods for time delay estimation.

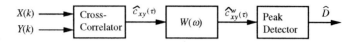

Figure 8.2   Smoothing of the cross-correlation function

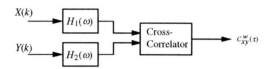

Figure 8.3   Prefiltering of the received signals to obtain smooth cross-correlation

Roth Window

$$W(\omega) = \frac{1}{C_2^x(\omega)} \tag{8.16}$$

where $C_2^x(\omega)$ is the power spectrum of $\{X(k)\}$, defined as

$$C_2^x(\omega) = F\{c_2^x(\tau)\} \tag{8.17}$$

with $c_2^x(\tau)$ being the autocorrelation of $\{X(k)\}$. Utilizing the assumptions discussed in section 8.2.1, (8.1) can be transformed to the power spectrum domain as follows:

$$C_2^x(\omega) = C_2^s(\omega) + C_2^{w_1}(\omega), \tag{8.18}$$

where $C_2^s(\omega)$ and $C_2^{w_1}(\omega)$ are the power spectra of $\{S(k)\}$ and $\{W_1(k)\}$, respectively. For finite length observations, the estimated cross-spectrum, after the application of the Roth window, will be

$$C_{xy}^w(\omega) = C_{xy}(\omega)W(\omega) = \frac{\hat{C}_{xy}(\omega)}{C_2^x(\omega)} \tag{8.19}$$

where $\hat{C}_{xy}(\omega)$ is given by (8.14). Combining (8.19) and (8.18) with (8.3) we obtain

$$C_{xy}^w(\omega) = \frac{C_2^s(\omega)}{C_2^s(\omega) + C_2^{w_1}(\omega)} \exp\{j\omega D\}. \tag{8.20}$$

Equation (8.20) implies that $C_{xy}^w(\omega)$ is unattenuated when $C_2^s(\omega) \gg C_2^{w_1}(\omega)$, i.e., when the signal-to-noise ratio (SNR) is high, and is weighted by the SNR when the SNR is low. Note that

$$\text{SNR} \equiv \frac{C_2^s(\omega)}{C_2^{w_1}(\omega)}. \tag{8.21}$$

SCOT (The Smoothed Coherence Transform)

Knapp and Carter [1976] introduced the following smoothing window:

$$W(\omega) = \frac{1}{[C_2^x(\omega) \cdot C_2^y(\omega)]^{1/2}} = H_1(\omega) \cdot H_2(\omega) \tag{8.22}$$

where $C_2^x(\omega)$ and $C_2^y(\omega)$ are the power spectra of $\{X(k)\}$ and $\{Y(k)\}$, respectively, and

$$\begin{aligned} H_1(\omega) &= \frac{1}{[C_2^x(\omega)]^{1/2}} \\ H_2(\omega) &= \frac{1}{[C_2^y(\omega)]^{1/2}}. \end{aligned} \tag{8.23}$$

The transfer functions $H_1(\omega)$ and $H_2(\omega)$ are the prewhitening filters being applied on $\{X(k)\}$ and $\{Y(k)\}$, respectively, before the computation of the cross-correlation, as shown in Figure 8.3.

PHAT (The Phase Transform)

The smoothing window in this case is inversely proportional to the magnitude cross-spectrum, and has the form

$$W(\omega) = \frac{1}{|C_{xy}(\omega)|}. \tag{8.24}$$

Combining (8.3) and (8.29), we obtain

$$C_{xy}^w(\omega) = \frac{C_{xy}(\omega)}{|C_{xy}(\omega)|} = \exp\{j\omega D\}. \tag{8.25}$$

Apparently, the PHAT window flattens the magnitude of the cross-spectrum, or equivalently, results in a sharp peak in the cross-correlation domain. However, this sharp peak may turn out to be very sensitive to additive noise in practice.

Eckart window

The Eckart window has the form

$$W(\omega) = \frac{C_2^s(\omega)}{C_2^{w_1}(\omega) C_2^{w_2}(\omega)} \tag{8.26}$$

and gives small weight to the cross-spectrum in the regions where the SNR is low, while it gives large weight in those regions where the SNR is high. In other words, it amplifies the signal with respect to the noise in such a way that the peak of the cross-correlation is affected by noise as little as possible.

ML (Maximum Likelihood) or Hannan-Thompson Window

The window has the form

$$W(\omega) = \frac{1}{|C_{xy}(\omega)|} \cdot \frac{|\gamma_{xy}(\omega)|^2}{1 - |\gamma_{xy}(\omega)|^2} \tag{8.27}$$

where

$$|\gamma_{xy}(\omega)|^2 = \frac{|C_{xy}(\omega)|^2}{C_2^x(\omega) C_2^y(\omega)} \tag{8.28}$$

is the magnitude-squared coherence (MSC), taking values between zero and one.

Chan, et al. [1978] showed that in the case of zero-mean, Gaussian uncorrelated processors,

$$z(\omega) = \frac{|\gamma_{xy}(\omega)|^2}{1 - |\gamma_{xy}(\omega)|^2} \sim \frac{1}{\text{variance of phase of } \hat{C}_{xy}(\omega)}, \tag{8.29}$$

where the phase of $\hat{C}_{xy}(\omega)$ consists of the term (ωD) plus a random term caused by the noise sources $W_1(k)$ and $W_2(k)$, as well as by the effect of finite record length. In turn, this implies that the ML window weights the phase of $\hat{C}_{xy}(\omega)$ as the inverse of its standard deviation, after prewhitening, in contrast to the PHAT window which only whitens the cross-spectrum. The generalized cross-correlation with ML window takes the form

$$c_{xy}^w(\tau) = F^{-1}\{\frac{\hat{C}_{xy}(\omega)}{|C_{xy}(\omega)|} z(\omega)\} \tag{8.30}$$

which peaks at the estimated time delay $\hat{D}$. Knapp and Carter [1976] calculated the variance of the estimated time delay to be

$$\sigma_{\hat{D}}^2 = \frac{2\pi \int_{-\infty}^{\infty} |W(\omega)|^2 \omega^2 C_2^x(\omega) C_2^y(\omega)(1 - |\gamma_{xy}(\omega)|^2) d\omega}{T[\int_{-\infty}^{\infty} \omega^2 |C_{xy}(\omega)| \cdot W(\omega) d\omega]^2} \tag{8.31}$$

where T is the length of the observation interval, and $W(\omega)$ is a general smoothing window. For the ML window, substitution of (8.27) into (8.31), leads to [Carter, 1987]

$$\sigma_{\hat{D}}^2(ML) = \frac{1}{\frac{T}{2\pi}[\int_{-\infty}^{\infty} \omega^2 z(\omega) d\omega]} \tag{8.32}$$

which is also the Cramer-Rao bound.

Wiener Processor

In the previous discussion, to achieve a good time delay estimate (TDE), we have seen the application of smoothing windows on the cross-correlation function. These windows depend on input power spectra of the signal and/or noise that are generally unknown and thus have to be estimated. A different approach to the TDE was introduced by Chan, et al. [1980] using the discrete Wiener filter. The method is parametric in the sense that it estimates the parameters of a FIR filter from received signals $\{X(k)\}$ and $\{Y(k)\}$. In fact, the main idea is to pass $\{X(k)\}$ through a FIR filter with parameters $\{a_n\}$, and then choose the optimum values of the parameters such that the filter output is as close to $\{Y(k)\}$ as possible in the mean-square error (MSE) sense. *The resulting set of Wiener filter parameters corresponds to a smoothed cross-correlation function between $X(k)$ and $Y(k)$.* As such, the peak location of the smoothed cross-correlation gives the TDE. According to Figure 8.4, the filter output $Z(k)$ is related to the input $X(k)$ through

$$Z(k) = \sum_{n=-\infty}^{\infty} a(n) X(k-n). \tag{8.33}$$

In general, an infinite number of coefficients are needed to model the time delay. However, in practice, we truncate the summation to include $\{a(n), -p \leq n \leq p\}$, where p is sufficiently large to guarantee $-p \leq D \leq p$, where again D is the time delay parameter. The Wiener problem is to minimize the MSE

$$\begin{aligned} J &= E\{[Y(k) - Z(k)]^2\} \\ &= E\{[Y(k) - \sum_{n=-p}^{p} a(n) X(k-n)]^2\} \end{aligned} \tag{8.34}$$

with respect to filter parameters $\{a_n\}$. Taking the partial derivatives of J with respect to $a(n), n = -p, \ldots, p$; i.e.,

$$\frac{\partial J}{\partial a(i)} = 2E\{[Y(k) - \sum_{n=-p}^{p} a(n) X(k-n)] X(k-i)\} = 0, \tag{8.35}$$

and letting

$$c_{xy}(i) = E\{Y(k) X(k-i)\}, \quad \text{and} \tag{8.36}$$

$$c_2^x(i-n) = E\{X(k-n) X(k-i)\} \tag{8.37}$$

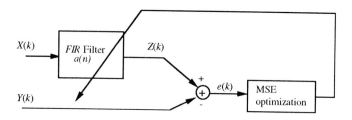

Figure 8.4 The Wiener processor for time delay estimation

for $i = -p, \ldots, p$, we obtain

$$\begin{bmatrix} c_{xy}(-p) \\ \vdots \\ c_{xy}(p) \end{bmatrix} = \begin{bmatrix} c_2^x(0) & c_2^x(-1) \cdots & c_2^x(-2p) \\ \vdots & & \vdots \\ c_2^x(2p) & \vdots & c_2^x(0) \end{bmatrix} \begin{bmatrix} a(-p) \\ \vdots \\ a(0) \\ \vdots \\ a(p) \end{bmatrix}$$

or

$$\mathbf{c} = \boldsymbol{\Phi}\boldsymbol{\alpha} \qquad (8.38)$$

where $\boldsymbol{\alpha}$ is a vector of length $(2p+1)$, $\boldsymbol{\Phi}$ is a $(2p+1) \times (2p+1)$ Toeplitz and symmetric matrix, and $\mathbf{c}$ is a vector of length $(2p+1)$. The system of equations in (8.38) can be solved using the original Levinson algorithm [Marple, 1987].

The time delay D is chosen to be the index of the parameter $\{a(n)\}$ which has the maximum value. However, in the case where the time delay is not an integer, the sampling function interpolation is performed between the values of the parameters to find the non-integer index at which the function has a maximum, [Chan, 1985]. In practical situations, the expectation operation $E\{\cdot\}$ in (8.36) and (8.37) is usually substituted by time average or ensemble average.

8.2.4 Conventional TDE Methods Based on Third-Order Statistics

In several practical applications, including passive and active sonar, the signal $\{S(k)\}$ in (8.1) and (8.2) may be regarded as a non-Gaussian process [Sato and Sasaki, 1977; Sasaki, et al., 1977], whereas the noise sources $\{W_1(k)\}$ and $\{W_2(k)\}$ may be regarded as zero-mean, stationary Gaussian processes. Under these signal conditions, there are certain advantages to comparing similarities between $\{X(k)\}$ and $\{Y(k)\}$ in higher-order spectrum domains, such as the bispectrum or trispectrum. Let us note that self-emitting signals from complicated mechanical systems may contain strong quasi-periodic components, and therefore, may be regarded as non-Gaussian signals [Sasaki, et al., 1977].

One of the fundamental properties of higher-order spectra that makes them attractive tools for time delay estimation is their ability to suppress additive Gaussian processes of unknown autocorrelation sequences [Nikias and Raghuveer, 1987; Nikias and Pan, 1988].

In the TDE problem described by (8.1) and (8.2), if the noise sources are spatially correlated and Gaussian, the generalized cross-correlation (GCC) techniques will fail to work well because they will also estimate the noise spatial correlation. On the other hand, if we "compare similarities" of $\{X(k)\}$ and $\{Y(k)\}$ in higher-order spectrum domains, the effect of spatially correlated Gaussian noises $\{W_1(k)\}$ and $\{W_2(k)\}$ will be suppressed.

Let $\{S(k)\}$ be a zero-mean, non-Gaussian stationary random process with a nonzero measure of skewness, i.e., $E\{S^3(n)\} \neq 0$. Assume that the two received signals are $\{X(k)\}$ and $\{Y(k)\}$ described by (8.1) and (8.2), respectively, and that $\{W_1(k)\}$ and $\{W_2(k)\}$ are zero-mean Gaussian, possibly correlated stationary random processes, statistically independent of $\{S(k)\}$. The cross-correlation of the noise sources (i.e., spatial correlation) is defined by

$$c_{w_1 w_2}(\tau) = E\{W_1(k)W_2(k+\tau)\} \tag{8.39}$$

and is assumed to be *unknown*. Note that these signal conditions arise in several practical situations including those of passive/active sonar. Based on the aforementioned assumptions, the cross-correlation of $\{X(k)\}$ and $\{Y(k)\}$ is given by

$$c_{xy}(\tau) = c_2^s(\tau - D) + c_{w_1 w_2}(\tau), \quad -\infty < \tau < \infty, \tag{8.40}$$

and therefore, all the generalized cross-correlation (GCC) methods will generally fail to work for this problem. On the other hand, the following relationships hold in the third-order moment or cumulant domain of the received signals:

$$\begin{aligned} m_3^x(\tau, \rho) &= E\{X(k)X(k+\tau)X(k+\rho)\} \\ &= m_3^s(\tau, \rho) \end{aligned} \tag{8.41}$$

and

$$\begin{aligned} m_3^{xyx}(\tau, \rho) &= E\{X(k)Y(k+\tau)X(k+\rho)\} \\ &= m_3^s(\tau - D, \rho), \end{aligned} \tag{8.42}$$

where

$$m_3^s(\tau, \rho) = E\{S(k)S(k+\tau)S(k+\rho)\}. \tag{8.43}$$

This is because third-order statistics (moments or cumulants) of a zero-mean Gaussian process are identical to zero.

Additional equivalent expressions are

$$m_3^y(\tau, \rho) = E\{Y(k)Y(k+\tau)Y(k+\rho)\}$$
$$= m_3^s(\tau, \rho) \qquad (8.44)$$

and

$$m_3^{xxy}(\tau, \rho) = E\{X(\tau)X(k+\tau)Y(k+\rho)\}$$
$$= m_3^s(\tau, \rho - D). \qquad (8.45)$$

In the bispectrum and cross-bispectrum domains, (8.41) and (8.42) respectively become

$$M_3^x(\omega_1, \omega_2) = M_3^s(\omega_1, \omega_2) \qquad (8.46)$$

$$M_3^{xyx}(\omega_1, \omega_2) = M_3^s(\omega_1, \omega_2) e^{j\omega_1 D} \qquad (8.47)$$

where $M_3^s(\omega_1, \omega_2)$ is the bispectrum of $\{S(k)\}$. We can also write

$$M_3^x(\omega_1, \omega_2) = |M_3^x(\omega_1, \omega_2)| e^{j\psi_3^x(\omega_1, \omega_2)} \qquad (8.48)$$

$$M_3^{xyx}(\omega_1, \omega_2) = |M_3^{xyx}(\omega_1, \omega_2)| e^{j\psi_3^{xyx}(\omega_1, \omega_2)}. \qquad (8.49)$$

The following set of ad hoc conventional methods have been developed for time delay estimation [Nikias and Pan, 1988].

Method I

Define

$$\Phi(\omega_1, \omega_2) = \Psi_3^{xyx}(\omega_1, \omega_2) - \Psi_3^x(\omega_1, \omega_2) \qquad (8.50)$$

$$I(\omega_1, \omega_2) = \exp\{j\Phi(\omega_1, \omega_2)\}. \qquad (8.51)$$

Combining (8.46) – (8.51), we obtain

$$I(\omega_1, \omega_2) = e^{j\omega_1 D}, \qquad (8.52)$$

and therefore, the function

$$T(\tau) = \int_{-\pi}^{\pi} \int_{-\pi}^{\pi} I(\omega_1, \omega_2) e^{-j\omega_1 \tau} d\omega_1 d\omega_2 = \int_{-\pi}^{\pi} d\omega_2 \int_{-\pi}^{\pi} d\omega_1 e^{j\omega_1(D-\tau)} \qquad (8.53)$$

peaks at $\tau = D$. If the signal $\{S(k)\}$ has a flat bispectrum, $I(\omega_1, \omega_2) = M_3^{xyx}(\omega_1, \omega_2)$, and thus, $T(\tau) = m_3^{xyx}(\tau, 0)$.

Method II

Similarly, we may also define

$$\Phi'(\omega_1, \omega_2) = \Psi_3^{xyx}(\omega_1, \omega_2) - \frac{1}{2}[\Psi_3^x(\omega_1, \omega_2) + \Psi_3^y(\omega_1, \omega_2)]$$

and

$$I(\omega_1,\omega_2) = \exp\{j\Phi'(\omega_1,\omega_2)\} \tag{8.54}$$

and $T(\tau)$ follows from (8.53).

Method III

[Sato, et al., 1977; Sasaki, et al., 1977]

$$I(\omega_1,\omega_2) = \frac{M_3^{xyx}(\omega_1,\omega_2)}{M_3^x(\omega_1,\omega_2)}. \tag{8.55}$$

Method IV

For $S(k)$ being a linear process

$$I(\omega_1,\omega_2) = \frac{|M_3^{xyx}(\omega_1,\omega_2)| \cdot \exp\{j\Phi'(\omega_1,\omega_2)\}}{\sqrt{|M_3^x(\omega_1,\omega_2)| \cdot |M_3^y(\omega_1,\omega_2)|}} \tag{8.56}$$

with $\Phi'(\omega_1,\omega_2)$ defined in (8.54). Oh, Kim, and Powers [1990] suggested the squared skewness weighting function which combines second- and third-order statistics. That is

$$P_3(\omega_1,\omega_2) = \frac{|M_3^{xyy}(\omega_1,\omega_2)|^2}{[M_2^x(\omega_1+\omega_2)M_2^y(\omega_1)M_2^y(\omega_2)]}. \tag{8.57.1}$$

The time delay D is estimated by generating

$$\hat{M}_3^{xyy}(\tau,\rho) = \int_{-\pi}^{+\pi} d\omega_1 \int_{-\pi}^{+\pi} d\omega_2 \cdot \exp\{j\Psi_3^{xyy}(\omega_1,\omega_2)\} \cdot P_3(\omega_1,\omega_2)\exp\{j(\omega_1\tau+\omega_2\rho)\}. \tag{8.57.2}$$

Note that ideally

$$M_3^{xyy}(\tau,\rho) = \delta(\tau-D,\rho-D). \tag{8.57.3}$$

In practice, we are always given finite length data records. Therefore, bispectrum and cross-bispectrum estimation procedures, such as those described in Chapter 4, have to be employed to estimate $I(\omega_1,\omega_2)$. Methods I–IV will generally give different estimates of $I(\omega_1,\omega_2)$. Since the implementation of Methods I–IV is based on conventional bispectrum estimation methods, we denote them as *Conventional TDE Methods*.

Zhang and Raghuveer [1991] extended the conventional Methods I and II for simultaneous estimation of two time delays in a situation involving three sensors. They showed that it is enough to compute just one cross-bispectrum, instead of two cross-bispectra as required by conventional Methods I–IV.

Example 8.1

The purpose of this example is to implement Method I as a signal processing algorithm. Let $\{X(0),\ldots,X(N-1)\}$ and $\{Y(0),\ldots,Y(N-1)\}$ be the available sets of data. The bispectrum conventional Method I can be implemented as follows:

a. Segment the data into K records of M samples each with 50% overlapping, i.e., $N = KM/2$.

b. Assuming that $\{X^{(i)}(0),\ldots,X^{(i)}(M-1)\}$ and $\{Y^{(i)}(0),\ldots,Y^{(i)}(M-1)\}$ are the data of the ith segment, apply a FFT on each segment, i.e.,

$$X^{(i)}(\omega) = \sum_{k=0}^{M-1} X^{(i)}(k) e^{-j\frac{2\pi k \omega}{M}}$$

$$Y^{(i)}(\omega) = \sum_{k=0}^{M-1} Y^{(i)}(k) e^{-j\frac{2\pi k \omega}{M}}.$$

c. Estimate the bispectra and cross-bispectra (direct approach)

$$M_3^{x,i}(\omega_1,\omega_2) = X^{(i)}(\omega_1) \cdot X^{(i)}(\omega_2) X^{(i)*}(\omega_1+\omega_2)$$

$$M_3^{xyx,i}(\omega_1,\omega_2) = X^{(i)}(\omega_1) Y^{(i)}(\omega_2) X^{(i)*}(\omega_1+\omega_2).$$

d. Average the bispectra over K records

$$\hat{M}_3^{xyx}(\omega_1,\omega_2) = \frac{1}{K} \sum_{i=1}^{K} M_3^{x,i}(\omega_1,\omega_2)$$

$$\hat{M}_3^{xyx}(\omega_1,\omega_2) = \frac{1}{K} \sum_{i=1}^{K} M_3^{xyx,i}(\omega_1,\omega_2).$$

e. Compute the principal values of the phases $\hat{M}_3^x$

$$\hat{\Psi}_3^x(\omega_1,\omega_2) = \tan^{-1}\left\{\frac{Im[\hat{M}_3^x(\omega_1,\omega_2)]}{Re[\hat{M}_3^x(\omega_1,\omega_2)]}\right\}$$

$$\hat{\Psi}_3^{xyx}(\omega_1,\omega_2) = \tan^{-1}\left\{\frac{Im[\hat{M}_3^{xyx}(\omega_1,\omega_2)]}{Re[\hat{M}_3^{xyx}(\omega_1,\omega_2)]}\right\}.$$

f. Compute
$$\hat{\Phi}(\omega_1,\omega_2) = \hat{\Psi}_3^{xyx}(\omega_1,\omega_2) - \hat{\Psi}_3^x(\omega_1,\omega_2)$$
and form
$$\hat{I}(\omega_1,\omega_2) = e^{j\hat{\Phi}(\omega_1,\omega_2)}.$$

g. Calculate
$$\hat{T}(\tau) = \sum_{\omega_1=0}^{M-1}\sum_{\omega_2=0}^{M-1} \hat{I}(\omega_1,\omega_2)e^{-j\omega_1\tau}.$$

h. Select $\hat{D}$ that maximizes $\hat{T}(\tau = \hat{D})$ as the estimate of time delay D. Conventional Methods II–IV can be easily implemented following similar steps. □

8.2.5 Parametric TDE Method in the Bispectrum Domain

From (8.1) we can always write $S(k - D) = X(k - D) - W_1(k - D)$, and thus from (8.2) we have
$$Y(k) = X(k-D) - W_1(k-D) + W_2(k) \qquad (8.58.1)$$
or, in a more general form,
$$Y(k) = \sum_{i=-\infty}^{+\infty} \alpha(i)X(k-i) + W_2(k) - W_1(k-D) \qquad (8.58.2)$$
where in theory $\alpha(i) = 0$ for all $\{i\}$ except $i = D$, and $\alpha(D) = 1$. In practice, D is always finite, and therefore,
$$Y(k) = \sum_{i=-p}^{+p} \alpha(i)X(k-i) + W_2(k) - W_1(k-D) \qquad (8.59)$$
where p is much greater than the largest possible delay we can expect. If we take (8.58.2) for time instant $k + \tau$, multiply both sides by $X(k)X(k + \rho)$ and take expectations, we obtain
$$\begin{aligned} E\{X(k)Y(k+\tau)X(k+\rho)\} &= \sum_{i=-p}^{+p} \alpha(i)E\{X(k)X(k-i+\tau)X(k+\rho)\} \\ &\quad + E\{X(k)W_2(k+\tau)X(k+\rho)\} \\ &\quad - E\{X(k)W_1(k+\tau-D)X(k+\rho)\} \end{aligned} \qquad (8.60)$$
or
$$m_3^{xyx}(\tau,\rho) = \sum_{i=-p}^{p} \alpha(i)m_3^x(\tau-i,\rho) + m_3^{xw_2x}(\tau,\rho) - m_3^{xw_1x}(\tau-D,\rho). \qquad (8.61)$$

However, $m_3^{xw_2x}(\tau,\rho)$ and $m_3^{xw_1x}(\tau-D,\rho)$ are identically zero for all (τ,ρ) because (i) the signal and noise sources are zero-mean and independent, and (ii) the noise sources are Gaussian. So, even in the case where the noise sources are spatially correlated with cross-correlation function described by (8.39), (8.61) becomes [Nikias and Pan, 1988]

$$m_3^{xyx}(\tau,\rho) = \sum_{i=-p}^{p} \alpha(i) m_3^x(\tau-i,\rho). \tag{8.62}$$

Selecting various integers for τ and ρ, we form an overdetermined system of equations. For example, if τ ranges from $-p$ to $+p$, and ρ is assigned to $\{-1,0,1\}$, a set of linear equations can be produced as follows:

$$\mathbf{m}^{xyx} = \mathbf{m}^x \cdot \mathbf{A} \tag{8.63}$$

where

$$\mathbf{m}^{xyx} = [m_3^{xyx}(-p,0), m_3^{xyx}(-p+1,0), \ldots, m_3^{xyx}(p,0), m_3^{xyx}(-p,1), \ldots,$$
$$m_3^{xyx}(p,1), m_3^{xyx}(-p,-1), \ldots, m_3^{xyx}(p,-1)]^T (6p+3) \times 1$$
$$\mathbf{A} = [\alpha(-p), \alpha(-p+1), \ldots, \alpha(0), \ldots, \alpha(p)]^T (2p+1) \times 1$$

$$\mathbf{m}^x = \begin{bmatrix} m_3^x(0,0) & \ldots & m_3^x(-2p,0) \\ \vdots & & \vdots \\ m_3^x(2p,0) & \ldots & m_3^x(0,0) \\ m_3^x(0,1) & \ldots & m_3^x(-2p,1) \\ \vdots & & \vdots \\ m_3^x(2p,1) & \ldots & m_3^x(0,1) \\ m_3^x(0,-1) & \ldots & m_3^x(-2p,-1) \\ \vdots & & \vdots \\ m_3^x(2p,-1) & \ldots & m_3^x(0,-1) \end{bmatrix} (6p+3) \times (2p+1).$$

The least-squares solution of (8.63) is given by

$$\mathbf{A} = [(\mathbf{m}^x)^T \mathbf{m}^x]^{-1} (\mathbf{m}^x)^T \mathbf{m}^{xyx}. \tag{8.64}$$

Lower rank approximation via singular value decomposition may also be used in (8.64). The time delay D is chosen to be the index of the parameter $\{\alpha(i)\}$ which has maximum value. However, in the case where the time delay is noninteger, interpolation is performed between the parameters of vector $\mathbf{A}$ to find the time instant at which the function is maximum. The parameter vector $\mathbf{A}$ is essentially the "smoothed" cross-cumulant vector $\mathbf{m}^{xyx}$. The number of equations in (8.63) may be increased by assigning more values to integer $\{\rho\}$.

The Wiener filtering approach for TDE introduced by Chan, et al. [1980] is based on second-order statistics and leads to biased TDE estimates even when the correlation data are exactly known and the noise sources are spatially uncorrelated. On the other hand, for exactly known third-order statistics, the TDE method described in (8.62) and (8.64) provides unbiased estimates, even for spatially correlated Gaussian noise sources.

Example 8.2

To illustrate the algorithmic implementation of the parametric TDE method in the bispectrum domain, we consider the following steps in this example.

a. Segment the data into K records with 50% overlapping. Each segment consists of M samples.

b. Assuming that $\{X^{(i)}(0), \cdots, X^{(i)}(M-1)\}$ and $\{Y^{(i)}(0), \cdots, Y^{(i)}(M-1)\}$ are the data of the ith segment, calculate the third-order statistics

$$\hat{m}_3^{x,i}(\tau,\rho) = \frac{1}{M} \sum_{k=s_1}^{s_2} X^{(i)}(k)X^{(i)}(k+\tau)X^{(i)}(k+\rho)$$

$$\hat{m}_3^{xyx,i}(\tau,\rho) = \frac{1}{M} \sum_{k=s_1}^{s_2} X^{(i)}(k)Y^{(i)}(k+\tau)X^{(i)}(k+\rho)$$

where

$$s_1 = \max(0, -\tau, -\rho)$$

$$s_2 = \min(M, M-\tau, M-\rho).$$

c. Average over all segments

$$\hat{m}_3^x(\tau,\rho) = \frac{1}{K} \sum_{i=1}^{K} \hat{m}_3^{x,i}(\tau,\rho)$$

$$\hat{m}_3^{xyx}(\tau,\rho) = \frac{1}{K} \sum_{i=1}^{K} \hat{m}_3^{xyx,i}(\tau,\rho).$$

d. Substitute the estimated third-order statistics into (8.63)

$$\hat{\mathbf{m}}^{xyx} = \hat{\mathbf{m}}^x \cdot \mathbf{A}.$$

e. Obtain the least-squares solution

$$\hat{\mathbf{A}} = [(\hat{\mathbf{m}}^x)^T \hat{\mathbf{m}}^x]^{-1} (\hat{\mathbf{m}}^x)^T \hat{\mathbf{m}}^{xyx}$$

where

$$\hat{A} = [\hat{\alpha}(-p), \cdots, \hat{\alpha}(0), \cdots, \hat{\alpha}(p)]^T.$$

f. Select $\hat{a}(d)$ as the maximum among $\{a(i)\}$ for $i = -p, \cdots, 0, \cdots, p$. The TDE is $\hat{D} = d$. A possibly more accurate time delay estimate, $\hat{D}$, may be obtained by applying the sampling interpolation formula on the estimated FIR parameters [Chan, et al., 1980]. □

Example 8.3

In this example, we present some computer simulation results on the performance of TDE techniques. Specifically, the conventional bispectrum Method I, as well as the parametric bispectrum method, are employed in the simulations. For comparison purposes, three generalized cross-correlation (GCC) methods (i.e., ROTH, SCOT, and ML) are also employed in the same simulations. Throughout these computer experiments, we segment the data into records, assuming 50% overlapping, and choose the time delay to be $D = 16$. The results will be demonstrated for different lengths of data and different signal-to-noise ratios (SNR), where SNR is defined as $20 \log_{10}(\sigma_s/\sigma_n)$, where σ_s and σ_n are the standard deviations of signal and noise, respectively. In this example, we assume that: (i) the random signal $\{S(k)\}$ in non-Gaussian with nonzero skewness, and (ii) the noise sources $\{W_1(k)\}$ and $\{W_2(k)\}$ are Gaussian. The range of search for the time delay estimate is chosen from $p = -30$ to $p = 30$ with a step increment equal to one. Two particular cases are discussed in this example depending on the power spectrum nature of signal $\{S(k)\}$, namely, white signal and colored signal in the presence of spatially correlated Gaussian noises [Nikias and Pan, 1988].

White Signal and Spatially Correlated Noises

In this case, the signal is white and the noise sources $\{W_1(k)\}$ and $\{W_2(k)\}$ are spatially correlated. The $\{W_1(k)\}$ was generated first, while $\{W_2(k)\}$ was computed from

$$W_2(k) = \sum_{i=0}^{10} b(i) W_1(k+i) \qquad (8.65)$$

where $\{b(i)\}$ take values $\{0.2, 0.4, 0.6, 0.8, 1, 1, 1, 0.7, 0.5, 0.3, 0.1\}$. Figure 8.5 (SNR = 0 dB) and Figure 8.6 (SNR = − 5dB) illustrate the TDE results of this case. Apparently, the parametric bispectrum method exhibits better performance than the conventional bispectrum method I. The suppression of Gaussian spatially correlated noise sources is a function of both the SNR and the length of data. However, from Figures 8.5 and 8.6, the noise suppression appears to be more effective when the parametric bispectrum method is utilized. As the SNR decreases, longer data lengths are required to reduce the effect of Gaussian noise sources. As expected, the GCC methods (ROTH, SCOT, ML) are incapable of suppressing the spatially correlated Gaussian noise sources and therefore fail to provide reliable results for this signal and noise environment.

Colored Signal and Spatially Correlated Colored Noises

The signal $\{S(k)\}$ is generated by passing a non-Gaussian white process (one-sided exponential) through a low-pass filter with cutoff frequency 0.1π. The noise sources are generated the same way by passing a Gaussian white process through the same low-pass filter. Thus, signal and noise sources have flat power spectra in the band $|\omega| \leq 0.1\pi$, and are assumed to be zero-mean. Assuming that the noise sources also satisfy (8.65), i.e., they are spatially correlated, the time delay between $\{X(k)\}$ and $\{Y(k)\}$ is estimated using the bispectrum and cross-correlation methods. Figure 8.7 (SNR = 0 dB) and Figure 8.8 (SNR = -5 dB) illustrate the TDE results of this case.

When the SNR = 0 dB, the parametric bispectrum method suppresses the correlated Gaussian noises even when the length of data is $64 \times 128 (K = 64, M = 128)$. On the other hand, the conventional bispectrum method I would achieve the same performance as the parametric for data lengths longer than 256×128. As the SNR decreases, the bispectrum methods require longer lengths of data to suppress the effect of correlated Gaussian noises. As expected again, GCC methods (ROTH, SCOT, and ML) do not work well in this case.
□

8.2.6 TDE Method Based on the Cross-Bicepstrum

Both bispectrum TDE methods described in section 8.2.5 can be characterized as indirect, in the sense that they provide time delay estimates by examining the results of a reference function. On the other hand, Nikias and Liu [1990] introduced a TDE method based on the cepstrum of a cross-bispectrum that provides time delay estimates directly and explicitly, combining the use of both second- and third-order statistics. From (8.46) and (8.47), if we use the 2-d Z transform, instead of the Fourier transform, for the bispectrum, $M_3^x(z_1, z_2)$, and cross-bispectrum, $M_3^{xyx}(z_1, z_2)$, we obtain the following ratio:

$$\frac{M_3^{xyx}(z_1, z_2)}{M_3^x(z_1, z_2)} = z_1^{-D}. \tag{8.66}$$

Applying the complex logarithm on (8.66) we obtain

$$\ln[M_3^{xyx}(z_1, z_2)] - \ln[M_3^x(z_1, z_2)] = -D\frac{1}{z_1}\ln[z_1]. \tag{8.67}$$

Operating the partial differentiation with respect to z_1, we have

$$\frac{1}{M_3^{xyx}(z_1,z_2)} \cdot \frac{\partial M_3^{xyx}(z_1,z_2)}{\partial z_1} - \frac{1}{M_3^{x}(z_1,z_2)} \cdot \frac{\partial M_3^{x}(z_1,z_2)}{\partial z_1} = \frac{D}{z_1}$$

which can also be written, in the time domain, as

$$m_3^x(m,n) * [m \cdot m_3^{xyx}(m,n)] m_3^{xyx}(m,n) * [m \cdot m_3^x(m,n)] = D m_3^{xyx}(m,n) * m_3^x(m,n). \quad (8.68)$$

Rewritting (8.68) in the Fourier transform domain, and solving for D, we obtain

$$D = \frac{F_2\{m \cdot m_3^{xyx}(m,n)\}}{F_2\{m_3^{xyx}(m,n)\}} - \frac{F_2\{m \cdot m_3^x(m,n)\}}{F_2\{m_3^x(m,n)\}}. \quad (8.69)$$

Note that time delay D is actually $D(\omega_1,\omega_2)$; i.e., we have the same time delay parameter at each frequency pair (ω_1,ω_2).

In practical situations $M_3^x(\omega_1,\omega_2)$ and $M_3^{xyx}(\omega_1,\omega_2)$ are estimated from $\{X(k)\}$ and $\{Y(k)\}$. Hence, what we obtain from (8.69) is actually a set of estimates $\hat{D}(\omega_1,\omega_2)$ of D at each point (ω_1,ω_2). The final estimate, $\hat{D}$, can be obtained by averaging

$$\hat{D} = \frac{1}{K} \sum_{\omega_1} \sum_{\omega_2} \hat{D}(\omega_1,\omega_2), \quad (8.70)$$

where K is the total number of lattice points in the plane (ω_1,ω_2).

8.2.7 TDE with the Mean Fourth-Cumulant (MFC) Criterion

Tugnait [1989] proposed an optimization approach where the time delay between two signals is estimated by minimizing a mean fourth-cumulant (MFC) criterion. Since the method is based on fourth-order cumulants, it requires that the signal $\{S(k)\}$ has nonzero kurtosis; i.e.,

$$\gamma_4^s = E\{S^4(k)\} - 3[\gamma_2^s]^2 \neq 0. \quad (8.71)$$

Hinich, et al. [1989] have reported that ship-radiated noise is indeed non-Gaussian and it satisfies (8.71).

Chan, et al. [1980] pointed out that the GCC approach with no weighting (i.e., rectangular window) is equivalent to choosing a delay D_0 which minimizes the performance index

$$J_2(D_0) = E\{[X(k - D_0) - Y(k)]^2\}. \quad (8.72)$$

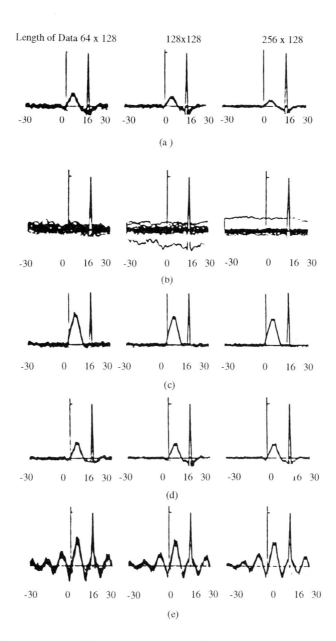

Figure 8.5 Non-Gaussian white signal and spatially correlated Gaussian noise source: time delay estimates using (a) bispectrum conventional method I (direct), (b) parametric bispectrum, (c) ROTH, (d) SCOT, (e) ML methods in SNR = 0 dB and for three different lengths of data: 64 × 128, 128 × 128 and 256 × 128. The true time delay is $D = 16$ ([Nikias and Pan, 1988]. © 1988 IEEE).

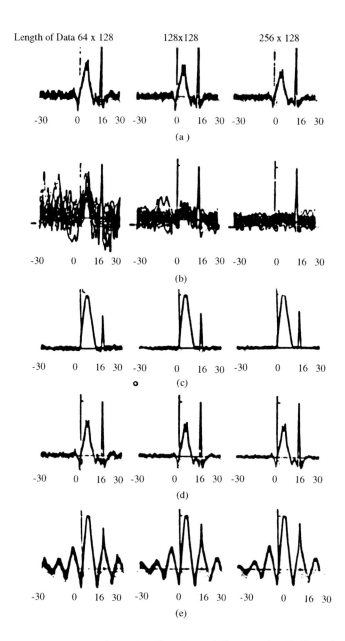

Figure 8.6 Non-Gaussian white signal and spatially correlated Gaussian noise source: time delay estimates using (a) bispectrum conventional method I (direct), (b) parametric bispectrum, (c) ROTH, (d) SCOT, (e) ML methods in SNR = −5 dB and for three different lengths of data: 64 × 128, 128 × 128 and 256 × 128. The true time delay is $D = 16$ ([Nikias and Pan, 1988]. ©1988 IEEE).

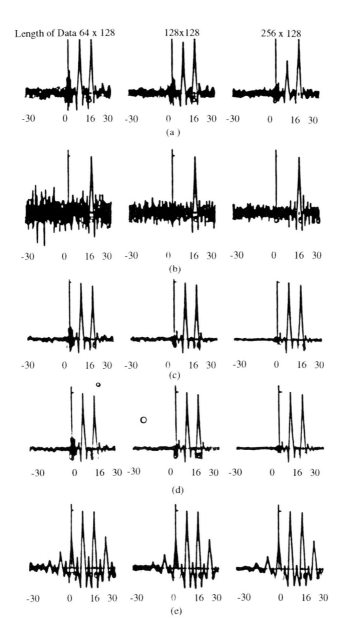

Figure 8.7 Non-Gaussian color signal and spatially correlated color Gaussian noise sources (directional noise): time delay estimates using (a) bispectrum conventional method I (direct), (b) parametric bispectrum, (c) ROTH, (d) SCOT, (e) ML methods in SNR = 0 dB and for three different lengths of data: 64 × 128, 128 × 128 and 256 × 128. The true time delay is $D = 16$ ([Nikias and Pan, 1988]. ©1988 IEEE).

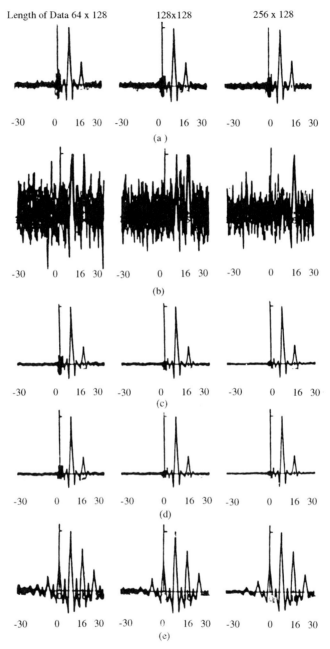

Figure 8.8 Non-Gaussian color signal and spatially correlated color Gaussian noise sources (directional noise): time delay estimates using (a) bispectrum conventional method I (direct), (b) parametric bispectrum, (c) ROTH, (d) SCOT, (e) ML methods in SNR = −5 dB and for three different lengths of data: 64 × 128, 128 × 128 and 256 × 128. The true time delay is $D = 16$ ([Nikias and Pan, 1988]. © 1988 IEEE).

Tugnait's approach [1989] is an extension of performance index $J_2(D_0)$ to fourth-order statistics domain; viz:

$$J_4(D_0) = E\{[X(k - D_0) - Y(k)]^4\} - 3(E\{[X(k - D_0) - Y(k)]^2\})^2. \tag{8.73}$$

Thus, the desired time delay D is defined as the one that minimizes the performance index $J_4(D_0)$ if $\gamma_4^s > 0$ or the one that maximizes $J_4(D_0)$ if $\gamma_4^s < 0$.

In practical situations, we are given $X(k)$ and $Y(k)$ for $k = 1, 2, \cdots, N$. Consequently, (8.73) becomes

$$\begin{aligned}
\hat{J}_4(D_0) &= \frac{1}{N}\sum_{k=1}^{N}[X(k - D_0) - Y(k)]^4 \\
&\quad - 3(\frac{1}{N}\sum_{k=1}^{N}[X(k - D_0) - Y(k)]^2)^2
\end{aligned} \tag{8.74}$$

where the kurtosis γ_4^s is estimated as follows [Tugnait, 1989]:

$$\hat{\gamma}_4^s = 0.5(A_x + A_y) \tag{8.75}$$

$$\begin{aligned}
A_x &= \frac{1}{N}[\sum_{k=1}^{N} X^4(k)] - 3[\frac{1}{N}\sum_{k=1}^{N} X^2(k)]^2 \\
A_y &= \frac{1}{N}[\sum_{k=1}^{N} Y^4(k)] - 3[\frac{1}{N}\sum_{k=1}^{N} Y^2(k)]^2
\end{aligned} \tag{8.76}$$

Example 8.4 [Tugnait, 1989]

In this example we present computer simulation results that demonstrate the performance of the "basic" GCC (e.g., rectangular window) and the TDE method with MFC criterion. The signal $S(k) = W(k) \cdot \mu(k)$ where $W(k)$ is Laplace i.i.d. and $\mu(k)$ is a Bernoulli process (i.e., $\mu(k)$ is i.i.d. with two possible values, 0 or 1, with probabilities 0.8 and 0.2, respectively).

This leads to $\gamma_4^s > 0$, so that we seek the minimum of $J_4(D_0)$. The noise sources $W_1(k)$ and $W_2(k)$ in (8.1) and (8.2), respectively, are spatially correlated, as described in Example 8.3, and (8.65). The SNR $= -$ 5dB and the true time delay $D = 16$. The length of the observed signals $\{X(k)\}$ and $\{Y(k)\}$ is $N = 2000$. Figure 8.9 illustrates $\hat{J}_2(D_0)$ and $\hat{J}_4(D_0)$ as functions of D_0 for 10 different realizations, and clearly demonstrates that the MFC criterion suppresses the spatially correlated noise sources at the expense of higher variability (i.e., lower output SNR).

.3 BEARING ESTIMATION WITH HIGHER-ORDER STATISTICS

The purpose of this section is to introduce the bearing estimation problem of a source

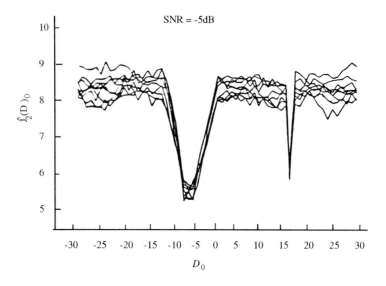

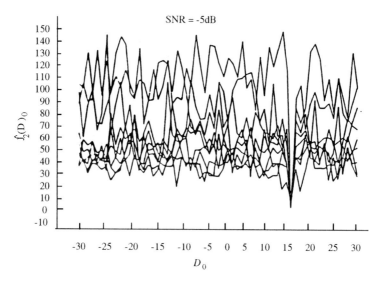

Figure 8.9 Time delay estimation using (a) GCC method with rectangular window, and (b) mean fourth-order cumulant (MFC) method. True time delay is $D = 16$ ([Tugnait, 1989] © 1989 Asilomar Conference Proceedings).

relative to a line array configuration. First, the bearing estimation problem is formulated and the properties of the spatial covariance matrix are discussed. Second, a brief review of array processing techniques based on the spatial covariance matrix is given. Third, the spatial fourth-order cumulant matrix and its properties are introduced, as well as array processing methods based on the fourth-order cumulant matrix. Finally, the bearing estimation problem in the bispectrum and trispectrum domains is addressed by presenting the cross-bispectrum beamforming and MUSIC-like methods, respectively. Several examples are given to demonstrate the performance of all of these methods. Note that the array processing problem with a line array configuration is the dual to the harmonic retrieval problem. Consequently, *the array processing techniques based on higher-order statistics described in this section can also be used for harmonic retrieval.*

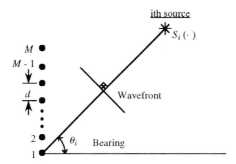

Figure 8.10 Linear array of sensors and bearing definition of a source

8.3.1 Problem Formulation

Let us consider a set of narrow-band incoherent plane waves of frequency $\omega_0 = 2\pi f_0$, and bearings $\theta_1, \theta_2, \ldots, \theta_I$, impinging on a linear array of M elements (as shown in Figure 8.10). Let us assume that: (a) the number of sources is less than the number of array elements $(I < M)$, and (b) the distance between the array elements is $d \leq \frac{\lambda_0}{2}$ where $\lambda_0 = \frac{c}{f_0}$ and c is the speed of propagation. For a linear array, the sensor signals are

$$Y_m(k) = \sum_{i=1}^{I} S_i(k) \exp\{-j\omega_0 (m-1) \frac{\sin \theta_i}{c} d\} + W_m(k); \quad m = 1, 2, \ldots, M; k = 1, 2, \ldots, N$$

(8.77)

where $\{S_i(k)\}$ is the signal of the ith wavefront and $\{W_m(\cdot)\}$ is additive noise at the mth sensor with variance γ_2^w, uncorrelated with $\{W_j(\cdot)\}$, for $m \neq j$ and N is the number of snapshots. In vector forms (8.77) becomes

$$\begin{aligned}
\mathbf{Y}(k) &= \mathbf{A} \ \mathbf{S}(k) + \mathbf{W}(k), \\
\mathbf{Y}(k) &= [Y_1(k), Y_2(k), \ldots, Y_M(k)]^T \quad (M \times 1) \\
\mathbf{S}(k) &= [S_1(k), S_2(k), \ldots, S_I(k)]^T \quad (I \times 1) \\
\mathbf{A} &= [\boldsymbol{\alpha}(\theta_1), \boldsymbol{\alpha}(\theta_2), \ldots, \boldsymbol{\alpha}(\theta_I)] \quad (M \times I) \\
\boldsymbol{\alpha}(\theta_i) &= \left[e^{-j\phi_1(\theta_i)}, \ldots, e^{-j\phi_{M-1}(\theta_i)}\right]^T \quad (M \times 1) \\
\phi_m(\theta_i) &= \omega_0 \frac{d}{c} m \sin\theta_i, \\
\mathbf{W}(k) &= [W_1(k), W_2(k), \ldots, W_M(k)]^T \quad (M \times 1).
\end{aligned}$$

The array processing problem is to estimate the number of sources, I, and their corresponding bearings $\theta_i, i = 1, 2, \ldots, I$ from the array snapshots $\mathbf{Y}(k), k = 1, 2, \ldots, N$. The number of columns of $\mathbf{A}$ equals the number of sources, and each column $\boldsymbol{\alpha}(\theta_i)$ corresponds to the steering vector of the line array pointing at the source bearing θ_i. Note that noise sources are assumed to be uncorrelated with incoming source signals.

In this section we describe array processing techniques that solve the bearing estimation problem using spatial second- or higher-order statistics of $\mathbf{Y}(k)$.

8.3.2 The Spatial Covariance Matrix

The spatial covariance matrix of the array signals, $\mathbf{R}$, is defined by

$$\mathbf{R} = E\{\mathbf{Y}(k)\mathbf{Y}^H(k)\}, \qquad (8.79)$$

where $\mathbf{R}$ is an $M \times M$ matrix, "H" denotes complex transpose operation, $\mathbf{Y}(k)$ is the kth snapshot, and $E\{.\}$ is the statistical expectation. In practice, the spatial covariance matrix can be estimated as follows [Johnson, 1982]:

$$\hat{\mathbf{R}} = \frac{1}{N} \sum_{k=1}^{N} \mathbf{Y}(k)\mathbf{Y}^H(k) \qquad (8.80)$$

where N is the total number of snapshots. The matrix $\mathbf{R}$ is Toeplitz and Hermitian. However, for finite N, the $\hat{\mathbf{R}}$ is *only* Hermitian. From (8.78) and the problem assumptions, it follows that

$$\mathbf{R} = \mathbf{S}_M + \mathbf{W}_M \qquad (8.81.1)$$

where $\mathbf{S}_M$ and $\mathbf{W}_M$ are the signal and noise covariance matrices, respectively, given by

$$\mathbf{S}_M = \mathbf{A} \ \mathbf{S} \ \mathbf{A}^H, \ (M \times M) \qquad (8.81.2)$$

where **A** was defined in (8.78),

$$\mathbf{S} = E\{\mathbf{S}(k)\mathbf{S}^H(k)\}, \ (I \times I) \tag{8.81.3}$$

and

$$\mathbf{W}_M = \gamma_2^w \mathbf{I}_M, \ (M \times M). \tag{8.81.4}$$

From (8.81.1) – (8.81.4) the following comments are in order:

(i) The noise covariance matrix $\mathbf{W}_M$ is full rank, i.e., rank $[\mathbf{W}_M] = M$.

(ii) For incoherent or partially correlated sources, rank $[S_M] = I$ ($I < M$).

(iii) If all source signals are fully correlated (coherent), rank $[S_M] = 1$.

(iv) Rank $[\mathbf{R}] = M$; this is always true because $\mathbf{W}_M$ is full rank.

The spatial covariance matrix admits the orthogonal decomposition [Bienvenu and Kopp 1981]

$$\mathbf{R} = \sum_{i=1}^{M} \rho_i \mathbf{V}_i \mathbf{V}_i^H \tag{8.82.1}$$

where ρ_i are the eigenvalues and $\mathbf{V}_i$ the corresponding eigenvectors with

$$\mathbf{V}_i^H \mathbf{V}_j = \begin{cases} 1, & i = j \\ 0, & i \neq j. \end{cases} \tag{8.82.2}$$

The first I eigenvectors, i.e., $\{\mathbf{V}_1, \mathbf{V}_2, \ldots, \mathbf{V}_I\}$ form the *signal subspace* of the matrix with corresponding eigenvalues, $\rho_i = \lambda_i + \gamma_2^w$ for $i = 1, 2, \ldots, I$. On the other hand, the eigenvectors $\{\mathbf{V}_{I+1}, \mathbf{V}_{I+2}, \ldots, \mathbf{V}_M\}$ form the *noise subspace* of the matrix with eigenvalues $\rho_i = \gamma_2^w$ for $i = I+1, \ldots, M$.

In other words

$$\mathbf{R} = \sum_{i=1}^{I} (\lambda_i + \gamma_2^w) \mathbf{V}_i \mathbf{V}_i^H + \gamma_2^w \sum_{i=I+1}^{M} \mathbf{V}_i \mathbf{V}_i^H. \tag{8.83}$$

Figure 8.11 illustrates the eigenvalues ρ_i versus $i = 1, 2, \ldots, M$. From this figure, it is apparent that the number of sources, I, can be easily identified, since it corresponds to a point where after I sources, all matrix eigenvalues are equal to each other. The number of sources, I, can also be estimated, in practical situations, by finding the minimum of the MDL function defined by [Wax 1985]

$$MDL(k) = -\log\left[\frac{G(k)}{A(k)}\right]^N + E(k) \tag{8.84}$$

for $k = 0, 1, \ldots, M - 1$, where

$$G(k) = \prod_{i=k+1}^{M} \rho_i,$$

$$A(k) = \left[\frac{1}{M-k} \sum_{i=k+1}^{M} \rho_i\right]^{M-k},$$

$$E(k) = \frac{1}{2}k(2M - k)\log N,$$

$$\rho_1 \geq \rho_2 \geq \ldots \geq \rho_M > 0,$$

with N being the number of snapshots and ρ_i the eigenvalues of **R**.

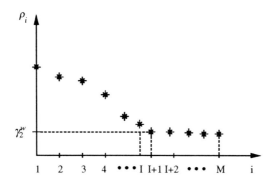

Figure 8.11 The eigenvalues of the spatial covariance matrix. Note that $\rho_{I+1} = \ldots = \rho_M = \gamma_2^w$, and thus the number of sources, I, can be easily identified.

8.3.3 Bearing Estimation Based on the Spatial Covariance Matrix

For a linear (or line) array that the sensor signals satisfy (8.77), the array steering vector is defined as

$$\mathbf{C}(\theta) = [1, \exp\{-j\phi_1(\theta)\} \ldots, \exp\{-j\phi_{M-1}(\theta)\}]^T \quad (8.85.1)$$

where

$$\phi_m(\theta) = \omega_0 \frac{d}{c} m \sin\theta, -90° \leq \theta \leq 90°. \quad (8.85.2)$$

The array processing methods based on the spatial covariance matrix are summarized below.

A. Conventional Method (Beamforming):

The distribution of power over bearings, i.e., $P(\theta)$ versus θ, is given in terms of the spatial covariance matrix $\mathbf{R}$ of (8.79) and the steering vector $\mathbf{C}(\theta)$ of (8.85) as

$$P(\theta) = \mathbf{C}^H(\theta)\mathbf{R}\mathbf{C}(\theta). \tag{8.86}$$

The number of sources and their bearings are given by the peak locations of $P(\theta)$. Although this method is simple to implement, it exhibits sidelobes and limited resolution capability.

Based on the orthogonal decomposition of the spatial covariance matrix [see (8.82)], array processing methods can be divided into signal subspace and noise subspace methods. The signal subspace methods are based on the first I eigenvalues and eigenvectors of $\mathbf{R}$, while the noise subspace methods are based on the last $M - I$ eigenvectors of $\mathbf{R}$. A brief description of these methods follows next.

B. Signal Subspace Methods:

From (8.83) and Figure 8.11, we see that the white noise variance γ_2^w can easily be identified. As such, the signal spatial covariance matrix can be reconstructed from

$$\hat{\mathbf{S}}_M = \sum_{i=1}^{I}(\rho_i - \gamma_2^w)\mathbf{V}_i\mathbf{V}_i^H, \tag{8.87}$$

and its pseudo-inverse by

$$\mathbf{S}_M^{-1} = \sum_{i=1}^{I}\frac{1}{(\rho_i - \gamma_2^w)}\mathbf{V}_i\mathbf{V}_i^H. \tag{8.88}$$

Array processing methods that belong to the signal subspace category are listed below [Marple, 1987; Gabriel, 1980].

Maximum Likelihood (ML) of Capon:

$$P(\theta) = \frac{1}{\mathbf{C}^H(\theta)\hat{\mathbf{S}}_M^{-1}\mathbf{C}(\theta)}. \tag{8.89}$$

This method is also known as Minimum Variance Distortionless technique.

Autoregressive (AR) or Yule-Walker:

$$P(\theta) = \frac{1}{|\mathbf{u}^T\hat{\mathbf{S}}_M^{-1}\mathbf{C}(\theta)|^2} \tag{8.90}$$

where $\mathbf{u}^T = [1, 0, \ldots, 0]$ is the unit steering vector.

Thermal Noise:

$$P(\theta) = \frac{1}{|\mathbf{u}^T \hat{\mathbf{S}}_M^{-H} \hat{\mathbf{S}}_M^{-1} \mathbf{C}(\theta)|^2} \tag{8.91}$$

where "H" denotes transpose conjugate and " $-H$ " is the inverse transpose conjugate.

C. Noise Subspace Methods:

From the orthogonal decomposition of **R**, the noise subspace methods are given by

$$P(\theta) = \frac{1}{\sum_{i=I+1}^{M} \alpha_i |\mathbf{C}^H(\theta) \mathbf{V}_i|^2} \tag{8.92}$$

where $\mathbf{V}_i, i = I+1, \ldots, M$ are the noise subspace eigenvectors of **R**. Based on how the α_i coefficients are chosen, the following methods exist:

Multiple Signal Classification (MUSIC):

$$\alpha_i = 1 \text{ for all } i = I+1, \ldots, M. \tag{8.93}$$

Eigenvector:

$$\alpha_i = \frac{1}{\rho_i}, i = I+1, \ldots, M. \tag{8.94}$$

Pisarenko:

$$\alpha_{I+1} = \ldots = \alpha_{M-1} = 0; \; \alpha_M = 1. \tag{8.95}$$

Signal subspace and noise subspace methods are high-resolution techniques capable of resolving closely spaced sources. The sizes of the signal subspace (I) and noise subspace $(M - I)$ are clearly determined by the number of sources (I). Consequently, in those array processing settings where the number of sources, I, is much less than $(M - I)$, the noise subspace methods are utilized. On the other hand, signal subspace methods are preferred when $I > (M - I)$.

D. ESPRIT:

Paulraj, et al. [1985] and Roy, et al. [1985, 1986] introduced a bearing estimation method based on the idea of estimating signal parameters via rotation invariance techniques (ESPRIT). It has been shown, however, that for line arrays ESPRIT is identical to the Toeplitz approximation method (TAM) originally introduced by Kung, et al. [1983]. The basic approach of ESPRIT is to divide the array into two subarrays of constant displacement and utilize the spatial covariance matrix and the cross-covariance matrix between the two subarrays.

For the line array described by (8.78), ESPRIT proceeds as follows:

The array signals $\mathbf{Y}(k)$ are divided into two subarrays, $\mathbf{Y}_1(k)$ and $\mathbf{Y}_2(k)$, of the form

$$\begin{aligned}\mathbf{Y}_1(k) &= [Y_1(k), Y_2(k), \ldots, Y_{M-1}(k)]^T \\ \mathbf{Y}_2(k) &= [Y_2(k), Y_3(k), \ldots, Y_M(k)]^T.\end{aligned} \quad (8.96)$$

Consequently, from (8.78) it follows that

$$\begin{aligned}\mathbf{Y}_1(k) &= \mathbf{A}\,\mathbf{S}(k) + \mathbf{W}_1(k) \\ \mathbf{Y}_2(k) &= \mathbf{A}\,\boldsymbol{\Phi}\,\mathbf{S}(k) + \mathbf{W}_2(k)\end{aligned} \quad (8.97)$$

where

$$\begin{aligned}\mathbf{S}(k) &= [S_1(k), \ldots, S_I(k)]^T, \\ \mathbf{W}_1(k) &= [W_1(k), \ldots, W_{M-1}(k)]^T, \\ \mathbf{W}_2(k) &= [W_2(k), \ldots, W_M(k)]^T,\end{aligned}$$

$$A = \begin{bmatrix} 1 & 1 & \cdots & 1 \\ e^{j\phi_1(\theta_1)} & e^{-j\phi_1(\theta_2)} & \cdots & e^{-j\phi_1(\theta_I)} \\ \vdots & \vdots & & \vdots \\ e^{-j\phi_{M-2}(\theta_1)} & e^{-j\phi_{M-2}(\theta_2)} & \cdots & e^{-j\phi_{M-2}(\theta_I)} \end{bmatrix}$$

and

$$\boldsymbol{\Phi} = \text{diag}\,\{\exp(j\phi_1(\theta_1)), \ldots \exp(j\phi_1(\theta_I))\}$$

where $\phi_m(\theta_i)$ is given by (8.78) or (8.85.2).

The spatial covariance matrix of $\mathbf{Y}_1(k)$ is given by

$$\mathbf{R}_{11} = E\{\mathbf{Y}_1(k)\mathbf{Y}_1^H(k)\} = \mathbf{A}\,\mathbf{S}\,\mathbf{A}^H + \mathbf{R}_{11}^w, \ (M-1) \times (M-1) \quad (8.98)$$

where $\mathbf{S} = \text{Diag}\,\{\gamma_2^{S_1}, \ldots, \gamma_2^{S_I}\}$ with $\gamma_2^{S_i}$ being the variance of $\{S_i(k)\}$ and $\mathbf{R}_{11}^w$ is the spatial covariance matrix of the noise sources which are assumed to be spatially white; i.e., $\mathbf{R}_{11}^w$ is a diagonal matrix.

On the other hand, the cross-covariance matrix between subarray data $\mathbf{Y}_1(k)$ and $\mathbf{Y}_2(k)$ is given by

$$\mathbf{R}_{12} = E\{\mathbf{Y}_1(k)\mathbf{Y}_2^H(k)\} = \mathbf{A}\,\mathbf{S}\,\boldsymbol{\Phi}^H\mathbf{A}^H + \mathbf{R}_{12}^w \ (M-1) \times (M-1). \quad (8.99)$$

Since the noise sources $\{W_i(k)\}, i = 1, 2, \ldots, M$ are assumed to be uncorrelated, $\mathbf{R}_{11}^w$ and $\mathbf{R}_{12}^w$ become

$$\mathbf{R}_{11}^w = \gamma_2^w - \mathbf{I}_{M-1}$$

and

$$\mathbf{R}_{12}^w = \begin{bmatrix} 0 \ldots & & 0 \\ \gamma_2^w \mathbf{I}_{M-2} & & \vdots \\ & & 0 \end{bmatrix} \quad (8.100)$$

where $\mathbf{I}_{M-1}, \mathbf{I}_{M-2}$ are the identity matrices of sizes $(M-1) \times (M-1)$ and $(M-2) \times (M-2)$, respectively. The algorithmic steps of ESPRIT are the following:

(i) Compute the eigenvalues and eigenvectors of $\mathbf{R}_{11}$ from (8.98); from (8.100) it follows that the noise subspace eigenvalues are equal to γ_2^w. Thus, $\hat{\mathbf{R}}_{11} = \mathbf{R}_{11} - \mathbf{R}_{11}^w$.

(ii) Form matrices $\mathbf{R}_{12} = \mathbf{R}_{12} - \mathbf{R}_{12}^w$ where $\mathbf{R}_{12}$ is given by (8.99).

(iii) The matrix pair $(\mathbf{R}_{11}, \mathbf{R}_{12})$ has I generalized eigenvalues at $\{\exp[j\phi_1(\theta_i)]\}, i = 1, 2, \ldots, I$.

As such, the ESPRIT algorithm estimates the bearings directly by computing the generalized eigenvalues of the matrix pair in step (iii), above.

8.3.4 The Spatial Fourth-Order Cumulant Matrix and Bearing Estimation

The array processing methods described in the previous section are all based on the spatial covariance matrix of the array signals, and make the assumption that additive noise sources are spatially white, i.e., each has a diagonal spatial covariance matrix. The sensitivity of these methods to spatially colored noise has yet to be quantified, especially in those cases where the noise spatial covariance matrix is unknown or cannot be estimated accurately. Array processing problems in the presence of spatially colored noise have been addressed assuming either a known covariance matrix [Reddy, et al., 1987], or estimated covariance from secondary inputs [Kelley, 1986].

In this section, we address the bearing estimation problem using the spatial fourth-order cumulant matrix of the array data. If the additive noise sources are spatially colored and Gaussian, they will be suppressed in the fourth-order cumulant domain, and it would not be necessary to know their covariance matrix [Pan and Nikias, 1988]. This property of fourth-order statistics constitutes the main motivation behind the development of array processing methods in the trispectrum domain.

The spatial *fourth-order cumulant matrix* of sensor signals given by (8.78) is defined by

$$\mathbf{C}_4 = C_{um} \left\{ \begin{bmatrix} Y_1(k)Y_1^*(k)Y_1(k) \\ \vdots \\ Y_M(k)Y_M^*(k)Y_M(k) \end{bmatrix} [Y_1^*(k), Y_2^*(k), \ldots, Y_M^*(k)] \right\} \quad (8.101)$$

where "*" denotes conjugation, and "Cum" is the abbreviation of cumulant. Note that the definition of cumulants of complex array data is not unique. However, definition (8.101) is a convenient equation to solve the problem. Substituting (8.78) into (8.101) we obtain, after some algebra,

$$\mathbf{C}_4 = \mathbf{A}\mathbf{\Gamma}\mathbf{A}^H \quad (8.102)$$

where

$$\mathbf{\Gamma} = \text{Diag}\{\gamma_1, \gamma_2, \ldots, \gamma_I\}. \quad (8.103)$$

The diagonal elements $\{\gamma_i\}$ is the kurtosis measure of the i-th source signal defined by

$$\gamma_i = \text{Cum}\{[S_i(k), S_i^*(k), S_i(k), S_i^*(k)]\}, \quad i = 1, 2, \ldots, I. \tag{8.104}$$

Comparing (8.102) with (8.81.1) and (8.81.2), we see that the spatial fourth-order cumulant matrix is free of Gaussian noise, white or colored. Consequently, all array processing techniques, described in the previous section, can also be applied on $\mathbf{C}_4$ of (8.102), and an estimate can be made of the number of sources and their bearings. As explained earlier, the advantage of this approach is that it can suppress spatially colored Gaussian noise sources of unknown covariance matrix $\mathbf{W}_M$.

The fourth-order cumulant matrix $\mathbf{C}_4$ admits the orthogonal decomposition

$$\mathbf{C}_4 = \sum_{i=1}^{M} r_i \mathbf{E}_i \mathbf{E}_i^H \tag{8.105}$$

where $r_1 \geq r_2 \geq \ldots \geq r_I > 0$, $r_{I+1} = \ldots = r_M = 0$ and $\mathbf{E}_i$ are the associated eigenvectors. In this case, $\mathbf{E}_1, \ldots, \mathbf{E}_I$ form the *signal subspace*. Note that $\mathbf{E}_i, i = I+1, \ldots, M$ are generalized eigenvectors corresponding to zero eigenvalues.

Array processing methods based on spatial fourth-order cumulant matrix $\mathbf{C}_4$ are described below [Pan and Nikias, 1988].

A. Cumulant-Based Beamforming Method:

We define the function $P(\theta)$ to determine source bearings:

$$P(\theta) = \mathbf{C}^H(\theta) \mathbf{C}_4 \mathbf{C}(\theta) \tag{8.106}$$

where $\mathbf{C}(\theta)$ is the steering vector described by (8.85.1) and (8.85.2). The function $P(\theta)$ reaches its maximum when θ coincides with one of the source bearings.

B. Cumulant-Based Signal Subspace Method:

The maximum likelihood of Capon, autoregressive, and thermal noise methods can also be expressed in terms of the pseudo-inverse of the fourth-order cumulant matrix, which is given by

$$\mathbf{C}_4^{-1} = \sum_{i=1}^{I} \frac{1}{r_i} \mathbf{E}_i \mathbf{E}_i^H. \tag{8.107}$$

These are:

Maximum Likelihood of Capon Method:

$$P(\theta) = \frac{1}{\mathbf{C}^H(\theta)\mathbf{C}_4^{-1}\mathbf{C}(\theta)} \qquad (8.108)$$

Autoregressive Method:

$$P(\theta) = \frac{1}{|\mathbf{u}^T \mathbf{C}_4^{-1}\mathbf{C}(\theta)|^2} \qquad (8.109)$$

where $\mathbf{u}^T = [1, 0, \ldots, 0]$

Thermal Noise Method:

$$P(\theta) = \frac{1}{|\mathbf{u}^T \mathbf{C}_4^{-H}\mathbf{C}_4^{-1}\mathbf{C}(\theta)|^2} \qquad (8.110)$$

where "$-H$" denotes inverse transpose conjugate.

Signal subspace methods exhibit higher-resolution than beamforming and are well suited for those array processing scenarios where the number of sources, I, is greater than or equal to $M/2$, where M is the number of sensors in the array.

C. Cumulant-Based Noise Subspace Methods:

The maximum search function for noise subspace methods is given by

$$P(\theta) = \frac{1}{\sum_{i=I+1}^{M} \alpha_i |\mathbf{C}^H(\theta)\mathbf{E}_i|^2} \qquad (8.111)$$

where $\alpha_i = 1, i = I+1, \ldots, M$ for MUSIC, and $\alpha_i = 0, i = I+1, \ldots, M-1; \alpha_M = 1$ for Pisarenko. A modified Pisarenko method based on fourth-order cumulants was also suggested by Swami and Mendel [1988].

Example 8.5

The performance of beamforming and MUSIC with spatial covariance and fourth-order cumulant matrices is demonstrated for a linear array with eight sensors ($M = 8$) and half wavelength spacing among sensors.

Case 1: A source is at bearing $\theta = -25°$ and the additive Gaussian noises are spatially correlated with correlation coefficient close to unity. The signal-to-noise ratio (sensor's

SNR) is 0 dB and the number of samples (snapshots) taken from the array is $16 \times 32 = 512$, i.e., 16 segments with 32 samples each. The results from the beamforming methods are shown in Figure 8.12(a) and Figure 8.12(b), while the results from the MUSIC methods are illustrated in Figures 8.12(c) and 8.12(d). The plots in Figures 8.12(a) and 8.12(c) are produced by covariance-based methods, whereas those of Figures 8.12(b) and (d) by fourth-order cumulant methods. Two peaks can be clearly seen in Figures 8.12(a) and (c): one peak corresponding correctly to the source signal and the other stemming from the additive correlated noises. It is obvious from Figures 8.12(b) and (d) that the fourth-order cumulant methods are effective in suppressing Gaussian noise and show only one peak at $\theta = -25°$.

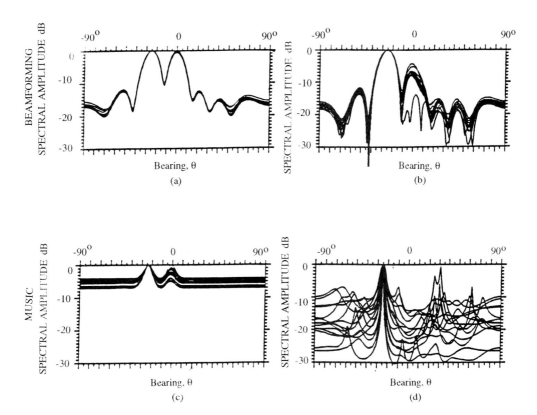

Figure 8.12 Bearing estimation of a source at $\theta = -25°$ in the presence of spatially correlated Gaussian noises using the following methods: (a) beamforming, (b) fourth-order cumulant-based beamforming, (c) MUSIC based on spatial covariance, and (d) MUSIC based on fourth-order cumulants.

Case 2: A similar scenario is used in this case except that an additional source is present at $\theta = -15°$ and the number of snapshots is now 16×256. The SNR is taken to be 5 dB.

The arrangement of plots in Figure 8.13 is the same as that in Figure 8.12. The results from the beamforming methods illustrate that neither covariance-based nor cumulant-based methods work well, as illustrated in Figures 8.13(a) and (b), due to the poor resolution of beamforming. However, the capability of suppressing Gaussian noises by the fourth-order cumulant methods is still apparent, if one compares Figure 8.13(b) with Figure 8.13(a). The cumulant-based MUSIC method appears to be superior to the covariance-based MUSIC method both in suppressing and additive colored noise and in resolving the two sources with bearings $\theta_1 = 15°$ and $\theta_2 = -25°$.

It is important to note that for the same number of snapshots the cumulant-based methods generally exhibit higher variance than the covariance-based methods, as it appears from the results of Figures (8.12) and (8.13). Consequently, cumulant-based methods need relatively longer data length than covariance-based methods to produce the same variance estimates. □

D. Cumulant-Based ESPRIT Method:

Chiang and Nikias [1989] reformulated the ESPRIT algorithm by utilizing fourth-order cumulant matrices instead of covariance matrices. The main motivation behind the extension of the ESPRIT algorithm to a fourth-order cumulant domain was to suppress spatially correlated additive Gaussian noise sources of unknown covariance matrix.

For the two subarrays, $\mathbf{Y}_1(k)$ and $\mathbf{Y}_2(k)$, defined by (8.96), the fourth-order cumulant and cross-cumulant matrices are respectively given by

$$\mathbf{C}_4^{11} = \text{Cum} \left\{ \begin{bmatrix} Y_1(k)Y_1^*(k)Y_1(k) \\ Y_2(k)Y_2^*(k)Y_2(k) \\ \vdots \\ Y_{M-1}(k)Y_{M-1}^*(k)Y_{M-1}(k) \end{bmatrix} [Y_1^*(k), Y_2^*(k), \ldots, Y_{M-1}^*(k)] \right\} \quad (8.112)$$

$$\mathbf{C}_4^{12} = \text{Cum} \left\{ \begin{bmatrix} Y_1(k)Y_1^*(k)Y_1(k) \\ Y_2(k)Y_2^*(k)Y_2(k) \\ \vdots \\ Y_{M-1}(k)Y_{M-1}^*(k)Y_{M-1}(k) \end{bmatrix} [Y_2^*(k), Y_3^*(k), \ldots, Y_M^*(k)] \right\} \quad (8.113)$$

where "Cum" is the abbreviation of cumulant.

Combining (8.97) with (8.112) and (8.113), we obtain

$$\mathbf{C}_4^{11} = \mathbf{A} \, \mathbf{\Gamma} \, \mathbf{A}^H, \; (M-1) \times (M-1) \quad (8.114.1)$$

$$\mathbf{C}_4^{12} = \mathbf{A} \, \mathbf{\Gamma} \, \mathbf{\Phi} \, \mathbf{A}^H, \; (M-1) \times (M-1), \quad (8.114.2)$$

even when $\mathbf{W}_1(k)$ and $\mathbf{W}_2(k)$ are spatially correlated Gaussian noise vectors. The matrix $\mathbf{\Gamma} = \text{diag} \{\gamma_1, \ldots, \gamma_I\}$ where γ_i is the kurtosis of the ith source signal $\{S_i(k)\}$. The matrix

Chapter 8

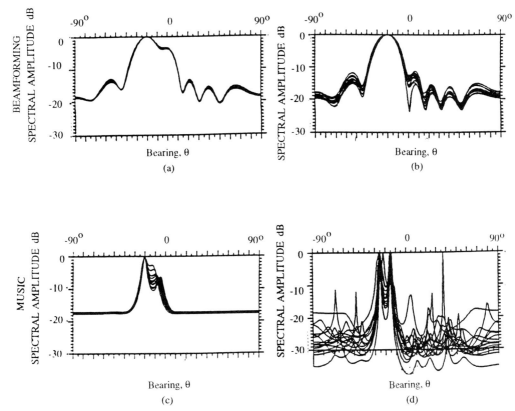

Figure 8.13 Bearing estimation of two sources at $\theta = -25°$ and $\theta = -15°$ in the presence of spatially correlated Gaussian noises using the following methods: (a) beamforming, (b) cumulant-based beamforming, (c) MUSIC based on spatial covariance, and (d) MUSIC based on fourth-order cumulants.

pair $(\mathbf{C}_4^{11}, \mathbf{C}_4^{12})$ has I generalized eigenvalues at $\exp\{j\phi_1(\theta_i)\}, i = 1, 2, \ldots, I,$, with $\phi_1(\theta_i) = \omega_0 \frac{d}{c} \sin \theta_i$. In practical situations, however, due to errors in the estimation of cumulants from finite length data records, the procedure with steps (i)–(iii) described at the end of Section 8.3.3 will also have to be adopted here using $\mathbf{C}_4^{11}$ and $\mathbf{C}_4^{12}$ instead of $\mathbf{R}_{11}$ and $\mathbf{R}_{12}$, respectively.

Example 8.6

The performance of ESPRIT with covariance and fourth-order cumulant matrices is tested for the case of additive noise sources that are spatially correlated and Gaussian.

Consider a uniform linear array with $M = 8$ sensors. Let us assume that there are $I = 2$ incoherent sources in the far field emitting non-Gaussian stationary signals with

$\omega_0 = 0.2\pi$ Hz and $\phi_1(\theta_1) = +0.8131$ and $\phi_1(\theta_2) = +1.3277$.

Figure 8.14 illustrates the results of the ESPRIT algorithms for 16×32 and 16×64 data lengths when SNR = 20 dB. From this figure it is apparent that although the ESPRIT algorithm based on covariances resolves the two sources, it does not suppress the spatially correlated colored Gaussian noise sources, which show up as an extra "source" at $\theta = 45°$ (i.e., high concentration of generalized eigenvalues close to the unit circle at approximately $\theta = 45°$). On the other hand, ESPRIT based on fourth-order cumulants resolves the two sources with much less concentration of extraneous (i.e., noise) generalized eigenvalues near the unit circle. □

8.3.5 Bearing Estimation with Cross-Bispectrum Beamforming

In this section a bearing estimation method is presented based on cross-bispectrum estimates of array output signals [Forster and Nikias, 1991]. The method is based on the asymptotic normality of cross-bispectrum estimates and employs maximum likelihood theory. It is demonstrated that when the additive noise sources are spatially correlated and Gaussian *with unknown cross-spectral matrix (CMS)*, the cross-bispectrum method can provide better bearing estimates than the stochastic maximum likelihood method.

The problem formulation is exactly the same as that of section 8.3.1. Equations (8.77) and (8.78) describe the relationship between array signals $\{Y(k)\}$, source signals $\{S(k)\}$ and noise sources $\{W(k)\}$. Source signals and noises are independent, stationary, and zero-mean processes. The source signals are assumed to be non-Gaussian with nonzero skewness, while the noises are Gaussian and possibly spatially correlated.

Cross-Bispectrum of the Array Output:

Let $m^Y_{\lambda\mu n}(\tau, \rho)$ be the third-order cross-moment sequence of the outputs of sensors $\lambda, \mu,$ and n:

$$m^Y_{\lambda\mu n}(\tau, \rho) \triangleq E\{Y_\lambda(k+\tau)Y_\mu(k+\rho)Y_n^*(k)\}. \tag{8.115}$$

Similarly, let $m^s_{iii}(\tau, \rho)$ be the third-order moment sequence of the signal $\{S_i(k)\}$ emitted by the ith source:

$$m^s_{iii}(\tau, \rho) = E\{S_i(k+\tau)S_i(k+\rho)S_i^*(k)\}. \tag{8.116}$$

Chapter 8

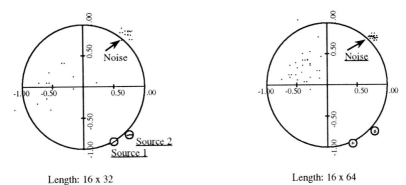

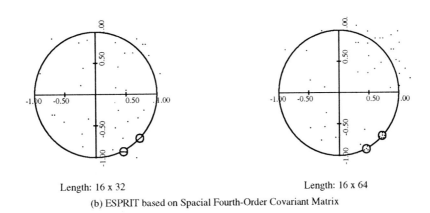

Figure 8.14 Bearing estimation of two sources in the presence of spatially correlated Gaussian noises using the following ESPRIT algorithms: (a) covariance-based, (b) cumulant-based.

From (8.78), (8.115), and the problem assumptions we obtain

$$m^Y_{\lambda\mu n}(\tau,\rho) = E\{Y_\lambda(k+\tau)Y_\mu(k+\rho)Y_n^*(k)\} = \sum_{i=1}^{I}[E\{S_i(k+\tau)S_i(k+\rho)S_i^*(k)\} \cdot$$
$$\exp\{-j[\phi_\lambda(\theta_i) + \phi_\mu(\theta_i) - \phi_n(\theta_i)]\} + E\{W_\lambda(k+\tau)W_\mu(k+\rho)W_n^*(k)\}. \quad (8.117)$$

Since the noise sources are zero-mean and Gaussian, the second term of the right side of (8.117) is zero. Consequently,

$$m^Y_{\lambda\mu n}(\tau,\rho) = \sum_{i=1}^{I} m^s_{iii}(\tau,\rho)\exp\{-j[\phi_\lambda(\theta_i) + \phi_\mu(\theta_i) - \phi_n(\theta_i)]\} \quad (8.118)$$

where $m^s_{iii}(\tau,\rho)$ is given by (8.116). If we consider in (8.118) all M^3 possible values of the triplet (λ,μ,n), we can form a vector $\mathbf{m}(\tau,\rho)$ whose components are $m^Y_{\mu\lambda n}(\tau,\rho)$ with λ,μ and n ranging from 1 to M.

Ordering the triplets (λ,μ,n) in lexicographical order, so that the ith component of $\mathbf{m}(\tau,\rho)$ is $m^Y_{\lambda\mu n}(\tau,\rho)$ for $i = (\lambda-1)M^2 + (\mu-1)M + n$, (8.116) can be written as

$$\mathbf{m}(\tau,\rho) = \sum_{i=1}^{I} m^s_{iii}(\tau,\rho)[\mathbf{C}(\theta_i) \otimes \mathbf{C}(\theta_i) \otimes \mathbf{C}^*(\theta_i)] \quad (M^3 \times 1) \quad (8.119)$$

where $\mathbf{C}(\theta_i)$ is defined by (8.85), and $\otimes$ denotes the Kronecker product.

Note that for matrices $\mathbf{A}$ of size $k \times \ell$ and $\mathbf{B}$ of size $m \times n$, the Kronecker product is a matrix $\mathbf{C}$ of size $km \times \ell n$ defined as

$$\mathbf{C} \triangleq \mathbf{A} \otimes \mathbf{B} = \begin{bmatrix} \alpha_{11}\mathbf{B} .... \alpha_{1\ell}\mathbf{B} \\ \alpha_{k_1}\mathbf{B} .... \alpha_{k\ell}\mathbf{B} \end{bmatrix}. \quad (8.120)$$

where a_{ij} is the ij-th element of $\mathbf{A}$.

Since the vector $[\mathbf{C}(\theta_i) \otimes \mathbf{C}(\theta_i) \otimes \mathbf{C}^*(\theta_i)]$ is encountered frequently, we adopt the simplifying notation

$$\mathbf{C}_\otimes(\theta_i) = [\mathbf{C}(\theta_i) \otimes \mathbf{C}(\theta_i) \otimes \mathbf{C}^*(\theta_i)]. \quad (8.121)$$

In the cross-bispectrum domain, (8.115) takes the form

$$M^Y_{\lambda\mu n}(\omega_1,\omega_2) = \sum_{\tau,\rho=-\infty}^{\infty} m^Y_{\lambda\mu n}(\tau,\rho) \cdot e^{-j2\pi(\omega_1\tau+\omega_2\rho)}. \quad (8.122)$$

Similarly, (8.116) in the bispectrum domain becomes

$$M^s_{iii}(\omega_1,\omega_2) = \sum_{\tau,\rho=-\infty}^{\infty} m^s_{iii}(\tau,\rho) \cdot e^{-j2\pi(\omega_1\tau+\omega_2\rho)} \quad (8.123)$$

where $M_{iii}^s(\omega_1,\omega_2)$ is the bispectrum of the ith-source signal. Combining (8.119), (8.122), and (8.123) we obtain the cross-bispectrum of the array output in terms of the bispectrum of source signals as follows:

$$\mathbf{M}(\omega_1,\omega_2) = \sum_{i=1}^{I} M_{iii}^s(\omega_1,\omega_2) \cdot \mathbf{C}_\otimes(\theta_i) \qquad (8.124)$$

with ω_1 and ω_2 in $[0,2\pi]$. $\mathbf{M}(\omega_1,\omega_2)$ is an $(M^3 \times 1)$ vector whose i^{th} element is $M_{\lambda\mu n}^Y(\omega_1,\omega_2)$ where $i=(\lambda-1)M^2+(\mu-1)M+n$. Since the source signals are complex, the only symmetry of their bispectra is

$$M_{iii}^s(\omega_1,\omega_2) = M_{iii}^s(\omega_2,\omega_1) \qquad (8.125)$$

and consequently, from (8.124)

$$\mathbf{M}(\omega_1,\omega_2) = \mathbf{M}(\omega_2,\omega_1). \qquad (8.126)$$

Thus, the cross-bispectrum needs only to be computed in the region

$$D = \{(\omega_1,\omega_2), 0 \leq \omega_2 \leq \omega_1 \leq 2\pi\}. \qquad (8.127)$$

In summary, the following model equations obey the linear array data:

Time domain:	(8.78)
Covariance domain:	(8.81.1)–(8.81.4)
Fourth-order cumulant domain:	(8.102)–(8.104)
Third-order moment domain:	(8.119)
Cross-bispectrum domain	(8.124).

The bearing estimation problem via cross-bispectrum beamforming is carried out using the model described by (8.124).

Estimation of the Cross-Bispectrum:

Let N be the number of snapshots. Then, the array output $\{\mathbf{Y}(1),\ldots,\mathbf{Y}(N)\}$ consists of K consecutive segments of L data samples each, i.e., $N = K \cdot L$. Denote by $\mathbf{Y}^{(\tau)}(0),\ldots,\mathbf{Y}^{(\tau)}(L-1)$ the data of the τth segment. Their Fourier transform will be

$$\mathbf{Y}^{(\tau)}(\omega) = \sum_{k=0}^{L-1} \mathbf{Y}^{(\tau)}(k)e^{-j\omega k} \qquad (8.128)$$

where $\omega = \frac{2\pi\ell}{L}, \ell = 0,\ldots L-1$. An estimate of the cross-bispectrum vector at $(\omega_1,\omega_2) = (\frac{2\pi\ell}{L}, \frac{2\pi m}{L})$ is obtained by using the direct method, described in Chapter 4, with frequency domain smoothing and time domain averaging over segments; viz,

$$\hat{M}(\omega_1,\omega_2) = \frac{1}{K(2J+1)^2 L} \sum_{\tau=1}^{K} \sum_{j_1,j_2=-J}^{J} Y^{(\tau)}(\ell+j_1) \\ \otimes Y^{(\tau)}(m+j_2) \otimes Y^{*(\tau)}(\ell+m+j_1+j_2), \quad (8.129)$$

with $\omega_1 = \frac{2\pi\ell}{L}$ and $\omega_2 = \frac{2\pi m}{L}$.

If we denote by $\Delta = \frac{2J+1}{L}$ the width of the square window over which the frequency domain smoothing is performed, then it has been shown that the cross-bispectrum estimate $\hat{M}(\omega_1,\omega_2)$ of (8.129) is unbiased and consistent when $N \to \infty$, as soon as Δ is chosen as a function that satisfies [Forster and Nikias, 1991]:

$$\lim_{N \to \infty} \Delta = 0 \quad (8.130.1)$$

$$\lim_{N \to \infty} \Delta^2 \cdot N = \infty. \quad (8.130.2)$$

Consistency is ensured by (8.130.2), while the estimate is ensured to be unbiased by (8.130.1).

Let us consider the cross-bispectrum grid S shown in Figure 8.15. Forster and Nikias [1991] established the following result, presented here in the form of a theorem.

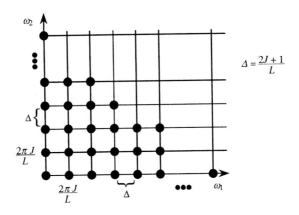

Figure 8.15 The grid S of cross-bispectrum estimates

Theorem:

Let

$$\mathcal{E}(\omega_1,\omega_2) = \hat{M}(\omega_1,\omega_2) - M(\omega_1,\omega_2) \quad (8.131)$$

be the error cross-bispectrum for $(\omega_1,\omega_2) \in S$. The normalized errors $\Delta\sqrt{N}\mathcal{E}(\omega_1,\omega_2)$ are *asymptotically complex Gaussian, uncorrelated, zero-mean* with covariance matrix

$$\lim_{N \to \infty} \Delta^2 N . E\{\mathcal{E}(\omega_1,\omega_2)\mathcal{E}^*(\omega_1,\omega_2)\} = \Gamma(\omega_1) \otimes \Gamma(\omega_2) \otimes \Gamma(\omega_1+\omega_2), \quad (8.132)$$

where $\boldsymbol{\Gamma}(\omega)$ is the cross-spectral matrix (CSM) of the array output. Again, for the sake of simplicity we establish the notation

$$\boldsymbol{\Gamma}_\otimes(\omega_1,\omega_2) = \boldsymbol{\Gamma}(\omega_1) \otimes \boldsymbol{\Gamma}(\omega_2) \otimes \boldsymbol{\Gamma}^*(\omega_1+\omega_2). \tag{8.133}$$

Bearing Estimation:

Let us assume that the CSM of the array data, i.e., $\boldsymbol{\Gamma}_\otimes(\omega_1,\omega_2)$, is known. From the asymptotic distribution of the error cross-bispectrum we obtain the log-likelihood function of grid S [Forster and Nikias, 1991]

$$\mathcal{L}(\mathbf{J}) = \sum_{(\omega_1,\omega_2)\in S} \mathcal{E}^H(\omega_1,\omega_2)\boldsymbol{\Gamma}_\otimes^{-1}(\omega_1,\omega_2)\mathcal{E}(\omega_1,\omega_2). \tag{8.134}$$

As we observe from (8.134), the vector $\mathbf{J}$ of the unknown parameters consists of the bispectral values of the I source signals defined over the grid S, and the source bearings $\boldsymbol{\Theta} = [\theta_1, \theta_2, \cdots \theta_I]^T$. Forster and Nikias [1991] have shown, however, that the maximization of the log-likelihood function with respect to $\mathbf{J}$ is equivalent to maximizing the following function of $\boldsymbol{\Theta}$:

$$P(\boldsymbol{\Theta}) = \sum_{(\omega_1,\omega_2)\in S} \| \boldsymbol{\Pi}(\omega_1,\omega_2,\boldsymbol{\Theta})\boldsymbol{\Gamma}_\otimes^{-1/2}(\omega_1,\omega_2)\hat{M}(\omega_1,\omega_2) \|^2 \tag{8.135}$$

where $\boldsymbol{\Gamma}_\otimes^{1/2}(\omega_1,\omega_2)\boldsymbol{\Gamma}_\otimes^{*/2}(\omega_1,\omega_2)$ is a factorization (e.g., Cholesky) of $\boldsymbol{\Gamma}_\otimes(\omega_1,\omega_2)$, and $\boldsymbol{\Pi}(\omega_1,\omega_2,\boldsymbol{\Theta})$ is the projector onto the space spanned by the set of I vectors

$$\mathbf{V}(\omega_1,\omega_2,\boldsymbol{\Theta}) = \{\boldsymbol{\Gamma}_\otimes^{-1/2}(\omega_1,\omega_2) \cdot \mathbf{C}_\otimes(\theta_1),\ldots,\boldsymbol{\Gamma}_\otimes^{-1/2}(\omega_1,\omega_2) \cdot \mathbf{C}_\otimes(\theta_I)\}. \tag{8.136}$$

with $\mathbf{C}_\otimes(\theta) \triangleq \mathbf{C}(\theta)_\otimes \mathbf{C}(\theta) \otimes \mathbf{C}^*(\theta)$ and $\mathbf{C}(\theta)$ defined in (8.85).

Single Source Case:

When only *one source* is assumed to be present, the projector $\boldsymbol{\Pi}(\omega_1,\omega_2,\boldsymbol{\theta})$ is simply given by

$$\boldsymbol{\Pi}(\omega_1,\omega_2,\boldsymbol{\theta}) = \frac{\boldsymbol{\Gamma}_\otimes^{-1/2}(\omega_1,\omega_2) \cdot \mathbf{C}_\otimes(\theta) \cdot \mathbf{C}_\otimes^H(\theta) \cdot \boldsymbol{\Gamma}_\otimes^{-*1/2}(\omega_1,\omega_2)}{\mathbf{C}_\otimes^H(\theta) \cdot \boldsymbol{\Gamma}_\otimes^{-1}(\omega_1,\omega_2) \cdot \mathbf{C}_\otimes(\theta)}. \tag{8.137}$$

For practical purposes, a consistent estimator of $\boldsymbol{\Gamma}_\otimes(\omega_1,\omega_2)$ is obtained by substituting an averaged periodogram estimate, $\hat{\boldsymbol{\Gamma}}(\omega)$, for the cross-spectral matrix, $\boldsymbol{\Gamma}(\omega)$, in

$$\hat{\boldsymbol{\Gamma}}_\otimes(\omega_1,\omega_2) = \hat{\boldsymbol{\Gamma}}(\omega_1) \otimes \hat{\boldsymbol{\Gamma}}(\omega_2) \otimes \hat{\boldsymbol{\Gamma}}^*(\omega_1+\omega_2), \tag{8.138}$$

where
$$\hat{\boldsymbol{\Gamma}}(\omega) = \frac{1}{K(2J+1)L} \sum_{\tau=1}^{K} \sum_{j=-J}^{J} \mathbf{Y}^{(\tau)}(\ell+j) \mathbf{Y}^{(\tau)*}(\ell+j) \qquad (8.139)$$

with $\omega = \frac{2\pi\ell}{L}, \ell = 0, 1, \ldots, L-1$.

After some algebraic manipulations, combination of (8.137) and (8.138) yields

$$P(\theta) = \sum_{(\omega_1,\omega_2) \in S} \frac{|\hat{\mathbf{M}}^H(\omega_1,\omega_2) \cdot \hat{\boldsymbol{\Gamma}}_\otimes^{-1}(\omega_1,\omega_2) \cdot \mathbf{C}_\otimes(\theta)|^2}{\mathbf{C}_\otimes^H(\theta) \cdot \hat{\boldsymbol{\Gamma}}_\otimes^{-1}(\omega_1,\omega_2) \cdot \mathbf{C}_\otimes(\theta)}. \qquad (8.140)$$

$P(\theta)$ will be referred to as the *Cross-Bispectrum Beamformer* (CBBF) output. The source is localized by sketching $P(\theta)$ versus θ and looking for the location of its maximum.

The Multiple Sources Case:

The algorithm for the maximization of $P(\theta)$ of (8.135), described in the sequel, is based on a technique called *alternating maximization* [Ziskind and Wax, 1988]. The main idea of this technique is to decompose the initial multidimensional maximization problem into a sequence of one-dimensional problems that can be easily handled. Let us assume that the number of sources I is known. The algorithm proceeds as follows. Let

$$\boldsymbol{\Theta}^{(k)} = [\theta_1^{(k)}, \ldots, \theta_I^{(k)}]^T \qquad (8.141)$$

be the estimated bearings at the end of the kth iteration. The $(k+1)^{\text{th}}$ iteration consists of maximizing $P(\boldsymbol{\Theta})$ with respect to the single parameter $\theta_i, i = k \cdot \text{mod}(I) + 1$, while $\theta_1 = \theta_1^{(k)}, \ldots, \theta_{i-1} = \theta_{i-1}^{(k)}, \theta_{i+1} = \theta_{i+1}^{(k)}, \ldots, \theta_I = \theta_I^{(k)}$ are kept fixed. Let $\theta_i^{(k+1)}$ be the value of θ_i that maximizes $P(\boldsymbol{\Theta})$ with respect to θ_i,

$$\theta_i^{(k+1)} = \arg\max_{\theta_i} P(\boldsymbol{\Theta}), \qquad (8.142)$$

where
$$\boldsymbol{\Theta} = [\theta_1^{(k)}, \ldots, \theta_{i-1}^{(k)}, \theta_i, \theta_{i+1}^{(k)}, \ldots, \theta_I^{(k)}]^T.$$

The estimated bearings $\boldsymbol{\theta}^{(k+1)}$ at the end of the $(k+1)^{\text{th}}$ iteration are defined as

$$\boldsymbol{\Theta} = [\theta_1^{(k)}, \ldots, \theta_{i-1}^{(k)}, \theta_i^{(k+1)}, \theta_{i+1}^{(k)}, \ldots, \theta_I^{(k)}]^T.$$

The problem that arises now is to solve (8.142). When $\boldsymbol{\Theta} = [\theta_1^{(k)}, \ldots, \theta_{i-1}^{(k)}, \theta_i^{(k)}, \theta_{i+1}^{(k)}, \ldots, \theta_I^{(k)}]^T$, the projector $\Pi(\omega_1, \omega_2, \boldsymbol{\Theta})$ of (8.137) can be decomposed into

$$\boldsymbol{\Pi}(\omega_1, \omega_2, \boldsymbol{\Theta}) = \boldsymbol{\Pi}^{(k)}(\omega_1, \omega_2) + \frac{\mathbf{u}(\omega_1, \omega_2, \theta_i) \mathbf{u}^H(\omega_1, \omega_2, \theta_i)}{\|\mathbf{u}(\omega_1, \omega_2, \theta_i)\|^2}, \qquad (8.143)$$

where $\boldsymbol{\Pi}^{(k)}(\omega_1,\omega_2)$ is the projector onto the space spanned by the $(I-1)$ vectors

$$\mathbf{V}^{(k)}(\omega_1,\omega_2) = \{\hat{\boldsymbol{\Gamma}}_\otimes^{-1/2}\mathbf{C}_\otimes(\boldsymbol{\theta})\} \text{ where } \boldsymbol{\theta} = \{\theta_1^{(k)},\ldots,\theta_{i-1}^{(k)},\theta_{i+1}^{(k)},\ldots,\theta_I^{(k)}\}$$

and $\mathbf{u}(\omega_1,\omega_2,\theta_i) = \{\mathbf{I} - \boldsymbol{\Pi}^{(k)}(\omega_1,\omega_2)\}\hat{\boldsymbol{\Gamma}}_\otimes^{-1/2}\mathbf{C}_\otimes(\theta_i)$ is the residual of $\hat{\boldsymbol{\Gamma}}_\otimes^{-1/2}\mathbf{C}_\otimes(\theta_i)$ when projected onto the space spanned by $\mathbf{V}^{(k)}(\omega_1,\omega_2)$. Substituting (8.134) into (8.135) we obtain

$$P(\boldsymbol{\Theta}) = \sum_{(\omega_1,\omega_2)\in S} \|\boldsymbol{\Pi}^{(k)}(\omega_1,\omega_2)\hat{\mathbf{M}}_w(\omega_1,\omega_2)\|^2 + \sum_{(\omega_1,\omega_2)\in S} \frac{|\mathbf{u}^H(\omega_1,\omega_2,\theta_i)\hat{\mathbf{M}}_w(\omega_1,\omega_2)|^2}{\|\mathbf{u}(\omega_1,\omega_2,\theta_i)\|^2} \quad (8.144)$$

where

$$\hat{\mathbf{M}}_w(\omega_1,\omega_2) = \boldsymbol{\Gamma}_\otimes^{-1/2}(\omega_1,\omega_2)\hat{\mathbf{M}}(\omega_1,\omega_2). \quad (8.145)$$

In (8.144), the first term of the right-hand side is independent of θ_i. Thus, maximization of $P(\boldsymbol{\Theta})$ with respect to θ_i is equivalent to maximizing $F(\theta_i)$ defined as

$$F(\theta_i) \triangleq \sum_{(\omega_1,\omega_2)\in S} \frac{|\mathbf{u}^H(\omega_1,\omega_2,\theta_i)\hat{\mathbf{M}}_w(\omega_1,\omega_2)|^2}{\|\mathbf{u}(\omega_1,\omega_2,\theta_i)\|^2}. \quad (8.146)$$

The quantity $F(\theta_i)$ is a one-dimensional function of θ_i, hence the maximization is carried out by seeking its maximum over a grid of values of θ_i. A detailed description of the algorithm is provided by Forster and Nikias [1991].

Special Case for a White Source Signal:

Consider, the case of a single source signal with power Q, skewness β, and location θ in spatial white noise of power γ_2^w. For a linear array of $\mathbf{M}$ equi-spaced sensors with spacing equal to one half wavelength at the center frequency, Forster and Nikias [1991] showed that

$$N \cdot \text{var}[\hat{\theta}] \approx 4\beta^{-2} \cdot (\frac{\gamma_2^w}{Q \cdot M})(1 + \frac{\gamma_2^w}{Q \cdot M})^2(\frac{1}{\pi \cdot M \cdot \cos\theta})^2. \quad (8.147)$$

The Cramer-Rao Lower Bound (CRLB) on $\hat{\theta}$, assuming $\theta, Q,$ and γ_2^w to be unknown, based on the Gaussian asymptotic distribution of the Fourier transform of the vectors of the array output is

$$N \cdot \text{var}[\hat{\theta}]_{CRLB} \approx 6 \cdot (\frac{\gamma_2^w}{Q \cdot M}) \cdot (1 + \frac{\gamma_2^w}{Q \cdot M}) \cdot (\frac{1}{\pi \cdot M \cdot \cos\theta})^2. \quad (8.148)$$

Combining (8.147) and (8.148), we obtain

$$\text{var}[\hat{\theta}] = \frac{2}{3}(\beta^{-2})(1 + \frac{\gamma_2^w}{Q \cdot M})\text{var}[\hat{\theta}]_{CRLB}. \quad (8.149)$$

As long as the source skewness satisfies

$$\beta > \sqrt{\frac{2}{3}(1 + \frac{\gamma_2^w}{Q \cdot M})}, \tag{8.150}$$

the cross-bispectrum beamformer (CBBF) will provide a better bearing estimate than any cross-spectrum-based method. Note, however, that the lower the signal-to-noise ratio Q/γ_2^w is, the higher the skewness β must be to satisfy inequality (8.150).

Example 8.7

Consider a linear array of equi-spaced sensors, with spacing equal to a half wavelength at the center frequency. The steering vector of the array, $\mathbf{C}(\theta)$, is given by (8.85.1) and (8.85.2). The source signals $\{S_i(k)\}$ are independent, zero-mean, white processes, distributed according to a one-sided exponential distribution.

The noise sources $\mathbf{W}(k)$ in (8.78) are zero-mean, Gaussian with spatial covariance matrix

$$\mathbf{W}_M = E\{\mathbf{W}(k)\mathbf{W}^H(k)\}. \tag{8.151}$$

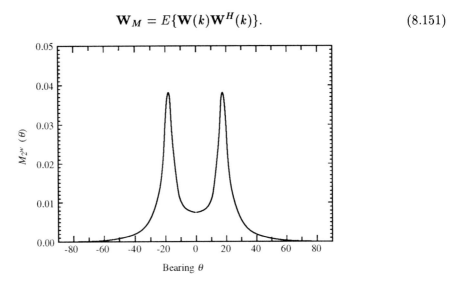

Figure 8.16 Noise spatial power spectral density (PSD) for Example 8.7 ([Forster and Nikias, 1991]. © 1991 IEEE).

The signals of each sensor have been divided into segments of $L = 128$ samples each, and the cross-bispectrum has been estimated over the grid S shown in Figure 8.15 according to (8.129), with $J = 5$.

In this example there are $M = 8$ sensors and $I = 2$ incoherent sources of power 0 dB at $\theta_1 = 14°$ and $\theta_2 = 18°$. The noise spatial power spectral density (PSD) per degree, $M_2^W(\theta)$, is illustrated in Figure 8.16. It has been computed from

$$M_2^W(\theta) = \frac{1}{|\mathbf{u}^T \mathbf{W}_M^{-1} \mathbf{C}(\theta)|^2}, \quad (8.152)$$

i.e., it is autoregressive (AR) type where $\mathbf{u}^T = [1, 0, \ldots, 0]$. Figures 8.17(a) and (b) illustrate the response of the cross-bispectrum beamformer (CBBF) computed from (8.140), to 4×128 data and 32×128 data, respectively. It is interesting to note that the CBBF resolves the two sources although the angular separation between the two sources is only one third the beamwidth of the conventional beamformer. As a comparison, the response of the MUSIC method based on spatial covariance matrix has been computed. Figures 8.18(a) and 8.18(b) display the results obtained by MUSIC, *after noise spatial whitening*, to the same 4×128 data and 32×128 data, respectively.

Note that the results of Figures 8.18(a) and (b) require the noise spatial covariance W_M *to be known exactly*, but this information is usually not available. Figures 8.19(a) and (b) show the results obtained by MUSIC assuming the noise to be spatially white. The two sources are not resolved, and there is a MUSIC spectral peak at $-20°$, although there is no source at that location. However, the MUSIC algorithm based on spatial covariance matrix does what it is supposed to do, i.e., it correctly estimates the directional distribution of the signal plus colored noise. □

8.3.6 MUSIC-Like Method Based on Cumulants

The problem of high-resolution direction finding based on cumulants was also addressed by Porat and Friedlander [1990] in a MUSIC-like framework. This method is based on the eigendecomposition of the fourth-order cumulant matrix. It turns out that there is a subspace of eigenvectors of the cumulant matrix which is orthogonal to the array manifold vector $\boldsymbol{\alpha}(\theta)$. As such, the MUSIC-like algorithm estimates the bearings by searching for peaks of a function that involves the projection of the array manifold vector on this subspace.

Again, the problem is formulated as before. There are I, zero-mean, non-Gaussian signals that impinge on an array of M sensors ($M > I$). The signals emitted at different time instances are statistically independent. The received signals at the array will be described by (8.78), with $\mathbf{W}(k)$ being a complex white zero-mean Gaussian noise vector with unknown spatial covariance matrix. The I-dimensional vector $\boldsymbol{\alpha}(\theta_i)$ in (8.78) is known as the array manifold. The sources are divided into G groups, such that the gth group consists of I_g sources ($\sum_{g=1}^{G} I_g = I$), and within a group sources are dependent, while sources belonging to different groups are independent.

Based on the above assumption, the problem is reformulated as follows:

$$\begin{aligned} \mathbf{Y}(k) &= \sum_{g=1}^{G} \mathbf{A}_g \mathbf{S}_g(k) + \mathbf{W}(k) \\ &= \mathbf{X}_g(k) + \mathbf{W}(k) \end{aligned} \quad (8.153)$$

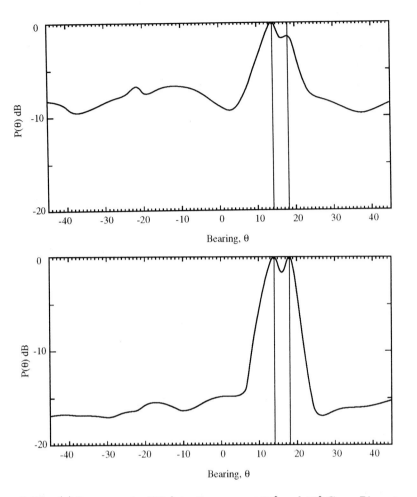

Figure 8.17 (a) 8 sensors, 4 × 128 data, 2 sources at 14° and 18° Cross-Bispectrum Beamformer (CBBF). (b) 8 sensors, 32 × 128 data, 2 sources at 14° and 18°, Cross-Bispectrum Beamformer (CBBF) ([Forster and Nikias, 1991] © 1991 IEEE).

where $\mathbf{A}_g$ is a submatrix of $\mathbf{A}$, which was defined in (8.78), and corresponds to the gth group of sources; $\mathbf{S}_g(k)$ is a subvector of $\mathbf{S}(k)$, which was also defined in (8.78), and corresponds to the same group g.

For some applications such as QAM signals, due to symmetries, only the even-order moments are different than zero. For this type of signals the fourth-order moments are given by [Porat and Friedlander, 1990]:

$$m_4^y(k_1, k_2, \ell_1, \ell_2) = E\{Y_{k_1}(k)Y_{k_2}(k)Y_{\ell_1}^*(k)Y_{\ell_2}^*(k)\}. \tag{8.154}$$

Due to symmetries of QAM signals, it holds that

$$E\{Y_{k_1}(k)Y_{\ell_1}(k)\} = 0, \tag{8.155}$$

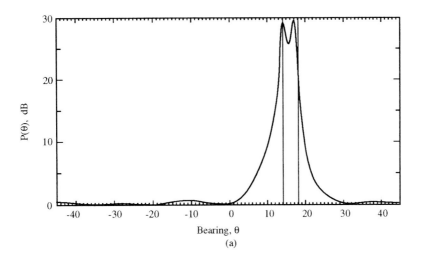

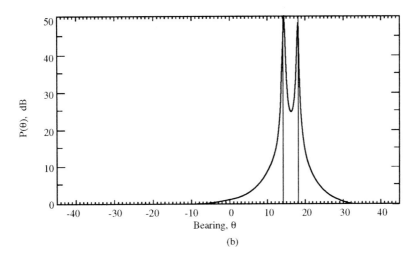

Figure 8.18 (a) 8 sensors, 4 × 128 data, 2 sources at 14° and 18°, MUSIC after noise spatial whitening. (b) 8 sensors, 32 × 128 data, 2 sources at 14° and 18°, MUSIC after noise spatial whitening. ([Forster and Nikias, 1991] © 1991 IEEE).

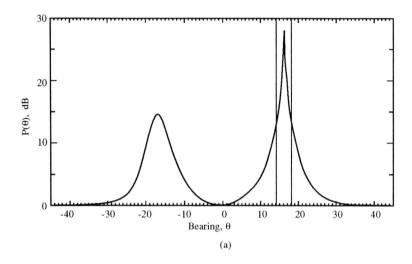

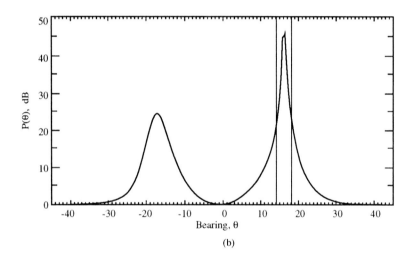

Figure 8.19 (a) 8 sensors, 4 × 128 data, 2 sources at 14° and 18° MUSIC, based on spatial covariance matrix. (b) 8 sensors, 32 × 128 data, 2 sources at 14° and 18°, MUSIC based on spatial covariance matrix. ([Forster and Nikias, 1991] © 1991 IEEE).

and therefore the fourth-order cumulants are given by

$$c_4^y(k_1, k_2, \ell_1, \ell_2) = m_4^y(k_1, k_2, \ell_1, \ell_2) - m_2^y(k_1, \ell_1) \cdot m_2^y(k_2, \ell_2) \\ - m_2^y(k_1, \ell_2) m_2^y(k_2, \ell_1). \tag{8.156}$$

Let the vector $\mathbf{Z}_g(k)$ be defined by

$$\mathbf{Z}_g(k) = \mathbf{X}_g(k) \otimes \mathbf{X}_g^*(k) \tag{8.157}$$

where $\mathbf{X}_g$ is defined by (8.153) and $\otimes$ denotes the Kronecker product. Then

$$\begin{aligned} \mathbf{Z}_g(k) &= (\mathbf{A}_g \mathbf{S}_g(k)) \otimes (\mathbf{A}_g^* \mathbf{S}_g^*(k)) \\ &= (\mathbf{A}_g \otimes \mathbf{A}_g^*) \cdot (\mathbf{S}_g(k) \otimes \mathbf{S}_g^*(k)). \end{aligned} \tag{8.158}$$

The fourth-order cumulants matrix of $\mathbf{Z}_g(k)$, i.e., $\mathbf{C}_g^z$, is found to be [Porat and Friedlander, 1990]

$$\mathbf{C}_g^z = (\mathbf{A}_g \otimes \mathbf{A}_g^*) \mathbf{C}_g^s (\mathbf{A}_g \otimes \mathbf{A}_g^*)^H \tag{8.159.1}$$

where

$$\mathbf{C}_g^s = E\{(\mathbf{S}_g(k) \otimes \mathbf{S}_g^*(k))(\mathbf{S}_g(t) \otimes \mathbf{S}_g^*(k))^H\} - E\{\mathbf{S}_g(k)\mathbf{S}_g^4(k)\} \otimes E\{\mathbf{S}_g(k)\mathbf{S}_g^*(k)\}. \tag{8.159.2}$$

Let $\mathbf{C}^y$ be the matrix of fourth-order cumulants of $\mathbf{Y}(k)$. Then, we have

$$\mathbf{C}^y = \sum_{g=1}^G \mathbf{C}_g^z = \sum_{g=1}^G (\mathbf{A}_g \otimes \mathbf{A}_g^*) \mathbf{C}_g^s (\mathbf{A}_g \otimes \mathbf{A}_g^*)^H. \tag{8.160}$$

Since the components of $\mathbf{S}_g(k)$ are statistically dependent, $\mathbf{C}_g^s$ has full rank, i.e., the rank of $\mathbf{C}_g^s$ is I_g^2. Hence, the rank of $\mathbf{C}_g^z$ and the rank of $\mathbf{C}^y$ is $\sum_{g=1}^G I_g^2$ in general. $\mathbf{C}^y$ is a $(M^2 \times M^2)$ matrix, and having rank $\sum_{g=1}^G I_g^2$ means that it has $M^2 - \sum_{g=1}^G I_g^2$ eigenvalues that are zero.

Computing the singular value decomposition (SVD) of $\mathbf{C}^y$

$$\mathbf{C}_0^y = [\mathbf{U}_1 \mathbf{U}_2] \begin{bmatrix} \wedge & 0 \\ 0 & 0 \end{bmatrix} \begin{bmatrix} \mathbf{V}_1^H \\ \mathbf{V}_2^H \end{bmatrix} \tag{8.161}$$

where $\wedge$ is a $(\sum_{g=1}^G I_g^2) \times (\sum_{g=1}^G I_g^2)$ diagonal matrix of singular values. From (8.160), $\mathbf{C}^y$ can be expressed as

$$\mathbf{C}^y = [(\mathbf{A}_1 \otimes \mathbf{A}_1^*), \ldots, (\mathbf{A}_G \otimes \mathbf{A}_G^*)] \begin{bmatrix} \mathbf{C}_1^s & & 0 \\ & \ddots & \\ 0 & & \mathbf{C}_G^s \end{bmatrix} \begin{bmatrix} (\mathbf{A}_1 \otimes \mathbf{A}_1^*)^H \\ \vdots \\ (\mathbf{A}_G \otimes \mathbf{A}_G^*)^H \end{bmatrix}. \tag{8.162}$$

As in the case of the original MUSIC algorithm, all the columns of all the matrices $\{(\mathbf{A}_g \otimes \mathbf{A}_g^*), 1 \leq g \leq G\}$ are orthogonal to all columns of $\mathbf{U}_2$, which consists of the left eigenvectors of $\mathbf{C}^y$ that correspond to the zero singular values.

Let $\theta_{i,g}$ be the bearing of the ith source in the gth group. Then, the vector $[\boldsymbol{\alpha}(\theta_{i,g}) \otimes \boldsymbol{\alpha}^*(\theta_{i,g})]$ is the $[(i-1)I_g + i]^{\text{th}}$ column of $(\mathbf{A}_g \otimes \mathbf{A}_g^*)$. So, all $\{\boldsymbol{\alpha}(\theta_i) \otimes \boldsymbol{\alpha}^*(\theta_i), 1 \leq i \leq I)\}$ are orthogonal to the range space of $\mathbf{U}_2$. Porat and Friedlander [1990] used the function

$$d(\theta) = \| [\boldsymbol{\alpha}(\theta) \otimes \boldsymbol{\alpha}^*(\theta)]^H \mathbf{U}_2 \|^2, \tag{8.163}$$

and estimated the bearings at the locations where $d(\theta)$ has minima.

It is important to note that the MUSIC-like algorithm becomes the cumulant-based MUSIC algorithm described in section 8.3.4 if each of the G groups consists of $I_g = 1$ source only.

8.4 PARAMETER ESTIMATION OF TRANSIENT SIGNALS

In this section we discuss the estimation of parameters of exponentially damped sinusoidal signals in white or colored Gaussian noise using higher-order statistics.

8.4.1 Problem Formulation

Let us assume that the observed signal $x(k)$ can be represented by a finite sum of complex exponentials of the form

$$x(k) = \sum_{m=1}^{L} h_m \exp\{b_m k\}, \ k = 0, 1, 2, \ldots \tag{8.164}$$

where the complex constants are defined as $h_m = d_m e^{j\theta_m}$ and $b_m = -c_m + j\omega_m$, $c_m > 0$. Notice that d_m and θ_m are the amplitude and initial phase of the mth exponential, respectively; c_m is its damping factor and ω_m its frequency.

The problem of estimating the parameters $\{h_m, b_m\}$ when $x(k)$ is observed in the presence of additive Gaussian noise has been previously approached by the following procedures: (a) the exact maximum likelihood (ML) estimation of the signal parameters, (b) Prony's method [Marple, 1987], and (c) using the minimum norm method which is also known as the Kumaresan and Tufts (KT) method [1982]. The maximum likelihood procedure, though optimum, requires the solution of a nonlinear system of equations. As such, it has not been given a lot of attention in practical situations. On the other hand, Prony's method and the least-squares Prony method, although simple, are very sensitive to the presence of additive noise, even when the SNR is relatively high [Marple, 1987]. The KT method has been shown to be asymptotically equivalent to the maximum likelihood procedure when the additive noise is white Gaussian. However, for finite length data, the KT method performs well in moderate SNR environments if the noise is white. A parameter estimation method using cumulants has been proposed by Swami and Mendel [1991]. They showed that the diagonal slice of the fourth-order cumulants of signals defined in (8.164) behave like autocorrelation functions and satisfy autoregressive equations; hence, linear prediction based procedures can be used with this diagonal slice.

In this section we discuss the parameter estimation of complex exponentials using higher-order statistics, and especially the case of additive colored Gaussian noise. In particular, the minimum norm (or KT) method is reformulated in the third-order and fourth-order statistics domain.

8.4.2 The Third-Order Recursion Equation

One of Prony's important contributions was to show that for (8.164) there exists a set of complex coefficients $\{a(i)\}, i = 1, 2, \ldots, L$ such that [Marple, 1987]:

$$x(k) = \sum_{i=1}^{L} a(i)x(k-i) + \epsilon(k) \tag{8.165}$$

where $\epsilon(k)$ is a zero-mean white process. If we now multiply both sides of (8.165) by the double product $x^*(k-\ell)x^*(k-\ell)$ and take expectations, we obtain the following third-order recursion equation:

$$\sum_{i=0}^{L} a(i) m_3^x(i-\ell, i-\ell) = 0; \quad a(0) = -1, \tag{8.166}$$

for all ℓ, where $m_3^x(.,.)$ is the third-order moment sequence of the energy signal $x(k)$. Identity (8.166) implies that the set of complex coefficients in (8.165) also satisfies the third-order recursion. However, both (8.165) and (8.166) require *noise-free* data and *exact* knowledge of the number, L, of complex exponentials.

8.4.3 Parameter Estimation Method

The estimation procedure for obtaining the $\{b_m\}$ parameters from exactly known third-order moments is described in this section, assuming that the number of complex exponentials, L, is unknown. The utilization of this method when the transient signal is observed in the presence of noise is also discussed. We begin by forming the linear system of equations

$$\mathbf{m}_3^x \cdot \hat{\boldsymbol{\alpha}} = \mathbf{0} \tag{8.167}$$

where
$$\hat{\boldsymbol{\alpha}} = [\hat{\alpha}(K), \hat{\alpha}(K-1), \ldots, \hat{\alpha}(0)] \text{ is } (K+1) \times 1, \quad K \geq L$$
and $\mathbf{m}_3^x$ is the Hankel matrix of size $(K+1) \times (K+1)$ defined as
$$\mathbf{m}_3^x = \begin{bmatrix} m_3^{x*}(0,0) & \cdots\cdots & m_3^{x*}(-K,-K) \\ m_3^{x*}(-1,-1) & \cdots\cdots & m_3^{x*}(-K-1,-K-1) \\ \vdots & & \\ m_3^{x*}(-K,-K) & \cdots\cdots & m_3^{x*}(-2K,-2K) \end{bmatrix},$$

Let us define
$$A(\ell) \triangleq \sum_{m=1}^{L} \sum_{v=1}^{L} h_\ell^* h_m h_v \cdot \frac{1}{(1-\exp\{b_\ell^* + b_m + b_v\})}. \tag{8.168}$$

Papadopoulos and Nikias [1990] showed that if $A(\ell) \neq 0$ for $\ell = 1, 2, \ldots, L$ then the matrix $\mathbf{m}_3^x$ has rank L and the Kth order polynomial
$$A(z) = \sum_{i=0}^{K} \hat{\alpha}(K-i) z^{-i}, \quad K > L \tag{8.169}$$

has L roots at locations $\exp\{-b_\ell^*\}, \ell = 1, 2, \ldots, L$ which are outside the unit circle. The extraneous $(K - L)$ zeros of $A(z)$ lie inside the unit circle if (i) $K \geq L$ and (ii) $\hat{\boldsymbol{\alpha}}$ is the *minimum norm* solution of (8.167) [Kumaresan and Tufts, 1982].

The sufficient conditions under which $A(\ell) \neq 0$ for $\ell = 1, 2, \ldots, L$ are [Papadopoulos and Nikias, 1990]:
$$0 \leq \gamma' \leq \pi - 2\sin^{-1}(e^{-3c_i}) \tag{8.170.1}$$
where γ' is such that
$$\mu' \leq \arg[h_m \cdot h_v] \leq \mu' + \gamma' < \mu' + \pi \tag{8.170.2}$$
i.e., γ' corresponds to the spread of initial phases $(\theta_m + \theta_v)$ for $v = 1, 2, \ldots, L$ and $m = 1, 2, \ldots, v$, and $c_i = \min(c_1, \ldots, c_L)$.

Equation (8.167) can be rewritten as follows:
$$\mathbf{m}_3^{xc} \hat{\boldsymbol{\alpha}} = -\mathbf{m}^x \tag{8.171}$$

where
$$\hat{\boldsymbol{\alpha}} = [\hat{\alpha}(K-1), \hat{\alpha}(K-2), \ldots, \hat{\alpha}(0)]^T$$
$$\mathbf{m}_3^{xc} = \begin{bmatrix} m_3^{x*}(-2,-2) & \cdots & m_3^{x*}(-K-1,-K-1) \\ m_3^{x*}(-3,-3) & \cdots & m_3^{x*}(-K-2,-K-2) \\ \vdots & & \vdots \\ m_3^{x*}(-K-1,-K-1) & \cdots & m_3^{x*}(-2K,-2K) \end{bmatrix}$$
$$\mathbf{m}^x = [m_3^{x*}(-1,-1), \ldots, m_3^{x*}(-K,-K)]^T.$$

The minimum norm solution of (8.171) is then given within the constant $\hat{\alpha}(k)$ as:

$$\alpha' = -[(\mathbf{m}_3^{xc})^H \mathbf{m}_3^{xc}]^{-1} (\mathbf{m}_3^{xc})^H \mathbf{m}. \tag{8.172}$$

Let us now consider the case where the signal $x(k)$ is observed in additive noise

$$Y(k) = x(k) + W(k), \quad k = 0, \ldots, N-1 \tag{8.173}$$

where $\{W(k)\}$ is a zero-mean colored or white complex Gaussian process with identically distributed real and imaginary parts which are independent of each other and the signal, $x(k)$.

In the third-order moments domain it holds that

$$\begin{aligned} m_3^y(\tau,\tau) &= m_3^x(\tau,\tau) + m_3^w(\tau,\tau) + m_2^{w*w*}(0) \cdot m_1^x \\ &\quad + 2m_2^w(\tau)(m_1^x)^* + m_2^{x*x*}(0) \cdot m_1^w \\ &\quad + 2m_2^x(\tau)(m_1^w)^* \end{aligned} \tag{8.174}$$

Assuming multiple realizations of $Y(k)$ are available, averaging the third-order moment of each realization, the cross-term (8.174) disappears and what we are left with is: $m_3^y(\tau,\tau) = m_3^x(\tau,\tau)$. Equation (8.171) can be rewritten as

$$\mathbf{m}_3^{yc} \alpha = -\mathbf{m}^y. \tag{8.175}$$

where $\mathbf{m}^y$ is defined analogous to $\mathbf{m}^x$. The parameter estimation algorithm can be summarized as follows [Papadopoulos and Nikias, 1990]:

Let $\{Y^{(i)}(0), \ldots, Y^{(i)}(N-1)\}$ be the given ensemble data set of the form:

$$Y^{(i)}(k) = x(k - D_i) + W^{(i)}(k), \quad i = 1, 2, \ldots, M$$

where $\{W^{(i)}(k)\}$ are different noise realizations of identical statistical properties and D_i are integer linear phase shifts.

Step 1: Subtract the average value of each record (i).

Step 2: Estimate the ensemble average third-order moment sequence:

$$m_3^y(-\tau,-\tau) = \frac{1}{M} \sum_{i=1}^{M} m_3^{y(i)}(-\tau,-\tau) \tag{8.176.1}$$

where

$$m_3^{y(i)}(\tau,\tau) = \frac{1}{N} \sum_{k=s_1}^{s_2} Y^{(i)}(k) Y^{(i)*}(k+\tau) Y^{(i)*}(k+\tau) \tag{8.176.2}$$

with
$$s_1 = \max(0, -\tau)$$
$$s_2 = \min(N-1, N-1-\tau).$$

Step 3: Form the $\mathbf{m}_3^y$ matrix and use a SVD algorithm to decide on the rank of the matrix.

Step 4: Let $\hat{L}$ be the rank of the matrix $\mathbf{m}_3^y$. Compute the least-squares solution of (8.175). Because of variance in the estimates due to noisy data, use SVD to solve (8.175); i.e., compute $\hat{\boldsymbol{\alpha}}$ by

$$\boldsymbol{\alpha}' = -\sum_{m=1}^{\hat{L}} t_m^{-1} \mathbf{q}_m (\mathbf{p}_m^*)^T \cdot \mathbf{m}^y \qquad (8.177)$$

where $t_m, m = 1, 2, \ldots, \hat{L}$ are the singular values of $\mathbf{m}_3^{yc}$ and $\mathbf{q}_m$ and $\mathbf{p}_m, m = 1, 2, \ldots, \hat{L}$ are the eigenvectors of $((\mathbf{m}_3^{yc})^H \cdot \mathbf{m}_3^{yc})$ and $(\mathbf{m}_3^{yc} \cdot (\mathbf{m}_3^{yc})^H)$, respectively. The solution of (8.177) is a truncated least-squares solution, based on the eigenvectors that correspond to the $\hat{L}$ largest eigenvalues of $\mathbf{m}_3^{yc}$. Since the effective rank of the matrix is now known, when the transient signal is observed in additive noise, and third-order moment estimates are used, a statistical criterion for rank determination could be employed, such as the one introduced by Konstantinides and Yao [1988].

8.4.4 Parameter Estimation with Fourth-Order Moments

If fourth-order moments are considered, $m_4^{x*}(-\tau, -\tau, -\tau)$ can be easily obtained as follows [Papadopoulos and Nikias, 1990]:

$$\begin{aligned} m_4^{x*}(-\tau, -\tau, -\tau) &= \sum_{k=\tau}^{\infty} x^*(k) x^*(k-\tau)(x(k-\tau))^2 \\ &= \sum_{\ell=1}^{L} B(\ell) \exp\{b_\ell^* \cdot \tau\} \end{aligned} \qquad (8.178.1)$$

where

$$B(\ell) = \sum_{q=1}^{L} \sum_{m=1}^{L} \sum_{v=1}^{L} h_\ell^* h_q^* h_m h_v \frac{1}{1 - \exp\{b_\ell^* + b_q^* + b_m + b_v\}}. \qquad (8.178.2)$$

Assuming that
$$\begin{aligned} 0 &\leq \gamma'' \leq \pi - [2\sin^{-1}(e^{-4} c_i)] \\ \mu'' &\leq \arg[h_q^* h_m h_v] \leq \mu'' + \gamma'' < \mu'' + \pi \end{aligned} \qquad (8.179)$$

where $c_i = \min(c_1, \ldots, c_L)$, the rank of the fourth-order matrix $\mathbf{m}_x^4$, which is similar to the structure of $\mathbf{m}_3^x$ in (8.167), becomes equal to the number of complex damped sinusoids. As such, the procedure described in the previous section for parameter estimation can also be applied using the fourth-order moments.

Example 8.8

The purpose of this example is to demonstrate the performance of the parameter estimation methods based on third-order moments and fourth-order cumulants of transient signals observed in noise. In this example, N is the number of data samples, $\{c_i\}$ and $\{\omega_i\}$ are the damping factors and frequencies, respectively, and γ_2^w is the variance of the additive complex Gaussian noise. The SNR is defined as $\text{SNR} \triangleq (1/\gamma_2^w)$. Note that $\gamma_2^w = 2\gamma_2^{wr}$, where γ_2^{wr} is the variance of the real and imaginary parts of the complex noise process. The experiment that we present here is described by the equation [Papadopoulos and Nikias, 1990]:

$$Y(k) = e^{b_1 k} + e^{b_2 k} + W(k), \quad k = 0, 1, 2, \ldots, 63$$

where

$$\begin{aligned} b_1 &= -0.2 + j2\pi(0.42) \\ b_2 &= -0.1 + j2\pi(0.52) \end{aligned}$$

and $W(k)$ is colored noise generated by passing a complex white Gaussian process through a FIR filter with impulse response

$$h(k) = \sum_{i=0}^{15} a(i)\delta(k-i),$$

$\mathbf{a} = [a(0), \ldots, a(15)]^T = [0.5, 0.6, -0.7, 0.8, 0.7, 0.6, 0.5, 0, 0, 0.5, 0.6, 0.7, 0.8, 0.7, 0.6, 0.5]^T$.

Figure 8.20 illustrates the estimated signal zeros obtained from 40 independent noise realizations. For each noise realization the transient signal is kept the same. The experiment is repeated for two different SNR's, using third-order moments [Figure 8.20(a)] and fourth-order cumulants [Figure 8.20(b)], as well as the original Kumaresan and Tufts (KT) method [Figure 8.20(c)] which is based on second-order statistics. From Figure 8.20, it is apparent that the fourth-order cumulants give a better estimation of the signal zeros than the third-order moments or KT methods for this particular example. □

5   DETECTION OF TRANSIENT SIGNALS

In this section we present two different ad hoc detection schemes for the detection of a transient signal in additive noise. The first scheme, by Hinich [1990], requires that the noise

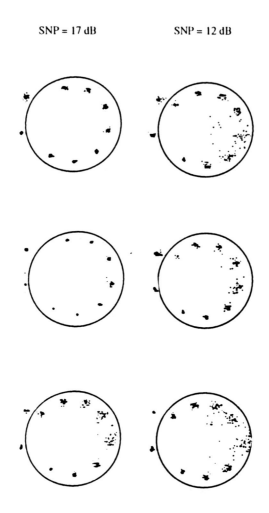

Figure 8.20 Additive colored Gaussian noise: Estimated signal zeros from 40 independent runs using $N = 64$ samples and filter order $K = 9$, in SNR = 17 dB and 12 dB. (a) Third-order moments-based method, (b) Fourth-order cumulants-based method, and (c) Kumaresan-Tufts method [Papadopoulos and Nikias, 1990] © 1990 IEEE).

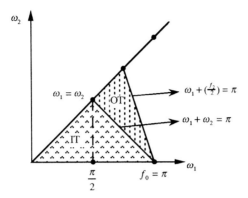

Figure 8.21 Discrete-time principal domain of the bispectrum

is stationary and possibly non-Gaussian, while the second detection scheme, by Giannakis and Tsatsanis [1990], requires stationary Gaussian noise.

8.5.1 Detection of Transient Signals Using the Bispectrum

The method proposed by Hinich [1990] for the detection of a deterministic signal of finite duration in broad-band noise is described in this section. This method can be used for the detection of transients and makes no assumptions about the functional form of transient signals. The only information which is required about the signal is that its frequency band is in the interval $(0, \omega_0)$.

The detection scheme is based on the following key observation reported by Hinich and Wolinsky [1988]. The continuous time bispectrum of a band-limited process is a proper subset of the principal domain of the bispectrum of the corresponding sampled process. In particular, Hinich and Wolinsky [1988] showed that if the received signal is sampled at the Nyquist rate, then the bispectrum of a stationary sampled process must be zero in the triangle OT of Figure 8.21, i.e., in the region defined by the triangle

$$OT = \{\omega_1, \omega_2 : \omega_2 \leq \omega_1, \pi \leq \omega_1 + \omega_2 \leq 2\pi - \omega_1\}.$$

The bispectrum in the OT region will be nonzero if the process is nonstationary.

The test statistics: The two hypotheses are:

H_0 : only noise is present $Y(k) = W(k)$
H_1 : Signal plus noise $Y(k) = W(k) + x(k)$; $k = 0, \ldots, N-1$

where $W(k)$ is the observed noise. Note that the noise is not assumed to be Gaussian; the only requirement about $W(k)$ is that of a sixth-order stationary, zero-mean random process, with bounded moment functions of all orders.

Let $\hat{M}_3^y(\omega_1, \omega_2)$ be a smoothed estimator of the bispectrum of the observed data $\{Y(k)\}$. In Chapter 4 different methods to obtain smooth bispectrum estimates are described. The distribution of this estimate for large N is approximately complex Gaussian, with variance $V(\gamma_2^w)^3/(N \cdot \Delta_N^2)$, where V is the energy of the smoothing window used in the computation of $\hat{M}_3^y(\omega_1, \omega_2)$, γ_2^w is the noise variance assuming $W(k)$ is white, and Δ_N is the bispectrum bandwidth.

Null Hypothesis:

The bispectrum of the observed signal is zero in the triangle OT of Figure 8.21. The mean of $\hat{M}_3^y(\omega_1, \omega_2)$ is zero even if the noise is not Gaussian. Hence, the distribution of the statistic

$$\text{CHI}(j,k) = 2N\Delta_N^2 |\hat{M}_3^y(\omega_j, \omega_k)|^2 / V(\gamma_2^w)^3 \tag{180}$$

is approximately central chi-squared with two degrees of freedom for (ω_j, ω_j) in OT. Thus, the sum of CHI(j,k) denoted by CHISUM is approximately central chi-squared with $2p$ degrees of freedom, i.e., χ_{2p}^2, where p is the number of points (ω_j, ω_k) in OT. Assuming $\Delta_N = \frac{1}{\sqrt{N}}, p \approx \frac{N}{48}$ then CHISUM can be transformed into an approximate Gaussian statistic Z with zero-mean and unit variance as follows [Hinich, 1990]:

$$Z = [\frac{1}{9p}]^{-1/2}[(\text{CHISUM}/2p)^{1/3} - (1 + 1/9p)]. \tag{8.181}$$

Thus, Z is the test statistic.

Signal is Present:

The test threshold C_α for a given probability of false alarm α is determined from $\alpha = Pr(Z > C_\alpha/H_0)$ which can be found in most statistics books. Under the H_1, hypothesis, $Y(\omega) = X(\omega) + W(\omega)$, and in this case the mean of $\hat{M}_3^y(\omega_1, \omega_2)$ is different from zero in the triangle OT. This is because the deterministic signal $x(k)$ makes $\{Y(k)\}$ a nonstationary process. Hinich [1990] suggested that under the H_1, hypothesis, the test statistic can be approximated as

$$Z \approx V^{-1/3}(\frac{3}{16})^{1/2} N^{5/6} \rho \tag{8.182}$$

where $\rho = \gamma_2^x/\gamma_2^w$ is the SNR, γ_2^x is the energy of $x(k)$, and V is the energy of the smoothing window used for the estimation of $\hat{M}_3^y(\omega_1, \omega_2)$. This approximation implies that the probability of signal detection is nearly one if

$$(3/16)^{1/2} N^{5/6} \gg 1/\rho. \tag{8.183}$$

The performance evaluation of this detector was developed by Hinich [1990].

8.5.2 Signal Detection Using Matched Filter Operations

When the additive noise is white, or has a known spectral density, the matched filter receiver is the optimum receiver in the sense that it maximizes the signal-to-noise ratio (SNR) and minimizes the detection error probability.

Giannakis and Tsatsanis [1990] combined matched filtering and higher-order statistics in order to come up with a detection scheme that is tolerant to signal shifts and deals with the additive noise of unknown spectral characteristics. Their proposed algorithm constructs a hypothesis test using the zeroth lag of higher-order correlations of the matched filter output. The signal can be either deterministic or random non-Gaussian, and the noise is assumed to be zero-mean Gaussian with unknown covariance sequence.

Let $s(k), k = 0, \cdots, N$ be a deterministic FIR signal. Then the matched filter (MF) has impulse response $h(k) = s(N - k)$. Hence, when the MF is excited by $x(k)$, the third-order moment sequence of its output, $y(k)$, is

$$\begin{aligned} m_3^y(\tau, \rho) &= \sum_{k,\ell=-N}^{N} m_3^x(k, \ell) m_3^h(\tau - k, \ell - \rho) \\ &= \sum_{k,\ell=-N}^{N} m_3^x(k, \ell) m_3^s(k - \tau, \rho - \ell). \end{aligned} \tag{8.185}$$

In the noise-free case, $x(k) = s(k)$. Rewritting (8.185) in terms of inverse Fourier transforms and after some algebra, we obtain

$$\begin{aligned} m_3^y(\tau, \rho) &= \frac{1}{(2\pi)^2} \int \int M_3^y(\omega_1, \omega_2) e^{-j\omega_1 \tau - j\omega_2 \rho} d\omega_1 d\omega_2 \\ &= \frac{1}{(2\pi)^2} \int \int |M_3^s(\omega_1, \omega_2)|^2 e^{-j\omega_1 \tau - j\omega_2 \rho} d\omega_1 d\omega_2. \end{aligned} \tag{8.186}$$

Applying the Cauchy-Schwartz inequality on (8.186), it follows that

$$\begin{aligned} |m_3^y(\tau, \rho)| &\leq \frac{1}{(2\pi)^2} \int \int M_3^y(\omega_1, \omega_2) d\omega_1 d\omega_2 \\ &= m_3^y(0, 0). \end{aligned} \tag{8.187}$$

The zeroth lag $m_3^y(0, 0)$ computed from that MF output, $y(k)$, as

$$m_3^y(0, 0) = \sum_{k=0}^{N} y^3(k) \tag{8.188}$$

is the basis of the detection algorithm proposed by Giannakis and Tsatsanis [1990].

Detection of Deterministic Signals in Additive Gaussian Noise

To decide between the hypotheses

$$H_0 : X(k) = W(k); \quad k = 0, 1, \cdots, T-1 \tag{8.189.1}$$

$$H_1 : X(k) = s(k) + W(k) : \quad k = 0, 1, \cdots, T-1 \tag{8.189.2}$$

where $T > N$ and $W(k)$ is zero-mean Gaussian noise, the detection statistic defined by (8.188) is used, which is the sum of the cubes of the MF output.

Under H_0, the detection statistic becomes

$$H_0 : \frac{1}{T+N} m_3^y(0,0) = \frac{1}{T+N} \sum_{k,\ell=-N}^{N} m_3^w(k,\ell) m_3^s(k,\ell) \xrightarrow[\text{as } T \to \infty]{} 0 \tag{8.190}$$

since for stationary $W(k)$ and sufficiently large T, $m_3^w(k,\ell) \approx 0$.

Under H_1 the detection statistic becomes

$$\begin{aligned}
\frac{1}{T+N} m_3^y(0,0) &= \frac{1}{T+N} \sum_{k=0}^{T+N-1} y^3(k) \\
&= \frac{1}{T+N}(m_3^s(0,0) + m_3^w(0,0) + 3\sum_{k=0}^{T+N-1} y_s^2(k)Y_w(k) \\
&\quad + 3\sum_{k=0}^{T+N-1} y_s(k)Y_w^2(k))
\end{aligned} \tag{8.191}$$

where $y_s(k), Y_w(k)$ are the MF outputs due to $s(k)$ and $W(k)$ respectively. As in (8.190), $m_3^w(0,0)$ vanishes as $T \to \infty$. Since $E\{W(k)\} = 0$, the third term in (8.191) is eliminated if averaged over independent records, while the fourth term converges to $M_2^s(0)E\{Y_w^2(k)\}$. Hence, by averaging over R independent records we get

$$H_1 : \frac{1}{T+N} \frac{1}{R} \sum_{r=1}^{R} m_3^{y,r}(0,0) \xrightarrow[R \to \infty]{} \frac{1}{T+N} m_3^s(0,0) + \frac{3}{T+N} E\{Y_w^2(k)\} M_2^s(0). \tag{8.192}$$

If the signal $s(k)$ has zero-mean, then $M_2^s(0) = 0$ and the detection statistic is proportional to $m_3^s(0,0)$.

The detection algorithm is summarized in the following steps:

Step 1: Design the MF $h(k) = s(N - k)$ for the known signal $s(k)$, $k = 0, \cdots, N - 1$.

Step 2: Generate the MF outputs of R independent records $\{Y^{(r)}(k), r = 1, \cdots, R\}$. For each record compute $m_3^{y,r}(0,0) = \sum_{k=0}^{T+N-1} [Y^{(r)}(k)]^3$, and compare it to a threshold.

The test can be extended for detection of random non-Gaussian signals in additive Gaussian noise [Giannakis and Tsatsanis, 1990]. The asymptotic performance of the detection algorithm was provided analytically and experimentally by Giannakis and Tsatsanis [1990]. Through simulation results, the algorithm was shown to behave better than the conventional matched filtering at low SNR.

8.6 SUMMARY

In this chapter, we studied a number of important signal processing problems utilizing higher-order statistics. In particular, we described time delay estimation methods based on cross third-order statistics, direction of arrival estimation methods based on spatial fourth-order cumulant matrices, beamforming techniques based on cross-bispectrum and its asymptotic normality properties, as well as MUSIC-like methods based on cumulants. We also discussed an extension of the minimum norm method to third- and fourth-order statistics domains for parameter estimation of damped exponentials in additive Gaussian noise. Finally, we discussed two different detection schemes for transient signals in noise, which form their test statistics from bispectrum estimates.

A tutorial paper on time delay estimation (TDE) using second-order statistics was published by Carter [1987]. TDE methods based on cross third- and fourth-order statistics are treated by Nikias and Pan [1988], Nikias and Liu [1990], Tugnait [1989], Zhang and Raghuveer [1991], and Oh, et al. [1990].

The tutorial papers by Johnson [1982] and Gabriel [1980] as well as the publications by Marple [1987], Schmidt [1979 and 1981], Wax [1985], Bienvenu and Kopp [1981], and Paulraj, et al. [1985] provide discussions on direction of arrival estimation using second-order statistics. Array processing methods based on third- or fourth-order statistics have been published by Pan and Nikias [1988], Forster and Nikias [1991], Porat and Friedlander [1990], and Swami and Mendel [1988]. Finally, detection schemes for transient signals in noise using the bispectrum have been treated by Hinich [1990] and Giannakis and Tsatsanis [1990].

REFERENCES

Bienvenu, G., and L. Kopp, "Adaptive High Resolution Spatial Discrimination of Passive Sources," **Underwater Acoustics and Signal Processing**, pp. 509–515, D. Reidel Publishing Co., 1981.

Brillinger, D., and M. Rosenblatt, "Asymptotic Theory of K-th Order Spectra," pp. 153–183, **Spectral Analysis of Time Series**, B, Harris (ed.), New York: Wiley, 1967.

Carter, G. C., "Coherence and Time Delay Estimation," *Proceedings of the IEEE*, **75**(2), February, 1987.

Chan, Y. T., "Notes on: Time Delay Estimation, ARMA Processes, Tracking Filters," Department of Electrical Engineering, Royal Military College of Canada, Kingston, Ontario, CA K7L2W3, April, 1985.

Chan, Y. T., R. V. Hattin, and J. B. Plant, "The Least Squares Estimation of Time Delay and Its Use in Signal Detection," *IEEE Trans. Acoustics, Speech and Signal Processing*, **ASSP-26**(3), June, 1978.

Chan, Y. T., J. M. Riley, and J. B. Plant, "A Parameter Estimation Approach to Time-Delay Estimation and Signal Detection," *IEEE Trans. Acoustics, Speech and Signal Processing*, **ASSP-28**(1), February, 1980.

Chiang, H. H., and C. L. Nikias, "The ESPRIT Algorithm with Higher-Order Statistics," *Proceedings Workshop on Higher-Order Spectral Analysis*, Vial, Co., pp. 163–168, June 28-30, 1989.

Forster, P., and C. L. Nikias, "Bearing Estimation in the Bispectrum Domain," *IEEE Trans. on Signal Processing*, **39**(9), pp. 1994–2006, September, 1991.

Gabriel, W. F., "Spectral Analysis and Adaptive Array Super-resolution Techniques," *Proceedings of IEEE*, **68**, pp. 654–666, June, 1980.

Giannakis, G. B., and N. K. Tsatsanis, "Signal Detection and Classification Using Matched Filtering and Higher-Order Statistics," *IEEE Trans. Acoustics, Speech and Signal Processing*, **38**(7), July, 1990.

Hassab, J. C., **Underwater Signal and Data Processing**, Boca Raton, Florida: CRC Press, 1989.

Hinich, M. J., "Testing for Gaussianity and Linearity of a Stationary Time Series," *J. Times Series Anal.*, **3**(3), pp. 169–176, 1982.

Hinich, M. J., "Detecting a Transient Signal by Bispectral Analysis," *IEEE Trans. Acoustics, Speech and Signal Processing*, **38**, 7, pp. 1277–1283, July, 1990.

Hinich, M. J., and M. A. Wolinsky, "A Test for Aliasing Using Bispectral Analysis," *Journal of American Statistical Association*, **83**, 402, pp. 499–502, June, 1988.

Hinich, M. J., D. Marandino, and E. J. Sullivan, "Bispectrum of Ship-Radiated Noise," *Journal of Acoust. Soc. American*, **85**, pp. 1512–1517, April, 1989.

Johnson, D. H., "The Application of Spectral Estimation Methods to Bearing Estimation Problems," *Proceedings of the IEEE*, **9**, September, 1982.

Kelley, E. J., "An Adaptive Detection Algorithm," *IEEE Trans. Aerospace and Electronic Systems*, **AES-22**(1), pp. 115–127, March, 1986.

Knapp, C. H., and G. C. Carter, "The Generalized Correlation Method for Estimation of Time Delay," *IEEE Trans. Acoustic, Speech and Signal Processing*, **ASSP-24**, pp. 320–327, August, 1976.

Konstantinides, R., and K. Yao, "Statistical Analysis of Effective Singular Values in Matrix Rank Determination," *IEEE Trans. Acoustics, Speech and Signal Processing*, **ASSP-36**, pp. 757–763, May, 1988.

Kumaresan, R., and D. W. Tufts, "Estimating the Parameters of Exponentially Damped Sinusoids and Pole-Zero Modeling in Noise," *IEEE Trans. Acoustics, Speech and Signal Processing*, **ASSP-30**, pp. 833–840, December, 1982.

Kung, S. Y., K. S. Arun, Rao, and D.V. Bhaskar, "State-space and Singular Value Decomposition-based Approximation Methods for the Harmonic Retrieval Problem," *Proc. 2nd ASSP Workshop on Spectral Estimation*, Tampa, FL, November, 1983.

Lii, K. S., and M. Rosenblatt, "Deconvolution and Estimation of Transfer Function, Phase and Coefficients for Non-Gaussian Linear Processes," *Annals of Statistics*, pp. 1195-1208, 1982.

Marple, S. L., **Digital Spectral Analysis with Applications**, Englewood Cliffs, NJ: Prentice Hall, 1987.

Nikias, C. L., "Higher-Order Spectral Analysis," **Advances in Spectrum Analysis and Array Processing**, S. Haykin (ed.), Vol. I, pp. 326–365, Englewood Cliffs, NJ: Prentice-Hall, 1991.

Nikias, C. L., and F. Liu, "Bicepstrum Computation Based on Second- and Third-Order Statistics with Applications," *Proc. ICASSP'90*, pp. 2381–2386, Albuquerque, New Mexico, April, 1990.

Nikias, C. L., and R. Pan, "Time Delay Estimation in Unknown Gaussian Spatially Correlated Noise," *IEEE Trans. on Acoustics, Speech and Signal Processing*, **36**(11), pp. 1706–1714, November, 1988.

Nikias, C. L., and M. R. Raghuveer, "Bispectrum Estimation: A Digital Processing Framework," *Proc. IEEE*, **75**, pp. 869–891, July, 1987.

Oh, W.T., S. B. Kim, and E. J. Powers, "The Squared Skewness Processor for Time Delay Estimation in the Bispectrum Domain," **Signal Processing V: Theory and Applications**, L. Torres, E. Masgrau, and M. A. Lagunas (eds.), pp. 111–114, Elsevier Science Publishers, R.V., 1990.

Pan, R., and C. L. Nikias, "Harmonic Decomposition Methods in Cumulant Domains," *Proceedings ICASSP'88*, pp. 2356–2359, New York, 1988.

Papadopoulos, C. K., and C. L. Nikias, "Parameter Estimation of Exponentially Damped Sinusoids Using Higher-Order Statistics," *IEEE Trans. on Acoustics, Speech, and Signal Processing*, **38**(8), pp. 1424–1445, August, 1990.

Paulraj, A., R. Roy, and T. Kailath, "Estimation of Signal Parameters via Rotational Invariance Techniques – ESPRIT," *Proc. 19th Asilomar Conf. Circuits, Syst. Comput.*, Asilomar, CA, November, 1985.

Porat, B., and B. Friedlander, "Direction Finding Algorithms Based on Higher-Order Statistics," *Proceedings ICASSP'90*, **5**, pp. 2675-2678, Albuquerque, NM, April 3–6, 1990.

Reddy, V. U., A. Paulraj, and T. Kailath, "Performance Analysis of the Optimum Beamformer in the Presence of Correlated Sources and Its Behavior Under Spatial Smoothing," *IEEE Trans. Acoustics, Speech and Signal Processing*, **ASSP-35**, pp. 927–936, July, 1987.

Rosenblatt, M., **Stationary Sequences and Random Fields**, Boston: Birkhauser, 1985.

Roy, R., A. Paulraj, and T. Kailath, "Direction-of-Arrival Estimation by Subspace Rotation Methods – ESPRIT," *Proc. ICASSP'86*, pp. 2495–2498, Tokyo, Japan, April, 1985.

Roy, R., A. Paulraj, and T. Kailath, "ESPRIT – A Subspace Rotation Approach to Estimation of Parameters of Cisoids in Noise," *IEEE Trans. Acoust., Speech and Signal Processing*, **ASSP-34**(5), pp. 1340–1342, October, 1986.

Sasaki, K., T. Sato, and Y. Makamura, "Holographic Passive Sonar," *IEEE Trans. Sonics Ultrason*, **SU-24**, pp. 193–200, May, 1977.

Sato, T., and K. Sasaki, "Bispectral Holography," *Journal of Acoust. Soc. Amer.*, **62**, pp. 404–408, 1977.

Schmidt, R. O., "Multiple Emitter Location and Signal Parameter Estimation," *Proc. RADC Spectral Estimation Workshop*, pp. 243–258, 1979.

Schmidt, R. O., "A Signal Subspace Approach to Multiple Emitter Location and Spectral Estimation," Ph.D. Dissertation, Stanford Univ., Stanford, CA, 1981.

Subba, T. R., and M. M. Gabr, "A Test for Linearity of Stationary Time Series," *J. Time Series Anal.*, **2**, pp. 145–158, 1980.

Swami, A., and J. M. Mendel, "Cumulant Based Approach to the Harmonic Retrieval Problem," *Proc. ICASSP'88*, pp. 2264–2266, New York, April, 1988.

Swami, A., and J. M. Mendel, "Cumulant-based Approach to the Harmonic Retrieval and Related Problems," *IEEE Trans. on Signal Processing*, **39**(5), pp. 1099–1109, May, 1991.

Tugnait, J. K., "Time Delay Estimation in Unknown Spatially Correlated Gaussian Noise Using Higher-Order Statistics," *Proc. 23rd Asilomar Conf. Signals, Systems, Computers*, pp. 211–215, Pacific Grove, CA, 1989.

Wax, M., "Detection and Estimation of Superimposed Signals," Ph.D. Dissertation, Stanford University, Stanford, CA, March, 1985.

Zhang, W., and M. R. Raghuveer, "Nonparametric Bispectrum-based Time-delay Estimators for Multiple Sensor Data," *IEEE Trans. on Signal Processing*, **39**(13), pp. 770–774, March, 1991.

Ziskind, I., and M. Wax, "Maximum Likelihood Localization of Multiple Sources by Alternating Projection," *IEEE Trans. on ASSP*, **ASSP-36**, pp. 1553–1560, October, 1988.

9

ADAPTIVE FILTERING WITH HIGHER-ORDER STATISTICS

9.1 INTRODUCTION

There are signal processing problems in practice where signals have time-varying characteristics, but knowledge of the signal characteristics is not available. Adaptive filtering algorithms are self-adjusting or self-designing techniques that can be applied to the analysis of signals with unknown, or time-varying, statistics. The adaptive algorithm starts with a set of initial conditions which, after successive iterations, converges to an optimum solution, provided that the signal is stationary. On the other hand, if the signal has time-varying statistics (i.e., it is nonstationary), the adaptive algorithm exhibits tracking capabilities by following up the variations in the statistics of the signal. Applications of adaptive filtering algorithms can be found in system identification and equalization problems, as well as in speech analysis, spectrum estimation, detection of signals, echo cancellation, line enhancement, array processing, etc. See for example the excellent book by Haykin [1991] on adaptive filter theory and references therein.

As described by Haykin [1991], the recursive algorithms for the operation of adaptive filters can be identified in three distinct categories; namely, (i) approaches based on Wiener filter theory, (ii) the Kalman filter theory, and (iii) the method of least-squares. These algorithms are based on mean-square error (MSE) or least-squares criteria, and consequently, employ second-order statistics of the signals involved in the adaptation process. These algorithms are limited to tracking variations in the second-order statistics domain.

There is a fourth category of adaptive filtering algorithms that has received a lot of attention recently. These algorithms are well suited for problems where tracking of higher-order statistical variations is needed. This class of techniques is based on higher-order statistics (HOS), or nonlinear (e.g., Non-MSE) criteria.

Higher-order statistical variations arise in non-Gaussian signals and may contain important information that has to be extracted by the adaptive filter. Such information includes phase variations, as well as nonlinearities.

The purpose of this chapter is to describe adaptive filtering algorithms based on higher-order statistics. However, a brief review of adaptive algorithms based on second-order statistics is also provided. In particular, the Least-Mean Square (LMS) and Recursive Least-Squares (RLS) algorithms, which are based on second-order (or quadratic) MSE cost functions, are described. Extensions of these algorithms to adaptive techniques with fourth-order cost functions, the so-called LMF and RLF algorithms, are then described. Adaptive algorithms based on IIR, lattice and noncausal AR models, and higher-order statistics are also discussed. Finally, a description is given of adaptive delay estimation methods and adaptive blind equalization algorithms.

9.2 ADAPTIVE FILTERING WITH SECOND-ORDER STATISTICS

Let us consider the block diagram illustrated in Figure 9.1 representing a linear discrete-time adaptive filter with coefficients $\mathbf{W}^T(k) = [w_1(k), \ldots, w_M(k)]$. The input and output sequences to the filter are $\mathbf{u}^T(k) = [u(k), u(k-1), \ldots, u(k-M-1)]$ and $\{\tilde{X}(k)\}$, respectively. The goal of adaptive filtering is to choose the values of the coefficient vector, $\mathbf{W}(k)$, such that the filter output $\{\tilde{X}(k)\}$ is "as-close-as possible" to a given desired response $\{d(k)\}$. From Figure 9.1 we obtain the following equations:

$$\tilde{X}(k) = \mathbf{W}^H(k)\mathbf{u}(k) \qquad (9.1)$$

$$\begin{aligned} e(k) &= d(k) - \tilde{X}(k) \\ &= d(k) - \mathbf{W}^H(k)\mathbf{u}(k) \end{aligned} \qquad (9.2)$$

where "H" denotes the transpose conjugate, and $e(k)$ is the error between the desired and actual response of the adaptive filter. The objective is to minimize a cost function of the error with respect to the filter coefficients, so that $\{\tilde{X}(k)\}$ is as close to $\{d(k)\}$ as possible. We show in this section that if the cost function is the mean-square error (MSE), the filter coefficients are only calculated from second-order statistics of the data, i.e., from the autocorrelation matrix of $\mathbf{u}(k)$ and the cross-correlation vector of $\mathbf{u}(k)$ and $\{d(k)\}$. The autocorrelation matrix of $\mathbf{u}(k)$ is defined by

$$\mathbf{m}_2^\mathbf{u} = E\{\mathbf{u}(k)\mathbf{u}^H(k)\}, \quad (M \times M), \qquad (9.3.1)$$

whereas the cross-correlation vector is given as

$$\mathbf{m}_2^{\mathbf{u}d} = E\{\mathbf{u}(k) \cdot d^*(k)\}, \ (M \times 1). \tag{9.3.2}$$

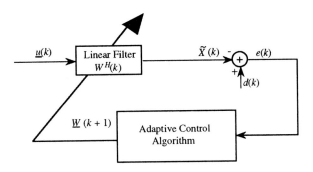

Figure 9.1 Block diagram of a linear adaptive filter

Assuming that $\mathbf{u}(k)$ and $d(k)$ are jointly stationary, the mean-square error (MSE) at time k is $J(k) = E\{|e(k)|^2\}$ or [Haykin, 1991]

$$J(k) = \gamma_2^d - \mathbf{W}^H(k)\mathbf{m}_2^{\mathbf{u}d} - (\mathbf{m}_2^{\mathbf{u}d})^H\mathbf{W}(k) + \mathbf{W}^H(k)\mathbf{m}_2^{\mathbf{u}}\mathbf{W}(k), \tag{9.4}$$

where $\gamma_2^d = E\{|d(k)|^2\}$ is the variance of $d(k)$.

The MSE cost function $J(k)$, viewed as a function of $\mathbf{W}(k)$, has a unique minimum. The value of $\mathbf{W}(k)$ that corresponds to this minimum, i.e., $\mathbf{W}_o$, is defined as

$$\mathbf{m}_2^{\mathbf{u}} \cdot \mathbf{W}_o = \mathbf{m}_2^{\mathbf{u}d}, \tag{9.5}$$

and at this point, $J(k)$ is

$$J_{\min} = \gamma_2^d - (\mathbf{m}_2^{\mathbf{u}d})^H \mathbf{W}_o \tag{9.6}$$

Since the input sequence is stationary, the task of the adaptive transversal filter is to find a solution, $\mathbf{W}(k)$, that satisfies (9.5).

9.2.1 The Steepest Descent Method

One approach to solving (9.5) is by matrix inversion, or by the use of the Levinson algorithm, since $\mathbf{m}_2^{\mathbf{u}}$ is Toeplitz and Hermitian [Haykin, 1991]. An adaptive approach, however, for solving (9.5) is the method of *Steepest Descent* [Murray, 1972], which may also be useful when the order of the adaptive filter is large. The steepest descent method is summarized below.

Let $\nabla(k)$ be the gradient vector at time k, and let $\mathbf{W}(k)$ be the tap-weight vector at k. The adaptation of $\mathbf{W}(k)$ is described by

$$\mathbf{W}(k+1) = \mathbf{W}(k) + \frac{1}{2}\mu[-\nabla(k)]. \tag{9.7}$$

where μ is a positive and real constant. Differentiating both sides of (9.4) with respect to $\mathbf{W}(k)$ we obtain

$$\nabla(k) = \frac{\partial J(k)}{\partial \mathbf{W}(k)} = -2\,\mathbf{m}_2^{\mathbf{u}d} + 2\,\mathbf{m}_2^{\mathbf{u}}\mathbf{W}(k). \tag{9.8}$$

Substitution of (9.8) into (9.7) yields

$$\mathbf{W}(k+1) = \mathbf{W}(k) + \mu[\mathbf{m}_2^{\mathbf{u}d} - \mathbf{m}_2^{\mathbf{u}}\mathbf{W}(k)]. \tag{9.9}$$

The latter equation corresponds to the mathematical formulation of the Steepest Descent algorithm, otherwise known as the deterministic gradient algorithm. Table 9.1 illustrates the description of the Steepest Descent algorithm.

TABLE 9.1   THE STEEPEST DESCENT ALGORITHM

Adaptive Filter	$\tilde{X}(k) = \mathbf{W}^H(k)\mathbf{u}(k)$
Error Equation	$e(k) = d(k) - \tilde{X}(k)$
Adaptive Control Equation	$\mathbf{W}(k+1) = \mathbf{W}(k) + \mu[\mathbf{m}_2^{\mathbf{u}d} - \mathbf{m}_2^{\mathbf{u}}\mathbf{W}(k)]$
Initialization	$\mathbf{W}(0) = \mathbf{0}$

The necessary and sufficient condition for convergence of the Steepest Descent algorithm is

$$0 < \mu < \frac{2}{\lambda_{\max}} \tag{9.10.1}$$

where $\lambda_{\max}$ is the maximum eigenvalue of the autocorrelation matrix $\mathbf{m}_2^u$. Since $\lambda_{\max} < M \cdot m_2^u(0)$, a tighter bound for μ is

$$0 < \mu < \frac{2}{M \cdot m_2^u(0)}, \tag{9.10.2}$$

TABLE 9.2   THE LMS ALGORITHM

Adaptive Filter	$\tilde{X}(k) = \hat{\mathbf{W}}^H(k)\mathbf{u}(k)$
Error Equation	$e(k) = d(k) - \tilde{X}(k)$
Adaptive Control Equation	$\hat{\mathbf{W}}(k+1) = \hat{\mathbf{W}}(k) + (\mu)\mathbf{u}(k) \cdot e^*(k)$
Initialization	$\hat{\mathbf{W}}(0) = \mathbf{0}$

where $m_2^u(0) = E\{|u(k)|^2\}$. As pointed out by Haykin [1991] and Widrow and Stearns [1985], the convergence behavior of the Steepest Descent algorithm is highly sensitive to the variation of μ and eigenvalue spread of the matrix $\mathbf{m}_2^\mathbf{u}$.

9.2.2   The LMS Algorithm

The Steepest Descent algorithm described in the previous section requires *exact* knowledge of the second-order statistics of the data. In practice this is not feasible, and the second-order statistics have to be estimated from the available data. An algorithm that updates the tap-weight vector based on second-order statistic estimates is the *Least-Mean Square* (LMS) algorithm [Widrow and Stearns, 1985].

According to the LMS algorithm, an estimate of the gradient vector is given by

$$\hat{\nabla}(k) = -2\hat{\mathbf{m}}_2^{\mathbf{u}d} + 2\hat{\mathbf{m}}_2^\mathbf{u}\hat{\mathbf{W}}(k) \qquad (9.11)$$

where $\hat{\mathbf{m}}_2^{\mathbf{u}d}$ and $\hat{\mathbf{m}}_2^\mathbf{u}$ are *single point unbiased estimates* of $\mathbf{m}_2^{\mathbf{u}d}$ and $\mathbf{m}_2^\mathbf{u}$, respectively, i.e.,

$$\begin{aligned} \hat{\mathbf{m}}_2^{\mathbf{u}d} &= \mathbf{u}(k)d^*(k) \\ \hat{\mathbf{m}}_2^\mathbf{u} &= \mathbf{u}(k)\mathbf{u}^H(k). \end{aligned} \qquad (9.12)$$

Substituting $\hat{\nabla}(k)$ of (9.11), (9.12) for $\nabla(k)$ in (9.7), we obtain the recursive relation

$$\hat{\mathbf{W}}(k+1) = \hat{\mathbf{W}}(k) + (\mu)\mathbf{u}(k)e^*(k) \qquad (9.13)$$

which is entirely based on the incoming data samples. The LMS algorithm is summarized in Table 9.2.

When the input data $\mathbf{u}(k)$ is stationary, the LMS algorithm has the following properties [Haykin, 1991; Widrow and Stearns, 1985]:

1. The algorithm is convergent in the mean, i.e. $E\{\hat{\mathbf{W}}(k)\} \to \mathbf{W}_0$ as $k \to \infty$, if the step size μ satisfies $0 < \mu < 2/\lambda_{\max}$.

2. The LMS is convergent in the mean-square, i.e.,

$$E\{J(k)\} = J_{\min} + E\{J_{ex}(\infty)\}, \text{ if } 0 < \mu < \frac{2}{\sum_{i=1}^{M}\lambda_i}$$

where $J(k)$ is given in (9.4), $\{\lambda_i\}$ are eigenvalues of $\mathbf{m}_2^{\mathbf{u}}$, and $E\{J_{ex}(\infty)\}$ is the excess MSE given by

$$E\{J_{ex}(\infty)\} = \frac{\mu J_{\min} \sum_{i=1}^{M}\lambda_i}{2 - \mu \sum_{i=1}^{M}\lambda_i}.$$

3. If the LMS is convergent in the mean-square, it is also convergent in the mean. This follows from the fact that $\lambda_{\max} \leq \sum_{i=1}^{M}\lambda_i$.

4. The step-size μ is the "memory" of the LMS algorithm. If μ is small, the LMS has slow adaptation and $E\{J_{ex}(\infty)\}$ will be small. On the other hand, if μ is large, the algorithm exhibits fast adaptation at the expense of large $E\{J_{ex}(\infty)\}$.

5. The convergence rate of the algorithm is slow if the autocorrelation matrix of the incoming data has large eigenvalue spread. The eigenvalue spread is defined as $(\lambda_{\max}/\lambda_{\min})$.

6. The LMS algorithm exhibits misadjustment at steady-state. That is, $\mathcal{M} = E\{J_{ex}(\infty)\}/J_{\min}$ which increases approximately linearly with μ.

9.2.3 The RLS Algorithm

The RLS algorithm may be viewed as the deterministic counterpart of Kalman filtering. The cost function of the RLS, $\mathcal{E}(k)$, is given by

$$\mathcal{E}(k) = \sum_{i=1}^{k} \beta(k,i)|e(i)|^2 \qquad (9.14)$$

where k is the variable length of the observation data, $\beta(k,i)$ is the forgetting factor, $0 < \beta(k,i) < 1$ for all values of $\{i\}$, and $e(i)$ is defined by

$$e(i) = d(i) - \mathbf{W}^H(k)\mathbf{u}(i). \tag{9.15}$$

Note that usually $\beta(k,i) = \lambda^{k-i}$ where $0 < \lambda < 1$. The values of the tap-weight vector that minimize $\mathcal{E}(k)$ are given by the set of normal equations

$$\mathbf{m}_2^{\mathbf{u}}(k) \cdot \hat{\mathbf{W}}(k) = \mathbf{m}_2^{\mathbf{u}d}(k) \tag{9.16}$$

where $\mathbf{m}_2^{\mathbf{u}}(k)$ is the $(M \times M)$ autocorrelation matrix defined as

$$\mathbf{m}_2^{\mathbf{u}}(k) = \sum_{i=1}^{k} \lambda^{k-i} \mathbf{u}(i)\mathbf{u}^H(i) \tag{9.17}$$

and $\mathbf{m}_2^{\mathbf{u}d}(k)$ is the $(M \times 1)$ cross-correlation vector between input and the desired response defined by

$$\mathbf{m}_2^{\mathbf{u}d}(k) = \sum_{i=1}^{k} \lambda^{k-i} \mathbf{u}(i) d^*(i). \tag{9.18}$$

Equation (9.16) can be solved by matrix inversion for $\mathbf{W}(k)$. However, this approach is not computationally efficient because an inverse matrix computation should be performed for each instant k. An alternative solution of (9.16) can be obtained recursively by the RLS algorithm, which is based on the matrix inversion lemma [Haykin, 1991]. The RLS algorithm proceeds by computing the gain vector

$$\mathbf{K}(k) = \frac{\lambda^{-1}\mathbf{P}(k-1)\mathbf{u}(k)}{1 + \lambda^{-1}\mathbf{u}^H(k)\mathbf{P}(k-1)\mathbf{u}(k)}, \quad (M \times 1) \tag{9.19}$$

and the a priori error

$$a(k) = d(k) - \hat{\mathbf{W}}^H(k-1)\mathbf{u}(k) \tag{9.20}$$

and then the tap-weight vector

$$\hat{\mathbf{W}}(k) = \hat{\mathbf{W}}(k-1) + \mathbf{K}(k)a^*(k). \tag{9.21}$$

The matrix $\mathbf{P}(k)$ is the inverse of the autocorrelation matrix $\mathbf{m}_2^{\mathbf{u}}(k)$, which is updated by the following equation

$$\mathbf{P}(k) = \lambda^{-1}\mathbf{P}(k-1) - \lambda^{-1}\mathbf{K}(k)\mathbf{u}^H(k)\mathbf{P}(k-1). \tag{9.22}$$

Table 9.3 describes the RLS algorithm equations. The properties of RLS can be summarized as follows [Haykin, 1991].

TABLE 9.3   THE RLS ALGORITHM

Adaptive Filter	$\tilde{X}(k) = \hat{\mathbf{W}}^H(k)\mathbf{u}(k)$
A priori Error	$a(k) = d(k) - \hat{\mathbf{W}}^H(k-1)\mathbf{u}(k)$
A posteriori Error	$e(k) = d(k) - \hat{\mathbf{W}}^H(k)\mathbf{u}(k)$
Gain Vector	$\mathbf{K}(k) = \frac{\lambda^{-1}\mathbf{P}(k-1)\mathbf{u}(k)}{1+\lambda^{-1}\mathbf{u}^H(k)\mathbf{P}(k-1)\mathbf{u}(k)}$
Adaptive Control Equation	$\hat{\mathbf{W}}(k) = \hat{\mathbf{W}}(k-1) + \mathbf{K}(k)a^*(k)$
Matrix Inversion Lemma	$\mathbf{P}(k) = \lambda^{-1}\mathbf{P}(k-1) - \lambda^{-1}\mathbf{K}(k)\mathbf{u}^H(k)\mathbf{P}(k-1)$
Initialization	$\hat{\mathbf{W}}(0) = \mathbf{0}$ $\mathbf{P}(0) = \delta^{-1}\mathbf{I}\ (\delta > 0)$

1. The RLS is convergent in the mean, i.e., $E\{\hat{\mathbf{W}}(k)\}\ \mathbf{W}_0$ as $k \to \infty$.

2. The algorithm is convergent in the mean-square, i.e.,

$$E\{J(k)\} \approx J_{\min}(1 + \frac{M}{k}) \text{ as } k \to \infty.$$

   where $J(k) = (\hat{\mathbf{W}}(k) - \mathbf{W}_0)(\hat{\mathbf{W}}^H(k) - \mathbf{W}_0^H)$.

3. The RLS algorithm exhibits no misadjustment and its convergence is independent of the eigenvalue spread of the autocorrelation matrix.

4. The convergence rate of RLS is significantly higher than that of the LMS. For example, if the LMS requires 20M iterations to converge in the mean-square, the RLS for the same example would only require 2M iterations.

5. The RLS requires $O(M^2)$ multiplications per iteration, whereas the LMS requires $O(M)$ multiplications.

Fast RLS algorithms that require $O(M)$ multiplications per iteration are discussed by Haykin [1991].

9.3   THE LMF AND RLF ALGORITHM

The LMS algorithm estimates the tap weights by minimizing the expected value of the squared error $E\{|e(k)|^2\}$ between the estimated and the desired filter response. This choice of a quadratic performance measure is due to its simplicity and convenience of analysis, and is motivated by the assumption that the underlying processes are Gaussian. In this section, the LMS algorithm is shown to be a special case of a more general adaptive scheme that minimizes $E\{e^{2N}(k)\}$ for some arbitrary value of N. This particular cost function was proposed by Walach and Widrow [1984]; for $N = 1$ gives the already studied LMS algorithm, while the case for $N = 2$ gives the Least-Mean-Fourth (LMF) error algorithm.

The advantage of the aforementioned general algorithm, against the LMS, is found in the case where the observed error is contaminated with additive noise. In this case, when both the LMF and LMS are set to have the same time constants for the weight adaptation process, the LMF will have substantially lower weight noise than the LMS.

Many practical problems, such as plant modeling and channel equalization, can be modeled as shown in Figure 9.2. The output of the channel with transfer function $H(z)$ is corrupted by additive noise $n(k)$. The objective is to find a channel model $\hat{H}(z)$ in some adaptive way by minimizing $E\{e^{2N}(k)\}, N = 1, 2, \ldots$ with respect to $\hat{H}(z)$ coefficients. The adaptive model of the channel $H(z)$ is a tap-delay line of length M, where M is assumed to be equal to the order of the channel $H(z)$.

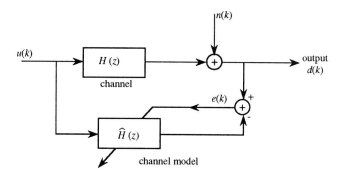

Figure 9.2 Adaptive channel modeling

The input vector at time k is given by $\mathbf{u}^T(k) = [u(k), u(k-1), \ldots, u(k-M+1)]$. For the subsequent analysis, it will be assumed that the input $u(k)$ is independent of the noise $n(k)$, and that both are *distributed symmetrically around zero*. This implies that all odd moments of $u(k)$ and $n(k)$ are zero.

9.3.1 The LMF Algorithm

Let $\hat{\mathbf{W}}^T(k) = [\hat{w}_1(k), \hat{w}_2(k), \ldots \hat{w}_M(k)]$ and $\mathbf{W}^T = [w_1, w_2, \ldots, w_M]$ be the weight vectors for the channel model $\hat{H}(z)$ and channel $H(z)$, respectively. Then the instantaneous error at time k is given by

$$\begin{aligned} e(k) &= d(k) - \hat{\mathbf{W}}^T(k)\mathbf{u}(k) = \mathbf{W}^T\mathbf{u}(k) + n(k) - \hat{\mathbf{W}}^T(k)\mathbf{u}(k) \\ &= n(k) - \mathbf{c}^T(k)\mathbf{u}(k) \end{aligned} \quad (9.23)$$

where

$$\mathbf{c}(k) = \hat{\mathbf{W}}(k) - \mathbf{W} \quad (9.24)$$

is the difference between the current vector and the optimum solution. Taking the $2N^{th}$ power of $e(k)$ and differentiating with respect to $\hat{\mathbf{W}}(k)$, we find that the *instantaneous gradient* of $e^{2N}(k)$ is

$$\nabla(e^{2N}(k)) = -2Ne^{2N-1}\mathbf{u}(k). \quad (9.25)$$

From (9.7) and (9.25) the weight adaptation rule becomes

$$\hat{\mathbf{W}}(k+1) = \hat{\mathbf{W}}(k) + \mu N e^{2N-1}(k) \cdot \mathbf{u}(k) \quad (9.26)$$

where μ is the step-size parameter. At the point of convergence, $E\{\hat{\mathbf{W}}(k+1)\} = E\{\hat{\mathbf{W}}(k)\}$ and, therefore, from (9.26) and (9.23) we obtain

$$E\{e^{2N-1}(k)\mathbf{u}(k)\} = E\{[d(k) - \hat{\mathbf{W}}^T(k)\mathbf{u}(k)]^{2N-1}\mathbf{u}(k)\} = 0. \quad (9.27)$$

Equation (9.27) breaks down into a system of M polynomials of degree $2N - 1$ each, with unknown the weights $\hat{w}_i(k), i = 1, \ldots, M$. The function $E\{e^{2N}(k)\}$ is a convex function of the weight vector and, therefore, cannot have local minima. For the special case of LMS ($N = 1$), the system of (9.27) is linear in $\hat{\mathbf{W}}(k)$ and has only one solution. Next, the convergence in the mean of $\mathbf{c}(k)$ is studied.

In this analysis, only the simple case of small deviations from the optimum solution, $\mathbf{W}$, are considered, i.e., when $\mathbf{c}(k)$ is close to zero. For a more general convergence analysis the reader is referred to the original publication by Walach and Widrow [1984]. Combining (9.24) and (9.26) we obtain

$$\mathbf{c}(k+1) = \mathbf{c}(k) + \mu \ Ne^{2N-1}(k) \cdot \mathbf{u}(k). \quad (9.28)$$

Substituting (9.23) into (9.28) we find

$$\mathbf{c}(k+1) = \mathbf{c}(k) + \mu N \mathbf{u}(k) \sum_{i=0}^{2N-1} \binom{2N-1}{i} n^i(k)(-\mathbf{u}^T(k)\mathbf{c}(k))^{2N-1-i}$$

$$\approx \mathbf{c}(k) + \mu N \mathbf{u}(k)(n^{2N-1}(k) - (2N-1)n^{2N-2}(k)\mathbf{u}^T(k)\mathbf{c}(k))$$
(9.29)

where in the second part of (9.29) the high powers of $\mathbf{c}(k)$ were neglected since we assume that $\mathbf{c}(k)$ is already close to zero. Initially, it was assumed that $n(k)$ and $u(k)$ are independent. Furthermore, for small values of μ, it can be assumed [Walach and Widrow 1984] that $n(k)$ is also independent of $\mathbf{c}(k)$. Hence, expectations on both sides of (9.29) yield

$$E\{\mathbf{c}(k+1)\} = [\mathbf{I} - \mu N(2N-2)E\{n^{2N-2}(k)\}\mathbf{m}_2^{\mathbf{u}}]E\{\mathbf{c}(k)\} = \mathbf{Q}E\{\mathbf{c}(k)\}, \quad (9.30)$$

where $\mathbf{m}_2^{\mathbf{u}}$ is given by (9.31) and $\mathbf{Q}$ can be inferred from (9.30). To guarantee convergence, μ is chosen so that

$$0 < \mu < \frac{2}{N(2N-1)E\{n^{2N-2}(k)\}\lambda_{\max}} \quad (9.31.1)$$

where $\lambda_{\max}$ is the maximum eigenvalue of the input autocorrelation matrix $\mathbf{m}_2^{\mathbf{u}}$.

An easily applied sufficient condition for convergence in the mean can be found if we replace the maximum eigenvalue of the input autocorrelation matrix by the trace of the same matrix, i.e., $tr\{\mathbf{m}_2^{\mathbf{u}}\} = ME\{u^2(k)\}$. Then, the condition for the step-size parameter μ becomes

$$0 < \mu < \frac{2}{NM(2N-1)E\{n^{2N-2}(k)\}E\{u^2(k)\}}. \quad (9.31.2)$$

Walach and Widrow [1984] have also shown that in the steady state the weight errors $\mathbf{c}(k)$ are uncorrelated with each other.

The misadjustment $\mathcal{M}$ for some arbitrary N of the Least-Mean $2N^{th}$-order algorithm is defined as [Walach and Widrow, 1984]:

$$\mathcal{M}(N) = \frac{\text{(error power due to weight noise)}}{\text{(optimal error power)}}$$

$$= \frac{E\{[\mathbf{c}^T(k)\mathbf{u}(k)]^2\}}{E\{n^2(k)\}} \quad (9.32)$$

$$= \frac{E\{n^{4N-2}(k)\}}{2(2N-1)^2 E\{n^2(k)\}(E\{n^{2N-2}(k)\})^2} \sum_{i=1}^{M} \frac{1}{\tau_i}$$

where τ_i are the relaxation time constants for the M different modes of convergence of $E\{\mathbf{c}(k)\}$ given by

$$\tau_i = \frac{1}{\mu N(2N-1)E\{n^{2N-2}(k)\}\lambda_i}, \quad i = 1, 2, \ldots, M \tag{9.33}$$

and λ_i's are the eigenvalues of $\mathbf{m}_2^{\mathbf{u}}$.

Note that for $N = 1$ (LMS), $\mathcal{M}(1) = \frac{1}{2}\sum_{i=1}^{M}\frac{1}{\tau_i}$ is the LMS misadjustment for small μ [Haykin, 1991].

This $2N^{th}$-order algorithm performs better when $\mathcal{M}$ is lower. By defining

$$\alpha(N) = \frac{\mathcal{M}(1)}{\mathcal{M}(N)} = \frac{(2N-1)^2 E\{n^2(k)\}(E\{n^{2N-2}(k)\})^2}{E\{n^{4N-2}(k)\}} \tag{9.34}$$

we can see that when $\alpha(N) > 1$ it is better to use algorithms with $N > 1$ than the LMS algorithm ($N = 1$). This means that when $\alpha(N) > 1$, the algorithm with $N > 1$ leads to lower misadjustment than the LMS for the same speed of convergence. The optimum value for N can be obtained if we know the moments of the noise $n(k)$. Four noise densities have been selected for their practical importance and they are illustrated in Figure 9.3(a)–(d), i.e., the Gaussian density, the uniform density, the density of the sinusoidal signal, and the density of a square wave. The corresponding values for $\alpha(N)$ are given in Table 9.4 for $N = 1, 2, 3,$ and 4.

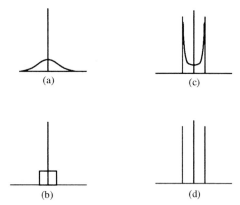

Figure 9.3 Four noise probability densities: (a) Gaussian, (b) uniform, (c) density of sine wave, and (d) density of square wave. ([Walach and Widrow, 1984] © 1984 IEEE)

The Least Mean Fourth-Order (LMF) algorithm is a special case of the general $2N^{th}$-order algorithm for $N = 2$. Table 9.5 illustrates the mathematical description of the LMF algorithm.

TABLE 9.4 VALUES OF $\alpha(N)$ FOR THE NOISE PROBABILITY DENSITIES OF FIGURE 9.3

	Gaussian	Uniform	Sine Wave	Square Wave
N=1	1	1	1	1
N=2	0.6	2.3	3.6	9
N=3	0.24	3.69	7.14	25
N=4	0.08	5	11.4	49

Example 9.1 [Walach and Widrow, 1984]

If the noise $n(k)$ is uniformly distributed between -1 and 1, then from Table 9.4 we see that $\alpha(2) = 2.3$. Consequently, in this case we expect that the use of the LMF will result in an improvement of about 3dB over the LMS algorithm. For the same speed of convergence, we expect 3 dB less noise in the weights, using the LMF algorithm.

The LMS and LMF algorithms are employed to estimate the weights $\mathbf{W}^T = (0.1, 0.2, 0.3, 0.4, 0.5, 0.4, 0.3, 0.2, 0.1)$. The input signal is white with unit power. The noise is simulated as uniformly white of power 100. A 10-weight tap-delay adaptive model is used.

TABLE 9.5 THE LMF ALGORITHM

Adaptive Filter	$\tilde{X}(k) = \hat{\mathbf{W}}^T(k)\mathbf{u}(k)$
Error Equation	$e(k) = d(k) - \hat{\mathbf{W}}^T(k)\mathbf{u}(k)$
Adaptive Control Equation	$\hat{\mathbf{W}}(k+1) = \mathbf{W}(k) + 2\mu \cdot e^3(k)\mathbf{u}(k)$
Initialization	$\hat{\mathbf{W}}(0) = \mathbf{0}$

The weights are initialized by adding to each of the true weights a random component drawn from a population having power equal to 0.56. Results of 10 independent experiments are averaged.

For the LMS, μ was chosen to be 18×10^{-4}, and the misadjustment is measured to be 0.959×10^{-2} (theoretical value 0.9×10^{-2}). For the LMF, μ is chosen to be 3×10^{-6}, and the misadjustment is 0.445×10^{-2} (theoretical value 0.386×10^{-2}).
□

9.3.2 The RLF Algorithm

The generalization of the RLS algorithm to the Nth-order cost function was presented by Figueras-Vidal, et al. [1988] and is based on the minimization of the sum of arbitrarily weighted N^{th} powers of the observed error. The advantage of this generalized scheme is as in the case of the LMF, i.e., less weight noise when compared to the quadratic case.

The performance index to be minimized is

$$\mathcal{E}(k) = \sum_{i=1}^{k} \lambda^{k-i} |e(i/k)|^N \tag{9.35}$$

where $0 < \lambda < 1$ is the forgetting factor and $e(i/k)$ is the error computed at time i using the parameters at time k as follows:

$$\begin{aligned} e(i/k) &= d(i) - \mathbf{u}^T(i)\hat{\mathbf{W}}(k) \\ &= \mathbf{u}^T(i)\mathbf{W} + n(i) - \mathbf{u}^T(i)\hat{\mathbf{W}}(k) \end{aligned} \tag{9.36}$$

with $\mathbf{u}(i), d(i)$ being the input vector and output scalar at time i, respectively; $\hat{\mathbf{W}}(k)$ is the weight vector at time k, $\mathbf{W}$ is the channel vector, and $n(i)$ is the additive noise (again the reader is referred to Figure 9.2).

Setting the derivative of $\mathcal{E}(k)$ with respect to $\hat{\mathbf{W}}(k)$ to zero, we obtain the general solution of the coefficient vector at time k :

$$\hat{\mathbf{W}}(k) = \mathbf{S}^{-1}(k) \cdot \mathbf{P}(k) \tag{9.37}$$

where

$$\mathbf{S}(k) = \sum_{i=1}^{k} \lambda^{k-i} \mathbf{u}(i) \mathbf{u}^T(i) / |e(i/k)|^{N-2} \tag{9.38}$$

and

$$\mathbf{P}(k) = \sum_{i=1}^{k} \lambda^{k-i} \mathbf{u}(i) [\mathbf{u}^T(i)\mathbf{W} + n(i)] |e(i/k)|^{N-2}. \tag{9.39}$$

Note that for $N = 2$, (9.37) reduces to the classical RLS of (9.16). Equation (9.38) can be written as

$$\mathbf{S}(k) = \mathbf{u}(k)\mathbf{u}^T(k) |e(k/k)|^{N-2} + \lambda \mathbf{S}(k-1) \tag{9.40}$$

from which, by applying the matrix inversion Lemma, we obtain

$$\mathbf{S}^{-1}(k) = \frac{1}{\lambda}[\mathbf{S}^{-1}(k-1) - \mathbf{g}(k)\mathbf{u}^T(k)\mathbf{S}^{-1}(k-1)] \tag{9.41.1}$$

with
$$g(k) = \frac{|e(k/k)|^{N-2}S^{-1}(k-1)u(k)}{\lambda + |e(k/k)|^{N-2}u^T(k)S^{-1}(k-1)u(k)}. \quad (9.41.2)$$

Note that in (9.41.2) the term $e(k/k)$ appears, and hence we cannot compute (9.41.1) recursively. Figueras-Vidal, et al. [1988] suggested the approximation $e(k/k) \approx e(k/k-1)$ in (9.41.2). As a result, (9.41.1) can be computed by utilizing the past values of $S^{-1}(k-1)$ and $\hat{W}(k-1)$, and the new input $u(k)$.

Substituting (9.41.1) into (9.37) we obtain the weight adaptation rule

$$\hat{W}(k) = \hat{W}(k-1) + g(k) \cdot e(k/k-1). \quad (9.42)$$

Table 9.6 summarizes the RLF algorithm which has been demonstrated to exhibit similar improvements over RLS with those of the LMF algorithm over LMS.

9.4 ADAPTIVE IIR ALGORITHMS BASED ON HOS

In this section we describe the adaptive algorithms introduced by Friedlander and Porat [1989] for the estimation of MA and ARMA processes. These algorithms can be used for adaptive IIR filtering of non-Gaussian processes, and they are based on higher-order statistics (HOS).

9.4.1 An Adaptive Algorithm for MA Processes Based on Third-Order Cumulants

Let the observed process be

$$Y(k) = \sum_{n=0}^{q} b(n)u(k-n); \; b(0) = 1 \quad (9.43)$$

where $\{u(k)\}$ is zero-mean, non-Guassian, third-order white noise, with skewness γ_3^u and variance γ_2^u. Recalling from Chapter 7 that the "diagonal" cumulants of $\{Y(k)\}$ are

$$c_3^y(\tau, \tau) = E\{Y(k)Y^2(k+\tau)\} = \gamma_3^u \sum_n b(n)b^2(n+\tau), \quad (9.44)$$

and the second-order cumulants

$$c_2^y(\tau) = E\{Y(k)Y(k+\tau)\} = \gamma_2^u \sum_n b(n)b(n+\tau), \quad (9.45)$$

we saw that Giannakis and Mendel [1989] established the following equation

TABLE 9.6 THE RLF ALGORITHM ($N = 4$)
NOTE THAT $\mathbf{P}(k) = S^{-1}(k)$

Adaptive Filter	$\tilde{X}(k) = \hat{\mathbf{W}}^T(k) \cdot \mathbf{u}(k)$
A priori Error	$e(k/k) = d(k) - \hat{\mathbf{W}}^T(k)\mathbf{u}(k)$
A posteriori Error	$e(k/k-1) = d(k) - \hat{\mathbf{W}}^T(k-1)\mathbf{u}(k)$
Gain Vector	$\mathbf{g}(k) = \dfrac{\|e(k/k-1)\|^2 \mathbf{P}(k-1)\mathbf{u}(k)}{\lambda + \|e(k/k-1)\|^2 \mathbf{u}^T(k)\mathbf{P}(k-1)\mathbf{u}(k)}$
Adaptive Control Equation	$\hat{\mathbf{W}}(k) = \hat{\mathbf{W}}(k-1) + \mathbf{g}(k)e(k/k-1)$
Matrix Inversion Lemma	$\mathbf{P}(k) = \lambda^{-1}\mathbf{P}(k-1) - \lambda^{-1}\mathbf{g}(k)\mathbf{u}^T(k)\mathbf{P}(k-1)$
Initialization	$\hat{\mathbf{W}}(0) = \mathbf{0}$ $\mathbf{P}(0) = \delta^{-1}\mathbf{I} (\delta > 0)$

$$\sum_{n=0}^{q} b^2(n) c_2^y(\tau - n) = \epsilon \sum_{n=0}^{q} b(n) c_3^y(\tau - n, \tau - n), \quad -q \leq \tau \leq 2q \tag{9.46}$$

with $\epsilon = \gamma_2^u / \gamma_3^u$, which can also be written in a matrix form as

$$\mathbf{M}\boldsymbol{\theta} = \mathbf{r} \tag{9.47}$$

where

$$\boldsymbol{\theta} = \left[\epsilon, \epsilon b(1), \ldots, \epsilon b(q), -b^2(1), \ldots, -b^2(q)\right]^T$$

$$\mathbf{M} = \begin{bmatrix} c_3^y(-q,q) & & \cdots & 0 & 0 & \cdots & 0 \\ c_3^y(-q+1,-q+1) & & & & c_2^y(-q) & \cdots & 0 \\ \vdots & & & c_3^y(-q,-q) & \vdots & & \\ & c_3^y(q,q) & & c_3^y(-q+1,-q+1) & c_2^y(q) & \cdots & c_2^y(-q) \\ & & & \vdots & & & \\ & 0 & \cdots & c_3^y(q,q) & 0 & \cdots & c_2^y(q) \end{bmatrix}$$

$$\underbrace{}_{q+1} \quad \underbrace{}_{q}$$

is a $(3q+1) \times (2q+1)$ matrix and

$$\mathbf{r} = [c_2^y(-q), \cdots, c_2^y(q), 0, \cdots, 0]^T$$

is a $(3q+1) \times 1$ vector.

When we are using estimates of $c_2^y(\tau)$ and $c_3^y(\tau,\tau)$, instead of their true values, (9.47) has the least-squares solution [Giannakis and Mendel, 1989]:

$$\hat{\boldsymbol{\theta}} = (\mathbf{M}^T\mathbf{M})^{-1}\mathbf{M}^T\mathbf{r}. \tag{9.48}$$

An adaptive way of solving (9.47) for $b(k)$'s has been introduced by Friedlander and Porat [1989]. Instead of $\mathbf{M}$ and $\mathbf{r}$, Friedlander and Porat proposed the use of the following matrices:

$$\hat{\mathbf{M}}(k) = \mathbf{Z}^T(k)[\mathbf{V}(k)|\mathbf{W}(k)];$$
$$\hat{\mathbf{r}} = \mathbf{Z}^T(k)\mathbf{y}(k), \tag{9.49}$$

where

$$\mathbf{V}(k) = \begin{bmatrix} 0 & & 0 \\ \vdots & \ddots & \vdots \\ Y^2(0) & & \vdots \\ \vdots & \ddots & Y^2(0) \\ \vdots & & \vdots \\ Y^2(k-q) & \cdots & Y^2(k-2q) \end{bmatrix} (k+1) \times (q+1), \tag{9.50.1}$$

$$\mathbf{Z}(k) = \begin{bmatrix} Y(0) & \cdots & 0 \\ \vdots & & \vdots \\ \vdots & & Y(0) \\ \vdots & \cdots & \\ Y(k) & \cdots & Y(k-3q) \end{bmatrix} (k+1) \times (3q+1), \tag{9.50.2}$$

$$\underbrace{}_{3q+1}$$

$$\mathbf{W}(k) = \begin{bmatrix} 0 & & 0 \\ \vdots & \ddots & \\ 0 & & 0 \\ Y(0) & & \vdots \\ \vdots & \ddots & Y(0) \\ \vdots & & \vdots \\ Y(k-q-1) & \cdots & Y(k-2q) \end{bmatrix}, \; Y(k) = \begin{bmatrix} 0 \\ \vdots \\ 0 \\ Y(0) \\ \vdots \\ Y(k-q) \end{bmatrix} (k+1) \times 1.$$

(9.51)

From the definitions of $\mathbf{Z}(k)$, $\mathbf{V}(k)$, and $\mathbf{W}(k)$, it turns out that the $(i,j)^{th}$ element of $\{\mathbf{Z}^T(k)\mathbf{V}(k)\}$ is

$$[\mathbf{Z}^T(k)\mathbf{V}(k)]_{i,j} = \sum_{n=\max(i,q+j)}^{k} Y(n-i)Y^2(n-q-j) \qquad (9.52)$$

and the one of $\{\mathbf{Z}^T(k)\mathbf{W}(k)\}$ is

$$[\mathbf{Z}^T(k)\mathbf{W}(k)]_{i,j} = \sum_{n=\max(i,q+j+1)}^{k} Y(n-i)Y(n-q-j-1) \qquad (9.53)$$

Equations (9.52) and (9.53) indicate that $[\mathbf{Z}^T(k)\mathbf{V}(k)]_{i,j}$ is a consistent estimate of $c_3^y(i-j-q, i-j-q)$, while $[(Z^T(k)\mathbf{W}(k)]_{i,j}$ is a consistent estimate of $c_2^y(i-j-q-1)$. Similarly,

$$[\mathbf{Z}^T(k)\mathbf{y}(k)]_i = \sum_{n=\max(i,q)}^{k} Y(n-i)Y(n-q) \qquad (9.54)$$

which is a consistent estimate of $c_2^y(i-q)$.

The least-squares solution of (9.48) becomes

$$\hat{\boldsymbol{\theta}}(k) = [\mathbf{X}^T(k)\mathbf{Z}(k)\mathbf{Z}^T(k)\mathbf{X}(k)]^{-1}\mathbf{X}^T(k)\mathbf{Z}(k)\mathbf{Z}^T(k)\mathbf{Y}(k) \qquad (9.55.1)$$

where

$$\mathbf{X}(k) = [\mathbf{V}(k)|\mathbf{W}(k)] \qquad (9.55.2)$$

A time-recursive algorithm based on (9.55) would converge to the exact solution of (9.47) as $k \to \infty$.

Friedlander [1984] proposed a recursive algorithm, the so-called ORIV (overdetermined recursive instrumental variable) algorithm for the updating of estimators which can be expressed in the form of (9.55). The ORIV algorithm based on third-order cumulants is summarized in Table 9.7.

The main advantage of using (9.55) instead of (9.47) and (9.48) is that (9.55) leads to faster updating when the number of data points increases.

The algorithm described in Table 9.7 can be extended to work with input white noise $\{u(k)\}$ that has symmetric probability density function. In order to achieve this, fourth-order cumulants, instead of third-order, should be utilized [Friedlander and Porat, 1989].

9.4.2 Adaptive Algorithms for ARMA Processes

The MA algorithms presented in the previous section have also been extended by Friedlander and Porat [1989] to ARMA processes. Let

$$Z(k) = -\sum_{n=1}^{p} a(n)Z(k-n) + \sum_{n=0}^{q} b(n)u(k-n); \ b(0) = 1, \quad (9.56)$$

where $\{u(k)\}$ is a zero-mean, non-Gaussian, third-order white noise process. The AR parameters can be estimated from the second-order sample moments. Let $\hat{a}(n)$ be the estimated AR parameters. Based on $\hat{a}(n)$ we may construct the process

$$Y(k) = Z(k) + \sum_{n=1}^{p} \hat{a}(n)Z(k-n) \quad (9.57)$$

and use it to estimate the MA parameters by utilizing the algorithm discussed in section 9.4.1. The adaptive ARMA filtering scheme is illustrated in Figure 9.4.

Example 9.2 [Friedlander and Porat, 1989]

The adaptive MA parameter estimation method is tested for a specific example. The ORIV algorithm is initialized exactly, i.e., $3q + 1$ data points are used to generate true initial values for $L(0), P(0)$, and $\theta(0)$ in Table 9.7. Once estimates for θ are obtained, the parameters $\{b(n), 1 \le n \le q\}$ are computed in one of the following ways:

(a) $\hat{b}^{(1)}(n) = \widehat{\epsilon b(n)}/\hat{\epsilon}$, where $\epsilon = \frac{\gamma_2^u}{\gamma_3^u}$, or

(b) $\hat{b}^{(2)}(n) = \sqrt{\widehat{b^2(n)}} sgn[\hat{b}^{(1)}(n)]$.

For this specific example, the data are generated using the simple MA(1) model $B(z) = 1 - 1.25z^{-1}$. The input noise is taken to be exponentially distributed with zero-mean, $\gamma_2^u = 1$, and $\gamma_3^u = 2$ (i.e., $\epsilon = 0.5$). One hundred Monte-Carlo simulations are performed with 400 data points each, and the time histories for $\hat{\epsilon}, \widehat{\epsilon b(1)}$, and $\hat{b}^2(1)$ means and $\pm \sigma$ bounds) are shown in Figure 9.5. In this figure, variance bounds are plotted versus iteration time index k. In Figure 9.6, $\hat{b}^{(1)}(1)$ and $\hat{b}^{(2)}(1)$ arc plotted vs. iteration index k.

The sample statistics after 400 data points are -1.2918 ± 0.1731 for $\hat{b}^{(1)}(1)$ and -1.266 ± 0.1397 for $\hat{b}^{(2)}(1)$. In Figure 9.6, $\hat{b}^{(2)}(1)$ is observed to converge better than $\hat{b}^{(1)}(1)$. This is accounted to the fact that $\hat{b}^{(1)}(1)$ is sensitive to errors in ϵ when $|\epsilon|$ is small. □

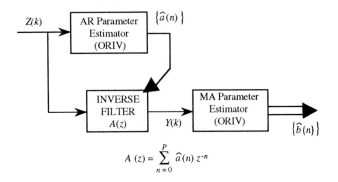

Figure 9.4 Adaptive ARMA filtering by two ORIV algorithms; one for AR and another for MA parameter estimation

9.5 ADAPTIVE LATTICE LINEAR PREDICTION USING CUMULANTS

A variation of the recursive instrumental variable method, originally introduced by Young [1984], for adaptive linear prediction was proposed by Swami and Mendel [1988]. By using nonconventional orthogonality conditions in the linear prediction problem, cumulant-based linear prediction equations were derived. This development led to a double lattice structure with one lattice excited by the observed process $\{Y(k)\}$ and the other by the instrumental process $\{Z(k)\}$. The lattices are coupled with order- and time-update equations, which can be related to the AR parameters.

Consider the forward linear prediction model for the AR process $\{Y(k)\}$. The linear prediction coefficients $a_{m,n}, \{n = 1, \ldots, m\}$ have to be calculated so that the error sequence

$$f_m(k) = Y(k) + \sum_{n=1}^{m} a_{m,n}(k) Y(k-n) \qquad (9.58)$$

is optimal in some sense. The prediction error sequence can be written in the component form

$$f_m(k) = \mathbf{a}_m^H(k) \mathbf{Y}_m(k); \quad a_{m,0}(k) = 1 \qquad (9.59)$$

where $\mathbf{a}_m(k) = [1, a_{m,1}(k), \ldots, a_{m,m}(k)]^T$ is the prediction error filter and $\mathbf{Y}_m(k) = [Y(k), \ldots, Y(k-m)]^T$ is the vector that contains the $m+1$ latest samples of $\{Y(k)\}$.

The prediction error filter is chosen so that

$$< Z(k-i), f_m(k) > = 0, \ i = 1, \ldots, m \qquad (9.60)$$

where $\{Z(k)\}$ is instrumental for $\{Y(k)\}$ (see [Young, 1984] for definitions). The

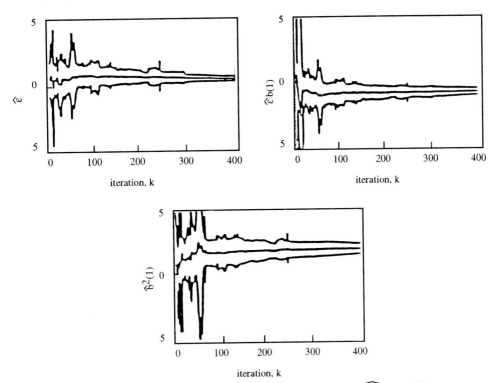

Figure 9.5 The mean and the ± variance bounds for $\hat{\epsilon}, \widehat{\epsilon b(1)}$ and $\hat{b}^{(2)}(1)$ of Example 9.2. ([Friedlander and Porat, 1989], © IEEE)

symbol $< \cdot >$ denotes time-averaging. Let

$$F_m(k) = < Z(k), f_m(k) > \tag{9.61}$$

and

$$\boldsymbol{\Phi}_{m+1}(k) = < \mathbf{Z}_m(k), \mathbf{Y}_m(k) > = \sum_{i=1}^{k} \lambda^{k-i} \mathbf{Z}_m(i) \mathbf{Y}_m^H(i) \tag{9.62}$$

where $\mathbf{Y}_m(k) = [Y(k), Y(k-1), \ldots, Y(k-m)]^T$, $\mathbf{Z}_m(k) = [Z(k), Z(k-1), \ldots, Z(k-m)]^T$, λ is the exponential forgetting factor ($0 < \lambda < 1$), and $\boldsymbol{\Phi}_{m+1}(k)$ is the deterministic cross-correlation matrix.

Combining (9.59), (9.60), and (9.61) with (9.62), we obtain

$$\boldsymbol{\Phi}_{m+1}(k) \mathbf{a}_m(k) = \begin{bmatrix} F_m(k) \\ \mathbf{0}_m \end{bmatrix} \tag{9.63}$$

which gives the prediction error filter of order m for the data $\{Y(k)\}$, so that the error sequence $f_m(k)$ is orthogonal to the associated process $\{Z(k)\}$. Similarly, we consider the m-coefficient backward linear predictor

$$\mathbf{c}_m(k) = [c_{m,0}(k), \ldots, c_{m,m-1}(k), 1]^T$$

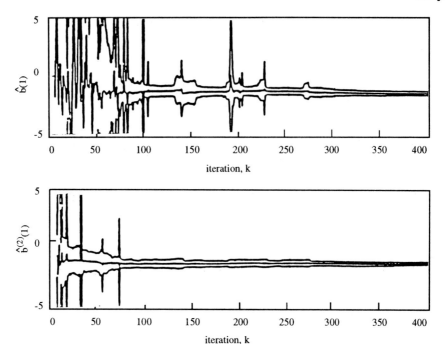

Figure 9.6 The mean and the ± variance bounds for $\hat{b}(1)$ and $\hat{b}^{(2)}(1)$ of Example 9.2 ([Friedlander and Porat, 1989] © 1989 IEEE).

for the data $\{Y(k)\}$. The backward prediction error is given by

$$b_m(k) = \mathbf{c}_m^H \mathbf{Y}_m(k); \quad c_{m,m}(k) = 1. \tag{9.64}$$

Again, the prediction error filter is chosen so that

$$< Z(k-i), b_m(k) > = 0; \ i = 0, \cdots, m-1. \tag{9.65}$$

Let

$$B_m(k) = < Z(k-m), b_m(k) > . \tag{9.66}$$

Combining the last two equations we obtain the following system of linear equations

$$\boldsymbol{\Phi}_{m+1}(k)\mathbf{c}_m(k) = \begin{bmatrix} \mathbf{0}_m \\ B_m(k) \end{bmatrix} \tag{9.67}$$

which gives the prediction error filter of order m for the data $\{Y(k)\}$, so that the error sequence is orthogonal to the associated instrumental process $\{Z(k)\}$.

Generally, $\{Z(k)\}$ is different from $\{Y(k)\}$. Swami and Mendel [1988] also posed the following problem: forward and backward linear prediction of $Z(k)$ subject to the error sequences being orthogonal to $Y(k)$. Note that if $Z(k) = Y(k)$, then this new problem is the same with the already described one.

TABLE 9.7 THE ORIV ALGORITHM BASED ON THIRD ORDER STATISTICS

Initialization	
$\mathbf{S}(0) = \mu[\mathbf{I}_{2q+1}\mathbf{0}]$	$(2q+1) \times (3q+1)$
$\mathbf{L}(0) = \mathbf{0};$	$(3q+1) \times 1$
$\mathbf{P}(0) = \mu^{-2} \cdot \mathbf{I}_{2q+1}$	$(2q+1) \times (2q+1)$
$\hat{\theta}(0) = \mathbf{0}$	$(2q+1) \times 1$
for $k = 0, 1, 2, \cdots$	
$\mathbf{Z}(k+1) = [Y(k+1) \cdots Y(k-3q+1)]^T$	$(3q+1) \times 1$
$\mathbf{X}(k+1) = [Y^2(k-q+1) \cdots Y^2(k-2q+1) \vert Y(k-q) \cdots Y(k-2q+1)]^T$	$(2q+1) \times 1$
$\mathbf{w}(k+1) = \mathbf{S}(k)\mathbf{z}(k+1)$	$(2q+1) \times 1$
$\mathbf{S}(k+1) = \lambda\mathbf{S}(k) + \mathbf{x}(k+1) \cdot \mathbf{z}^T(k+1)$	$(2q+1) \times (3q+1)$
$\Phi(k+1) = [\mathbf{w}(k+1)\vert\mathbf{X}(k+1)]$	$(2q+1) \times 2$
$\Lambda(k+1) = \begin{bmatrix} -\mathbf{z}^T(k+1)\mathbf{z}(k+1) & \lambda \\ \lambda & 0 \end{bmatrix}$	2×2
$\mathbf{K}(k+1) = \mathbf{P}(k)\Phi(k+1)[\Lambda(k+1) + \Phi^T(k+1)\mathbf{P}(k)\Phi(k+1)]^{-1}$	$(2q+1) \times 2$
$\mathbf{P}(k+1) = \lambda^{-2}[\mathbf{P}(k) - \mathbf{K}(k+1)\Phi^T(k+1)\mathbf{P}(k)]$	$(2q+1) \times (2q+1)$
$\mathbf{v}(k+1) = \begin{bmatrix} \mathbf{z}^T(k+1)\mathbf{L}(k) \\ Y(k-q+1) \end{bmatrix}$	(2×1)
$\mathbf{L}(k+1) = \lambda\mathbf{L}(k) + \mathbf{Z}(k+1)Y(k-q+1)$	$(3q+1) \times 1$
$\hat{\theta}(k+1) = \hat{\theta}(k) + \mathbf{K}(k+1)(\mathbf{v}(k+1) - \Phi^T(k+1)\hat{\theta}(k))$	$(2q+1) \times 1$

For the forward prediction of $Z(k)$ we seek a prediction error filter $\tilde{\mathbf{a}}_m(k)$ subject to the error sequence $\tilde{f}_m(k)$ being orthogonal to $Y(k)$. Hence,

$$\tilde{f}_m(k) = \tilde{\mathbf{a}}_m(k)\mathbf{Z}(k); \qquad \tilde{a}_{m,0}(k) = 1 \tag{9.68.1}$$

$$\Phi_m^H(k)\tilde{\mathbf{a}}_m(k) = \begin{bmatrix} \tilde{F}_m(k) \\ \mathbf{0}_m \end{bmatrix} \tag{9.68.2}$$

$$\tilde{F}_m(k) = <Y(k), \tilde{f}_m(k)> . \tag{9.68.3}$$

The backward prediction error filter for $Z(k)$ seeks $\tilde{\mathbf{c}}_m(k)$ subject to the error sequence $\tilde{b}_m(k)$ being orthogonal to $Y(k)$. This leads to

$$\tilde{b}_m(k) = \tilde{\mathbf{c}}_m^H(k)\mathbf{Z}_m(k); \qquad c_{m,m}(k) = 1 \tag{9.69.1}$$

$$\Phi_{m+1}(k)\tilde{c}_m(k) = \begin{bmatrix} 0_m \\ \tilde{B}_m(k) \end{bmatrix} \qquad (9.69.2)$$

$$\tilde{B}_m(k) = <Y(k-m), \tilde{b}_m(k)> . \qquad (9.69.3)$$

Note that since $Y(k)$ and $Z(k)$ have been interchanged in the two forward and the two backward prediction problems, the coefficient matrices in (9.64) and in (9.69.2) and the coefficient matrices in (9.67) and (9.69.2) are conjugate-transpose of one another.

If $Z(k) = Y(k)$ then $\Phi_{m+1}(k)$ is the autocorrelation matrix, the two prediction problems are identical, and the orthogonality conditions lead to the least-squares solution. If $Z(k) = Y(k)Y(k+k_0)$, then $\Phi_{m+1}(k)$ contains one-dimensional slices of the third-order cumulants. If $Z(k) = Y^3(k) - \gamma_2^y(k) \cdot Y(k)$, $\Phi_{m+1}(k)$ is formed using the diagonal slices of fourth-order cumulants.

For the second-order statistics case, i.e., $Z(k) = Y(k)$, there exist order-recursive adaptive lattice algorithms described in Haykin [1991], for obtaining time-varying estimates of the linear prediction coefficients. These algorithms exploit the Hermitian nature of the autocorrelation matrix. Generally, for $Z(k) \neq Y(k)$, $\Phi_{m+1}(k)$ will not be Hermitian. However, Swami and Mendel [1988] proposed a lattice structure to obtain the linear prediction coefficients. This structure consists of two lattices, one excited by $Y(k)$ and the other by $Z(k)$. The two lattices are coupled via time update equations for the reflection coefficients. For the derivation of the update equations the reader is referred to Swami and Mendel [1988]. The lattice algorithm is based on a single cumulant slice; hence it cannot guarantee identifiability. A p-slice lattice algorithm have been proposed by Fonollosa [1991]. The double lattice algorithm is summarized in Table 9.8 for an arbitrary choice of $Z(k)$.

Table 9.8 illustrates the adaptive lattice linear prediction algorithm, as well as reasonable choices for the instrumental variable $Z(k)$.

9.6 ADAPTIVE MA MODEL ESTIMATION VIA NONCAUSAL AR APPROXIMATIONS

An adaptive deconvolution and system identification scheme is presented in this section, which is applicable to the case of a non-Gaussian, zero-mean white noise driven, linear nonminimum phase FIR system. It is based on approximating the FIR system by noncausal AR models and using higher-order cumulants of the system output. The set of updated AR parameters is obtained by employing a gradient-type algorithm, and by using higher-order cumulants instead of time samples, of the signals involved in the adaptation process.

Assuming that the received signal $X(k)$ is the output of an unknown FIR system described by

$$X(k) = \sum_{i=-L_1}^{L_2} c(i)W(k+i) \qquad (9.70)$$

where $W(k)$ is non-Gaussian, white, noise with $E\{W(k)\} = 0, E\{W(k)W(k+\tau)\} = \gamma_2^w \delta(\tau)$ and $E\{W(k)W(k+\tau)W(k+\rho)\} = \gamma_3^w \delta(\tau,\rho)(\gamma_3^w \neq 0)$. The problem is to trace the magnitude and phase response of the system at each iteration from $X(k)$ as well as to reconstruct the input $W(k)$. Let us assume that the FIR system is nonminimum phase and it has no zeros on the unit circle. The magnitude, $|C(\omega)|$, and the phase, $\phi_c(\omega)$, of the FIR system are defined by

$$C(\omega) = |C(\omega)|e^{j\phi_c(\omega)}, \ |\omega| \leq \pi \qquad (9.71.1)$$

where

$$C(\omega) = \sum_{i=-L_1}^{L_2} c(i)e^{j\omega i}, \ |\omega| \leq \pi. \qquad (9.71.2)$$

Chiang and Nikias [1990;a] introduced an adaptive parametric method for both the reconstruction of the input sequence $W(k)$ and the identification of $|C(\omega)|$ and $\phi_c(\omega)$. This method employs a noncausal autoregressive (AR) model approximation of the FIR system and third-order cumulants of the signal $X(k)$. The magnitude and phase response estimates at each iteration are expressed directly in terms of the updated parameters of the noncausal AR model. As such, the input sequence may easily be reconstructed by deconvoluting the output sequence.

Given the received sequence, $X(k)$, we can always design a noncausal AR filter for deconvolution of the form

$$W(k) = X(k) + \sum_{i=-\infty}^{-1} a(i)X(k+i) + \sum_{i=1}^{\infty} a(i)X(k+i) \qquad (9.72.1)$$

where

$$C(z) = \frac{1}{\sum_{i=-\infty}^{\infty} a(i)z^i} = \sum_{i=-L_1}^{L_2} c(i)z^i \qquad (9.72.2)$$

Assuming $X(k)$ is stationary, from (9.72.1) we obtain the following third-order recursion equation for the noncausal AR filter:

$$\sum_{i=-\infty}^{\infty} a(i)c_3^x(m-i, n-i) = c_3^{wxx}(m,n) \qquad (9.73.1)$$

where

$$c_3^{wxx}(m,n) = E\{W(k)X(k+m)X(k+n)\} \qquad (9.73.2)$$

TABLE 9.8 THE ADAPTIVE LATTICE LINEAR PREDICTION ALGORITHM

for $k = 1, 2 \cdots$
for $m = 1, 2, \cdots M$

$$\Delta_{m-1}^{f}(k) = \lambda \Delta_{m-1}^{f}(k-1) + \frac{f_{m-1}^{*}(k)\tilde{b}_{m-1}(k-1)}{\gamma_{m-1}^{*}(k-1)} \quad (1)$$

$$\Delta_{m-1}^{b}(k) = \lambda \Delta_{m-1}^{b}(k-1) + \frac{b_{m-1}^{*}(k-1)\tilde{f}_{m-1}(k)}{\gamma_{m-1}^{*}(k-1)} \quad (2)$$

$$\Gamma_{f,m}(k) = -\frac{\Delta_{m-1}^{f}(k)}{B_{m-1}(k-1)} \quad (3)$$

$$\Gamma_{b,m}(k) = -\frac{\Delta_{m-1}^{b}(k)}{F_{m-1}(k)} \quad (4)$$

$$\mu(k) = \frac{F_{m-1}^{*}(k)}{B_{m-1}^{*}(k-1)} \quad (5)$$

$$f_m(k) = f_{m-1}(k) + \Gamma_{f,m}^{*}(k) b_{m-1}(k-1) \quad (6)$$

$$b_m(k) = b_{m-1}(k-1) + \Gamma_{b,m}^{*}(k) f_{m-1}(k) \quad (7)$$

$$\tilde{f}_m(k) = \tilde{f}_{m-1}(k) + \mu(k) \Gamma_{b,m}(k) \tilde{b}_{m-1}(k-1) \quad (8)$$

$$\tilde{b}_m(k) = \tilde{b}_{m-1}(k+1) + \mu^{-1}(k) \Gamma_{f,m}(k) \tilde{f}_{m-1}(k) \quad (9)$$

$$F_m(k) = F_{m-1}(k) - \frac{\Delta_{m-1}^{f}(k) \Delta_{m-1}^{b}(k)}{B_{m-1}(k-1)} \quad (10)$$

$$B_m(k) = B_{m-1}(k-1) - \frac{\Delta_{m-1}^{f}(k) \Delta_{m-1}^{b}(k)}{F_{m-1}(k)} \quad (11)$$

$$\gamma_m(k-1) = \gamma_{m-1}(k-1) - \frac{\tilde{b}_{m-1}^{*}(k-1) b_{m-1}(k-1)}{B_{m-1}^{*}(k-1)} \quad (12)$$

Initializations

$$\Delta_{m-1}^{f}(0) = \Delta_{m-1}^{b}(0) = 0 \quad (13)$$

$$F_{m-1}(0) = B_{m-1}(0) = \delta \ll 1 \quad (14)$$

$$f_0(k) = b_0(k) = Y(k) \quad (15)$$

$$\tilde{f}_0(k) = \tilde{b}_0(k) = Z(k) \quad (16)$$

$$F_0(k) = B_0(k) = \gamma - F_0(k-1) + Z(k) Y^{*}(k) \quad (17)$$

$$\gamma_0(k) = 1 \quad (18)$$

Choices for $Z(k)$

$$Z(k) = Y(k) \quad (19)$$

$$Z(k) = Y(k) Y(k + k_0) \quad (20)$$

$$Z(k) = Y^3(k) - \gamma_2^y \cdot Y(k) \quad (21)$$

$$c_3^x(m-i, n-i) = E\{X(k+i)X(k+m)X(k+n)\}. \qquad (9.73.3)$$

From (9.70) and (9.73) it follows that $c_3^{wxx}(m,n) = 0$ if $(m,n)\epsilon S$, where $S = \{m > L_1,$ or $m < -L_2,$ or $n > L_1,$ or $n < -L_2\}$. Hence, (9.73.1) becomes

$$\sum_{i=-\infty}^{\infty} a(i)c_3^x(m-i, n-i) = 0, \text{ for } (m,n)\epsilon S \qquad (9.74.1)$$

or, by assuming $a(0) = 1$ (no loss of generality),

$$c_3^x(m,n) - \hat{c}_3^x(m,n) = e(m,n) \qquad (9.74.2)$$

where

$$\hat{c}_3^x(m,n) = \sum_{i=-M}^{-1} a(i)c_3^x(m-i, n-i) - \sum_{i=1}^{N} a(i)c_3^x(m-i, n-i)$$

and

$$e(m,n) = -\sum_{i=-\infty}^{-(M+1)} a(i)c_3^x(m-i, n-i) - \sum_{i=N+1}^{\infty} a(i)c_3^x(m-i, n-i).$$

From (9.74.2), it follows that the input estimates $\hat{W}(k)$ are obtained by the deconvolution

$$\hat{W}(k) = \sum_{i=-M}^{N} a(i)X(k+i), \ a(0) = 1. \qquad (9.75)$$

The values M and N should be chosen to be sufficiently large [depending, of course, on the zero location of the polynomial in (9.72.2)], so that the errors $e(m,n)$ and thus the error $\mathcal{E}(k) = W(k) - \hat{W}(k)$ are small. The coefficients $\{a(i)\}, i = -M, \ldots, N$ can be estimated by minimizing a criterion of "goodness" which is defined as

$$\mathcal{E} = \sum_m \sum_n e^2(m,n); \ (m,n)\epsilon S_1 \subseteq S \qquad (9.76.1)$$

Figure 9.7 illustrates the rectangular region where $c_3^{wxx}(m,n) \neq 0$, its complement region S, and the subregion S_1 defined as

$$S_1 = \{(m,n) : m = L_1 + i, n = -L_2 + i + j, \ i = 1, 2, \ldots, N \ j = 0, 1, \ldots, N_s - 1\}$$

$$\{(m,n) : m = -L_2 - i, n = L_1 - i - j, i = 1, 2, \ldots, M, j = 0, 1, \ldots, N_s - 1\}$$

where N_s is an integer with value $N_s \leq (L_1 + L_2 + 1)$. Let us note that the maximum number of slices in Figure 9.7 is $N_s = L_1 + L_2 + 1$. On each slice, the maximum number of lags is $N_1 = L_1 + N$ for there are at most $(L_1 + L_2 + 1)(L_1 + L_2 + N + M)$ nontrivial equations in $(N + M)$ unknowns. In the sequel, we use $(M + N)N_s$ equations only. Thus, the criterion of goodness, $\mathcal{E}$, becomes

$$\mathcal{E} = (\mathbf{c}_3^x + \hat{\mathbf{c}}_3^x \mathbf{A})^T (\mathbf{c}_3^x + \hat{\mathbf{c}}_3^x \mathbf{A}) \qquad (9.76.2)$$

$$\hat{\mathbf{c}}_3^x = \begin{pmatrix} R_1(L_1, -L_2; -L_2, L_1) \\ R_1(L_1, -L_2 + 1; -L_2, L_1 - 1) \\ R_1(L_1, -L_2 + 2; -L_2, L_1 - 2) \\ \vdots \\ R_1(L_1, -L_2 + N_s - 1; -L_2, L_1 - N_s + 1) \end{pmatrix}, [N_s(N+M)] \times (N+M),$$

$$(9.77.1)$$

and $A = [a(N), a(N-1), \ldots, a(1), a(-1), \ldots, a(M)]^T$ is $(N+M) \times 1$,

$$\mathbf{c}_3^x = \begin{pmatrix} r_1(L_1, -L_2; -L_2, L_1) \\ r_1(L_1, -L_2 + 1; -L_2, L_1 - 1) \\ r_1(L_1, -L_2 + 2; -L_2, L_1 - 2) \\ \vdots \\ r_1(L_1, -L_2 + N_s - 1; -L_2, L_1 - N_s + 1) \end{pmatrix}, [N_s(N+M)] \times 1 \qquad (9.77.2)$$

$$R_1(g,h;u,v) = \begin{pmatrix} c_3^x(g,h) & \cdots & c_3^x(g+N-1,h+N-1) \\ c_3^x(g-1,h-1) & \cdots & c_3^x(g+N-2,h+N-2) \\ \vdots & & \vdots \\ c_3^x(g-N+1,h-N+1) & \cdots & c_3^x(g,h) \\ c_3^x(u-N-1,v-N-1) & \cdots & c_3^x(u-2,v-2) \\ \vdots & & \vdots \\ c_3^x(u-N-M+1,v-N-M+1) & \cdots & c_3^x(u-M,v-M) \\ c_3^x(u-N-M,v-N-M) & \cdots & c_3^x(u-M-1,v-M-1) \end{pmatrix}$$

$$\begin{pmatrix} c_3^x(g+N+1,h+N+1) & \cdots & c_3^x(g+N+M,h+N+M) \\ c_3^x(g+N,h+N) & \cdots & c_3^x(g+N+M-1,h+N+M-1) \\ \vdots & & \vdots \\ c_3^x(g+2,h+2) & \cdots & c_3^x(g+M+1,h+M+1) \\ c_3^x(u,v) & \cdots & c_3^x(u+M-1,v+M-1) \\ \vdots & & \vdots \\ c_3^x(u-M+2,v-M+2) & \cdots & c_3^x(u+1,v+1) \\ c_3^x(u-M+1,v-M+1) & \cdots & c_3^x(u,v) \end{pmatrix}, (N+M) \times (N+M).$$

$$(9.77.3)$$

$$r_1(g,h;u,v) = \begin{bmatrix} c_3^x(g+N,h+N) \\ c_3^x(g+N-1,h+N-1) \\ \vdots \\ c_3^x(g+1,h+1) \\ c_3^x(u-1,v-1) \\ \vdots \\ c_3^x(u-M+1,v-M+1) \\ c_3^x(u-M,v-M) \end{bmatrix}, (N+M) \times 1, \qquad (9.77.4)$$

and $r_1(g,h;u,\nu)$ = the gradient of $\mathcal{E}$ is

$$\nabla = \frac{\partial \mathcal{E}}{\partial \mathbf{A}}. \tag{9.78}$$

In practice all cumulants, $c_3^x(m,n)$, have to be replaced by their estimates at each iteration k, i.e., $\hat{c}_3^x(m,n;k)$. Consequently, the gradient of $\mathcal{E}$ becomes $\nabla(k)$ and the parameter update equation takes the form

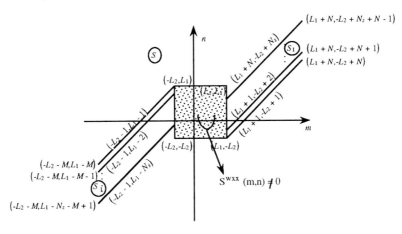

Figure 9.7 The region when $c_3^{wxx}(m,n) \neq 0$, its complement region S and the subregion S_1, which is chosen for the criterion of goodness

$$\hat{\mathbf{A}}(k+1) = \hat{\mathbf{A}}(k) - \mu(k)\hat{\nabla}(k) \tag{9.79}$$

where $\mu(k)$ is the step-size, properly chosen to guarantee convergence.

As long as the orders of the noncausal AR model (M,N) are chosen sufficiently large, the choice of (M,N) does not present any serious problem for the adaptive deconvolution scheme. However, the adaptive scheme may be sensitive to the choice of the MA orders L_1 and L_2. This is because (L_1, L_2) specify the region where $c_3^{wxx}(m,n) \neq 0$, and therefore, the criterion of goodness. Model order selection criteria for estimating (L_1, L_2) (assuming (M,N) have been chosen) based on third-order cumulants were discussed in Chapter 7.

Assuming we have the data $\{X(0), \cdots, X(k-1), X(k), X(k+1), \cdots, X(k+I_{\text{lag}}\}$ $I_{\text{lag}} = \max\{L_1 + N + M, N_s + N + M - L_2 - 1\}$, at the kth iteration, the third-order cumulants are estimated from

$$\hat{c}_3^x(m,n;k) = \frac{1}{k+1} \sum_{i=0}^{K} f^{k-i} X(i - I_{\text{lag}}) X(i - I_{\text{lag}} + m) X(i - I_{\text{lag}} + n) \tag{9.80}$$

where $\{0 < f \leq 1\}$ is a forgetting factor which controls the shape of the window of data taken into account at each iteration.

The adaptive deconvolution and system identification (ADSI) method is summarized in Table 9.9. The main advantage of the ADSI algorithm lies in its ability to estimate the "inverse system" directly, which can be used to perform the deconvolution operation directly on the output data. The price paid, however, is that the ADSI method cannot identify zeros on the unit circle.

For the nonstationary case (i.e., slowly time-varying FIR system), we may follow the variation of the coefficients by properly choosing the forgetting factor $\{f\}$ and applying the ADSI algorithm to track the coefficients of the noncausal AR adaptive filter.

The convergence analysis of the ADSI algorithms is discussed by Chiang and Nikias [1990a]. It is shown that all parameters converge to their true values as the number of iterations increases, provided that (M, N) are chosen large enough. Complexity issues are also discussed in the same paper.

Example 9.3

The performance of the ADSI algorithms is demonstrated for a specific example, for both stationary and nonstationary cases. In both cases the driving noise $\{W(k)\}$ is zero-mean, non-Gaussian (one-sided exponentially distributed) white.

Stationary Case (f=1)

The received data $\{X(k)\}$ are generated from (9.70) assuming a MA(1,1) process with parameters $\{c(1) = -0.7, c(0) = 1.21, c(-1) = -0.3\}$. Figure 9.8(a) illustrates the noncausal filter coefficients $a(i; k)\{i = 3, 2, 1, -1, -2\}$, of 5 different trials using the ADSI scheme with $L_1 = L_2 = 1, N = M = 10, N_s = 3$ and $\mu(k) = 0.9/tr[\hat{c}_3^x(k)]^T[\hat{c}_3^x(k)]$. Their average values are shown in Figure 9.8(b).

The true values of $\{\hat{a}(3; k), \hat{a}(2; k), \hat{a}(1; k), \hat{a}(-1; k), \hat{a}(-2; k)\}$ are $\{0.343, 0.49, 0.7, 0.3, 0.09\}$. From Figure 9.8(b) it appears that approximately 2000 iterations are required for the ADSI method to converge.

Nonstationary Case

The MA(1,1) model is employed in this case too, but with coefficients $c(i), \{i = 1, 0, -1\}$ that slowly vary with time, i.e.,

$$c(1; k) = -0.7 + 0.000025(k-1)$$

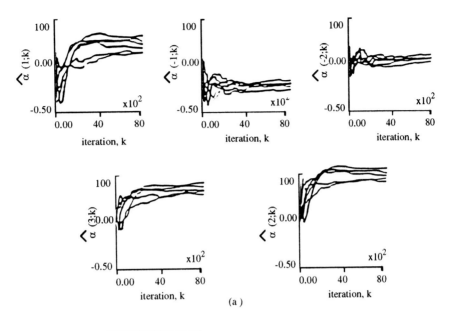
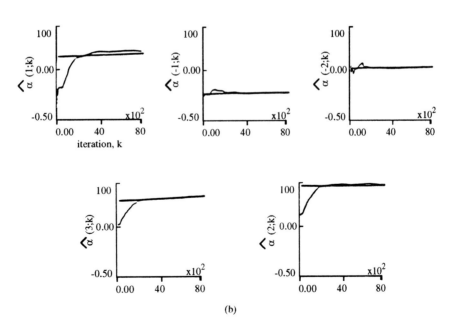

Figure 9.8 (a) The noncausal AR filter coefficients $\hat{a}(i;k)$, $\{i = 3, 2, 1, -1, -2\}$ of 5 different trials, (b) the average values of the coefficients $\hat{a}(i;k)$, $\{i = 3, 2, 1, -1, -2\}$.

$$c(-1;k) = -0.3 - 0.000025(k-1)$$

$$c(0;k) = 1. + c(1;k)c(-1;k)$$

In Figure 9.9 it is apparent that the ADSI method with $L_1 = L_2 = 1, N = M = 10, N_s = 3$ and $f = 0.999$, tracks very well the linear variation of the filter coefficients.

The ADSI method is easily extended to inputs $\{W(k)\}$ with symmetric p.d.f. (i.e., zero-skewness) by employing fourth-order instead of third-order cumulants. □

9.7 ADAPTIVE TIME DELAY ESTIMATION

The problem of estimating and tracking the time delay of a signal between two sensor measurements arises in many application fields such as sonar, radar, biomedicine, geophysics, communications, etc.

The conventional cross-correlation methods for estimating the time delay were discussed in Chapter 8. Parameter estimation approaches were also proposed, based on autocorrelations and cross-correlations, such as the least-squares and Wiener filtering methods [Chan, et al., 1980]. The least-mean-square (LMS) adaptive filter was applied to determine the time delay in a signal between two split-array outputs [Feintuch, et al., 1981]. A recursive least-squares algorithm and a peak detection algorithm have also been suggested for adaptive time delay estimation [Chan, et al., 1981]. Another adaptive technique that has been proposed was based on the assumption that the cross-correlation function is unimodal or periodically unimodal [Etter and Stearns, 1981]. Each of the aforementioned techniques has certain advantages and limitations depending upon the nature of the signal and noise sources.

In those practical application problems where the signal can be regarded as non-Gaussian and the additive noises can be regarded as zero-mean, stationary Gaussian processes, the time delay can be estimated in higher-order spectrum domains, such as the cross-bispectrum or cross-trispectrum, as it was discussed in Chapter 8.

In this section, we present the adaptive time delay estimation method proposed by Chiang and Nikias [1990b], which is based on parametric modeling of higher-order cross-cumulants, and as such, can be applied in non-Gaussian signal environments. Let $X(k)$ and $Y(k)$ be the two sensor measurements satisfying

$$\begin{aligned} X(k) &= S(k) + W_1(k) \\ Y(k) &= S(k - D(k)) + W_2(k) \end{aligned} \qquad (9.81)$$

where $S(k)$ is the unknown signal, $D(k)$ is the delay that changes slowly with time, and $W_1(k)$ and $W_2(k)$ are unknown noises. The problem assumptions are that $S(k)$

TABLE 9.9 SUMMARY OF THE ADSI METHOD: ADAPTIVE DECONVOLUTION AND SYSTEM IDENTIFICATION METHOD

Initialization

for $(g, h) = (L_1, -L_2), (L_1, -L_2 + 1), \cdots, (L_1, -L_2 + N_s - 1)$ and

$i = N - 1, N - 2, \cdots, -N - M,$

$c_3^x(g - i, h - i) = 0;$

for $(u, v) = (-L_2, L_1), (-L_2, L_1 - 1), \cdots, (-L_2, L_1 - N_s + 1)$ and

$i = N + M, N + M - 1, \ldots, -M + 1$

$c_3^x(u - i, v - i) = 0;$

$\hat{\mathbf{A}}(0) = \mathbf{A}_I$, where $\mathbf{A}_I$ is an initial guess $(N + M) \times 1$.

For $k = 0, 1, 2, \ldots; k_1 = k - \max[L_1 + N + M, N_s + N + M - L_2 - 1]$

for $(g, h) = (L_1, -L_2), (L_1, -L_2 + 1), \ldots, (L_1, -L_2 + N_s - 1)$ and $i = N - 1, N - 2, \ldots, -N - M$

$c_3^x(g - i, h - i) = f c_3^x(g - i, h - i) + X(k_1)X(k_1 + g - i)X(k_1 + h - i) = \hat{c}_3^x(g - i, h - i; k)$

for $(u, v) = (-L_2, L_1), (-L_2, L_1 - 1), \ldots, (-L_2, L_1 - N_s + 1)$ and

$i = N + M, N + M - 1, \cdots, -M + 1$

$c_3^x(u - i, v - i) = f c_3^x(u - i, v - i) + X(k_1)X(k_1 + u - i)X(k_1 + v - i) = \hat{c}_3^x(u - i, v - i; k).$

Form $\hat{\mathbf{c}}_3^x(k)$ and $\mathbf{c}_3^x(k)$ using (9.77.1) and (9.77.2), respectively.

$\hat{\nabla}(k) = 2[[\hat{\mathbf{c}}_3^x(k)]^T [\hat{\mathbf{c}}_3^x(k)] \hat{\mathbf{A}}(k) + [\hat{\mathbf{c}}_3^x(k)]^T \cdot \hat{\mathbf{c}}_3^x(k)]$

$\hat{\mathbf{A}}(k+1) = \hat{\mathbf{A}}(k) - \mu(k)\hat{\nabla}(k), 0 < \mu(k) < \frac{1}{\lambda_{\max}} \lambda_{\max}$: maximum eigenvalue of $[\hat{\mathbf{c}}_3^x(k)]^T \hat{\mathbf{c}}_3^x(k)$

---

is a zero-mean non-Gaussian stationary random process with nonzero skewness, and $W_1(k)$ and $W_2(k)$ are zero-mean Gaussian, possibly correlated stationary random processes, independent of $S(k)$. The goal is to track the time delay $D(k)$ at each time instant from the two measurements $X(k)$ and $Y(k)$.

For the stationary case, we established in Chapter 8 that equation (8.62) holds. Consequently, the adaptive scheme, to be presented, estimates the time delay $D(k)$ based on the adaptive parametric model equation

$$c_3^{yxx}(\tau, \rho; k) = \sum_{i=-P}^{P} a(i; k) c_3^x(\tau + i, \rho + i; k) \qquad (9.82)$$

where k is the iteration index, $c_3^{yxx}(\tau, \rho; k)$ and $c_3^x(\tau, \rho; k)$ are the cross third-order cumulants and third-order cumulants at iteration k, respectively.

The gradient-type algorithm is then employed to develop the adaptive time delay estimation scheme as follows. First we take the criterion

$$\hat{\xi} = \sum_{n=0}^{N_s-1} \sum_{\rho=-P}^{P} \left[\sum_{i=-P}^{P} \hat{a}(i,k) \hat{c}_3^x(\rho+n+i, \rho+i; k) - \hat{c}_3^{yxx}(\rho+n, \rho; k) \right]^2 \quad (9.83)$$

at the k-th iteration, where the model parameters form the vector

$$\hat{\mathbf{A}}(k) = [\hat{a}(-P; k), \hat{a}(-P+1; k), \ldots \hat{a}(0; k), \ldots, \hat{a}(P-1; k), \hat{a}(P; k)]^T,$$

$\hat{a}(i; k)$ is the estimate of $a(i)$ at the k-th iteration, and $\hat{c}_3^x(\tau, \rho), \hat{c}_3^{yxx}(\tau, \rho)$ are estimates of $c_3^x(\tau, \rho)$ and $c_3^{yxx}(\tau, \rho)$ at the k-iteration, given by

$$\hat{c}_3^x(\tau, \rho; k) = \frac{1}{k+1} \sum_{i=0}^{k} f^{k-i} X(i+2P) X(i+2P+\tau) X(i+2P+\rho) \quad (9.84.1)$$

$$\hat{c}_3^{yxx}(\tau, \rho; k) = \frac{1}{k+1} \sum_{i=0}^{k} f^{k-i} Y(i+2P) X(i+2P+\tau) X(i+2P+\rho) \quad (9.84.2)$$

NONSTATIONARY CASE

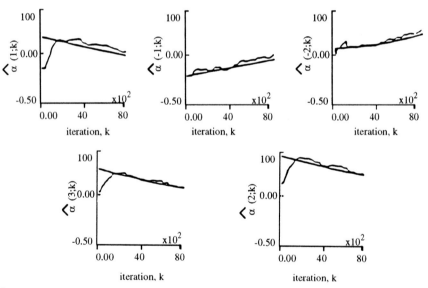

Figure 9.9 Nonstationary Case: The noncausal AR filter coefficients $\hat{a}(i; k), \{i = 3, 2, 1, -1, -2\}$; averaged over 5 different trials.

where $\{0 < f \leq 1\}$ is the forgetting factor. We choose $f = 1$ for the stationary case.

The parameter update equation takes the form

$$\hat{\mathbf{A}}(k+1) = \hat{\mathbf{A}}(k) - \mu(k) \cdot \hat{\nabla}(k) \quad (9.85)$$

where $\mu(k)$ is properly chosen to satisfy the convergence requirement and $\hat{\nabla}(k) = \partial \hat{\xi}(k)/\partial \hat{\mathbf{A}}(k)$ is the gradient vector.

Table 9.10 summarizes the cumulant-based adaptive time delay estimation (CATDE) method for both stationary and nonstationary cases. It also provides simple recursive expressions for the computation of cumulants at each iteration. Its block diagram configuration is illustrated in Figure 9.10.

The CATDE method is the LMS algorithm with the third-order cumulant estimates as its inputs instead of the data itself. Therefore, the convergence rate of the CATDE method depends on the step size, $\mu(k)$, the order of the parametric model, P, and the eigenvalue spread.

Example 9.4

The performance of the CATDE algorithm is demonstrated for some specific simulation examples. The performance of the cross-correlation-based adaptive time delay estimation (CCATDE) method will also be considered. For the reader's convenience the CCATDE method is summarized in Table 9.11.

The signal $s(k)$ is zero-mean non-Gaussian (one-sided exponentially distributed) white process, generated by the GGEXN subroutine of the IMSL library with $E\{S^2(k)\} = 1$ and $E\{S^3(k)\} = 2$. The noise processes are Gaussian correlated. The noise $W_1(k)$ is white, zero mean with unit variance, and $W_2(k)$ is generated from $W_1(k)$ using the FIR system equation.

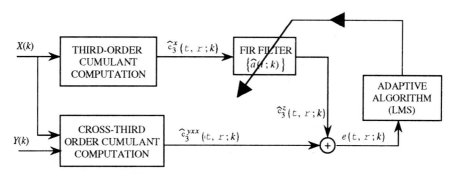

Figure 9.10 The configuration of the cumulant-based adaptive time delay estimation (CATDE) method

The range of search for the time delay is from -30 to $30(30)$ with step one. The signal $S(k)$ is a zero mean non-Gaussian (one-sided exponentially distributed) white process, generated by the GGEXN subroutine of the IMSL library with $E\{S^2(k)\} = $

1 and $E\{S^3(k)\} = 2$. The noise processes are Gaussian and spatially correlated. The noise $W_1(k)$ is white, zero-mean with unit variance, and $W_2(k)$ is generated from $W_1(k)$ using the FIR system equation

$$W_2(k) = \sum_{i=0}^{10} b(i)W(k-i)$$

with $\{b(i), i = 0, \ldots, 10\} = \{0.0337, 0.0675, 0.1012, 0.1349, 0.1514,$ $0.9466, 0.1514, 0.1181, 0.0843, 0.0506, 0.0169\}$.

The time delay $D(k)$ varies slowly with time, i.e.,

$$D(k) = \begin{cases} 16, & 0 \leq k \leq 4000 \\ 17, & 4000 < k \leq 8000 \\ 14, & 8000 < k \leq 12000. \end{cases}$$

The results obtained by the CATDE method with $f = 0.995$ are illustrated in Figures 9.11(a), (b) and (c), while the corresponding results obtained by the CCATDE method with the same f are shown in Figures 9.11(d), (e) and (f). From Figure 9.11 we see that the CATDE and CCATDE methods do follow the variation of the time delay. However, the CCATDE method does not suppress the correlated Gaussian noises and exhibits a stationary peak due to the nonzero value of $\hat{a}(5; k)$.

For comparison purposes the required complexity for the CATDE and the CCATDE methods is illustrated in Table 9.12. Apparently, the CATDE method requires $3N_s$ times more multiplications than the CCATDE method. Hence, the improved performance of the CATDE against the cross-correlation-based method is achieved at the expense of more computations. □

9.8 ADAPTIVE BLIND EQUALIZATION ALGORITHMS

The blind deconvolution, or equalization problem, deals with the reconstruction of the input sequence given the output of a linear system and statistical information about the input. Blind deconvolution algorithms are essentially adaptive filtering algorithms designed in such a way that they do not need the external supply of a desired response to generate the error signal in the output of the adaptive equalization filter. In other words, the algorithm is blind to the desired response. However, the algorithm itself generates an estimate of the desired response by applying a nonlinear transformation on sequences involved in the adaptation process. There are three important families of blind equalization algorithms depending on where the nonlinear transformation is being applied on the data. These are:

(i) The Bussgand algorithms, where the nonlinearity is in the output of the adaptive equalization filter,

(ii) The Polyspectra algorithms, where the nonlinearity is in the input of the adaptive equalization filter, and

(iii) The algorithms where the nonlinearity is inside the equalization filter, i.e., nonlinear filter (e.g., Volterra) or neural network.

This section gives on overview of the first two families of blind equalization algorithms and their properties. However, emphasis in the discussion is placed on the polyspectra-based blind equalizers.

Problem Definition

Let us consider a discrete-time linear transmission channel with impulse response, $f(k)$, which is unknown and possibly time-varying. The input data, $X(k)$, are assumed to be independent and identically distributed (i.i.d.) random variables with a non-Gaussian probability density function, with zero-mean and variance, $E\{X^2(k)\} = \gamma_2^x$. Initially the noise will not be taken into account in the output of the channel. Then, the received sequence, $Y(k)$, (see Figure 9.12) is

$$Y(k) = f(k) * X(k) = \sum_i X(k-i)f(i). \tag{9.86}$$

The problem is to restore $X(k)$ from the received sequence $Y(k)$, or equivalently, to identify the inverse filter (equalizer), $u(k)$, of the channel.

From Figure 9.12, we see that the output sequence $\tilde{X}(k)$ of the equalizer is given by

$$\begin{aligned}\tilde{X}(k) &= u(k) * Y(k) \\ &= u(k) * (f(k) * X(k)) \\ &= u(k) * f(k) * X(k).\end{aligned} \tag{9.87}$$

To achieve

$$\tilde{X}(k) = X(k-D)e^{j\Theta} \tag{9.88}$$

where D is a constant delay and Θ is a constant phase shift, it is required that

$$u(k) * f(k) = \delta(k-D)e^{j\Theta}, \tag{9.89}$$

where $\delta(k)$ is the delta function. Taking the Fourier transform of (9.89) we obtain

$$U(\omega)F(\omega) = e^{j(\Theta-\omega D)}.$$

TABLE 9.10 THE CUMULANT-BASED ADAPTIVE TIME DELAY ESTIMATION (CATDE) METHOD

Initialization:

$c_3^x(\rho + k, \rho) = 0; k = 0, 1, \ldots, N_s - 1; \rho = -2P, \ldots, 2P$

$c_3^{yxx}(\rho + k, \rho) = 0; k = 0, 1, \ldots, N_s - 1; \rho = -P, \ldots, P$

$\hat{\mathbf{A}}(0) = \mathbf{A}_I$, $\mathbf{A}_I$: initial guess

for $k = 0, 1, 2, \ldots, k_1 = k + 2P$

for $n = 0, 1, \ldots, N_s - 1; \rho = -2P, \ldots, 2P$
$c_3^x(\rho + n, \rho) = fc_3^x(\rho + n, \rho) + X(k_1)X(k_1 + \rho + n)X(k_1 + \rho)$

$c_3^x(\rho + n, \rho; k) = \frac{1}{k+1} c_3^x(\rho + n, \rho)$

for $n = 0, 1, \cdots, N_{s-1}, \rho = -P, \ldots, P$

$c_3^{yxx}(\rho + n, \rho) = fc_3^{yxx}(\rho + n, \rho) + Y(k_1)Y(k_1 + \rho + n)X(k_1 + \rho)$

$c_3^{yxx}(\rho + n, \rho; k) = \frac{1}{k+1} c_3^{yxx}(\rho + k, \rho)$

$\mathbf{c}_3^x(k) = [\hat{\mathbf{R}}_1^T(0), \hat{\mathbf{R}}_1^T(1), \ldots, \hat{\mathbf{R}}_1^T(N_s - 1)]^T$

$\hat{\mathbf{R}}_1(n) = \begin{bmatrix} \hat{c}_3^x(n, 0; k) & \hat{c}_3^x(n+1, 1; k) & \cdots & \hat{c}_3^x(n+2P, 2P; k) \\ \hat{c}_3^x(n-1, -1; k) & \hat{c}_3^x(n, 0; k) & \cdots & \hat{c}_3^x(n+2P-1, 2P-1; k) \\ \vdots & \vdots & \vdots & \vdots \\ \hat{c}_3^x(n-2P, -2P; k) & \hat{c}_3^x(n-2P+1, -2P+1; k) & \cdots & \hat{c}_3^x(n, 0; k) \end{bmatrix}$

$\mathbf{c}_3^{yxx}(k) = [\hat{\mathbf{R}}_2^T(0), \hat{\mathbf{R}}_2^T(1), \ldots, \hat{\mathbf{R}}_2^T(N_s - 1)]^T$

where $\hat{\mathbf{R}}_2(n) = [\hat{c}_3^{yxx}(n+P, P; k), \hat{c}_3^{yxx}(n+P-1, P-1; k), \ldots, \hat{c}_3^{yxx}(n-P, -P; k)]^T$

$\hat{\mathbf{V}}(k) = 3[[\hat{\mathbf{c}}_3^x(k)]^T [\hat{\mathbf{c}}_3^x(k)] \hat{\mathbf{A}}(k) - [\hat{\mathbf{c}}_3^x(k)]^T \cdot \hat{\mathbf{c}}_3^{yxx}(k)]$

$\hat{\mathbf{A}}(k + 1) = \hat{\mathbf{A}}(k) + \mu(k)\hat{\mathbf{V}}(k), 0 < \mu(k) < 1/tr[(\hat{\mathbf{c}}_3^x(k))^T \mathbf{c}_3^x(k)]$

---

*$N_s = 3$ for the calculation

TABLE 9.11 THE CROSS-CORRELATION-BASED ADAPTIVE TIME DELAY ESTIMATION (CCATDE) METHOD

Initialization:

$c_2^x(\tau) = 0; \tau = 0, 1, \ldots, 2P$

$c_2^{xy}(\tau = 0; \rho = -P, -P+1, \ldots, P$

$\hat{\mathbf{A}}(0) = \mathbf{A}_I$, $\mathbf{A}_I$: initial guess $(2+p) \times 1$,

for $k = 0, 1, 2, \ldots; k_1 = k + 2P$,

$c_2^x(\tau) = fc_2^x(\tau) + X(k_1)X(k_1 - \tau)$, $\tau = 0, 1, \ldots, 2P$

$c_2^{xy}(\tau) = fc_2^{xy}(\tau) - X(k_1)Y(k_1 - \tau)$, $\tau = -P, -P+1, \ldots, P$

$c_2^x(\tau; k) = \frac{1}{k+1} c_2^x(\tau)$, $\rho = 0, 1, \ldots, 2P$

$c_2^{xy}(\rho; k) = \frac{1}{k+1} c_2^{xy}(\rho)$, $\rho = -P, -P+1, \ldots, P$

$$\hat{\mathbf{c}}_2^x(k) = \begin{bmatrix} c_2^x(0;k) & c_2^x(1;k) & \cdots & c_2^x(2P;k) \\ c_2^x(1;k) & c_2^x(0;k) & & c_2^x(2P-1;k) \\ \vdots & \vdots & \vdots & \vdots \\ c_2^{xy}(2P;k) & c_2^x(2P-1;k) & \vdots & c_2^x(0;k) \end{bmatrix}$$

$\hat{\mathbf{c}}_2^{xy} = [c_2^{xy}(-P;k)c_2^{xy}(-P+1;k), \ldots, c_2^{xy}(P,k)]^T$
$\nabla(k) = 2[\hat{\mathbf{c}}_2^x(k)\hat{\mathbf{A}}(k) - \hat{\mathbf{c}}_2^{xy}(k)]$

$\mathbf{A}(k+1) = \hat{\mathbf{A}}(k) - \mu(k)\nabla(k)$, $0 < \mu(k) < \frac{1}{tr[\hat{\mathbf{C}}_2^x(k)]}$

TABLE 9.12 THE NUMBER OF MULTIPLICATIONS PER ITERATION OF THE CCATDE AND CATDE METHODS

Method	Multiplications/Iteration	$P = 30$ (for example)
CCATDE (2nd-order statistics)	$(2P+1)^2 + 3(2P+1) + 3$	3,907
CATDE* (3rd-order statistics)	$3N_s(2P+1)^2 + (12N_s+1) \cdot (2P+1)^{-3}$	35,743

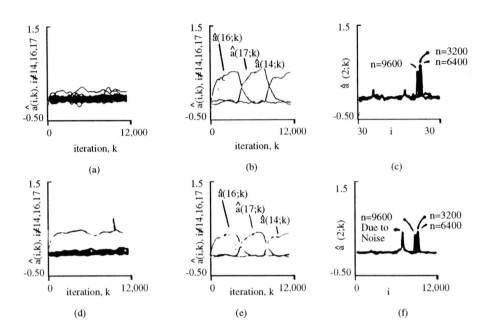

Figure 9.11 Time-Varying Time Delay: white signal in spatially correlated Gaussian noises. The values of the coefficients $\hat{a}(i;k)\{i = -30,\ldots,0,\ldots 30\}$ for each iteration k for (a) the CATDE method (d) the CCATDE method. The values of $\hat{a}(16;k), \hat{a}(17,k)$ and $\hat{a}(14,k)$ for each iteration k for (b) the CATDE method and (e) the CCATDE method. The values of $\{a(i;k)\}$ at $k = 3200, 6400$ and 9600 for (c) the CATDE method and (f) the CCATDE method. ([Chiang and Nikias, 1990b] © 1990 IEEE).

Hence, the objective of the equalizer is to achieve a transfer function

$$U(\omega) = \frac{1}{F(\omega)} e^{j(\Theta - \omega D)}. \tag{9.90}$$

In general, D and Θ are unknown. However, the constant delay D does not affect the reconstruction of the original input sequence $X(k)$. The constant phase Θ can be removed by a decision device. As such, in the sequel it will assumed that $D = 0$ and $\Theta = 0$.

9.8.1 The Bussgang Techniques

These algorithms are iterative deconvolution schemes that apply a memoryless non-linearity in the output of the equalization filter in order to generate the "desired response" at each iteration [Bellini, 1986; Bellini and Rocca, 1986; Benveniste and Goursat, 1984; Godfrey and Rocca, 1981].

Let us assume that an initial guess for the impulse response of the inverse filter

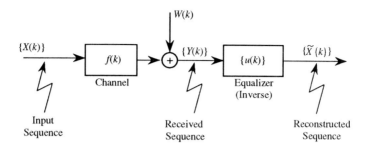

Figure 9.12 Block diagram of a baseband communication system subject to additive noise

(equalizer), $u_g(k)$, has been selected. Then, from (9.89) we have

$$u_g(k) * f(k) = \delta(k) + \epsilon(k) \tag{9.91}$$

where $f(k)$ is the impulse response of the transmission channel and $\epsilon(k)$ accounts for the difference (error) between the actual value, $u(k)$, and the initial guess $u_g(k)$. Convolving $u_g(k)$ with the received sequence, $Y(k)$, we obtain

$$\tilde{X}(k) = Y(k) * u_g(k) = X(k) * f(k) * u_g(k). \tag{9.92}$$

Combination of (9.92) with (9.91) yields

$$\begin{aligned} \tilde{X}(k) &= X(k) * [\delta(k) + \epsilon(k)] \\ &= [X(k) * \delta(k)] + [X(k) * \epsilon(k)] \\ &= X(k) + n(k) \end{aligned} \tag{9.93}$$

where $n(k) = X(k) * \epsilon(k)$ is the "convolutional noise," namely, the residual intersymbol interference (ISI) arising from the difference between $u_g(k)$ and $u(k)$. The problem now is in finding the "best" estimate of $X(k)$, namely $d(k)$, by utilizing the deconvolved sequence $\tilde{X}(k)$. The mean-square error (MSE) criterion may be employed to determine the best estimate of $X(k)$ given $\tilde{X}(k)$. Since the transmitted sequence $X(k)$ obeys a non-Gaussian probability density function, the MSE estimate is a nonlinear transformation of $\tilde{X}(k)$. Generally, the "best" estimate, $d(k)$, is given by

$$d(k) = g[\tilde{X}(k)] \text{ (memoryless)} \tag{9.94}$$

$$d(k) = g[\tilde{X}(k), \tilde{X}(k-1), \ldots, \tilde{X}(k-1)](m-\text{th order memory}) \tag{9.95}$$

where $g[\,\cdot\,]$ is a nonlinear function with or without memory. The $d(k)$ is feedback to the adaptive equalization filter as shown in Figure 9.13. From this figure, it is apparent that the nonlinearity $g[\,\cdot\,]$ appears in the output of the equalization filter. In summary, a well-treated classical estimation problem is formulated as follows.

$$\tilde{X}(k) = X(k) + n(k) \tag{9.96}$$

where

(i) $n(k)$ is Gaussian. Note that if the memory of $\epsilon(k)$ is long enough, the central limit theorem makes the Gaussianity assumption about $n(k)$ reasonable.

(ii) $X(k)$ are i.i.d., generally non-Gaussian, with known pdf. In digital communications the $\{X(k)\}$ are usually equi-probable discrete signal points.

(iii) $X(k), n(k)$ are assumed to be independent.

Given $\tilde{X}(k)$, the MSE estimate of $X(k)$ is the mean of the a posteriori density [Van Tress, 1968], i.e.,

$$d(k) = \int_{-\infty}^{\infty} x P_{x/\tilde{x}}(X/\tilde{X}) dx = E\{X(k)/\tilde{X}(k)\} \tag{9.97.1}$$

where

$$P_{x/\tilde{x}}(X/\tilde{X}) = \frac{P_{\tilde{x}/x}(\tilde{X}/X) \cdot P_x(X)}{P_{\tilde{x}}(\tilde{X})} \tag{9.97.2}$$

is the a posteriori density and $P_{x/\tilde{x}}(\tilde{X}/X) \sim N(X(k), \gamma_2^n)$ with γ_2^n being the variance of $n(k)$. The a priori density, $P_x(X)$, is the pdf of $X(k)$, and $P_{\tilde{x}}(\tilde{X})$ behaves as a normalization constant in the integral of (9.97). If $X(k)$ is zero-mean Gaussian with variance γ_2^x, i.e., $P_x(X) \sim N(0, \gamma_2^x)$, (9.97) reduces to

$$d(k) = \frac{\gamma_2^x}{\gamma_2^x + \gamma_2^n} \tilde{X}(k), \tag{9.98}$$

which, in turn, implies that $g[\,\cdot\,]$ is a constant. In general, for non-Gaussian $X(k), g[\,\cdot\,]$ is a memoryless nonlinear function that can be determined exactly from (9.94) and (9.98).

Table 9.13 illustrates the general form of Bussgang algorithms which are all based on LMS adaptation formulas. The main difference between the Bussgang type algorithms lies in the choice of the memoryless nonlinearity. Special cases of the Bussgang family algorithms include methods introduced by Sato [1975], Benveniste-Goursat [1984], Godard [1980], and Picchi and Prati [1987]. The algorithm by Picchi and Prati is known as the Stop-and-Go method. The blind equalizer most widely tested in practice is probably the Godard, also known as the Constant Modulus Algorithm (CMA) [Treichler and Agee, 1983].

From Table 9.13 and Figure 9.13 it is apparent that the output sequence of the nonlinear function, i.e., $d(k)$, "plays the role" of the desired response or training sequence. It is also apparent the Bussgang algorithms are simple to implement and understand, and it may be viewed as a modification of the original LMS algorithm (the desired response of the original LMS adaptation is a memoryless transformation of the transversal filter output). As such, it is expected that the algorithms will have convergence that will depend on the eigenvalue spread of the autocorrelation matrix of the observed data $\{Y(k)\}$.

In Table 9.13, the LMS adaptation equation for the equalizer coefficients is given by equation (4). If we compute the expected values of both sides of (4) in Table 9.13, we have

$$E\{\mathbf{u}(k+1)\} = E\{\mathbf{u}(k)\} + \mu \cdot E\{\mathbf{Y}(k)g^{*(k)}[\tilde{X}(k)]\} - \mu \cdot E\{\tilde{Y}(k)\tilde{X}^*(k)\} \quad (9.99)$$

TABLE 9.13 SUMMARY OF ADAPTIVE BUSSGANG TECHNIQUES
(LMS Type)

$\mathbf{u}(k) = [u_1(k),\ldots,u_N(k)]^T$	equalizer taps	
$\mathbf{Y}(k) = [Y(k),\ldots,Y(k-N+1)]^T$	Input to the equalizer block of data	
At iteration $(k), k = 1,2\ldots$		
$\tilde{X}(k) = \mathbf{u}^H(k)\mathbf{Y}(k)$	Equalizer output or reconstructed sequence	(1)
$d(k) = g^{(k)}[\tilde{X}(k)] = g^{(k)}[\mathbf{u}^H(k)\mathbf{Y}(k)]$	output of nonlinearity	(2)
$e(k) = d(k) - \tilde{X}(k)$	error sequence	(3)
$\mathbf{u}(k+1) = \mathbf{u}(k) + \mu\mathbf{Y}(k)e^*(k)$	LMS adaptation	(4)
Algorithm	Nonlinearity $g[\tilde{X}(k)]$	
Decision Directed (DD)	$\hat{X}(k)$	
Sato	$\gamma \cdot csgn[\tilde{X}(k)]$	
Benveniste-Goursat	$\tilde{X}(k) + k_1(\hat{X}(k) - \tilde{X}(k)) +$	
	$k_2\|\hat{X}(k) - \tilde{X}(k)\|(\gamma \cdot csgn[\tilde{X}(k)] - \tilde{X}(k))$	
Godard $(p=2)$	$\frac{\tilde{X}(k)}{|\tilde{X}(k)|}\{|\tilde{X}(k) + R_p|\tilde{X}(k)|^{p-1} - |\tilde{X}(k)|^{2p-1}\}$	
Stop and Go	$\tilde{X}(k) + \frac{1}{2}A(\hat{X}(k) - \tilde{X}(k)) + \frac{1}{2}B(\hat{X}(k) - \tilde{X}(k))^*$	

The adaptive algorithm converges in the mean when

$$E\{Y(k)g^{*(k)}[\tilde{X}(k)]\} = E\{\mathbf{Y}(k)\tilde{X}^*(k)\} \quad \text{(equilibrium)} \quad (9.100.1)$$

and it converges in the mean-square when

$$\begin{aligned}E\{\mathbf{u}^H(k)\mathbf{Y}(k)g^{*(k)}[\tilde{X}(k)]\} &= E\{\mathbf{u}^H(k)\mathbf{Y}(k)\tilde{X}^*(k)\}\\ &= E\{\tilde{X}^*(k)g^{(k)}[\tilde{X}(k)]\} = E\{\tilde{X}(k)\tilde{X}^*(k)\}.\end{aligned} \quad (9.101)$$

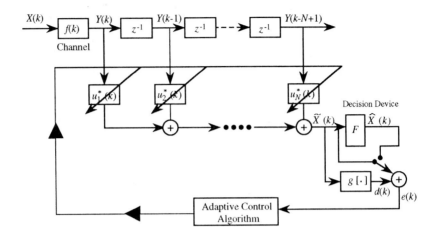

Figure 9.13 The linear adaptative blind equalization filter

Thus, it is required that the equalizer output, $\tilde{X}(k)$, be Bussgang at equilibrium, i.e., (9.100.2) states that the autocorrelation of $\tilde{X}(k)$ equals the cross-correlation between $\tilde{X}(k)$ and a memoryless nonlinear transformation of $\tilde{X}(k)$. Processes that satisfy (9.100.2) are said to be Bussgang [Bussgang, 1952].

The Bussgang algorithms are based on the minimization of a nonconvex cost function and therefore global convergence is not generally guaranteed. In fact, the possibility of ill-convergence of the Godard algorithm, as well as of the Sato and Benveniste-Goursat algorithms, has been established analytically and confirmed with specific examples [Ding, et al., 1989; Haykin, 1991; Shalvi and Weinstein, 1990].

The basic motivation behind Godard's algorithm [1980] was to find a cost function that characterizes the amount of Intersymbol Interference (ISI) at the equalizer output independently of the carrier phase. Since the input sequence $\{X(k)\}$ is i.i.d., the cost function that satisfies the aforementioned conditions is

$$J(p) = E\{(\tilde{X}(k)|^p - |X(k)|^p)^q\}. \tag{9.102}$$

This cost function cannot be used in practice because it assumes knowledge of the input sequence, $\{X(k)\}$, which is inaccessible. To avoid this difficulty, Godard [1980] suggested the use of a dispersion function

$$D(p) = E\{(|\tilde{X}(k)|^p - R_p)^q\} \tag{9.103.1}$$

where

$$R_p = \frac{E\{|X(k)|^{2p}\}}{E\{|X(k)|^p\}} \tag{9.103.2}$$

which was shown to behave like the cost function, $J(p)$, and yet it is independent of the input sequence. In this criterion, only the current equalizer output, $\tilde{X}(k)$, is involved and thus the resulting nonlinear function is memoryless. That is

$$g[\tilde{X}(k)] = \frac{\tilde{X}(k)}{|\tilde{X}(k)|}\{|\tilde{X}(k)| + R_p|\tilde{X}(k)|^{p-1} - |\tilde{X}(k)|^{2p-1}\}. \tag{9.104}$$

Let us recall that the ultimate goal of blind equalization is to achieve identity of equation (9.88). Since the transmitted symbols $X(k), \{k = 1, 2, \ldots\}$ are independent from each other, the $\tilde{X}(k)$ at perfect equalization should also be independent from each other. Since in most constellations $E\{X(k)\} = 0$, the following memory criterion, as an extension of the Godard dispersion index, has been proposed by Chen, Nikias, and Proakis [1991]:

$$M(p) = W_0 E\{(|\tilde{X}(k)|^p - R_p)^2\} + \sum_{i=1}^{M} W_i |E\{\tilde{X}(k)\tilde{X}^{(k)*}(k-i)\}|^2 \tag{9.105}$$

where W_i's are weights, R_p was defined in (9.103), M determines the size of the memory, and $\tilde{X}^{(k)}(k-i)$ is the equalizer output at time $k-i$ using the equalizer coefficients at time k. At perfect equalization, $E\{\tilde{X}(k)\tilde{X}^{(k)*}(k-i)\} = E\{X(k)X^*(k-i)\} = 0$ for $k \neq 0$. Consequently, the absolute minimum of $M(p)$ is identical to that of $D(p)$. Criterion (9.105) is called CRIMNO (i.e., criterion with memory nonlinearity) and its resulting equalizer has been shown to achieve a faster convergence rate than the Godard algorithm [Chen, Nikias, and Proakis, 1991].

9.8.2 The Polyspectra Algorithms

Blind equalization algorithms based on higher-order statistics perform nonlinear transformations on the input of the equalizer filter. This nonlinear transformation is a memory nonlinearity and it is identical to the generation of higher-order cumulants of the received channel data [Mendel, 1991; Nikias and Raghuveer, 1987]. In this section two blind equalization schemes will be discussed in detail.

(i) The Tricepstrum Equalization Algorithm (TEA) [Hatzinakos and Nikias, 1991] that estimates the equalizer impulse response by using the complex cepstrum of the fourth-order cumulants (tricepstrum) of the synchronously sampled received signal, and

(ii) parametric blind equalization schemes for QAM signals based on second- and fourth-order statistics.

A. The Tricepstrum Equalization Algorithm (TEA)

$$Y(k) = f(k) * X(k) + W(k) \text{ (time domain)} \qquad (9.106.1)$$

Consider a communication system subject to intersymbol interference (ISI) and additive noise. The received sequence after being demodulated, low-pass filtered, and synchronously sampled is where $f(k)$ is the transition channel and is assumed to be FIR and slowly time-varying, $X(k)$ is generally complex non-Gaussian white i.i.d. with $E\{X(k)\} = 0, E\{X^3(k)\} = 0$ and $E\{X^4(k)\} - 3(E\{X^2(k)\})^2 = \gamma_4^x \neq 0$. The additive noise $W(k)$ is zero-mean complex Gaussian independent of $X(t)$.

In the fourth-order cumulant spectrum domain it holds

$$c_4^y(\tau_1, \tau_2, \tau_3) = c_4^f(\tau_1, \tau_2, \tau_3) \cdot \gamma_4^x \quad (\text{fourth - order cumulant domain}) \qquad (9.106.2)$$

$$t_y(m, n, l) = t_f(w, n, l), \ (m, n, l)(0, 0, 0) \text{ (tricepstrum domain)}, \qquad (9.106.3)$$

where $c_4^y(\tau_1, \tau_2, \tau_3)$ is the fourth-order cumulant of $Y(k)$ and $t_y(\tau_1, \tau_2, \tau_3) = F_3^{-1}\{\ln C_4^y(\omega_1, \omega_2, \omega_3)\}$, $t_f(m, n, l) = F_3^{-1}\{\ln C_4^f(\omega_1, \omega_2, \omega_3)\}$ are the tricepstra of $Y(k)$ and $f(k)$, respectively. In addition, the following identity holds between the fourth-order cumulants $c_4^y(\tau_1, \tau_2, \tau_3)$ and the tricepstrum $t_y(\tau_1, \tau_2, \tau_3)$ [Pan and Nikias, 1988]:

$$\sum_{i=1}^{p} A^{(i)}[c_4^y(\tau_1 - i, \tau_2, \tau_3) - c_4^y(\tau_1 + i, \tau_2 + i, \tau_3 + i)] +$$
$$\sum_{j=1}^{q} B^{(j)}[c_4^y(\tau_1 - j, \tau_2 - j, \tau_3 - j) - c_4^y(\tau_1 + j, \tau_2, \tau_3)] = -m \cdot c_4^y(\tau_1, \tau_2, \tau_3)$$

$$(9.107.1)$$

where

$$i \cdot t_y(i, 0, 0) = \begin{cases} -A^{(i)}, & i = 1, \ldots, p \\ B^{(-i)}, & i = -1, -2, \ldots, -q \end{cases}$$

and $A^{(i)}, B^{(i)}$ are the minimum and maximum phase cepstral coefficients described by (5.11) in Chapter 5. The cepstrum parameters are exponentially decaying functions of i. Hence, although in theory $p, q \to \infty$, in practice p and q are chosen to be some finite numbers. A rule of thumb is to define $w = \max(p, q)$, $z \leq \frac{w}{2}$, $s \leq z$ and then choose $\tau_1 = -w, \ldots, -1, 1, \ldots, w$, $\tau_2 = -z, \ldots, 0, \ldots, z$ and $\tau_3 = -s, \ldots, 0, \ldots, s$ to form the overdetermined system of equations

$$\mathbf{P} \, \mathbf{a} = \mathbf{p} \qquad (9.107.2)$$

where $\mathbf{P}$ is an $(N_p \times (p+q))$ matrix, $N_p = 2w(2z+1)(2s+1)$ with entries of the form $\{c_4^y(\tau_1, \tau_2, \tau_3) - c_4^y(\tau_1', \tau_2', \tau_3')\}$; $\mathbf{a} = [A^{(1)}, \ldots, A^{(p)}, B^{(1)}, \ldots, B^{(q)}]^T$ is the $((p+q) \times 1)$ vector of unknown cepstrum coefficients and $\mathbf{P}$ is an $(N_p \times 1)$ vector with entries of the form $\{-\tau_1 c_4^y(\tau_1, \tau_2, \tau_3)\}$. In practice, the fourth-order cumulants $c_4^y(\cdot)$ in (9.107) need to be substituted by their estimates $\hat{c}_4^y(\cdot)$, obtained from a finite length window of the observed samples $Y(k)$.

The TEA algorithm solves the overdetermined system of equations of (9.107) adaptively, using an LMS adaptation algorithm. At each iteration, an estimate of the cepstrum parameters, $A^{(i)}$ and $B^{(i)}$, is computed. The equalizer coefficients at each iteration can be computed based on (5.4.3, 5.20.1, and 5.20.2) and using the negative of the cepstrum parameters $A^{(i)}$ and $B^{(i)}$.

TEA Algorithm for a Linear Equalizer with order $N_1 + N_2 + 1 = N$

The notation that appears in the TEA algorithm is:

$\{Y(k)\}$:	The received zero-mean synchronously sampled communication signal.
N_1, N_2 :	Lengths of minimum and maximum phase components of the equalizer, respectively.
p, q :	Length of minimum and maximum phase cepstral parameters, respectively.
$\hat{m}_4^y(\tau_1, \tau_2, \tau_3; k)$:	Estimated fourth-order moments of $\{Y(k)\}$ at iteration (k).
$\hat{m}_2^y(\tau; k)$:	Estimated second-order moments of $\{Y(k)\}$ at iteration (k).
$\hat{c}_4^y(\tau_1, \tau_2, \tau_3, k)$:	Estimated fourth-order cumulants of $\{Y(k)\}$ at iteration (k).
$\{u(i; k)\}$:	The equalizer at iteration k.
$\tilde{X}(k)$:	The equalizer output.

In general for 1-D (e.g., PAM) or 2-D (e.g., QAM) signaling with symmetric constellations, the following holds:

$$\hat{c}_4^y(\tau_1, \tau_2, \tau_3) = \hat{m}_4^y(\tau_1, \tau_2, \tau_3) - \hat{m}_2^y(\tau_1) \cdot \hat{m}_2^y(\tau_3 - \tau_2) - \hat{m}_2^y(\tau_2) \cdot \hat{m}_2^y(\tau_3 - \tau_1) - \hat{m}_2^y(\tau_3) \cdot \hat{m}_2^y(\tau_1 - \tau_2). \quad (9.108)$$

For symmetric L^2- QAM constellations it holds:

$$\hat{c}_4^y(\tau_1, \tau_2, \tau_3) = \hat{m}_4^y(\tau_1, \tau_2, \tau_3) \quad (9.109)$$

The TEA algorithm proceeds as follows:

Step 1:

Estimate adaptively the $\hat{c}_4^y(\tau_1, \tau_2, \tau_3; k)$ for $-M \leq \tau_1, \tau_2, \tau_3 \leq M$, from a finite length window of $Y(k)$. M should be sufficiently large so that $\hat{c}_4^y(\tau_1, \tau_2, \tau_3) \approx 0$ for $|\tau_1|, |\tau_2|, |\tau_3| > M$. Assuming that at iteration (0) we have received the samples $\{Y(1), \ldots, Y(I_{\text{lag}})\}$, the cumulants at each iteration are estimated as follows:

Step 1(a) Stationary Case with Growing Rectangular Window:

Compute

$$\begin{aligned}\hat{m}_4^y(\tau_1, \tau_2, \tau_3; k) &= (1 - n(k))\hat{m}_4^y(\tau_1, \tau_2, \tau_3; k-1) \\ &+ n(k)Y(s_4^k)Y(s_4^k + \tau_1)Y(s_4^k + \tau_2)Y(s_4^k + \tau_3)\end{aligned} \quad (9.110.1)$$

$$\hat{m}_2^y(\tau; k) = (1 - n(k))\hat{m}_2^y(\tau; k-1) + n(k) \cdot Y(s_2^k)Y(s_2^k + \tau) \quad (9.110.2)$$

where $n(k) = \frac{1}{k + I_{\text{lag}}}$, $s_4^k = \min(k + I_{\text{lag}}, k + I_{\text{lag}} - \tau_1, k + I_{\text{lag}} - \tau_2 + 2, k + I_{\text{lag}} - \tau_3)$, $s_2^k = \min(k + I_{\text{lag}}, k + I_{\text{lag}} - \tau)$ and substitute (9.110.1) and (9.110.2) into (9.108).

Step 1(b) Nonstationary Case (for Symmetric L^2-QAM Signaling):

Since in this case the second-order moment $m_2^y(\tau) = 0$, we can use $m_4^y(\tau_1, \tau_2, \tau_3)$ with a forgetting factor $\lambda, 0 < \lambda < 1$ as follows:

$$\begin{aligned}(k + I_{\text{lag}})\hat{m}_4^y(\tau_1, \tau_2, \tau_3; k) &= \lambda(k - 1 + I_{\text{lag}}) \cdot \hat{m}_4^y(\tau_1, \tau_2, \tau_3; k-1) + \\ &\quad Y(s_4^k)Y(s_4^k + \tau_1)Y(s_4^k + \tau_2)Y(s_4^k + \tau_3),\end{aligned} \quad (9.111)$$

Then with s_4^k defined as in (9.110.1), substitute (9.111) into (9.109).

Step 2:

Select p, q arbitrarily large, so that $A^{(i)} = 0$ and $B^{(j)} = 0$ for $i > p$ and $j > q$.
Consequently, in (9.107) we set

$$w = \max(p, q), \quad z = \frac{w}{2}, \quad s \leq z.$$

Step 3:

Based on (9.107.1) with $\hat{c}_4^y(\tau_1,\tau_2,\tau_3;k)$, $A^{(i;k)}$ and $B^{(j;k)}$ instead of $c_4^y(\tau_1,\tau_2,\tau_3)$, $A^{(i)}$ and $B^{(j)}$, respectively, form the overdetermined system of equations

$$\hat{\mathbf{P}}(k)\hat{\mathbf{a}}(k) = \hat{\mathbf{p}}(k), \ k = 0, 1, \ldots$$

In analogy to (9.107.2), $\hat{\mathbf{P}}(k)$ is an $(N_p \times (p+q))$ matrix with entries $\{\hat{c}_4^y(\tau_1,\tau_2,\tau_3;k) - \hat{c}_4^y(\tau_1',\tau_2',\tau_3';k)\}$, $\hat{\mathbf{a}}(k) = [A^{(1;k)}, \ldots, A^{(p;k)}, B^{(1;k)}, \ldots, B^{(q;k)}]^T$ is the $((p+q) \times 1)$ vector of the unknown cepstrum parameters and $\hat{\mathbf{p}}(k)$ is an $(N_p \times 1)$ vector with entries $\{-\tau_1 \hat{c}_4^y(\tau_1,\tau_2,\tau_3;k)\}$.

Step 4:

Assuming that $\mathbf{a}(0) = [0,\ldots,0]^T$, update $\hat{\mathbf{a}}(k)$ as follows:

$$\hat{\mathbf{a}}(k+1) = \hat{\mathbf{a}}(k) + \mu(k) \cdot \hat{\mathbf{p}}^H(k)\hat{e}(k), \qquad (9.112.1)$$

$$\mathbf{e}(k) = \hat{\mathbf{p}}(k) - \hat{\mathbf{P}}(k)\hat{\mathbf{a}}(k) \qquad (9.112.2)$$

with

$$0 < \mu(k) < \frac{2}{tr\{\hat{P}^H(k)\hat{P}(k)\}}. \qquad (9.112.3)$$

Step 5:

Calculate the equalizer within a scalar constant from the cepstrum coefficients. Initialize $\hat{i}_{\text{inv}}(0;k) = \hat{o}_{\text{inv}}(0;k) = 1$, then compute $\hat{i}_{\text{inv}}(i;k)$ and $\hat{o}_{\text{inv}}(i,k)$ as follows

$$\hat{i}_{\text{inv}}(i;k) = -\frac{1}{i}\sum_{n=2}^{i+1}[-\hat{A}^{(n-1;k)}]\,\hat{i}_{\text{inv}}(i-n+1;k), \ i=1,\ldots,N_1 \qquad (9.113.1)$$

$$\hat{o}_{\text{inv}}(i;k) = \frac{1}{i}\sum_{n=i+1}^{0}[-\hat{B}^{(1-n;k)}]\,\hat{o}_{\text{inv}}(i-n+1;k), \ i=-1,\ldots,-N_2 \qquad (9.113.2)$$

where $i_{\text{inv}}(\cdot)$ is a minimum phase sequence and $o_{\text{inv}}(\cdot)$ is a maximum phase sequence. Then

$$\hat{u}(i;k) = \hat{i}_{\text{inv}}(i;k) * \hat{o}_{\text{inv}}(i;k), i = -N_2,\ldots,0,\ldots,N_1$$

as defined in (9.90) at iteration k, where '*' denotes linear convolution. Note that because of (9.89), the cepstrum coefficients of the equalizer are the negative of the coefficients of the transmission channel.

Step 6:

Estimate the gain factor $G(k)$ as follows. In Step 1 we have calculated

$$\hat{c}_4^y(0,0,0;k) \simeq \gamma_4^x \cdot \sum_i |f(i)|^4$$

$$\hat{m}_2^y(0;k) =\simeq \gamma_2^x \sum_i |f(i)|^2$$

where $\gamma_2^x = E\{X^2(k)\}$ and $\gamma_4^x = E\{X^4(k)\} - 3(\gamma_2^x)^2$ are known.
Also

$$\hat{i}(\ell;k) = -\frac{1}{\ell} \sum_{n=2}^{\ell+1} \hat{A}^{(n-1;k)} \, \hat{i}(\ell-n+1;k), \ell = 1, 2, \ldots, p \qquad (9.114.1)$$

$$\hat{o}(\ell;k) = \frac{1}{\ell} \sum_{n=\ell+1}^{0} \hat{B}^{(\ell-n;k)} \, \hat{o}(\ell-n+1;k), \; \ell = -1, -2, \ldots, -q \qquad (9.114.2)$$

and $\hat{f}(\ell;k) = \hat{i}(\ell;k) * \hat{o}(\ell;k)$, $\gamma_2^{\hat{f}}(k) = \sum_\ell (\hat{f}(\ell;k))^2$ and $\gamma_4^{\hat{f}}(k) = \sum_\ell (\hat{f}(\ell;k))^4$. The gain $\frac{1}{\hat{G}(k)}$ can be computed in such a way that the power (or kurtosis) at the output of the equalizer is adjusted to the desired data power (or kurtosis) level. Hence,

for PAM signaling:

$$\left|\frac{1}{\hat{G}(k)}\right| \simeq \left(\frac{\gamma_2^x \cdot \gamma_2^{\hat{f}}(k)}{\hat{m}_2^y(0;k)}\right)^{1/2} \qquad (9.115.1)$$

for QAM signaling:

$$\frac{1}{\hat{G}(k)} \simeq \left(\frac{\gamma_2^x \cdot \gamma_4^{\hat{f}}(k)}{\hat{c}_4^y(0,0,0;k)}\right)^{1/4} = |\gamma_4^x|^{1/4} e^{j\pi/4} \left(\frac{\gamma_4^{\hat{f}}(k)}{\hat{c}_4^y(0,0,0;k)}\right)^{1/4} \qquad (9.115.2)$$

Step 7:
Let $\mathbf{Y}(k) = [Y(k+N_2), \ldots, Y(k-N_1)]^T$ and $\hat{\mathbf{u}}^T(k) = [\hat{u}(-N_2;k), \ldots, u(N_1;k)]^T$.
The output of the TEA equalizer is

$$\tilde{X}(k) = \frac{1}{\hat{G}(k)} \cdot \hat{\mathbf{u}}^T(k)\mathbf{Y}(k). \qquad (9.116)$$

Remarks: While cases of misconvergence have been encountered with most of the Bussgang blind equalization algorithms, the TEA algorithm, designed as described above, is a more reliable alternative, as it guarantees convergence. The properties of the gradient descent algorithm have been extensively analyzed and thus considered known.

Since Gaussian noise is suppressed in the fourth-order cumulant domain, the identification of the channel impulse response does not take into account the additive output noise. Consequently, the proposed equalizer works under the zero forcing (ZF) constraint.

The ability of TEA to identify separately the maximum and minimum phase components of the channel makes possible the design and implementation of a number of different equalization structures [Hatzinakis and Nikias, 1991].

In the recursive formulas (9.113.1) and (9.113.2) it was taken into account that a channel and its inverse have cepstrum coefficients of opposite signs. This is a unique characteristic that allows TEA to perform simultaneously channel identification and equalization.

The TEA algorithm is based only on the second- and fourth-order statistics of the received sequence, $Y(k)$, and does not take into account the decisions, $\hat{X}(k)$, at the equalizer output (see Figure 9.13). Consequently, wrong decisions do not affect the convergence of the algorithm.

Instead of using the LMS algorithm to solve adaptively the system of equations in (9.107.2), one may employ a recursive least-squares (RLS) algorithm which will converge faster at the expense of even more computations per iteration.

Example 9.5

The effectiveness of the TEA is demonstrated in this example by means of Monte-Carlo simulations, and is compared to the Benveniste-Goursat (BG) and Stop-and-Go (SG) algorithms. A short description of the BG and SG algorithms was given in Table 9.13.

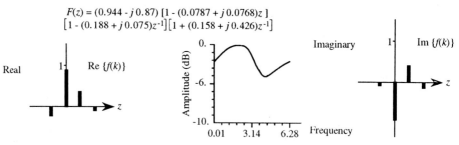

Figure 9.14 The impulse response and corresponding amplitude spectrum of a FIR channel

Let the channel be a nonminimum phase FIR filter as illustrated in Figure 9.14. For symmetric $L^2 - QAM$ signaling, $X(k) = X_r(k) + jX_I(k)$ with $X_r(k)$ and $X_I(k)$ two independent and identically distributed L-ary PAM sequences. The noise $W(k)$ is additive, complex white, and Gaussian. In all cases the SNR is measured at the input of the equalizer.

The TEA was applied as described in Steps 1–7. The fourth-order cumulants were calculated by taking $M = 3$ in Step 1. Other parameters were $p = q = 6$, $w = 6$, $s = 0$, $\mu(k) = 1/tr\{\hat{\mathbf{P}}^H(k)\hat{\mathbf{P}}(k)\}$.

The following performance metrics were considered in the simulations:

(i) the $MSE(k) = E\{|X(k-d) - \tilde{X}(k)|^2\}$ where $\tilde{X}(k)$ is the output of the equalizer at iteration (k), and $X(k-d)$ is the corresponding desired value;

(ii) the transitional symbol error rate (SER) which indicates the percentage of wrongly detected symbols in consecutive intervals of 500 samples (iterations). The $MSE(k)$ and SER were calculated by averaging over 10 experiments with independent signal and noise realizations. For the $MSE(k)$, time averaging over 100 samples was performed for each realization. The eye patterns of iteration (k) were obtained by drawing $\tilde{X}(k)$ for all 10 independent realizations and for a specific number of samples symmetrically located around (k).

Figures 9.15 and 9.16 illustrate the mean-square error learning curve, symbol error rate, as well as the eye diagrams generated by each blind equalizer. A close examination of these figures reveals the following:

1. *The objective of blind equalization is to open sufficiently the eye pattern of the distorted signal so that the equalizer can switch and operate successfully in the faster and simpler classical MSE decision-directed (DD) algorithm.* From these and other results obtained, as expected, the convergence rate for all algorithms slows down as the number of signal levels increases and the

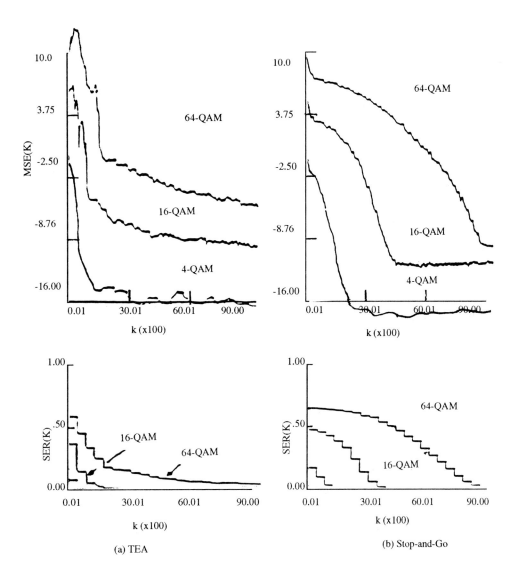

Figure 9.15 Mean-square error learning curve (top) and symbol error rate (bottom) versus number of iterations for $N = 31$ tap linear equalizer and SNR = 20 dB. (a) TEA and (b) Stop-and-Go.

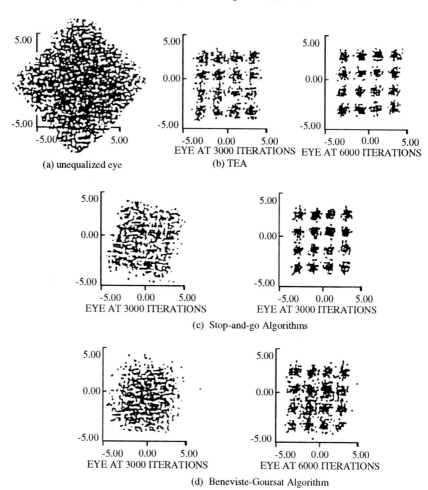

Figure 9.16 (a) Discrete eye pattern of a 16-QAM cancellation distorted by the channel example. (b) Equalized using TEA, (c) equalized using Stop-and-Go algorithm, (d) equalized using the Benveniste-Goursat algorithm ([Hatzinakos and Nikias, 1991] © 1991 IEEE).

channel distortion becomes more severe. However, we can easily see that the TEA opens the eye pattern faster and exhibits lower SER than the other two methods. For the simulation examples reported here, the TEA has a clear advantage in low-level QAM, schemes such as 4-QAM and 16-QAM, and has comparable convergence with the Stop-and-Go algorithm for 64-QAM.

2. The step-size in an adaptive algorithm controls both its transient behavior and stability. With large step-size values, high convergence rates can be achieved. However, if the step-size value is too large, convergence may not be achieved. While for the TEA the step-size ($\mu(k)$) is well determined, this is not the case for the step-size (γ) of the other Bussgang-type algorithms. In addition, the Benveniste-Goursat algorithm employs a few other parameters (k_1, k_2), which affect indirectly the choice of the step-size.

3. From Figure 9.15, we note that the TEA has fast initial convergence and slows down later. On the other hand, the Stop-and-Go algorithm has slow initial convergence (when the eye of the signal is closed) and speeds up later (when the eye starts to open). This is because the Stop-and-Go algorithm is becoming similar to the Decision-Directed (DD) algorithm when the eye pattern is open [Hatzinakos and Nikias, 1991]. Thus, one possible hybrid scheme is to use the TEA for fast initial convergence and then switch to the Stop-and-Go algorithm. □

Two extensions of the TEA have also been reported in the literature. The first one is the power cepstrum and tricoherence equalization algorithm (POTEA) which recovers the Fourier magnitude of the equalizer using autocorrelations and the power cepstrum, and its Fourier phase from fourth-order cumulants and the cepstrum of the tricoherence [Bessios and Nikias, 1991]. The second approach is an extension of TEA to the multichannel case using the cross-cumulants of the observed signals. It was thus designated as the Cross-Trispectrum Equalization Algorithm [CTEA] [Brooks and Nikias, 1991].

B. Blind Parametric Equalization Using Higher-Order Moments

Two blind equalization algorithms, based on second- and fourth-order moments of the received data sequence, were proposed by Porat and Friedlander [1991]. The proposed equalizers were developed for the general case of QAM signals. The first is based on the least-squares solution of a linear set of equations, while the second is based on nonlinear minimization of a certain cost function.

Let $\{X(k)\}$ be the transmitted data sequence. The symbols $X(k)$ are assumed to be i.i.d., each occurring with equal probability. The symbol constellation is assumed to be symmetric so that the odd-order moments are zero. The channel, $f(k)$, is assumed to be linear and time-invariant during each observation interval. Generally, $f(k)$ can be approximated by a finite impulse response of order $q = q_1 + q_2 + 1$, hence the observed data are:

$$Y(k) = \sum_{n=-q_1}^{q_2} f(n) X(k-n). \tag{9.117}$$

Without loss of generality we can assume that $f(0) = 1$.

Linear Least Squares Estimation Algorithm

Let $m_2^y(\tau)$ be the second-order moment of $Y(k)$, defined as

$$m_2^y(\tau) = E\{Y^*(k) Y(k+\tau)\} \tag{9.118}$$

and let $m_4^y(\tau, \tau, \tau)$ be the diagonal slice of the fourth-order moment of $Y(k)$ defined as

$$m_4^y(\tau) = m_4^y(\tau, \tau, \tau) = E\{Y^*(k) Y^*(k+\tau) Y^2(k+\tau)\}. \tag{9.119}$$

The corresponding slice of the fourth-order cumulant will be

$$c_4^y(\tau) = c_4^y(\tau, \tau, \tau) = m_4^y(\tau) - 2 m_2^y(0) m_2^y(\tau). \tag{9.120}$$

It can be easily shown that

$$m_2^y(\tau) = \gamma_2^x \sum_{k=-\infty}^{+\infty} f^*(k) f(k+\tau) \tag{9.121.1}$$

$$c_4^y(\tau) = \gamma_4^x \sum_{k=-\infty}^{+\infty} f^*(k) f^*(k+\tau) f^2(k+\tau) \tag{9.121.2}$$

where γ_2^x is the variance of $X(k)$ and γ_4^x is the kurtosis of $X(k)$. Taking the z transform of (9.121.1) and (9.121.2), we obtain:

$$M_2^y(z) = \gamma_2^x \sum_r \sum_k f^*(k) f(k+\tau) z^{-\tau} = \gamma_2^x F^*(z^{-1}) F(z) \tag{9.122.1}$$

$$C_4^y(z) = \gamma_4^x \sum_\tau \sum_k f^*(k) f^*(k+\tau) f^2(k+\tau) z^{-\tau} = \gamma_4^x F^*(z^{-1}) F_3(z) \tag{9.122.2}$$

where $F(z)$ is the transfer function of the channel, and $F_3(z)$ is the Z transform of $\{f^*(k) f^2(k); -q_1 \leq k \leq q_2\}$.

Combining (9.122.2) with (9.122.2), we obtain

$$C_4^y(z)F(z) + \epsilon F_3(z)M_2^y(z) = 0 \qquad (9.123)$$

where $\epsilon = -\gamma_4^x/\gamma_2^x$. Taking into account the finite extent of $f(k)$ and the fact that $f(0) = 1$, (9.123) can be written as

$$\sum_{k=-q_1}^{-1} f(k)c_4^y(\ell-k) + \sum_{k=1}^{q_2} f(k)c_4^y(\ell-k) + \sum_{k=-q_1}^{q_2} n(k)m_2^y(\ell-k) = -c_4^y(\ell-k)$$
$$-(2q_1+q_2) \le \ell \le (2q_2+q_1) \qquad (9.124)$$

where $n(k) = \epsilon f^*(k)f^2(k); -q_1 \le k \le q_2$. Treating $n(k)$ as if it were independent of $f(k)$, (9.124) can be viewed as an overdetermined system of $3q+1$ equations with $2q+1$ unknowns, where $q = q_1 + q_2$. In practice the quantities $c_4^y(\tau)$ and $m_2^y(\tau)$ have to be estimated. When $Y(k)$ is embedded in additive complex Gaussian noise, $W(k)$, then $c_4^y(\tau)$ and $m_2^y(\tau)$ can be estimated as follows:

$$\hat{m}_2^y(\tau) = (\hat{m}_2^y(-\tau))^* = \frac{1}{N}\sum_k Y^*(k)Y(k+m) - \gamma_2^w\delta(\tau),\ 0 \le \tau \le q \qquad (9.125.1)$$

$$\hat{m}_4^y(\tau) = \frac{1}{N}\sum_k Y^*(k)Y^*(k+\tau)Y^2(k+\tau),\ -q \le \tau \le q \qquad (9.125.2)$$

$$\hat{c}_4^y(\tau) = \hat{m}_4^y(\tau) - 2\hat{m}_2^y(\tau),\ q \le \tau \le q \qquad (9.125.3)$$

where N is the length of the observed data. Substituting (9.125) into the system of (9.124), we obtain the system

$$\mathbf{A}\ \mathbf{x} = -\mathbf{b} \qquad (9.126)$$

where $\mathbf{x} = [f(-q_1),\ldots,f(-1),\ldots,f(q_2),n(-q_1),\ldots,n(q_2)]^T$ is a $(2q+1) \times 1$ vector and the entries of $\mathbf{A}$ and $\mathbf{b}$ can be inferred from (9.124). The system of (9.126) has the least-squares solution

$$\mathbf{x} = -(\mathbf{A}^H\mathbf{A})^{-1}\mathbf{A}^H\mathbf{b} \qquad (9.127)$$

and the channel estimates can be extracted from the components of $\mathbf{X}$.

The algorithm of obtaining the channel estimate is basically an extension of the approach followed by Giannakis and Mendel [1989], modified to include the complex noncausal $f(k)$.

Nonlinear Least-Squares Estimation Algorithm

The linear least estimation algorithm just described is not optimal in any sense, but it performs well for the case of small intersymbol interference and low-order channels or when the channel varies slowly so that many data points can be collected for the channel estimation procedure. For the case where the intersymbol interference is severe, or the number of data samples is small, Porat and Friedlander [1991] suggested the asymptotically minimum variance nonlinear least squares algorithm which is described in the sequel.

Let $\boldsymbol{\Theta} = [f(-q_1), \ldots, f(q_2), \gamma_2^x, \gamma_2^4]^T$ be the vector of the unknown parameters and let s be a vector that consists of a subset of the second-order moments $\{m_2^y(\tau), 0 \leq \tau \leq q\}$ and the slice of fourth-order cumulants $\{c_4^y(\tau); -q \leq \tau \leq q\}$. It is required that the size of s is larger than the size of $\boldsymbol{\theta}$. Employing the moment and cumulant estimates of (9.125) we can similarly define the estimate of $\hat{\mathbf{s}}$, i.e.,

Let $sum(\boldsymbol{\Theta})$ be the asymptotic normalized covariance matrix of $\hat{\mathbf{s}}$, i.e.,

$$sum(\boldsymbol{\Theta}) = \lim_{N \to \infty} N \cdot \{E(\hat{\mathbf{s}} - \mathbf{s})(\hat{\mathbf{s}} - \mathbf{s})^H\}. \tag{9.128}$$

Porat and Friedlander [1989] have shown that an asymptotically minimum variance estimate of $\hat{\boldsymbol{\Theta}}$, $\hat{\boldsymbol{\Theta}}$ is the value for which the function

$$V(\boldsymbol{\Theta}) = (\hat{\mathbf{s}}(\boldsymbol{\Theta}) - \hat{\mathbf{s}})^H \boldsymbol{\Sigma}^{-1}(\boldsymbol{\Theta})(\hat{\mathbf{s}}(\boldsymbol{\Theta}) - \hat{\mathbf{s}}) \tag{9.129}$$

attains a global minimum (if a global minimum exists). The estimation of $\boldsymbol{\Theta}$ based on the global minimization of $V(\boldsymbol{\Theta})$ is not feasible in practice since the dependence of $\boldsymbol{\Sigma}(\Theta)$ on $\boldsymbol{\Theta}$ is difficult to express analytically.

Again, Porat and Friedlander [1991] have shown that if we can find some consistent estimate of $\boldsymbol{\Sigma}(\Theta)$, i.e., $\hat{\boldsymbol{\Sigma}}(\Theta)$, directly computed from the data $\{Y(k), 0 \leq k \leq N\}$ using an approximation of (9.128), then an asymptotically minimum variance estimate of $\boldsymbol{\Theta}$, is given by the global minimizer of the cost function

$$\hat{V}(\mathbf{s}(\boldsymbol{\Theta}) = (\hat{\mathbf{s}})^H \hat{\boldsymbol{\Sigma}}^{-1}(\boldsymbol{\Theta})(\mathbf{s}(\boldsymbol{\Theta}) = \hat{\mathbf{s}}). \tag{9.130}$$

In the same paper, they showed that under some regularity conditions on $\boldsymbol{\Sigma}(\Theta)$, the estimate of $\boldsymbol{\Theta}$, obtained from the global minimization of (9.130), achieves the same asymptotic variance of $\hat{\Theta}$ obtained as the global minimizer of (9.129).

The nonlinear least-squares estimation algorithm consists of the following steps: (i) Compute the estimated moments from (9.125). (ii) Compute $\hat{\boldsymbol{\Sigma}}(\boldsymbol{\Theta}$ using an approximation of (9.128). (iii) Apply the linear least-squares algorithm described in the previous subsection to obtain the initial estimates $\{\hat{f}(k), \hat{\gamma}_2^x, \hat{\gamma}_4^x\}$. Based on these estimates construct the vector $\boldsymbol{\Theta}_0$, and use it as an initial condition to the nonlinear minimization problem. (iv) Use a nonlinear minimization procedure (such as Newton-Gauss) to minimize $\hat{V}(\boldsymbol{\Theta})$ with respect to $\boldsymbol{\Theta}$.

Clearly, the computational complexity of the nonlinear least-squares algorithm is higher than the one of the linear least-squares estimation algorithm. The performance of the nonlinear least-squares estimation method is demonstrated by a simple example.

Example 9.6

The channel impulse response is taken to be $\{2-0.4j, 1.5+1.8j, 1, 1.2-1.3j, 0.8+1.6j\}$. The channel has a severe ISI of $f(k)$ 12dB, the signal is 4-QAM, and the SNR = 40 dB. The equalizer is chosen to be a linear transversal filter whose coefficients $\{\hat{u}(k), -K \leq k \leq K\}$ minimize the mean-square error (MSE). For the minimum mean-square error equalizer the $\hat{\mathbf{u}} = [\hat{u}(-K), \ldots, \hat{u}(K)]^T$ is given by [Clark, 1985]:

$$\hat{\mathbf{u}} = \left(\frac{\gamma_2^w}{\gamma_2^x}\mathbf{I} + \hat{\mathbf{F}}^H\hat{\mathbf{F}}\right)^{-1}\hat{\mathbf{F}}^H\mathbf{e}$$

where $\hat{\mathbf{F}}$ is a $((2K+q+1) \times (2K+1))$ matrix with each (i,j)-th element being $\hat{f}(i-j-q_1)$, or zero if $(i-j-q_1)$ is outside the interval $[-q_1, q_2]$, $\mathbf{e}$ is a $((2K+q+1) \times 1)$ vector with all its elements zero except the one of row $(K+q_1+1)$, and γ_2^w, γ_2^x are the variances of the noise and the signal respectively.

Figure 9.17 (a) and (b) illustrate the received and the equalized symbol constellations, respectively, out of a single simulation. For this test 15,000 data points were used to estimate the channel. The reconstructed constellation appears rotated due to imperfect channel reconstruction. Figure 9.18 shows the frequency response of the channel before and after equalization. The response of the equalized channel is almost flat, except at frequency 0.47. □

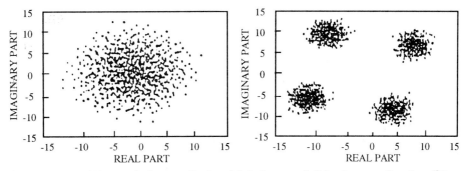

Figure 9.17 The symbol constellation (a) before, and (b) after equalization ([Porat and Friedlander, 1991] © 1991 IEEE).

9.9 SUMMARY

In this chapter we presented adaptive signal processing algorithms based on higher-order statistics and discussed their strengths and limitations. A brief overview of the

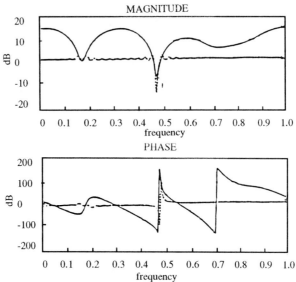

Figure 9.18 The frequency responses of the channel before and after equalization ([Porat and Friedlander, 1991] © IEEE)

popular LMS and RLS algorithms was first given within the framework of second-order statistics. The LMF and RLF methods were discussed and shown to be based on fourth-order statistics. Adaptive algorithms based on cumulants for MA, noncausal AR, ARMA, and AR lattice filters were described and their application to nonminimum phase system identification problem demonstrated. Finally, we introduced adaptive time delay estimation and blind equalization methods that are based on higher-order statistics and we showed their structural properties and performance.

Adaptive signal processing methods based on second-order statistics, including the LMS and RLS algorithms, can be found in Haykin [1991] and Widrow and Stearns [1985]. The LMF algorithm was introduced by Walach and Widrow [1984] and the RLF method by Figueras-Vidal, et al. [1988]. Friedlander and Porat [1989, 1990] introduced adaptive system identification algorithms based on cumulants for MA, AR, and ARMA processes whereas Chiang and Nikias [1990a,b] developed similar methods for MA system identification and time delay estimation. Swami and Mendel [1988] introduced an adaptive lattice method by combining cumulants and instrumental variables. Finally, a discussion on Bussgang techniques for blind equalization can be found in Haykin [1991], Bellini [1986], and Godfrey and Rocca [1981]. Polyspectra-based blind equalizers have been introduced by Hatzinakos and Nikias [1991], Porat and Friedlander [1991], Bessios and Nikias [1991], and Brooks and Nikias [1991].

REFERENCES

Bellini, S., " Bussgang Techniques for Blind Equalization," *Proceedings of IEEE-Globecom '86*, pp. 46.1.1–46.1.7, 1986.

Bellini, S., and F. Rocca, "Blind Deconvolution: Polyspectra or Bussgang Techniques," in **Digital Communications**, E. Biglieri and G. Prati (eds.), North Holland, pp. 251–262, 1986.

Benveniste, A., and M. Goursat, "Blind Equalizers," *IEEE Trans. on Communications*, **COM-32**, pp. 871–883, August, 1984.

Bessios, A. G., and C. L. Nikias, "POTEA: The Power Cepstrum and Tricoherence Equalization Algorithm," *Proceedings SPIE Conference*, San Diego, CA, July, 1991.

Brooks, D. H., and C. L. Nikias, "Cross-Bicepstrum and Cross-Tricepstrum Approaches to Multichannel Deconvolution," *Proceedings Int. Signal Processing Workshop on Higher-Order Statistics*, Chamrousse, France, July, 1991.

Bussgang, J. J., "Crosscorrelation Functions of Amplitude-Distorted Gaussian Signals," M.I.T. Technical Report, No. 216, March, 1952.

Chan, Y. T., J. M. Riley, and J. B. Plant, "Modeling of Time Delay and Its Application to Estimation of Nonstationary Delay," *IEEE Trans. Acoust., Speech, Signal Processing*, **ASSP-29**, pp. 577–581, June, 1981.

Chen, Y., C. L. Nikias, and J. G. Proakis, "CRIMNO: Criterion with Memory Nonlinearity for Blind Equalization," *Proceedings Int. Signal Processing Workshop on Higher-Order Statistics*, Chamrousse, France, pp. 57–90, July, 1991.

Chiang, H. H., and C. L. Nikias, "Adaptive Deconvolution and Identification of Nonminimum Phase FIR Systems Based on Cumulants," *IEEE Trans. on Automatic Control*, **35**(1), pp. 36–47, January, 1990a.

Chiang, H. H., and C. L. Nikias, "A New Method for Adaptive Time Delay Estimation for Non-Gaussian Signals," *IEEE Trans. on Acoustics, Speech, and Signal Proc.*, **38**(2), pp. 209–219, February, 1990b.

Clark, A. P., **Equalizers for Digital Modems**, London: Pentech, 1985.

Ding, Z., C. R. Johnson, R. A. Kennedy, B. D. O. Anderson, "On the Ill-Convergence of Godard Blind Equalizers in Data Communication Systems," *Proceedings of CISS'89*, pp. 538–543, March, 1989.

Etter, D. M., and S. D. Stearns, "Adaptive Estimation of Time Delays in Sampled Data Systems," *IEEE Trans. Acoust., Speech, Signal Processing*, **ASSP-29**, pp. 582–587, June, 1981.

Feintuch, P. L., N. J. Bershad, and F. A. Reed, "Time Delay Estimation Using the LMS Adaptive Filter-Dynamic Behavior," *IEEE Trans. Acoust., Speech, Signal Processing*, **ASSP-29**, pp. 571–576, June, 1981.

Figueras-Vidal, A. R., J. M. Paez-Borrallo, and F. Lorenzo-Speranzini, "Non-Quadratic Recursive Algorithms for Transversal Plant Identification," *Proceeding of ICASSP*, pp. 1395–1398, New York, April, 1988.

Fonollosa, J. A. R., "Adaptive System Identification Based on Higher-Order Statistics," *Proc. ICASSP 1991*, pp. 3437–3440, Toronto, Canada, May, 1991.

Friedlander, B., "The Overdetermined Recursive Instrumental Variable Method," *IEEE Trans., Autom. Contr.*, **AC-29**, pp. 353–356, April, 1984.

Friedlander, B., and B. Porat, "Adaptive IIR Algorithms Based on High-Order Statistics," *IEEE Trans. of Acoustics, Speech and Signal Processing*, **37(4)**, April, 1989.

Friedlander, B., and B. Porat, "Asymptotically Optimal Estimation of MA and ARMA Parameters of Non-Gaussian Processes from High-Order Moments," *IEEE Trans. Automat. Contr.*, **35(1)**, pp. 27–35, January, 1990.

Giannakis, G. B., and J. M. Mendel, "Identification of Nonminimum Phase Systems Using Higher Order Statistics," *IEEE Trans. on Acoustics, Speech and Signal Processing*, **37** (3), March, 1989.

Godard, D. N., "Self-Recovering Equalization and Carrier Tracking in Two-Dimensional Data Communication Systems," *IEEE Trans. on Communications*, **COM-28**, pp. 1867–1875, November, 1980.

Godfrey, R., and F. Rocca, "Zero Memory Nonlinear Deconvolution," *Geophysical Prospecting*, Vol. **29**, pp. 189–228, 1981.

Hatzinakos, D., and C. L. Nikias, "Blind Equalization Using a Tricepstrum Based Algorithm," *IEEE Trans. on Communications*, pp. 669–682, May, 1991.

Haykin, S., **Adaptive Filter Theory**, Second Edition, Englewood Cliffs, N.J.: Prentice Hall, 1991.

Mendel, J. M., "Tutorial on Higher-Order Statistics (Spectra) in Signal Processing and System Theory: Theoretical Results and Some Applications," *Proceedings IEEE*, **9**, pp. 278–305, March, 1991.

Murray, W. (ed.), **Numerical Methods for Unconstraint Optimization**, New York: Academic Press, 1972.

Nikias, C. L., "ARMA Bispectrum Approach to Nonminimum Phase System Identification," *IEEE Trans. Acoust., Speech, Signal Processing*, **36**, pp. 513–524, April, 1988.

Nikias, C. L., and R. Pan, "ARMA Modeling of Fourth-Order Cumulants and Phase Estimation," *Circuits Syst. Signal Processing*, **7**(3), pp. 291–325, 1988.

Nikias, C. L., and M. R. Raghuveer, "Bispectrum Estimation: A Digital Signal Processing Framework," *Proceedings IEEE*, **75**, No. 7, pp. 869–891, July, 1987.

Pan, R., and C. L. Nikias, "The Complex Cepstrum of Higher-Order Cumulants and Nonminimum Phase System Identification," *IEEE Trans. Acoustics, Speech and Signal Processing*, **ASSP-36**, pp. 186–205, February, 1988.

Picchi, G., and G. Prati, "Blind Equalization and Carrier Recovery Using a 'Stop-and-Go' Decision-Directed Algorithm," *IEEE Trans. on Communications*, **COM-35**(9), pp. 877–887, September, 1987.

Porat, B., and B. Friedlander, "Performance Analysis of Parameter Estimation Based on Higher-Order Moments," *Int. J. Adaptive Contr. Signal Processing*, **3**, pp. 191–229, 1989.

Porat, B., and B. Friedlander, "Blind Equalization of Digital Communications Channels Using Higher-Order Moments," *IEEE Trans. on Signal Processing*, **39**, pp. 522–526, February, 1991.

Sato, Y., "A Method for Self-Recovering Equalization for Multilevel Amplitude-Modulation Systems," *IEEE Trans. on Communications*, **COM-23**, pp. 679–682, June, 1975.

Shalvi, O., and E. Weinstein, "New Criteria for Blind Equalization of Nonminimum Phase Systems (Channels)," *IEEE Trans. on Information Theory*, Vol. **36**, pp. 312–321, March, 1990.

Swami, A., and J. M. Mendel, "Adaptive System Identification Using Cumulants," *Proceedings of ICASSSP*, pp. 2248–2251, New York, 1988.

Treichler, J. R., and B. G. Agee, "A New Approach to Multipath Correction of Constant Modulus Signals," *IEEE Trans. on Acoustics, Speech, and Signal Processing*, **ASSP-31**(2), pp. 459–471, April, 1983.

Walach, E., and B. Widrow, "The Least Mean Fourth (LMF) Adaptive Algorithm and Its Family," *IEEE Trans. on Information Theory*, **IT-30**(2), March, 1984.

Widrow, B., and S. D. Stearns, **Adaptive Signal Processing**, Englewood Cliffs, N.J.: Prentice Hall, 1985.

Wong, L. Y., and E. Polak, "Identification of Linear Discrete Time Systems Using the Instrumental Variable Method," *IEEE Trans. Automat. Contr.*, **AC-12**, pp. 707–718, December, 1967.

Van Tress, H. L., **Detection, Estimation, and Modulation Theory**, Part I, New York: Wiley, 1968.

Young, P., **Recursive Estimation and Time-Series Analysis**, New York: Springer-Verlag, 1984.

10

DETECTION AND CHARACTERIZATION OF NONLINEARITIES IN TIME SERIES

The main reason why time series analysis relies so heavily on linear models is the convenience in studying them and the availability of linear statistical tools. Time series found in physical sciences and other real life situations are far from conforming to linear models. Indeed, a stationary non-Gaussian time series is usually the result of some nonlinear operation on a Gaussian input process.

Higher-Order Spectra (HOS) are quite natural tools to analyze the nonlinearity of a system operating under a random input. General relations for arbitrary stationary random data passing through arbitrary linear systems have been studied extensively. On the other hand, general relations are not available for arbitrary random data passing through arbitrary nonlinear systems, and each type of nonlinearity has to be investigated as a special case.

Recently, third-order cumulant-based methods have been designed to detect certain types of nonlinearities in a time series. Rao and Gabr [1980] presented a test for nonlinearity using a sample estimate of the bispectrum of the time series. Hinich [1982] proposed a nonparametric test that uses the sample bispectrum and the asymptotic properties of bispectrum estimates.

Despite the fact that progress has been established in developing the theoretical properties of nonlinear models, only a few statistical methods for the detection and characterization of nonlinearities from a finite set of observations are available today. The procedure of a nonlinear system identification method consists of the following steps: (i) Detection of nonlinearity, (ii) Identification of candidate models, and (iii) Diagnostic checks to discover if the data fit to the considered model.

In this chapter we discuss the general formulation of a nonlinear Volterra system and provide up-to-date results on parameter identification of certain types of Volterra systems. We also present the quadratic and cubic phase coupling phenomena and their applications.

10.1 GENERAL VOLTERRA SYSTEMS

A general nonlinear (N/L) time series $Y(k)$ can be considered to be the output of a nonlinear system whose input is a stationary random process, $X(k)$. Typical representations of N/L systems are the Volterra series expansions [Schetzen, 1980]. The discrete p^{th}-order Volterra system is represented by the first $p+1$ terms of a Volterra series. If $Y(k)$ is the response of a discrete time-invariant p^{th}-order Volterra filter whose input is $X(k)$ then [Alper, 1965]:

$$\begin{aligned} Y(k) &= h_o + \sum_{i=1}^{p} H_i[X(k)] \\ &= h_o + \sum_i \sum_{(\tau_1, \tau_2, \ldots \tau_i)} h_i(\tau_1, \tau_2, \ldots \tau_i) X(k-\tau_1) \ldots X(k-\tau_i) \end{aligned} \qquad (10.1)$$

where $H_i[\cdot]$ denotes the ith-order Volterra operator, while $h_i(\tau_1, \tau_2, \ldots \tau_i)$ are the Volterra kernels of the system, which are bounded and discrete at each τ_i, and are symmetric functions of their arguments; for causal systems, $h_i(\tau_1, \tau_2, \ldots, \tau_i) = 0$ for any $\tau_i < 0$. When $X(k)$ is a pure random process with zero mean, $H_1[\cdot]$ corresponds to a general linear model, and the successive terms, $H_i[\cdot], i > 1$, are usually referred to as the "quadratic" ($H_2[\cdot]$), "cubic" ($H_3[\cdot]$) components, etc.

10.2 SECOND-ORDER VOLTERRA FILTERS

Suppose a time series is represented by the 2nd-order Volterra model

$$Y(k) = \sum_{\tau_1} h_1(\tau_1) X(k-\tau_1) + \sum_{\tau_1} \sum_{\tau_2} h_2(\tau_1, \tau_2) X(k-\tau_1) X(k-\tau_2) \qquad (10.2)$$

where $X(k)$ is a stationary random process with zero mean. The identification problem is to determine the impulse response, $h_1(\tau)$, and the kernel $h_2(\tau_1, \tau_2)$. Note that (10.2) can be viewed as a parallel connection of a linear system $\{h_1(\tau)\}$ and a quadratic system $\{h_2(\tau_1, \tau_2)\}$, as illustrated in Figure 10.1.

Chapter 10

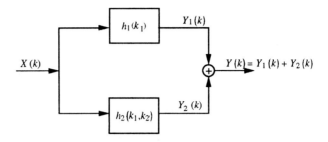

Figure 10.1 Second-Order Volterra System: linear and quadratic parts in a parallel structure

10.2.1 Identification Assuming Gaussian Input and Access to Both Input and Output [Tick, 1961]

Let $X(k)$ be a stationary zero-mean Gaussian process. From (10.2) and Figure 10.1 we obtain

$$Y(k) = Y_1(k) + Y_2(k) \qquad (10.3)$$

where

$$Y_1(k) = \sum_{\tau_1} h_1(\tau_1) X(k - \tau_1)$$

$$Y_2(k) = \sum_{\tau_1} \sum_{\tau_2} h_2(\tau_1, \tau_2) X(k - \tau_1) X(k - \tau_2).$$

The mean value of the output is given by

$$\begin{aligned} m_1^y &= E\{Y(k)\} = E\{Y_1(k)\} + E\{Y_2(k)\} \\ &= \sum_{\tau_1} \sum_{\tau_2} h_2(\tau_1, \tau_2) E\{X(k-\tau_1) X(k-\tau_2)\} \\ &= \sum_{\tau_1} \sum_{\tau_2} h_2(\tau_1, \tau_2) c_2^x(\tau_1 - \tau_2) \end{aligned} \qquad (10.4)$$

where $c_2^x(\tau)$ is the covariance (or autocorrelation) sequence of $X(k)$. Defining $H_2(\omega_1, \omega_2)$ to be Fourier transform of $h_2(\tau_1, \tau_2)$ i.e.,

$$H_2(\omega_1, \omega_2) = \sum_{\tau_1} \sum_{\tau_2} h_2(\tau_1, \tau_2) e^{-j(\omega_1 \tau_1 + \omega_2 \tau_2)} \qquad (10.5)$$

and $C_2^x(\omega)$ to be the Fourier transform of $c_2^x(\tau)$, i.e., the power spectrum of $X(k)$,

$$C_2^x(\omega) = \sum_{\tau} c_2^x(\tau) e^{-j\omega\tau}, \qquad (10.6)$$

(10.4) becomes

$$m_1^y = \frac{1}{(2\pi)^3} \sum_{\tau_1} \sum_{\tau_2} \int\int H_2(\omega_1,\omega_2) e^{j[\omega_1\tau_1+\omega_2\tau_2]} d\omega_1 d\omega_2 \int C_2^x(\omega) e^{j\omega(\tau_1-\tau_2)} d\omega$$

$$= \frac{1}{2\pi} \int H_2(\omega,-\omega) C_2^x(\omega) d\omega. \tag{10.7}$$

The cross-spectrum between the input and the output, $C_2^{xy}(\omega)$, is computed as follows: Let $c_2^{xy}(\tau)$ be the cross-covariance of the input and output processes. Then

$$\begin{aligned}c_2^{xy}(\tau) &= E\{X(k+\tau)[Y(k)-m_1^y]\} = E\{X(k+\tau)Y_1(k)\} \\ &= \sum_{\tau_1} h_1(\tau_1) c_2^x(\tau+\tau_1) \\ &= \frac{1}{2\pi} \int H_1(\omega) C_2^x(-\omega) e^{-j\omega\tau} d\omega\end{aligned} \tag{10.8}$$

where $H_1(\omega)$ is the Fourier transform of $h_1(\tau)$. Note that in (10.8), it was taken into account that the third-order cumulant sequence of the Gaussian process $X(k)$ is zero. The cross-spectrum is the Fourier transform of $c_2^{xy}(\tau)$, i.e.,

$$C_2^{xy}(\omega) = H_1(-\omega) C_2^x(\omega). \tag{10.9}$$

Hence, if a Gaussian process is input to the second-order Volterra system, the system's linear term can be identified by computing the power spectrum of the input and the cross-spectrum between input and output; viz,

$$H_1(-\omega) = \frac{C_2^{xy}(\omega)}{C_2^x(\omega)}. \tag{10.10}$$

The cross third-order cumulant between input and output is given by

$$\begin{aligned}c_3^{xxy}(\tau_1,\tau_2) &= E\{X(k+\tau_1)X(k+\tau_2)[Y(k)-m_1^y]\} \\ &= E\{X(k+\tau_1)X(k+\tau_2)Y_2(k)\} - m_1^y \cdot E\{X(k+\tau_1)X(k+\tau_2)\}\end{aligned} \tag{10.11}$$

and by utilizing (10.4) and the fact that if X_1, X_2, X_3, X_4 are zero-mean jointly Gaussian random variables, then

$$E\{X_1X_2X_3X_4\} = E\{X_1X_2\}E\{X_3X_4\}+E\{X_1X_3\}E\{X_2X_4\}+E\{X_1X_4\}E\{X_2X_3\},$$

we obtain [Schetzen, 1980]

$$\begin{aligned}c_3^{xxy}(\tau_1,\tau_2) &= 2\sum_{n_1}\sum_{n_2} h_2(n_1-\tau_1, n_2-\tau_2) c_2^x(n_1) c_2^x(n_2) \\ &= \frac{2}{(2\pi)^2} \int\int H_2(-\omega_1,-\omega_2) e^{j(\omega_1\tau_1+\omega_2\tau_2)} C_2^x(\omega_1) C_2^x(\omega_2) d\omega_1 d\omega_2.\end{aligned} \tag{10.12}$$

Hence, the cross-bispectrum, $C_3^{xxy}(\omega_1, \omega_2)$, is

$$C_3^{xxy}(\omega_1, \omega_2) = F_2\{c_3^{xxy}(\tau_1, \tau_2)\} = 2H_2(-\omega_1, -\omega_2)C_2^x(\omega_1)C_2^x(\omega_2), \qquad (10.13)$$

where $F_2\{\cdot\}$ denotes 2-d Fourier transform. The last equation indicates that the quadratic component of the second-order Volterra system can be identified by computing the power spectrum of the input and the cross-bispectrum between input and output; viz,

$$H_2(-\omega_1, -\omega_2) = \frac{C_3^{xxy}(\omega_1, \omega_2)}{2C_2^x(\omega_1)C_2^x(\omega_2)}. \qquad (10.14)$$

In summary, (10.10) and (10.14) provide the system identification formulas when the input is a zero-mean Gaussian process and both input and output are accessible for correlation and third-order cumulant computations.

10.2.2 Identification Assuming a Simple Linear Term and White Unobservable Input [Hinich, 1985]

A special class of quadratic filters was studied by Hinich [1985]. This class of systems is described by

$$Y(k) = X(k) + \sum_{m=1}^{\infty} \sum_{n=0}^{\infty} h_2(m, n) X(k-n) X(k-m-n) \qquad (10.15)$$

and can be viewed as a special case of (10.1), where $h_1(\tau) = \delta(\tau)$. The input process, $X(k)$, is assumed to be white, zero-mean, and *unobservable* (Gaussian or non-Gaussian).

The nonlinear series of (10.15) has many nonzero cumulants; however, $Y(k)$ is white noise if $[h_2(m,n) \cdot h_2(m, n+r)] = 0$ for $r \neq 0$. Hinich [1985] showed that for white input the filter coefficients, $h_2(m, n)$, of the model in (10.15) are related to the third-order cumulants of the output by

$$c_3^y(m, m+n) = (\gamma_2^x)^2 h_2(m, n) \qquad (10.16.1)$$

for all $m \geq 1$ and $n \geq 1$, and for $n = 0$ by

$$c_3^y(m, m) = 2(\gamma_2^x)^2 h_2(m, 0) \qquad (10.16.2)$$

where γ_2^x is the variance of the input process. The above expressions are fundamental because they allow estimation of the filter coefficients directly from cumulants without solving any system of equations. However, since $X(k)$ is unobservable, γ_2^x cannot be estimated. As a consequence, $h_2(m, n)$ can be computed up to the scale factor γ_2^x.

If the samples $\{Y(1),\ldots,Y(N)\}$ of the process $Y(k)$ are available and $\gamma_2^x = 1$, $h_2(m,n)$ can be estimated using (10.16) as follows:

$$\hat{h}_2(m,n) = \frac{1}{N}\sum_{k=0}^{N-1} Y(k)Y(k+m)Y(k+m+n). \qquad (10.17)$$

Note that (10.17) represents the estimation of the third-order cumulants of $Y(k)$, which are identical to the filter coefficients. This estimator is asymptotically normal. Its mean value is given by

$$E\{\hat{h}_2(m,0)\} = 2(\gamma_2^x)^2(1 - m/N)h_2(m,0), \ 1 \leq m < N$$

$$E\{\hat{h}_2(m,n)\} = (\gamma_2^x)^2[(1 - \tfrac{m+n}{N})h_2(m,n) \ + \ \tfrac{n}{N}h_2(N-m-n,m)$$
$$+ \ \tfrac{m}{N}h_2(n, N-m-n)], \ n \geq 1 \qquad (10.18)$$

$$E\{\hat{h}_2(m,n)\} = (\gamma_2^x)^2(1 - \frac{m+n}{N})h_2(m,n), \ m+n \leq \frac{N}{2} \text{ and } n \leq 1.$$

Assuming that $\sum_{r=1}^{\infty}\sum_{s=0}^{\infty} h_2^2(r,s) < \infty$, the estimator's variance is

$$\text{var}[\hat{h}^2(m,n)] = \frac{1}{N}(\gamma_2^y)^3 + O(N^{-2})$$

$$\text{var}[\hat{h}^2(m,0)] = \frac{1}{N}[c_4^y(0,0,0) + 3(\gamma_2^y)^2]\gamma_2^y + O(N^{-2}) \qquad (10.19)$$

where $c_4^y(\tau_1,\tau_2,\tau_3)$ denotes the fourth-order cumulant of $Y(k)$ at $\tau_1 = \tau_2 = \tau_3 = 0$. The signal-to-signal plus noise ratio (SSPN) of the process $Y(k)$ is defined as

$$SSPN \triangleq \frac{\gamma_2^x \sum_{m=1}^{\infty}\sum_{n=0}^{\infty} h_2^2(m,n)}{[1 + \gamma_2^x \sum_{m=1}^{\infty}\sum_{n=0}^{\infty} h_2^2(m,n)]^3}. \qquad (10.20)$$

SSPN has a maximum value of 4/27 or -8.29 dB, and measures the relative contribution of the quadratic term to the variance of the process $Y(k)$. The greater the value of the ratio, the greater is the quadratic nonlinearity contribution.

Example 10.1

Consider the following example of a simple quadratic system:

$$Y(k) = X(k) + bX(k-1)X(k-2), \ b \neq 0$$

where $X(k)$ is a white, zero-mean random process with variance γ_2^x. For this model, $h_2(m,n)$ is given by

$$h_2(m,n) = \begin{cases} b, & m = n = 1 \\ 0, & \text{otherwise.} \end{cases}$$

It can easily be proven that $Y(k)$ is a white noise process because

$$c_2^y(m) = E\{Y(k+m)Y(k)\} - (m_1^y)^2 = 0 \text{ for all } m \neq 0.$$

Although $\{Y(k)\}$ variables are uncorrelated, there is a relationship between $Y(k-2)$, $Y(k-1)$, and $Y(k)$ for all k such that

$$E\{Y(k-2)Y(k-1)Y(k)\} = b(\gamma_2^x)^2$$

or equivalently

$$c_3^y(1,2) = b(\gamma_2^x)^2$$

which was expected according to (10.16.2). □

10.2.3 Identification Assuming General Non-Gaussian Input: The Three-Wave Coupling Equation [Powers, et al., 1989]

Three-wave coupling phenomena are associated with turbulence in continuous media such as fluids and plasma. In such media, departures from thermodynamic equilibrium generate various unstable modes or waves. Initially these modes grow linearly, while later on, at sufficiently large amplitudes, they begin to interact nonlinearly. As a result of this nonlinear interaction, energy is transferred from the most unstable waves to damped waves. This process is referred to as energy cascading, and its final result is turbulence [Powers, et al., 1989].

Three-wave coupling results from a quadratic nonlinearity appearing in the governing equations of the system under consideration, and is described by the equation:

$$\frac{\partial \Phi(f_m, x)}{\partial x} = \Lambda_1(f_m)\Phi(f_m, x) + \sum_{f_i}\sum_{f_j} \Lambda_2(f_i, f_j)\Phi(f_i, x)\Phi(f_j, x) \qquad (10.21)$$

where $f_i + f_j = f_m$, $\Phi(f_m, x)$ is the Fourier transform of the fluctuation field $\Phi(t, x)$ measured at point x. $\Lambda_1(f_m)$ is given by

$$\Lambda_1(f_m) = \gamma(f_m) + jk(f_m) \qquad (10.22)$$

where $\gamma(f_m)$ is the linear growth rate and $k(f_m)$ is the dispersion relation. Finally, $\Lambda_2(f_i, f_j)$ is the three-wave coupling coefficient and is a quantity we wish to estimate from time series records of the fluctuation field observed at two spatial points.

The interpretation of (10.21) is the following: The left-hand side term corresponds to the rate of change of the Fourier amplitude of the wave at frequency f_m as it propagates through space. The first term of the right-hand side corresponds to the linear effects (or "dc" term) of this rate of change, while the second term models the quadratically nonlinear contributions. In this last term, all possible three-wave interactions that add or subtract to yield f_m are taken into account. The strength of each such interaction is the value of the coefficient $\Lambda_2(f_i, f_j)$.

Since the Fourier amplitude of the wave at frequency f_m changes as it propagates through space, the power spectral density, $C_2^\phi(f_m, x)$, also changes.

It can be shown that

$$\frac{\partial C_2^\phi(f_m, x)}{\partial x} = 2\gamma(f_m) \cdot C_2^\phi(f_m, x) + \sum_{f_1}\sum_{f_j} T(f_i, f_j) \qquad (10.22)$$

where $T(f_i, f_j)$, $f_1 + f_j = f_m$ is the three-wave energy transfer function given by

$$T(f_i, f_j) = 2Re[\Lambda_2(f_i, f_j) C_3^\phi(f_i, f_j)] \qquad (10.23)$$

where $C_3^\phi(f_i, f_j)$ is the bispectrum of the fluctuation field measured at point x. According to Kim and Powers [1979], the bispectrum, $C_3^\phi(f_i, f_j)$ is in fact measurable. In order to quantify the energy cascading, $T(f_i f_j)$, it is necessary to compute the coupling coefficients, $\Lambda_2(f_i, f_j)$, first. The determination of the coupling coefficients can be made as follows. The fluctuation field is monitored at two points closely spaced relative to a typical wavelength. The linear and quadratically nonlinear wave physics that occur between the two measurement points are modeled as depicted in Figure 10.2 where $X(f_m)$, and $Y(f_m)$ denote the Fourier transforms of the first and second measurements, respectively. Also, $H_1(f_m)$ denotes the linear transfer function while $H_2(f_i, f_j)$ denotes the quadratic one.

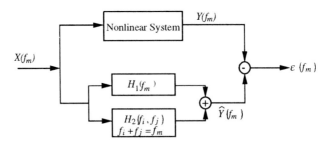

Figure 10.2 Frequency domain Volterra model of a quadratically nonlinear system

Comparing the three-wave coupling equation (10.21) and the second-order Volterra model shown in Figure 10.2 we see that $H_1(f_m)$ corresponds to $\Lambda_1(f_m)$ and

$H_2(f_i, f_j)$ corresponds to $\Lambda_2(f_i, f_j)$. Hence, the coefficients, $\Lambda_2(f_i, f_j)$, or equivalently, $H_2(f_i, f_j)$, can be computed from the cross-bispectrum of $X(f)$ and $Y(f)$, and the power spectrum of $X(f)$ from (10.14). Note that in order to apply (10.14), the process, $X(f)$, must be the Fourier transform of a zero-mean Gaussian signal.

If $X(f)$ is non-Gaussian, the linear and quadratic terms cannot be obtained separately. Kim and Powers [1988] suggested the following approach for the determination of the linear and quadratic transfer functions, in the case that the input is non-Gaussian.

The second-order Volterra filter of **Figure 10.2** is described by

$$Y(\lambda) = H_1(\lambda)X(\lambda) + \sum_{\omega_i}\sum_{\omega_2} H_2(\omega_1, \omega_2)X(\omega_1)X(\omega_2), \quad (10.24)$$

for $\omega_1 + \omega_2 = \lambda$, or in a vector form

$$Y(\lambda) = \mathbf{X}^T(\lambda)\mathbf{H}(\lambda) \quad (10.25)$$

where

$$\mathbf{X}(\lambda) = [X(\lambda), X(\frac{\lambda+1}{2})X(\frac{\lambda-1}{2}), X(\frac{\lambda+3}{2})X(\frac{\lambda-3}{2}), \cdots, X(\lambda+\frac{M}{4})X(\lambda-\frac{M}{4})]^T$$

$$\mathbf{H}(\lambda) = [H_1(\lambda), H_2(\frac{\lambda+1}{2}, \frac{\lambda-1}{2}), H_2(\frac{\lambda+3}{2}, \frac{\lambda-3}{2}), \cdots, H_2(\lambda+\frac{M}{4}, \lambda-\frac{M}{4})]^T.$$

Note that $\mathbf{X}(\lambda)$ and $\mathbf{H}(\lambda)$ are defined for odd values of λ. Similar definitions are also obtained for even values of λ. Multiplying both sides of (10.25) by $\mathbf{X}^*(\lambda)$ and taking expected values of both sides we get

$$E\{\mathbf{X}^*(\lambda)Y(\lambda)\} = E\{\mathbf{X}^*(\lambda)\mathbf{X}^T(\lambda)\}\mathbf{H}(\lambda). \quad (10.26)$$

The transfer function vector $\mathbf{H}(\lambda)$ is obtained by solving (10.26) via least-squares or a singular value decomposition algorithm.

10.3 THE IDENTIFICATION OF A PARTICULAR NONLINEAR TIME SERIES SYSTEM

Let us consider the particular nonlinear system illustrated in Figure 10.3 with the following structure: its output is the sum of a noise process and the result of passing an input signal through three cascade operations, namely, linear filtering, instantaneous functional composition, and again linear filtering. In this section we discuss a method proposed by Brillinger [1977] to estimate, up to a constant multiplier, the transfer functions of the linear filters, given a stretch of Gaussian

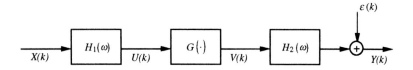

Figure 10.3 A particular nonlinear model

input series and their corresponding output. This method is based on a fundamental property of cumulants that moments do not share, which is discussed next.

Let cum $(U_1, \cdots, U_J)$ denote the joint cumulant of order J of the random variables $\{U_1, \cdots, U_J\}$. Based on a property of cumulants (see Chapter 2) that $\text{cum}(U_1, \cdots, U_J) = 0$ if the set of random variables can be divided into at least two independent groups, and that cum $(U_1, \cdots, U_J)$ is multilinear in its arguments, Brillinger [1977] presented the following lemma.

Lemma 10.1

Let $(U_1, \cdots, U_J, V_1, \cdots, V_K)$ be a set of multivariate Gaussian random variables with $U_1, \cdots, U_J$ being nondegenerate and statistically independent of each other. Let $cov(V_j, V_k)$ be the covariance of U_j and V_k, and let $G(u_1, \cdots, u_J)$ be a measurable function of $(u_1, \cdots, u_J)$ satisfying $E\{|G(U_1, \cdots, U_J)z_1 \cdots z_L|\} < \infty$ for $\{z_1, \cdots, z_L\}$ any subset of $(U_1, \cdots, U_J, V_1, \cdots, V_K)$. Then for $J, K \geq 1$,

$$\text{cum}\{G(U_1, \cdots, U_J), V_1, \cdots, V_K\}$$
$$= \sum_{j=1}^{J} \text{cum}\{G(U_1, ..., U_J), U_j..., U_j\} cov(U_j, V_1) \cdots cov(U_j, V_K)/(\gamma_2^{U_j})^K, \qquad (10.27)$$

where $\gamma_2^{U_j}$ denotes the variance of U_j. For the proof the reader is referred to Brillinger [1977].

Let $X(k)$ and $Y(k)$ denote the input and output series of the model illustrated in Figure 10.3. Then for $k = 0, \pm 1$,

$$\begin{aligned} Y(k) &= \mu + \sum_{u}[h_2(k-u)V(u)] + \epsilon(k) \\ V(k) &= G[U(k)] \\ U(k) &= \sum_{u} h_1(k-u)X(u) \end{aligned} \qquad (10.28)$$

Chapter 10   457

where μ is a constant, $\epsilon(k)$ is a stationary noise series independent of $X(k)$, and $G[\cdot]$ is a nonlinear function which maps real numbers to real numbers.

Suppose that $X(k)$ is stationary Gaussian. Then from Lemma 10.1 we obtain

$$c_2^{xy}(u) = L_1 \sum_w \sum_v h_1(w) h_2(v) c_2^x(w + v + u) \tag{10.29}$$

where $c_2^{xy}(u)$ is the cross-covariance of $X(k)$ and $Y(k)$, $L_1 = cov(v(0), V(0))/\gamma_2^{U(0)}$, and (10.29) holds provided that

$$\sum_u |c_2^x(u)|, \sum_u |h_1(u)|, \sum_u |h_2(u)|, E\{|G[U(k)]|\}, E\{|U(k+u)G[U(k)]|\} \tag{10.30}$$

are bounded.

The Fourier transform of (10.29) yields

$$C_2^{xy}(\omega) = L_1 H_1(-\omega) H_2(-\omega) C_2^x(\omega). \tag{10.31}$$

The above equation is meaningful when $L_1 \neq 0$. Obviously (10.30) is not sufficient to construct individual estimates of $H_1(\omega)$ and $H_2(\omega)$. More relationships involving these quantities can be defined using higher-order statistics of the input and output.

Let $c_3^{xxy}(u,v)$ be the cross-third-order cumulant of the input and output series. Again, applying Lemma 10.1 we obtain the relationship

$$c_3^{xxy}(u,v) = L_2 \sum_w \sum_x \sum_y h_1(x) h_1(y) m_2^x(x + u + w) m_2^x(y + v + w) h_2(w) \tag{10.32}$$

where $L_2 = cum\{U(0), U(0), V(0)\}/[\gamma_2^{U(0)}]^2$, provided that (10.30) is valid and $E\{|U(k+u)U(k+v)G[U(k)]|\} < \infty$. Taking the Fourier transform of (10.32) we obtain

$$C_3^{xxy}(\omega_1, \omega_2) = L_2 H_1(-\omega_1) H_1(-\omega_2) H_2(-\omega_1 - \omega_2) C_2^x(\omega_1) C_2^x(\omega_2). \tag{10.33}$$

From (10.31) and (10.33), it follows that

$$\frac{C_3^{xxy}(\omega_1, \omega_2) C_2^x(\omega_1 + \omega_2)}{C_2^{xy}(\omega_1 + \omega_2) C_2^x(\omega_1) C_2^x(\omega_2)} = \frac{L_2 \cdot H_1(-\omega_1) H_1(-\omega_2)}{L_1 H_1(-\omega_1 - \omega_2)}, \tag{10.34}$$

and setting $\omega_1 = -\omega_2$,

$$|H_1(\omega)|^2 \left(\frac{L_2}{L_1 H_1(0)}\right) = \frac{C_3^{xxy}(\omega, -\omega)}{[C_2^x(\omega)]^2} \cdot \frac{C_2^x(0)}{C_2^{xy}(0)}. \tag{10.35}$$

Hence, within a scalar constant, the Fourier magnitude of $H_1(\omega)$ is given by

$$|H_1(\omega)| = k \frac{[C_3^{xxy}(\omega, -\omega)]^{1/2}}{C_2^x(\omega)}, \quad k \neq 0 \tag{10.36}$$

where k is defined by comparing (10.35) and (10.36). If we let $\phi_1(\omega) = \arg\{H_1(\omega)\}$ and $\Psi(\omega_1,\omega_2) = \arg\{\frac{C_3^{xxy}(\omega_1,\omega_2)}{C_2^{xy}(\omega_1+\omega_2)}\}$, then from (10.34) we have

$$\phi_1(\omega_1+\omega_2) - \phi_1(\omega_1) - \phi(\omega_2) = \Psi(\omega_1,\omega_2) \bmod \pi \qquad (10.37)$$

and consequently, the formula

$$\phi_1(\omega) = \{\int_0^\omega \phi_1(a)da + \int_0^\omega \Psi(a,\omega-a)da\}/\omega \qquad (10.38)$$

is a recursive equation for obtaining $\phi_1(\omega)$ given $\phi_1(a)$ $(0 \leq a < \omega)$.

An alternative way of obtaining the Fourier phase of $H_1(\omega)$ is the following. Taking the conjugate of (10.31) at $\omega = \omega_1 + \omega_2$; and, multiplying the result with (10.33) we obtain

$$H_1(\omega_1)H_1(\omega_2)H_1^*(\omega+\omega_2)|H_2(\omega_1+\omega_2)|^2 = \frac{C_2^{xy*}(-\omega_1-\omega_2)C_3^{xxy}(-\omega_1,-\omega_2)}{L_1L_2C_2^x(-\omega_1-\omega_2)C_2^x(-\omega_1)C_2^x(-\omega_2)}. \qquad (10.39.1)$$

In other words,

$$\text{phase } \{C_3^{h_1}(\omega_1,\omega_2)\} = \text{phase } \{C_2^{xy*}(-\omega_1-\omega_2)C_3^{xxy}(-\omega_1,-\omega_2)\}. \qquad (10.39.2)$$

From the bispectral phase of $H_1(\omega)$, namely $\Psi_3^{h_1}(\omega_1,\omega_2)$, we can compute the phase of $H_1(\omega)$, i.e., $\phi_1(\omega)$, by using any of the nonparametric methods described in Chapter 6.

For the particular case where $G(u) = a_1u + a_2u^2$ (simple quadratic system), the results on system identification for Gaussian inputs, originally derived by Tick [1961] and presented in section 10.2, agree with the results obtained in this section.

Identification by Observing the Output Only

In the preceding section, the linear filters in Figure 10.3 were estimated, within a scalar constant, by computing cross-correlations between the input and the output of the system. The only assumption made about the nonlinear function $G[\cdot]$ was that it maps reals to reals. For the case of Gaussian white input, Rozario and Papoulis [1989] showed that the system illustrated in Figure 10.3 can be identified from knowledge of the output $Y(k)$ only, provided that (a) $H_1(\omega)$ and $H_2(\omega)$ are minimum phase, and (b) $G[\cdot]$ is monotonic. The identification procedure by Rozario and Papoulis [1989] is based on the following observation.

Let the process $V(k)$ be obtained from $U(k)$ by a memoryless nonlinear transformation, i.e., $V(k) = G[U(k)]$, where $G[\cdot]$ maps reals to reals. If $U(k)$ is Gaussian, then all polyspectra of $V(k)$ are real valued. The proof of this statement is briefly sketched by Rozario and Papoulis [1989]. Then the method proceeds as follows. Since $U(k)$ is Gaussian, its highest-order nonzero spectrum is the power spectrum which is a real function. The nonlinear function $G[\cdot]$ maps reals to reals, hence the phase of the polyspectrum of $Y(k)$ is entirely due to the phase of the linear system, $H_2(\omega)$. Therefore, after we estimate the phase of $Y(k)$, or equivalently, the phase of $H_2(\omega)$ based on the minimum phase assumption for $H_2(\omega)$, we can estimate the magnitude of $H_2(\omega)$ as described by Oppenheim and Schafer [1989]. Once $H_2(\omega)$ is determined, we can also determine $V(k)$ and its probability density function (pdf). The pdf of $U(k)$ is Gaussian, and in order to estimate $G[\cdot]$ by comparing the pdf of $V(k)$ and $U(k)$, we must assume that $G[\cdot]$ is monotonically increasing. Otherwise the solution is not unique. This can be seen from the following arguments:

If $G[\cdot]$ is monotonically increasing, then

$$P\{V(k) \leq \nu\} = P\{U(k) \leq u\} \qquad (10.40.1)$$

or

$$F_v(\nu) = F_u(u) \qquad (10.40.2)$$

where $P\{\cdot\}$ denotes probability and $F_v(\nu)$ and $F_u(u)$ are the probability distribution functions of $V(k)$ and $U(k)$, respectively.

If $G[\cdot]$ was not monotonic, for example if $V(k) = U^2(k)$, then

$$P\{V(k) \leq \nu\} = P\{-u \leq U(k) \leq u\} \qquad (10.41)$$

or

$$V_v(\nu) = F_u(u) - F_u(-u) \qquad (10.42)$$

which does not lead to a unique solution.

10.4 QUADRATIC PHASE COUPLING

There are situations where, because of interaction between two harmonic components of a process, there is contribution to the power at their sum and/or difference frequencies. Such a phenomenon which could be due to second-order nonlinearities gives rise to certain phase relations called quadratic phase coupling. A common example is amplitude modulation. Quadratic phase coupling can arise only among harmonically related components. Three frequencies are harmonically related when one of them is the sum or difference of the other two. A special case is when we have two components with one being at twice the frequency of the other. In certain applications it is necessary to find out if peaks at harmonically related positions in the power spectrum are in fact phase coupled. Since the power spectrum suppresses all phase relations, it cannot provide the answer.

Consider the process [Raghuveer and Nikias, 1985]

$$X(k) = \sum_{i=1}^{6} \cos(\lambda_i k + \phi_i) \tag{10.43}$$

where $\lambda_1 > \lambda_2 > 0$, $\lambda_4 > \lambda_5 > 0$, $\lambda_3 = \lambda_1 + \lambda_2$, $\lambda_6 = \lambda_4 + \lambda_5$, $\phi_1, \phi_2, \cdots, \phi_5$ are all independent, uniformly distributed random variables over $(0, 2\pi)$ and $\phi_6 = \phi_4 + \phi_5$.

In (10.43) while $(\lambda_1, \lambda_2, \lambda_3)$ and $(\lambda_4, \lambda_5, \lambda_6)$ are at harmonically related positions, only the component at λ_6 is a result of phase coupling between those at λ_4 and λ_5 while the one at λ_3 is an independent harmonic component. The power spectrum of the process consists of impulses at λ_i, $i = 1, 2, \cdots, 6$ as illustrated in Figure 10.4(a). Looking at the power spectrum one cannot say if the harmonically related components are in fact involved in quadratic phase coupling relationships. The third moment sequence, $c_3^x(\tau_1, \tau_2)$, of $X(k)$ can be easily obtained as

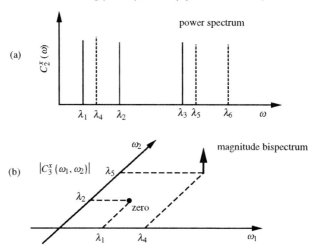

Figure 10.4 The effect of the phase coupling in the (a) power spectrum and (b) magnitude bispectrum domain

$$\begin{aligned}
c_3^x(\tau_1, \tau_2) &= \tfrac{1}{4}\{\cos(\lambda_5\tau_1 + \lambda_4\tau_2) + \cos(\lambda_6\tau_1 - \lambda_4\tau_2) \\
&+ \cos(\lambda_4\tau_1 + \lambda_5\tau_2) + \cos(\lambda_6\tau_1 - \lambda_5\tau_2) \\
&+ \cos(\lambda_4\tau_1 - \lambda_6\tau_2) + \cos(\lambda_5\tau_1 - \lambda_6\tau_2)\}.
\end{aligned} \tag{10.44}$$

It is important to note that in (10.44) only the phase coupled components appear. Hence, the bispectrum evaluated in the triangular region defined by the lines $\omega_2 = 0, \omega_1 = \omega_2$, and $\omega_1 + \omega_2 = \pi$, shows an impulse only at (λ_4, λ_5) indicating that only this pair is phase coupled (Figure 10.4b).

Thus, the fact that *only phase coupled components contribute to the third moment sequence* of a process is what makes the bispectrum a useful tool for detecting quadratic phase coupling and discriminating phase coupled components from those that are not.

A more general form of quadratic phase coupling can be seen by passing the signal $\{X(k) = A\cos(\lambda_1 k + \phi_1) + B\cos(\lambda_2 k + \phi_2)\}$ through the nonlinear quadratic system $\{G(x) = x + \epsilon x^2$ where $\epsilon > 0$ (ϵ small so that ϵ^3 is negligible)$\}$. The output of the quadratic system $\{Y(k) = X(k) + \epsilon X^2(k)\}$ contains cosinusoidal terms of the form $(\lambda_1, \phi_1), (\lambda_2, \phi_2), (2\lambda_2, 2\phi_2), (\lambda_1 + \lambda_2, \phi_1 + \phi_2)$ and $(\lambda_1 - \lambda_2, \phi_1 - \phi_2)$. Thus, quadratic phase coupling is the phenomenon which gives rise to phase relations that are of the same type of frequency relations in cosinusoidal signals.

Depending on the bispectrum estimation method employed, the techniques for the detection and quantification of quadratic phase coupling are divided into two categories: the *conventional* and the *parametric*. The conventional techniques are based on the bicoherence spectrum and they are better quantifiers of the phase coupling [Kim and Powers, 1978; 1979]. However, their resolution is limited by the "uncertainty principle" of the Fourier transform. On the other hand, the parametric techniques are based on the autoregressive (AR) modeling of the third-order cumulants. Although parametric AR methods are not good quantifiers, they possess a high resolution capability, much higher than the frequency resolution of conventional methods [Raghuveer and Nikias, 1985; 1986].

10.4.1 Conventional Techniques for Detection and Quantification of Quadratic Phase Coupling

The conventional techniques for bispectrum estimation were discussed extensively in Chapter 4. Kim and Powers [1978] considered the following equivalent definition of the bispectrum of a stochastic process, $X(k)$,

$$C_3^x(\omega_1, \omega_2) = E\{X(\omega_1)X(\omega_2)X^*(\omega_1 + \omega_2)\} \qquad (10.45)$$

where $X(\omega)$ is the Fourier transform of $X(k)$.

If the components present at λ_1, λ_2 and $(\lambda_1 + \lambda_2), (\lambda_1 - \lambda_2)$ in the power spectrum of the signal are spontaneously excited modes, each will be characterized by statistically independent random phases. Thus, through the statistical averaging in (10.45), the bispectrum will take a zero value. On the other hand, if the sum and/or difference component is generated through some nonlinear interaction, then phase coherency exists and the statistical averaging will not lead to a zero bispectrum value.

Let the magnitude-squared bicoherency be defined as

$$|b(\omega_1,\omega_2)|^2 \triangleq \frac{|C_3^x(\omega_1,\omega_2)|^2}{C_2^x(\omega_1)C_2^x(\omega_2)C_2^x(\omega_1+\omega_2)} \qquad (10.46)$$

where $C_2^x(\omega)$ is the power spectrum of $X(k)$. For a high degree of phase coherency, $b(\omega_1,\omega_2)$ at (λ_1,λ_2) frequency pairs is close to unity, which indicates that quadratic interaction has taken place. On the other hand, a value of $b(\omega_1,\omega_2)$ close to zero indicates low degree of phase coherency, or in other words, the power spectrum domain modes at $\lambda_1, \lambda_2, \lambda_1+\lambda_2, \lambda_1-\lambda_2$ are probably spontaneously excited independently, rather than being quadratically coupled modes.

Example 10.2 [Kim and Powers, 1978]

The detection and quantification of the quadratic phase coupling phenomenon using the bicoherence spectrum is demonstrated through an application to plasma fluctuation data. The plasma density fluctuations were recorded at two probes placed at a distance ΔS apart. Let $X(k)$ and $Y(k)$ be the recorded data at the two probes, and let $C_2^{xy}(\omega) = X(\omega)Y^*(\omega) = |C_2^{xy}(\omega)|e^{j\theta_{xy}(\omega)}$ be the cross-spectrum of $X(k)$ and $Y(k)$ when $X(\omega)$, and $Y(\omega)$ are the Fourier transforms of $X(k)$ and $Y(k)$, respectively. Of particular interest is the phase spectrum, $\theta_{xy}(\omega)$, which may be interpreted as the phase shift that each spectral component ω undergoes in traveling the distance ΔS between the two probes. Hence, $\theta_{xy}(\omega) = k(\omega)\Delta S$, where $k(\omega)$ is the wavenumber for a given frequency. Thus, from the peaks in $|C_2^{xy}(\omega)|$, one can determine the frequency of the modes present in plasma, and from the phase spectrum, one can determine the wavenumber, $k(\omega)$.

Plots of $|C_2^{xy}(\omega)|$ and $\theta_{xy}(\omega)$ versus frequency are displayed in Figure 10.5 for two different magnetic fields ($\omega = 2\pi f$).

The spectrum at $B = 570$ G in Figure 10.5(a) is of interest since it exhibits various peaks at $f_a = 28.5$ kHz, $f_b = 52.5$ kHz, $f_c = 81$ kHz, $f_d = 133.5$ kHz, and $f_e = 162$ kHz. Note that the following frequency relations hold:

$$f_a = f_c - f_b, \ f_d = f_c + f_b \text{ and } f_e = 2f_c.$$

From Figure 10.5(b), we can verify that for the corresponding wavenumbers the following relations hold:

$$k_a = k_c - k_b, \ k_d = k_c + k_b, \ k_e = 2k_c.$$

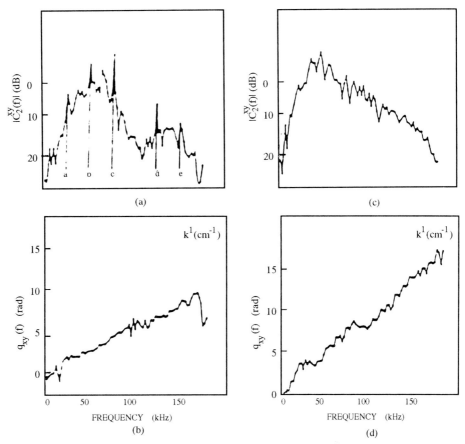

Figure 10.5 Cross-power spectra representative of the multimode regime at $B = 570$ G [plots (a) and (b)], and the turbulent regime at $B = 779$ G [plots (c) and (d)] ([Kim and Powers, 1978] © *Physics Fluids*).

Note that the right-hand vertical axis of the phase spectrum is calibrated in terms of wavenumbers.

In order to determine whether the a, b, c, d, and e modes are coupled modes or not, the magnitude-squared bicoherency is computed as follows. The measured data of length $N = 8192$ were divided into 64 records of 128 samples each. The fast Fourier transform was used to generate the Fourier transform of each record. The bispectrum was generated by forming the triple product of the Fourier transforms of each record at frequencies ω_1, ω_2 and $\omega_1 + \omega_2$, and then averaging the triple products corresponding to each record.

In Figure 10.6 the magnitude-squared bicoherency is demonstrated in the triangular region $0 \leq f_1 \leq f_N$, $f_2 \leq f_1$, $f_1 + f_2 \leq f_N$ where f_N is the Nyquist frequency. Perspective views and contour maps of the bicoherency are shown in Figures 10.6

(a) and (b). At $B = 570$ G, the bicoherency spectrum exhibits three peaks at (f_a, f_b), (f_b, f_c), and (f_e, f_c). Thus, we conclude that the primary modes, f_b and f_c, interact and generate a mode at $f_d = f_b + f_c$ and another mode at $f_a = f_c - f_b$. In addition the second harmonic of f_c is generated at f_e.

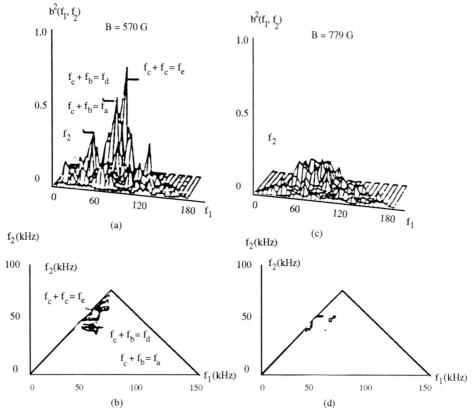

Figure 10.6 Perspective views and contour maps of the magnitude-squared bicoherency representative of the multimode regime at $B = 570$ G [Plots (a) and (b)] and the turbulent regime [plots (c) and (d)] ([Kim and Powers, 1978] ©*Physics Fluids*).

Considering the fluctuation data at $B = 779$ G, we observe no strong coherent peaks in the power spectrum. In such a turbulent situation the power spectrum contains many independent frequencies, characterized by statistically independent random phases. Thus, no phase coherency exists between them. Indeed, the magnitude-squared bicoherency spectrum [Figure 10.6 (c) and (d)] has very low value at the points which correspond to all combinations of f_1 and f_2. □

10.4.2 Parametric Methods for the Detection of Phase Coupling

The parametric techniques are based on autoregressive (AR) modeling of the third-order cumulants, and they are shown to be high-resolution quadratic phase coupling detectors. In the sequel, three AR parametric methods are discussed: (a) The Third-Order Recursion (TOR) method introduced by Raghuveer and Nikias [1985], (b) the Constrained Third-Order Mean (CTOM) [Raghuveer and Nikias, 1986] and (c) a least-squares extension of the TOR method, called the Optimized AR Method (OARM) suggested by An, Kim, and Powers [1988].

Third-Order Recursion (TOR) Method

This method is based on a parametric AR estimation of the bispectrum. Consider a pth order AR process, $X(k)$, described by

$$X(k) + \sum_{i=1}^{p} a(i)X(k-i) = W(k) \tag{10.47}$$

where $W(k)$ are i.i.d., non-Gaussian, with $E\{W(k)\} = 0$, $E\{W^2(k)\} = Q$, $E\{W^3(k)\} = \beta \neq 0$ and $X(k)$ independent of $W(\ell)$ for $k < \ell$. Since $W(k)$ is third-order stationary, $X(k)$ is also third-order stationary assuming it is a stable AR model. For the model of (10.47) we then have

$$c_3^x(-k,-\ell) + \sum_{i=1}^{p} a(i)c_3^x(i-k, i-\ell) = \beta \cdot \delta(k,\ell); \quad k, \ell \geq 0 \tag{10.48}$$

where $c_3^x(k,\ell)$ is the third-order moment or cumulant sequence of the AR process and $\delta(k,\ell)$ is the 2-d unit impulse function. From (10.48), which shall be referred to as the *third-order recursion equation*, it follows that $2p+1$ values of $c_3^x(k,\ell)$ along the line $k = \ell$ satisfy the matrix equation

$$\mathbf{R} \cdot \mathbf{a} = \mathbf{b} \tag{10.49}$$

where $\mathbf{a} \triangleq [1, a(1), \cdots, a(p)]^T$, $b \triangleq [\beta, 0, \cdots, 0]^T$ and

$$\mathbf{R} = \begin{vmatrix} c^x(0,0) & c_3^x(1,1) & \cdots & c_3^x(p,p) \\ c_3^x(-1,-1) & c_3^x(0,0) & \cdots & c_3^x(p-1,p-1) \\ \vdots & & & \\ c_3^x(-p,-p) & c_3^x(-p+1,-p+1) & \cdots & c_3^x(0,0) \end{vmatrix}.$$

The matrix **R** is Toeplitz and generally *not* symmetric. An essential condition for the existence of (10.49) is the stability of the AR filter

$$H(z) = 1/A(z) \qquad (10.50.1)$$

where

$$A(z) = 1 + \sum_{i=1}^{p} a(i)z^{-1}. \qquad (10.50.2)$$

A sufficient condition for the stability of $H(z)$ is that **R** is Toeplitz, symmetric, and positive definite [Makhoul, 1975]. However, this is not a necessary condition. Thus, for all those processes whose third-order cumulants satisfy the sufficient conditions stated above, stable AR representations of order p can be derived based on knowledge of the $2p + 1$ cumulants $\{c_3^x(-p,-p), \cdots, c_3^x(p,p)\}$.

The bispectrum of the AR process of (10.47) is given by

$$C_3^x(\omega_1, \omega_2) = \beta H(\omega_1) H(\omega_2) H^*(\omega_1 + \omega_2); \ |\omega_1|, |\omega_2| \le \pi. \qquad (10.51)$$

Parametric Bispectrum Estimation

Suppose now that we are given only a finite length of data and it is required to estimate the bispectrum of the underlying discrete random process. For this purpose, we propose a parametric method that involves the fitting of an AR model driven by non-Gaussian white noise. The method is based on (10.49) and (10.51), where we substitute estimated third moments in place of the true moments which are not known. Let $\{X(1), X(2), \cdots, X(N)\}$ be the given data set. Then the parametric bispectrum continuation proceeds as follows.

1. Form the biased third-order moment estimates, i.e.,

    (a) Segment the data into K records of M samples each, i.e., $N = K \cdot M$. Let $\{X^i(\ell); \ell = 1, 2, \cdots, M\}$ be the data samples of the ith record.

    (b) For each record, compute $c_3^{x,i}(m,n)$ as

    $$c_3^{x,i}(m,n) = \frac{1}{M} \sum_{\ell=\max(1,1-m,1-n)}^{\min(M,M-m,M-n)} X^i(\ell) X^i(\ell+m) X^i(\ell+n); \ i = 1,2,\cdots,K$$
    $$(10.52)$$

    (c) Average $c_3^{x,i}(m,n)$ over all records to obtain the overall estimate $\hat{c}_3^x(m,n)$ of $c_3^{x,i}(m,n)$; viz,

    $$\hat{c}_3^x(m,n) = \frac{1}{K} \sum_{i=1}^{K} c_3^{x,i}(m,n) \qquad (10.53)$$

2. Substitute the estimated moments (or cumulants) in place of the true ones in (10.49) to obtain

$$\hat{\mathbf{R}} \cdot \hat{\mathbf{a}} = \hat{\mathbf{b}} \qquad (10.54)$$

where

$$\hat{\mathbf{R}} = \begin{bmatrix} \hat{c}_3^x(0,0) & \hat{c}_3^x(1,1) & \cdots & \hat{c}_3^x(p,p) \\ \hat{c}_3^x(-1,-1) & \hat{c}_3^x(0,0) & \cdots & \hat{c}_3^x(p-1,p-1) \\ \vdots & & & \\ \hat{c}_3^x(-p,-p) & \hat{c}_3^x(-p+1,p+1) & \cdots & \hat{c}_3^x(0,0) \end{bmatrix}$$

$\{\hat{\mathbf{a}} = [1, \hat{a}(1), \cdots, \hat{a}(p)]^T; \hat{a}(i); i = 1, \cdots, p\}$ are the estimates of the AR parameters, $\{\hat{\mathbf{b}} = [\hat{\beta}, 0, \cdots, 0]^T$, and $\hat{\beta}$ is the estimate of the third-order moment of the driving noise.

3. Compute the normalized bispectrum estimate as

$$\frac{1}{\hat{\beta}} \hat{C}_3^x(\omega_1, \omega_2) = \hat{H}(\omega_1)\hat{H}(\omega_2)\hat{H}^*(\omega_1 + \omega_2) \qquad (10.55.1)$$

where

$$\hat{H}(\omega) = \frac{1}{1 + \sum_{i=1}^{p} \hat{a}(i) e^{-j\omega i}}, \quad |\omega| \leq \pi. \qquad (10.55.2)$$

Note that AR parameters obtained by second-order methods (i.e., Yule-Walker) contain only the power spectrum information, and thus cannot be used for bispectrum estimation.

The model that describes the bispectrum of a process is generally different from the one that describes its power spectrum [Raghuveer and Nikias, 1985]. Generally the appropriate model order for the presented method will be different from that for AR power spectrum estimation. Some of the well-known AR model order selection criteria such as AIC, FPE, and Parzen's CAT [Kay and Marple, 1981] depend on autocorrelations and hence cannot be used. However, the AR part of the criteria of Chow [1972] and Chan and Wood [1984] may be adapted to the third-order case.

The following theorem postulated by Raghuveer and Nikias [1985] provides a justification for the use of AR models for the detection of quadratic phase coupling.

Theorem 10.1

There always exist parameters $\{a(1), \cdots, a(6N)\}$ such that the third moment sequence of the process $\{X(k)\}$ consisting of a sum of cosinusoids of which N pairs are coupled, namely

$$X(k) = A \sum_{i=1}^{N} \sum_{\ell=1}^{3} \cos[\omega_\ell^{(i)} k + \phi_\ell^{(i)}] + \sum_{j=1}^{M} B_j \cos[\lambda_j k + \psi_j] \qquad (10.56)$$

where $\omega_1^{(i)}, \omega_2^{(i)}$ and $\omega_3^{(i)} = \omega_1^{(i)} + \omega_2^{(i)}$ are all distinct, $\phi_3^{(i)} = \phi_1^{(i)} + \phi_2^{(i)}$ for $i = 1, \cdots, N$ and $\phi_1^{(i)}, \phi_2^{(i)}, \psi_j$ are all independent and uniformly distributed over $(0, 2\pi)$, satisfies exactly the third-order recursion of (10.48) with $\mathbf{a} = [1, a(1), \cdots, a(6N)]^T$.

Proof:

Of the $3N + M$ cosinusoids in (10.56), only N pairs represented by the first sum are coupled to produce the N components at $\omega_3^{(i)}$. The third-order moment sequence, $c_3^x(m, n)$ of $X(k)$, is obtained as

$$\begin{aligned}
c_3^x(m, n) &= \frac{A^3}{4} \sum_{i=1}^{N} \{\cos(\omega_3^{(i)} m - \omega_1^{(i)} n) \\
&+ \cos(\omega_2^{(i)} m + \omega_1^{(i)} n) + \cos(\omega_3^{(i)} m - \omega_2^{(i)} n) \\
&+ \cos(\omega_1^{(i)} m + \omega_2^{(i)} n) + \cos(\omega_2^{(i)} m - \omega_3^{(i)} n) \\
&+ \cos(\omega_1^{(i)} m - \omega_3^{(i)} n)\}.
\end{aligned} \qquad (10.57)$$

Hence, the bispectrum of $X(k)$ in its principal triangular region ($\omega_2 = 0, \omega_1 = \omega_2, \omega_1 + \omega_2 = \pi$) has impulses at $(\omega_1^{(i)}, \omega_2^{(i)}), i = 1, \cdots, N$ (assuming $\omega_1^{(i)} > \omega_2^{(i)} > 0$). Define

$$\begin{aligned}
E(m, n) &\triangleq \sum_{i=1}^{N} \{\exp(j(\omega_3^{(i)} m - \omega_1^{(i)} n)) + \exp(j(\omega_2^{(i)} m + \omega_1^{(i)} n)) \\
&+ \exp(j(\omega_3^{(i)} m - \omega_2^{(i)} n)) + \exp(j(\omega_1^{(i)} m + \omega_2^{(i)} n)) \\
&+ \exp(j(\omega_2^{(i)} m - \omega_3^{(i)} n)) + \exp(j(\omega_1^{(i)} m - \omega_3^{(i)} n))\}.
\end{aligned} \qquad (10.58)$$

From (10.57) and (10.58) we get

$$c_3^x(m, n) = \frac{A^3}{8} \{E(m, n) + E^*(m, n)\} \qquad (10.59)$$

Consider the polynomial

$$P(z) = \prod_{i=1}^{N} \prod_{k=1}^{3} (z - \exp(j\omega_k^{(i)}))(z - \exp(-j\omega_k^{(i)})). \qquad (10.60)$$

Chapter 10   469

$P(z)$ has real coefficients and is of degree $6N$ and can therefore be written as

$$P(z) = \sum_{i=0}^{6N} a(i) z^i. \qquad (10.61)$$

It is easy to see that $a(0) = a(6N) = 1$. From (10.60) we have

$$P(\exp(\pm j\omega_k^{(i)})) = 0 \text{ for } i = 1, \cdots, N, \ k = 1, 2, 3. \qquad (10.62)$$

Now, making use of $\omega_3^{(i)} = \omega_1^{(i)} + \omega_2^{(i)}$,

$$\begin{aligned}
\sum_{i=0}^{6N} a(i) E(i-m, i-n) &= \sum_{i=1}^{N} \{\exp(-j(\omega_3^{(i)} m - \omega_1^{(i)} n)) \cdot \sum_{\ell=0}^{6N} a(\ell) \exp(j\omega_2^{(i)} \ell) \\
&+ \exp(-j(\omega_2^{(i)} m + \omega_1^{(i)} n)) \sum_{\ell=0}^{6N} a(\ell) \exp(j\omega_3^{(i)} \ell) \\
&+ \exp(-j(\omega_3^{(i)} m - \omega_2^{(i)} n)) \cdot \sum_{\ell=0}^{6N} a(\ell) \exp(j\omega_1^{(i)} \ell) \\
&+ \exp(-j(\omega_1^{(i)} m + \omega_2^{(i)} n)) \cdot \sum_{\ell=0}^{6N} a(\ell) \exp(j\omega_3^{(i)} \ell) \\
&+ \exp(-j(\omega_1^{(i)} m - \omega_3^{(i)} n)) \sum_{\ell=0}^{6N} a(\ell) \exp(-j\omega_1^{(i)} \ell) \\
&+ \exp(-j(\omega_1^{(i)} m - \omega_3^{(i)} n)) \cdot \sum_{\ell=0}^{6N} a(\ell) \exp(-j\omega_2^{(i)} \ell)]\}.
\end{aligned}$$
$$(10.63)$$

From (10.61) and (10.62) we can see that each of the $\sum_{\ell=0}^{6N} a(\ell) \exp(.)$ terms in (10.63) is an evaluation of $P(z)$ at one of its roots. Therefore

$$\sum_{\ell=0}^{6N} a(\ell) E(\ell - m, \ell - n) = 0. \qquad (10.64)$$

Similarly, it can be shown that

$$\sum_{\ell=0}^{6N} a(\ell) E^*(\ell - m, \ell - n) = 0. \qquad (10.65)$$

From (10.59), (10.64), and (10.65) it follows that

$$\sum_{\ell=0}^{6N} a(\ell) c_2^x(\ell - m, \ell - n) = 0. \tag{10.66}$$

In (10.56) the components that enter into quadratic phase coupling relationships, $(\omega_1^{(i)}, \omega_2^{(i)}, \omega_3^{(i)})$ are accounted for by the first summation. The second summation accounts for all other components including those that might be harmonically related but not phase coupled. The fact that there are no restrictions on the B_j's shows that the parameters $\{a(1), \cdots, a(6N)\}$ are determined solely by the coupled portion and the second term plays no role. If these parameters were substituted in (10.51), we would find impulses at all the positions where there are impulses in the true bispectrum of $X(k)$. If some of the $\omega_k^{(i)}$'s are identical, the third-order recursion is still satisfied, but with order less than $6N$. The process $\{X(k)\}$ in (10.43) is a special case of the one in (10.56) with $N = 1, M = 3$ and the corresponding AR parameters that satisfy the third-order recursion are:

$$a(1) = -2(\cos\lambda_4 + \cos\lambda_5 + \cos\lambda_6),$$

$$a(2) = 3 + 4(\cos\lambda_4\cos\lambda_5 + \cos\lambda_5\cos\lambda_6 + \cos\lambda_6\cos\lambda_4),$$

$$a(3) = -4(\cos\lambda_4 + \cos\lambda_5 + \cos\lambda_6 + 2\cos\lambda_4\cos\lambda_5\cos\lambda_6),$$

$$a(4) = a(2),\ a(5) = a(1),\ \text{and}\ a(6) = 1.$$

With these parameters the normalized parametric bispectrum shows an impulse only at (λ_4, λ_5). So far, the TOR method applies for the detection of quadratic phase coupling, but it does not provide any information about the degree of the coupling. In order to quantify the phase coupling in a situation that use of parametric methods is necessary (high resolution requirement could be a reason), the bicoherency can be estimated in a parametric way as follows:

$$\hat{b}_x(\omega_1, \omega_2) = \frac{\hat{C}_3^x(\omega_1, \omega_2)}{\sqrt{\hat{C}_2^x(\omega_1)\hat{C}_2^x(\omega_2)\hat{C}_2^x(\omega_1 + \omega_2)}} \tag{10.67}$$

where $\hat{C}_2^x(\omega)$ is the power spectrum estimate formed by using any of the known parametric methods (i.e., Yule-Walker) and $\hat{C}_3^x(\omega_1, \omega_2)$ is the AR bispectrum estimate. In (10.67) numerator and denominator involve parameter estimates of two completely different models and therefore $|\hat{b}_x(\omega_1, \omega_2)|$ could exceed unity, unlike some types of conventional bicoherency estimates. For this reason Raghuveer and Nikias [1985] pointed out that $|\hat{b}_x(\omega_1, \omega_2)|$ in (10.67) qualifies better as a detector rather than as a quantifier of phase coupling. Again, it may be used only for

Chapter 10

detection of quadratic phase coupling, especially in situations where conventional estimators completely fail to resolve closely spaced frequency components.

The Constrained Third-Order Mean Method (CTOM)

Let $\{X(1), \cdots, X(N)\}$ be zero-mean samples of a stationary process whose bispectrum is to be estimated. Consider

$$\hat{q}_m(k, i) = X(m-i)X^2(m-k); \ i = 1, \cdots, p. \tag{10.68}$$

We can easily verify that $E\{\hat{q}_m(k,i)\} = c_3^x(i-k, i-k)$ where $c_3^x(m,n)$ is the third-order moment (or cumulant) sequence of the process $X(k)$ whose samples are given above.

If we were given samples of $c_3^x(m,n)$ itself, then to fit a pth order AR model we would have to solve the system of equations [see (10.48)]:

$$E\{\hat{q}_m(k, 0) + \sum_{i=1}^{p} \hat{a}(i)\hat{q}_m(k,i)\} = 0; \ k = 1, \cdots, p. \tag{10.69}$$

If we denote the quantity within the braces by $\hat{C}(m,k)$, (10.69) becomes

$$E\{\hat{C}(m,k)\} = 0; \ k = 1, \cdots, p. \tag{10.70}$$

We refer to $\hat{C}(m,k)$ as the third-order recursion (TOR) process. With the given data samples we can obtain $N - p$ samples of $\hat{C}(m,k)$ for every value of k, i.e., we have $\hat{C}(m,k)$ for $m = p+1, p+2, \cdots, N; \ k = 1, \cdots, p$.

Instead of taking expectations as in (10.70) we now equate the sample mean of $\hat{C}(m,k)$ to zero and obtain p linear equations to be solved for the parameters:

$$\frac{1}{N-p} \sum_{m=p+1}^{N} \hat{C}(m,k) = 0; \ k = 1, \cdots, p. \tag{10.71}$$

In matrix form, (10.71) becomes

$$\hat{\mathbf{Q}} \cdot \hat{\mathbf{a}} = \hat{\mathbf{b}} \tag{10.72}$$

where

$$\hat{\mathbf{Q}} = \begin{bmatrix} \hat{q}_{11} & \cdots & \hat{q}_{1p} \\ \vdots & & \\ \hat{q}_{p1} & \cdots & \hat{q}_{pp} \end{bmatrix}$$

$$\hat{\mathbf{a}} = [\hat{a}(1), \cdots, \hat{a}(p)]^T; \ \mathbf{b} = [\hat{q}_{10}, \cdots, \hat{q}_{p0}]^T,$$

$$\hat{q}_{ij} \stackrel{\Delta}{=} \sum_{m=p+1}^{N} \hat{q}_m(i,j).$$

After solving for $\hat{a}(i)$'s, the bispectrum is formed as in (10.55).

Remarks

For simplicity of notation, $\hat{C}(m,k)$ was considered to be estimated from a single data record. However, we may choose to divide the given data into several records, with a different $\hat{C}(m,k)$ value for each record, and then average $\hat{C}(m,k)'s$ over all records.

Third-order ergodicity is a necessary condition for the method to give meaningful results. Asymptotically, Raghuveer and Nikias [1986] showed that the CTOM method is equivalent to the TOR for a given model order. The CTOM method provides consistent estimates of the AR parameters when the process satisfies the model assumption in (10.47).

Example 10.3

In this example the TOR and CTOM methods are compared to a conventional approach for the detection of quadratic phase coupling. Consider the real discrete process [Raghuveer and Nikias, 1986]

$$X(k) = \sum_i \cos(w_i k + \phi_i) + W(k); \quad i = a, b, \cdots, f$$

where $W(k)$ is taken to be -40dB white Gaussian noise and

$$\begin{array}{rclcrcl}
w_a &=& 2\pi(0.109375) &,& w_b &=& 2\pi(0.1935), \\
w_c &=& w_a + w_b &,& w_d &=& 2\pi(0.117375), \\
w_e &=& 2\pi(0.1875) &,& w_f &=& w_d + w_e.
\end{array}$$

ϕ_a, ϕ_b, ϕ_d, and ϕ_e are independent and uniformly distributed on $(0, 2\pi)$, $\phi_c = \phi_a + \phi_b$ and $\phi_f = \phi_d + \phi_e$. The true bispectrum magnitude has two impulses at (w_a, w_b) and (w_c, w_d). Sixty-four independent records of the process are obtained with each record containing 128 samples (the records are independent because ϕ_a, ϕ_b, ϕ_d, and ϕ_e are generated afresh for each record).

For comparison purposes, a conventional method is also employed for the detection of the quadratic phase coupling. For this particular conventional method the bispectrum is estimated as follows. The data are segmented into 64 records of

128 samples each. The DFT of each record is obtained and then the triple product of the DFT is formed. Finally, the bispectrum is taken to be the average of these triple products obtained from all records.

Figure 10.7(a) shows the conventional bispectrum magnitude estimate. Instead of two peaks at (w_a, w_b) and (w_d, w_e) we see just one midway between the correct locations. The conventional method has thus failed to resolve the two peaks. Figure 10.7(b) shows the estimate provided by the TOR method with a 30th order model. The two peaks are clearly resolved, but the estimated peak locations appear to be shifted from the true ones. Figure 10.7(c) shows the CTOM magnitude bispectrum estimate obtained by the CTOM method with a 20th order model. Now the peaks are resolved and correctly located. The important feature to observe in this example is the superiority in terms of resolution of the parametric methods against the conventional one. □

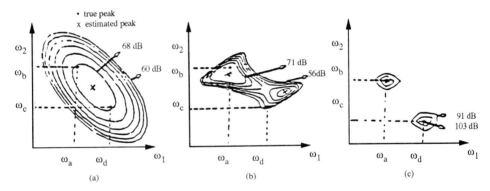

Figure 10.7 Magnitude bispectrum of the process in Example 10.3. (a) Conventional estimate, (b) TOR estimate, (c) CTOM estimate ([Raghuveer and Nikias, 1986], ©Elsevier Publishing Co.).

Example 10.4

We now consider performance of the different methods for short length data. Typically "short length data" for bispectrum estimation is longer than what the term implies for power spectrum estimation. We define the following process [Raghuveer and Nikias, 1986]

$$Y(k) = \sum_i \cos(\omega_i k + \phi_i) + W(k), \ i = a, \cdots, f$$

where

$$\omega_a = 2\pi(0.076125), \ w_b = 2\pi(0.09375), \ w_c = 2\pi(0.288875),$$
$$w_d = 2\pi(0.3045), \ w_e = w_a + w_c \ \text{and} \ w_f = w_b + w_d$$

$\phi_e = \phi_a + \phi_c$, $\phi_f = \phi_b + \phi_d$ and $\phi_a, \cdots, \phi_d$ are independent random variables uniformly distributed on $(0, 2\pi)$. The true bispectrum magnitude has impulses at (w_a, w_c) and (w_b, w_d). The level of the additive white Gaussian noise, $W(k)$, is set at -40dB. Figures 10.8(a), (b) and (c) show the bispectrum magnitude estimates provided by the conventional, TOR, and CTOM methods when 64 records of 128 samples each are considered. All approaches are successful in resolving the two peaks. The model orders for the TOR and the CTOM methods are 30 and 21 respectively.

Figures 10.8(d), (e), and (f) show the corresponding estimates when there are just 16 records with only 40 samples each. As shown, the conventional approach and the TOR approach have both failed to resolve the two peaks. The CTOM method is successful in resolving the peaks. A model order of 18 was employed for this purpose. □

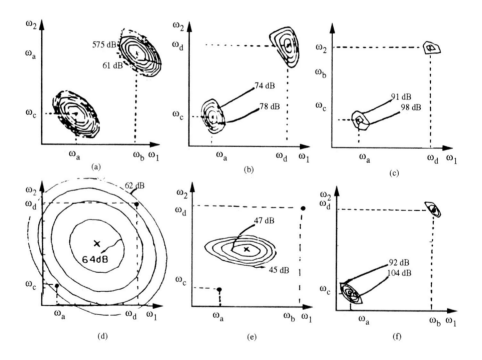

Figure 10.8 Magnitude bispectrum estimates for (i) 'long data': (a) Conventional, (b) TOR estimate, (c) CTOM estimate, and (ii) 'short data': (d) Conventional, (e) TOR, (f) CTOM estimates ([Raghuveer and Nikias, 1986] ©Elsevier Publishing Co.).

Chapter 10

The Optimized AR Method (OARM)

This method was proposed by An, Kim, and Powers [1988] and is in fact an extension of the TOR formulation to an overdetermined system of equations.

Under the same assumptions stated for the TOR method, the following third-order cumulant equation holds:

$$c_3^x(k,\ell) + \sum_{i=1}^{p} a(i) c_3^x(k-i, \ell-i) = \beta \cdot \delta(k, \ell), \qquad (10.73)$$

which was referred to as the third-order recursion (TOR). Considering k and ℓ to take the values $k = 0, 1, \cdots, s$ and $\ell = 0, 1, \cdots, s$ we can rewrite (10.73) in a matrix form

$$\mathbf{r} \cdot \mathbf{a} = \mathbf{b} \qquad (10.74)$$

where

$$\mathbf{r} = \begin{bmatrix} c_3^x(0,0) & c_3^x(-1,-1) & \cdots & c_3^x(-s,-s) \\ c_3^x(0,1) & c_3^x(-1,0) & \cdots & c_3^x(-s,-s+1) \\ \vdots & \vdots & & \vdots \\ c_3^x(0,s) & c_3^x(-1,s-1) & & c_3^x(-s,0) \\ c_3^x(1,0) & c_3^x(0,-1) & & c_3^x(-s+1,-s) \\ \vdots & \vdots & & \vdots \\ c_3^x(s,s) & c_3^x(s-1,s-1) & & c_3^x(0,0) \end{bmatrix} (s+1)^2 X (s+1),$$

$\mathbf{a} = [1, a(1), \cdots, a(p)]^T$ and $\mathbf{b} = [\beta, 0, \cdots, 0]^T$. Note that $\mathbf{r}$ now contains all third-order cumulants. Since (10.74) is an overdetermined system, its least-squares solution is

$$\hat{\mathbf{a}} = (\mathbf{r}^T \mathbf{r})^{-1} \mathbf{r}^T \cdot \mathbf{b}. \qquad (10.75)$$

Once the AR parameters are computed, the method proceeds as with the computation of the bispectrum in (10.51).

Example 10.5 [An, Kim, and Powers, 1988]

The performance of the OARM method is compared to the performance of the

TOR method for the detection of quadratic phase coupling for the process:

$$X(k) = \sum_{i=1}^{6} \cos(2\omega_i\, k + \phi_i) + W(k)$$

where $W(k)$ is white Gaussian noise, $\omega_1 = 2\pi(0.34)$ Hz, $\omega_2 = 2\pi(0.15)$ Hz, $\omega_3 = 2\pi(0.49)$ Hz, $\omega_4 = 2\pi(0.23)$ Hz, $\omega_5 = 2\pi(0.06)$ Hz, $\omega_6 = 2\pi(0.29)$ Hz, $\theta_1 + \theta_2 = \theta_3$ and $\theta_4 + \theta_5 = \theta_6$. The Nyquist frequency is 0.64 Hz. The total number of data points is $N = 8192$ and the data are divided into 64 segments of 128 samples each. For this process, the true bispectrum would exhibit one peak at $(\omega_1, \omega_2) = (2\pi(0.34), 2\pi(0.15))$. Figure 10.9(a), (b), and (c) demonstrate the bispectrum obtained from the OARM with model order 26, the TOR with model order 29 and a conventional FFT-based method, respectively, for 3dB SNR. The three techniques perform comparatively while the performance of the OARM method is slightly better.

Next, the total number of data points is reduced to 1024, i.e., 16 segment of 64 samples each. Figures 10.10(a), (b), and (c) show the bispectra of the noisy short duration data. Using the OARM with model order 26 to estimate the bispectrum we obtain a strong peak at $(\omega_1, \omega_2) = (2\pi(0.34), 2\pi(0.156))$, while using the TOR method, spurious peaks appear besides the correct peak. The conventional FFT approach also yields several peaks.

Hence for this specific example, the OARM method seems to perform better in detecting the quadratic phase coupling out of short length data when compared to the TOR method or the conventional approach. □

10.4.3 Time-Domain Approach to Quadratic Phase Coupling Estimation

As was already mentioned, the conventional methods are good quantifiers of the quadratic phase coupling, but are not high frequency resolution methods. On the other hand, the parametric approaches possess higher frequency resolving ability and are used mainly for the detection of quadratic phase coupling. Raghuveer [1990] proposed a different method for the estimation of the degree of the phase coupling using a high-resolution approach. The method he proposed can be applied, once the triplets $(\omega_1, \omega_2, \omega_3)$, such that $\omega_3 = \omega_1 + \omega_2$, have been identified in the power spectrum of the signal, and is well suited to be used in conjunction with the Pisarenko method for power spectrum estimation. The approach involves only "time-domain" sequences, i.e., only autocorrelation and third-order cumulants. No Fourier transforms are involved in this method.

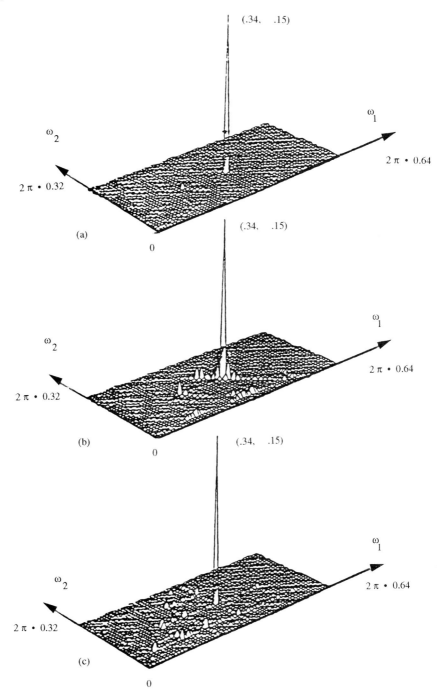

Figure 10.9 Magnitude bispectrum (a) optimized method (OARM), (b) TOR method, and (c) conventional method ([An, Kim, and Powers, 1988] © IEEE).

Consider the process

$$X(k) = \sum_{x}^{3} A_i \cos(\lambda_i k + \phi_i) \qquad (10.76)$$

where $\lambda_3 = \lambda_1 + \lambda_2$, the phase angles ϕ_1, and ϕ_2, are independent and uniformly distributed over $[0, 2\pi]$ and $\phi_3 = \phi_1 + \phi_2$.

The autocorrelation sequence of $X(k)$ is given by

$$c_2^x(k) = \sum_{i=1}^{3} P_i \cos(\lambda_i k) \qquad (10.77)$$

where P_i is the average power at frequency λ_i, i.e., $P_i = \frac{A_i^2}{2}$. Based on (10.44) the third-order cumulant of $X(k)$ is

$$c_3^x(k, \ell) = \frac{1}{4} A_1 A_2 A_3 \, G_1(\lambda_1, \lambda_2, k, \ell) \qquad (10.78)$$

where

$$\begin{aligned}
G_1(\lambda_1, \lambda_2, k, \ell) &= \cos(\lambda_1 k + \lambda_2 \ell) + \cos(\lambda_2 k + \lambda_1 \ell) \\
&+ \cos(\lambda_1 k - \lambda_3 \ell) + \cos(\lambda_2 k - \lambda_3 \ell) \\
&+ \cos(\lambda_3 k + \lambda_1 \ell) + \cos(\lambda_3 k - \lambda_2 \ell); \quad (\lambda_3 = \lambda_1 + \lambda_2).
\end{aligned}$$

Let $Q = \frac{A_1 A_2 A_3}{4}$. Then the magnitude-squared bicoherency, as defined in (10.46) at $\omega_1 = \lambda_1$ and $\omega_2 = \lambda_2$, will be

$$b^2(\lambda_1, \lambda_2) = \frac{2Q^2}{P_1 P_2 P_3} \qquad (10.79)$$

The P_i's ($i = 1, 2, 3$) can be computed from the autocorrelation samples $c_2^x(0), c_2^x(1)$, and $c_2^x(2)$ by solving three linear equations in P_i by utilizing (10.77). Q can be recovered from a sample of the third-order cumulant, say $c_3^x(0, 0)$ as

$$Q = c_3^x(0, 0) / G_1(\lambda_1, \lambda_2, 0, 0). \qquad (10.80)$$

Hence, the bicoherency at (λ_1, λ_2) can be computed from three autocorrelation samples and one third-order cumulant sample using (10.79) and (10.80). For (10.80) to apply, it must hold that $\lambda_1 = \lambda_2$. In the case that $\lambda_1 = \lambda_2$ (one frequency is the second harmonic of another) the bicoherency can still be estimated using (10.79), if P_2 and P_3 are replaced by P_1 and P_2, respectively. Now Q is recovered from $c_3^x(k, \ell)$ as

$$Q = \frac{A_1^2 A_2}{4} = \frac{2 c_3^x(0, 0)}{G_1(\lambda, \lambda, 0, 0)}. \qquad (10.81)$$

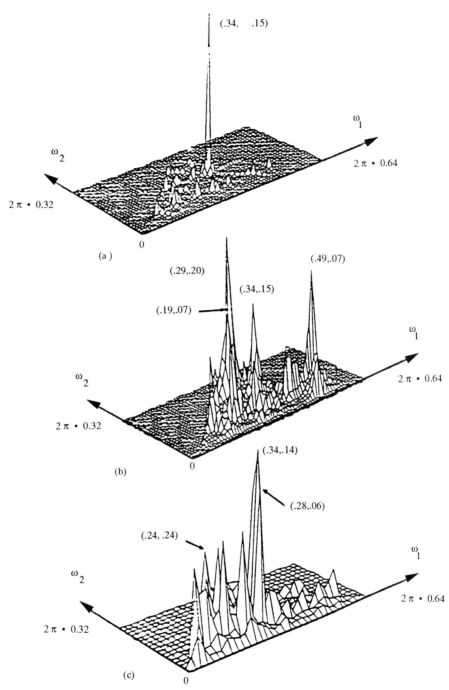

Figure 10.10 Magnitude bispectrum (a) optimized method (OARM), (b) TOR method, and (c) conventional method ([An, Kim, and Powers, 1988] © IEEE).

The method is next generalized for the case of multiple sinusoidal signals.

Let

$$X(k) = \sum_{i=1}^{J}\sum_{j=1}^{3} A_{ij}\cos(\lambda_{ij}k + \phi_{ij}) + \sum_{i=J+1}^{I}\sum_{j=1}^{2} A_{ij}\cos(j\lambda_i k + j\phi_i) \qquad (10.82)$$

where $\lambda_{i3} = \lambda_{i1}+\lambda_{i2}$; $\phi_{i3} = \phi_{i1}+\phi_{i2}$, $1 \leq i \leq J$, $\phi_{11}, \phi_{12}, \cdots, \phi_{J1}, \phi_{12}, \phi_{22}, \cdots, \phi_{J2}$, $\phi_{J+1}, \cdots, \phi_I$ and $\psi_1, \cdots, \psi_M$ are independent random variables uniformly distributed in $[0, 2\pi]$. Let $\lambda_{i1} = \lambda_{i2} = \lambda_i$ and $\lambda_{i3} = 2\lambda_i$ for $i = J+1, \cdots, I$. Hence, the power spectrum of $X(k)$ contains I harmonically related frequencies $(\lambda_{i1}, \lambda_{i2}, \lambda_{i3})$. For convenience, assume the following: (1) these are the only harmonically related frequencies in the power spectrum and they are distinct, and (2) none of the frequencies is in more than one harmonic relation. This means that there are no triples of a_m's forming harmonic relations; however, some of the a_m's can be equal to some of the λ's.

Given enough samples of the autocorrelation sequence of $X(k)$, the frequencies of the sinusoidals and their corresponding power can be determined, and one way toward their computation is the Pisarenko method [Marple, 1987].

Let $C_2^x(\lambda_{ik})$ $\{i = 1, \cdots, I; k = 1, 2, 3\}$ denote the power at frequency λ_{ik}. Once the frequency harmonic relations have been identified in the power spectrum, the bicoherency can be computed as follows.

The third-order cumulant of $X(k)$ can be found to be:

$$c_3^x(k, \ell) = \sum_{i=1}^{I} Q_i G(\lambda_{i1}, \lambda_{i2}, k, \ell) \qquad (10.83)$$

where

$$Q_i = \left\{ \begin{array}{l} A_{i1}A_{i2}A_{i3}/4, \ i = 1, 2, \cdots, J \\ A_{i1}^2 A_{i2}/4, \ i = J+1, \cdots, I \end{array} \right\}$$

and

$$G(\lambda_1, \lambda_2, k, \ell) = \left\{ \begin{array}{l} G_1(\lambda_1, \lambda_2, k, \ell), \ \lambda_1 \neq \lambda_2 \\ \dfrac{1}{2}G_1(\lambda_1, \lambda_1, k, \ell), \ \lambda_1 = \lambda_2. \end{array} \right.$$

Note that only the coupled components contribute to $c_3^x(k,l)$. Given I values of $c_3^x(k_i, \lambda_i), i = 1, \cdots, I$ we can compute $Q_i's$ by solving

$$\mathbf{G} \cdot \mathbf{q} = \mathbf{c} \qquad (10.84)$$

where G is an $I \times I$ matrix where each (i, j) element is $G(\lambda_{j1}, \lambda_{j2}, k_i, l_i)$, $\mathbf{q} = [Q_1, Q_2, \cdots, Q_I]^T$ and $\mathbf{c} = [c_3^x(k_1, l_1), c_3^x(k_2, l_2), \cdots, c_3^x(k_I, l_I)]^T$. The bicoherency at

$(\lambda_{i1}, \lambda_{i2})$ can then be determined as

$$b^2(\lambda_{i1}, \lambda_{i2}) = \frac{2Q_i^2}{C_2^x(\lambda_{i_1})C_2^x(\lambda_{i_2})C_2^x(\lambda_{i_3})}. \tag{10.85}$$

Estimation from Finite Samples

When a finite number of samples of $X(k)$ are given, i.e., $X(0), X(1), \cdots X(N-1)$, then the P_i's, and Q_i's can be obtained by substituting the estimated autocorrelation and the estimated third-order cumulant for the autocorrelation and the third-order cumulant, respectively.

An alternative approach was also suggested by Raghuveer [1990]. Assume we have available more than I samples of $c_3^x(k, l)$, $k = 0, 1, \cdots, K, l = 0, \cdots, k$. From (10.83) it follows that $Q_i's$ can be the solution to the least-squares problem:

$$\min_{T_i} \sum_{k=0}^{K} \sum_{\ell=0}^{k} [c_3^x(k, \ell) - \sum_{i=1}^{I} T_i G(\lambda_{i_1}, \lambda_{i_2}, k, \ell)]^2 \tag{10.86}$$

and $Q_i = T_i^*$ where T_i^* $\{i = 1, \cdots, I\}$ is the solution to (10.86).

The minimization of (10.88) yields

$$\mathbf{H} \cdot \mathbf{q} = \mathbf{d} \tag{10.87}$$

where $\mathbf{H}$ is a symmetric $I \times I$ matrix with the $(i, j)^{th}$ element equal to

$$\sum_{k=0}^{K} \sum_{\ell=0}^{k} G(\lambda_{i_1}, \lambda_{i_2} k, 1) G(\lambda_{i_1}, \lambda_{i_2}, k, \ell).$$

$\mathbf{q} = [Q_1, Q_2, \cdots, Q_I]^T$ and $\mathbf{d}$ is an $I \times 1$ vector whose i^{th} element equals

$$\sum_{k=0}^{K} \sum_{\ell=0}^{k} c_3^x(k, \ell) G(\lambda_{i_1}, \lambda_{i_2} k, \ell)$$

where $c_3^x(k, 1)$ can be substituted by its estimate when a finite number of samples is available.

The power at various frequencies can be computed in an analogous way. Let

$$X(k) = \sum_{i=1}^{I} A_i \cos(v_i k + \psi_i). \tag{10.88}$$

The autocorrelation of $X(k)$ is

$$c_2^x(k) = \sum_{i=1}^{I} P_i \cos(v_i k), \tag{10.89}$$

where P_i is the power at frequency v_i. If we have available the samples $c_2^x(k), k = K_1, \cdots, K_2$, and if we know the frequencies in $X(k)$, the power at these frequencies can be determined by solving the problem

$$\min_{T_i} \sum_{k=K_1}^{K_2} [c_2^x(k) - \sum_{i=1}^{I} T_i \cos(v_i k)]^2. \quad (10.90)$$

Then, $P_i = T_i^*$ is the solution to (10.90). Again $c_2^x(k)$ can be substituted by its estimate. Once P_i's and Q_i's have been estimated, (10.85) can be applied to yield an estimation of the degree of the phase coupling.

10.5 CUBIC PHASE COUPLING

The cubic phase coupling is a result of third-order (cubic) nonlinearities, and thus it can be detected with fourth-order cumulants [Swami and Mendel, 1988].

Let

$$X(k) = \sum_{i=1}^{N} A_i \cos(\omega_i k + \phi_i) + \sum_{i=1}^{N_q} \sum_{j=1}^{3} A_{ij}^q \cos(\omega_{ij}^q k + \phi_{ij}^q) + \sum_{i=1}^{N_c} \sum_{j=1}^{4} A_{ij}^c \cos(w_{ij}^c k + \phi_{ij}^c) \quad (10.91)$$

where $\omega_{i4}^c = \omega_{i1}^c + \omega_{i2}^c + \omega_{i3}^c$, $\phi_{i4}^c = \phi_{i1}^c + \phi_{i2}^c + \phi_{i3}^c$, $\omega_{i3}^q = \omega_{i1}^q + \omega_{i2}^q$, $\phi_{i3}^q = \phi_{i1}^q + \phi_{i2}^q$, the ω's are all distinct, and $\{\phi_{ij}^c\}_{j=1}^{3}, \phi_{i1}^q, \phi_{i2}^q$ and ϕ_i are all independent random variables uniformly distributed over $[-\pi, \pi]$. Then, the diagonal slice of the third-order cumulant sequence is

$$c_3^x(\tau, \tau) = \frac{1}{2} \sum_{i=1}^{N_q} [\prod_{j=1}^{3} A_{ij}^q \sum_{k=1}^{3} \cos(\omega_{ik}^q \tau)]. \quad (10.92)$$

Hence, the Fourier transform of $c_3^x(\tau, \tau)$ will exhibit peaks at each of the three frequencies involved in the quadratic phase coupling. The fourth-order cumulant is given by

$$c_4^x(\tau_1, \tau_2, \tau_3) = c_4^u(\tau_1, \tau_2, \tau_3) + \sum_{i=1}^{N_c} \sum_{k=1}^{4} A_{ik}^c \sum_{S_4} \cos(\omega_{i1}^c \tau_{j1} + \omega_{i2}^c \tau_{j2} + \omega_{i3}^c \tau_{j3} - \omega_{i4}^c \tau_{j4}) \quad (10.93)$$

where $c_4^u(\tau_1, \tau_2, \tau_3)$ is the fourth-order cumulant of $X(k)$ in (10.91) computed as if there were no phase coupling, i.e.,

$$c_4^u(\tau_1, \tau_2, \tau_3) = -\frac{1}{8} \sum_{k=1}^{N} A_k^4 \{ \cos(\omega_k \tau_1) \cos[\omega_k(\tau_2 - \tau_3)] + \\ \cos(\omega_k \tau_2) \cos[\omega_k(\tau_3 - \tau_1)] + \\ \cos(\omega_k \tau_3) \cos[\omega_k(\tau_1 - \tau_2)] \} \quad (10.94)$$

and $S_4 \triangleq \{(j_1, j_2, j_3, j_4) = \text{all permutations of } (1,2,3,4)\}$.

The diagonal slice of $c_4^x(\tau_1, \tau_2, \tau_3)$ is given by

$$c_4^x(\tau, \tau, \tau) = c_4^u(\tau, \tau, \tau) + \frac{3}{4} \sum_{i=1}^{N_c} \sum_{j=1}^{4} A_{ij}^c \sum_{k=1}^{4} \cos(\omega_{ik}^c \tau). \qquad (10.95)$$

Note that the trispectrum, i.e., the 3-d Fourier transform of (10.94), will have peaks at $(\lambda_1, \lambda_2, \lambda_3) = (\pm \omega_k, \pm \omega_k, \pm \omega_k)$ with the $(+,+,+)$ and $(-,-,-)$ combinations disallowed. The trispectrum will be zero over the first and eighth octants. Hence, the three-frequency terms $(\omega_{i1}^c, \omega_{i2}^c \omega_{i3}^c)$ are identifiable from the first octant of the trispectrum. On the other hand, the Fourier transform of (10.95) will show peaks at all the $\omega_i's$ coupled or otherwise.

10.6 SUMMARY

In this chapter we have introduced the general formulation of a second-order Volterra (or quadratic) filter and described three different methods for estimating the filter's parameters. Two of the methods assume access to both input and output of the filter, whereas the third method assumes white unobservable input. We have also discussed the identification of a particular nonlinear system consisting of three subsystems in cascade, namely two linear subsystems and one memoryless quadratic. The quadratic phase coupling phenomenon has been presented, as well as methods for its detection and quantification. In particular, conventional and parametric methods based on AR modeling of third-order cumulants have been described and their performance demonstrated with several examples. Finally, a brief discussion has been devoted to the more complex problem of cubic phase coupling.

The general representation of Volterra filters and their properties can be found in Schetzen [1980]. Tick [1961] introduced a method for estimating the linear and quadratic filter parameters assuming that the input is Gaussian, whereas Hinich [1985] describes a different approach based on the assumption that the input is white and unobservable. A general frequency-domain solution to the second-order Volterra filter identification problem, assuming access to both input and output, is given by Powers, et al. [1989]. On the other hand, Brillinger [1977] and Rozario and Papoulis [1989] provide identification solutions for a particular nonlinear time series system.

The detection and quantification of the quadratic phase coupling phenomenon using conventional bispectrum estimation methods has been studied by Kim and Power [1978, 1979, 1988] and Smith and Powers [1973]. Parametric methods based on AR modeling for the detection of quadratic phase coupling have been developed by Raghuveer and Nikias [1985, 1986], An, et al. [1988] and Raghuveer [1990]. Finally, a discussion on cubic phase coupling is given by Swami and Mendel [1988].

REFERENCES

Alper, P. "A Consideration of the Discrete Volterra Series," *IEEE Trans. Autom. Control*, **10**, pp. 322–327, 1965.

An, G. K., S. B. Kim, and E. J. Powers, "Optimized Parametric Bispectrum Estimation," *Proc. ICASSP'88*, pp. 2392–2395, New York, 1988.

Brillinger, D. R., "The Identification of a Particular Nonlinear Time Series System," *Biometrika* **64**, 3, pp. 509–515, 1977.

Chan, Y. T., and J. C. Wood, "A New Order Determination Technique for ARMA Processes," *IEEE Trans. Acoust., Speech, and Signal Processing*, **ASSP-32**, pp. 517–521, June 1984.

Chow, J. C., "On Estimating the Orders of an ARMA Process with Uncertain Observations," *IEEE Trans. Automat, Contr.*, **AC-17**, October, 1972.

Hinich, M. J., "A Testing for Gaussianity and Linearity of Stationary Time Series," *J. of Time Series Analysis* **3**, pp. 169–176, 1982.

Hinich, M. J., "Identification of the Coefficients in a Non-Linear Time Series of the Quadratic Type," *J. of Economics* **30**, pp. 269–288, 1985.

Kay, S. M., and L. Marple, "Spectrum Analysis: A Modern Perspective," *Proc. IEEE*, **69**(11), pp. 1380–1418, November 1981.

Kim, Y. C., and E. J. Powers, "Digital Bispectral Analysis and its Applications to Nonlinear Wave Interacting," *IEEE Trans. Plasma Science*, **PS-7**(2), pp. 120–131, June 1979.

Kim, Y. C., and E. J. Powers, "Digital Bispectral Analysis of Self-Excited Fluctuation Spectra," *Phys. Fluids*, **21**(8), pp. 1452–1453, August, 1978.

Kim, K. I., and E. J. Powers, "A Digital Method for Modeling Quadratically Nonlinear Systems with a General Random Input," *IEEE Trans. on Acoustics, Speech and Signal Processing*, **36**(11), pp. 1758–1769, November, 1988.

Makhoul, J., "Linear Prediction: A Tutorial Review," *Proc. IEEE*, **63**, pp. 561–580, April 1975.

Marple, S. L., "Digital Spectral Analysis with Application," Englewood Cliffs, N.J.: Prentice Hall, Signal Processing Series, 1987.

Oppenheim, A. V., and R. W. Schafer, "Discrete-Time Signal Processing," Englewood Cliffs, N.J.: Prentice Hall, 1989.

Powers, E. J., C. P. Ritz, C. K. An, S. B. Kim, R. W. Miksad, and S. W. Nam, "Applications of Digital Polyspectral Analysis to Nonlinear Systems Modeling and Nonlinear Wave Phenomena," Workshop on Higher-Order Spectral Analysis, pp. 73–77, Vail, Colorado, June, 1989.

Raghuveer, M. R., "Time Domain Approaches to Quadratic Phase Coupling Estimation," *IEEE Trans. on Automatic Control*, **35**(1), pp. 48–56, January, 1990.

Raghuveer, M. R., and C. L. Nikias, "Bispectrum Estimation: A Parametric Approach," *IEEE Trans. on Acoust. Speech and Signal Processing*, **ASSP-33**(4), October, 1985.

Raghuveer, M. R., and C. L. Nikias, "Bispectrum Estimation via AR Modeling," *Signal Processing* **10**, pp. 35–48, 1986.

Rozario, N., and A. Papoulis, "The Identification of Certain Nonlinear Systems by Only Observing the Output," *Workshop on Higher-Order Spectral Analysis*, pp. 78–82, Vail, Colorado, 1989.

Schetzen, M., **The Volterra and Wiener Theories of Nonlinear Systems**, New York: Wiley, 1980.

Smith, D. E., and E. J. Powers, "Experimental Determination of the Spectral Index of a Turbulent Plasma from Digitally Computed Power Spectra," *Phys. Fluids*, **16**, pp. 1373–1374, August, 1973.

Subba, Rao, T., and M. Gabr, "A Test for Linearity of Stationary Time Series," *Journal of Time Series Analysis*, **1**, pp. 145–158, 1980.

Swami, A., and J. M. Mendel, "Cumulant-Based Approach to the Harmonic Retrieval Problem," *International Conference on Acoustics, Speech, and Signal Processing, ICASSP'88*, pp. 2264–2267, New York, 1988.

Tick, L. J., "The Estimation of Transfer Functions of Quadratic Systems, *Technometrics*, **3**(4), pp. 562–567, November, 1961.

11

TIME-FREQUENCY DISTRIBUTIONS BASED ON HIGHER-ORDER SPECTRA

11.1 INTRODUCTION

The Fourier transform decomposes a signal into its frequency components. Looking at the energy spectrum we can identify these frequencies; however, we have no indication about their temporal localization. Time-frequency distributions (TFD) map a one-dimensional signal into a two-dimensional function of time and frequency, and describe how the spectral content of the signal is changing with time. Hence, TFDs are appropriate tools for non-stationary signal analysis, synthesis, and processing. The most well-known linear TFDs are the Short-Time Fourier Transform (STFT) [Allen, 1977; Rabiner and Shafer, 1978; Nawab and Quatieri, 1988], the Gabor representation [Gabor, 1946; Helstrom, 1966; Wexler, 1990; Friedlander and Porat, 1989], and the Wavelet Transform (WT) [Daubechies, 1990, 1991; Mallat, 1989; Rioul, 1991; Meyer, 1989]. Although linearity is a TFD's desirable property, quadratic TFDs have also been proposed and interpreted as time-frequency energy distributions, or instantaneous power spectra. Among the TFDs, Cohen's shift-invariant general class [Cohen, 1986] combines the concepts of the instantaneous power and the spectral energy density. Special cases of this general class are the Spectrogram, Rihaczek, Page, Wigner-Ville (WD) and Choi-Williams distributions [Cohen, 1989; Hlawatsch and Boudreaux-Bartels, 1992]. In particular, the WD has been of special interest since it satisfies a large number of important properties. It has been widely used for the analysis of phase modulated signals common in radar and sonar [Boudreaux-Bartels and Parks, 1986; Marinovic, Oklobdzija et al., 1988; Boashash and White, 1990]. Every member of Cohen's general class may be interpreted as a two-dimensional filtered WD [Hlawatsch and Boudreaux-Bartels, 1992]. A newly defined class of quadratic TFDs is the affine class [Flandrin, 1989; Bertrand and Bertrand, 1991] comprising all quadratic TFD's that preserve time scalings and time shifts.

A cubic TFD or the third-order Wigner distribution was first introduced by Gerr [Gerr, 1988]. An extension of the WD in a multilinear form, the Wigner Higher-Order Moment Spectra (WHOMS), was introduced by Follonosa and Nikias [1991, 1993]. WHOMS preserve most of the WD properties and extend them to the higher-order spectra domains. However, they are defined for deterministic signals only. A generalization of the third-order Wigner distribution, originally proposed by Gerr, has been developed by Swami [1991a]. Although the WD has been extended to random signals assuming quasi-stationarity, WHOMS of random signals based on a single data record demonstrate high variance. Multilinear representations that rely on the cyclostationarity of the data have been introduced by Gardner [1990], and Giannakis and Dantawate [1991], and Dantawate and Giannakis [1992]. One particular property of the WD, i.e., its optimal concentration for linear FM signals, has been extended to higher-order FM signals by means of a multilinear kernel by Boashash and Frazer [1992].

In this chapter, we present first the definition and properties of the WD. Based on these properties, the derivation of the WHOMS is discussed next along with their application to the detection of transient signals. Most of the material in this chapter is based on the publications by Fonollosa and Nikias [1991a,b, 1993].

11.2 WIGNER-VILLE DISTRIBUTION (WD)

The Wigner Distribution was initially defined in the context of quantum mechanics [Wigner, 1932], and was later extended by Ville to the Wigner Time-Frequency Distribution or Wigner-Ville Distribution (WD) [Ville, 1948].

Let us define the local autocorrelation function $m_2^x(t, \tau)$ of the signal $x(t)$ as:

$$m_2^x(t, \tau) = x^*(t - a)x(t + \tau - a) \tag{11.1}$$

where a is an arbitrary delay.

A time-frequency distribution can be derived by means of a Fourier transform of the local autocorrelation function, i.e.,

$$W_{1x}(t, f) = \int_\tau m_2^x(t, \tau)e^{-j2\pi f\tau}d\tau. \tag{11.2}$$

In order for a particular distribution to be interpreted as a distribution of the signal's energy in time and frequency, three two-fold properties must be satisfied. These are:

(a) Shift in time (or frequency) of the signal should result in a corresponding shift of the distribution.

(b) The integral of the distribution over all frequencies (or time) should be equal to the instantaneous power (or spectral density), i.e.,

$$\int W_{1x}(t,f)df = |x(t)|^2 \tag{11.3.1}$$

$$\int W_{1x}(t,f)dt = |X(f)|^2. \tag{11.3.2}$$

(c) The mean frequency (or time) of the distribution at each time (or frequency) should be equal to the instantaneous frequency (or group delay), i.e.,

$$E\{f\}_t = \frac{\int f \cdot W_{1x}(t,f)df}{\int W_{1x}(t,f)df} = \phi'(t) \tag{11.4.1}$$

where $x(t) = A(t)e^{j\phi(t)}, \phi'(t) = \frac{d\phi(t)}{dt}$, and

$$E\{f\}_t = \frac{\int t \cdot W_{1x}(t,f)dt}{\int W_{1x}(t,f)dt} = \phi'(f) \tag{11.4.2}$$

where $X(f) = A(f)e^{j\phi(f)}$ and $\phi'(f) = \frac{d\phi(f)}{df}$.

Note that properties (a) and (b) are satisfied by $W_{1x}(t,f)$ in (1.1) for any value of a. On the other hand, the third property requires the mean value of the arguments of $x(t)$ in $m_2^x(t,\tau)$ to be t; i.e., $m_2^x(t,\tau)$ would have to be a centered local autocorrelation function. This constraint, known as the lag-centering condition, implies that $a = \tau/2$ and yields the WD, i.e.,

$$W_{1x}(t,f) = \int x^*(t - \frac{\tau}{2})x(t + \frac{\tau}{2})e^{-j2\pi f\tau}d\tau, \tag{11.5.1}$$

or in terms of the spectral value of $x(t)$,

$$W_{1x}(t,f) = \int X^*(f + \frac{1}{2}\theta)X(f - \frac{1}{2}\theta)e^{-jt\theta}d\theta. \tag{11.5.2}$$

The Support of Wigner Distribution

From (11.5.1) we see that at time t, the terms of the summation consist of the product of past and future values of the signal, the time into the past equal to the time into the future. Hence, for a finite duration signal $x(t), t_1 \leq t \leq t_2$, the WD will be zero for $t < t_1$ or $t > t_2$. Similarly, by looking at (11.5.2), the WD will be zero for $f < f_1$ or $f > f_2$ if $X(f)$ is nonzero only in the interval $[f_1, f_2]$.

Example 11.1

For $x(t) = A_0 e^{j2\pi f_0 t}$, $X(f) = A_0 \delta(f-f_0)$, and thus, substitution of $X(f)$ in (11.5.2) yields the WD
$$W_{1x}(t,f) = A_0^2 \delta(f - f_0).$$

Hence, the impulse of the Fourier transform of the signal is mapped into an impulse centered at frequency f_0. □

Example 11.2

For chirp signals of the form
$$x(t) = e^{j(\beta t^2/2) + j(2\pi f_0 t)}$$

the WD becomes
$$W_{1x}(t,f) = \delta(2\pi f - 2\pi f_0 - \beta t),$$

which shows that the energy is concentrated along the instantaneous frequency.

Example 11.3

Consider the signal
$$x(t) = A_1 e^{j2\pi f_1 t} + A_2 e^{j2\pi f_2 t}.$$

The WD becomes
$$W_{1x}(t,f) = A_1^2 \delta(f-f_1) + A_2^2 \delta(f-f_2) + 2A_1 A_2 \delta[f - \frac{f_1 + f_2}{2}]\cos(2\pi(f_2-f_1)t)$$

and is illustrated in Figure 11.1.

Apparently, besides the peaks at $f = f_1$ and $f = f_2$, there are nonzero values of the WD along the line $f = (f_1 + f_2)/2$. This is the so-called cross-term effect, which

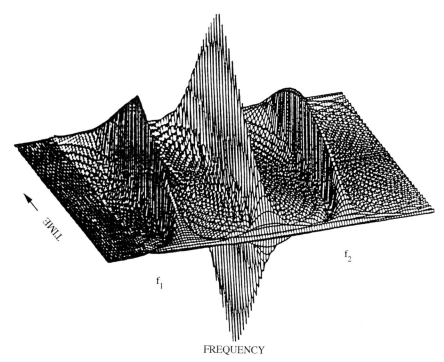

Figure 11.1 WD for the sum of two sine waves $x(t) = e^{j2\pi f_1 t} + e^{j2\pi f_2 t}$

is more dominant when the signal is the sum of N sine waves. In this case, there are $\frac{1}{2}N(N-1)$ terms which show as cross-term effects. The presence of nonzero values of the WD, in regions where zero values would be expected, is the main difficulty with the Wigner distribution. A way of reducing the cross-term effects in the WD due to multicomponent signals, while preserving most of the desirable mathematical properties, is discussed in section 11.3.1. □

11.3 COHEN'S GENERAL CLASS OF TIME-FREQUENCY DISTRIBUTIONS

Cohen's general class, defined by Cohen [1986] as

$$P_{GEN}(t,f) = \frac{1}{4\pi^2} \iiint e^{-j\theta t - j2\pi f\tau + j\theta u}\phi(\theta,\tau)x^*(u - \frac{\tau}{2})x(u + \frac{\tau}{2})du\,d\tau\,d\theta$$
(11.6.1)

has been used as a unified framework in time-frequency signal analysis. Depending on the form of the kernel $\phi(\theta,\tau)$, different distributions can be obtained. In fact, the Wigner distribution is a special case of (11.6.1) for $\phi(\theta,\tau) = 1$. Other known distributions are the Rihaczek and Page, for $\phi(\theta,\tau) = e^{j\theta\tau/2}$ and $e^{j\theta|\tau|/2}$, respectively. An alternative way of writing (11.6.1) is

$$P_{GEN}(t,f) = \int\int M_{GEN}(\theta,\tau)\exp(-j2\pi t\theta)\exp(-j2\pi f\tau)d\tau\, d\theta \qquad (11.6.2)$$

where $M_{GEN}(\theta,\tau)$ is the product of the kernel $\phi(\theta,\tau)$ with the inverse Fourier transform of the WD, i.e., the symmetrical ambiguity function,

$$M_{GEN}(\theta,\tau) = \phi(\theta,\tau)\int x^*(u-\frac{\tau}{2})x(u+\frac{\tau}{2})\exp(j2\pi u\theta).$$

Example 11.4

Let $s(t)$ be the signal (real and imaginary parts) illustrated in Figure 11.2. In the interval $(0,t_1)$ the signal is a sine wave of frequency f_1; in the interval (t_1,t_2) the signal is zero, and in the interval (t_2,t_3) is a sine wave of frequency, f_2. Figure 11.2 illustrates the Wigner distribution for this signal. As was expected, the WD distribution shows peaks at frequencies f_1 and f_2, corresponding to the intervals $(0,t_1)$ and (t_2,t_3) where the signal is non-zero.

One would expect the distributions to be zero in the interval (t_1,t_2) where the signal is zero. However, the Wigner distribution exhibits a peak in the middle of the range (t_1,t_2). □

11.3.1 Reduced Interference Distributions

The finite support property does not guarantee that the time-frequency distribution will be zero everywhere the signal or its spectrum is zero. Rather, the finite support property ensures only that a time-frequency distribution is zero outside the global time and frequency support of the signal. The nonzero intensities of the distribution at places where zero values would be expected, are called "artifacts," and are the main difficulty associated with many distributions. Choi and Williams [1989] devised a distribution that can reduce the artifacts caused by multicomponent signals. The distribution they proposed results from (11.6.1) by setting

$$\phi(\theta,\tau) = e^{-\theta^2\tau^2/\sigma} \qquad (11.7)$$

where σ is a constant. Hence, the Choi-Williams distribution is

$$P_{cw}(t,f) = \frac{1}{4\pi^{3/2}}\int\int \frac{1}{\sqrt{\tau^2/\sigma}}e^{-[(u-\tau)^2/(4\tau^2/\sigma)]-j2\pi f\tau}x^*(u-\frac{\tau}{2})x(u+\frac{\tau}{2})du\,d\tau.$$

$$(11.8)$$

Chapter 11   493

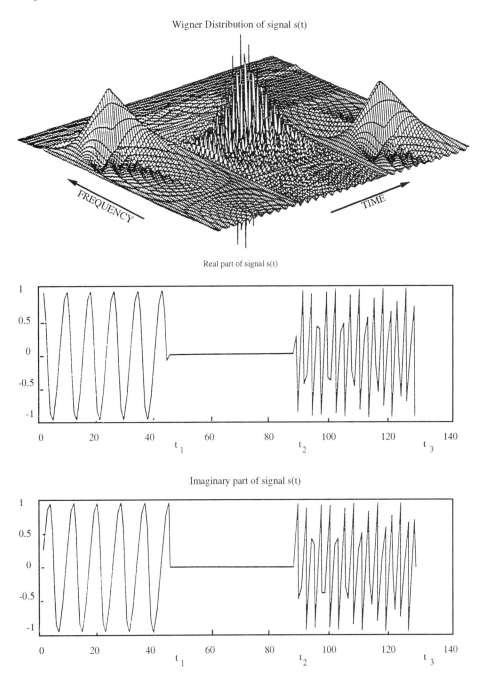

Figure 11.2 The Wigner distribution of the signal $s(t)$ of Example 11.4

Controlling the value of σ, the distribution in (11.8) can attenuate the cross-terms, as can be seen in the following examples.

Example 11.5 [Cohen, 1989]

The Choi-Williams distribution is computed for the signal

$$x(t) = A_1 e^{j2\pi f_1 t} + A_2 e^{j2\pi f_2 t},$$

i.e.,

$$P_{cw}(t,f) = A_1^2 \delta(f - f_1) + A_2^2 \delta(f - f_2)$$
$$+ 2A_1 A_2 \cos 2\pi (f_2 - f_1) t \frac{\exp[-\frac{(f - \frac{f_1+f_2}{2})^2}{4(f_1-f_2)^2/\sigma}]}{\sqrt{16\pi^3 (f_1 - f_2)^2/\sigma}}.$$

Note that as $\sigma \to \infty$, $P_{cw}(t, f)$ becomes the Wigner distribution. Since the cross-terms in $P_{cw}(t, f)$ are controlled by $\sqrt{\sigma}$, as σ becomes small, the cross-term effect becomes less dominant. This can be seen in Figure 11.3 where $P_{cw}(t, f)$ is plotted for different values of σ. □

11.4 WIGNER HIGHER-ORDER MOMENT SPECTRA: CONTINUOUS CASE

In this section we present the definition of a general class of time-frequency distributions in terms of higher-order moments. The definition of a general class requires, however, the formulation of a basic representation, i.e., the representation for which the kernel is equal to unity. In analogy to Cohen's general class, the Wigner Higher-Order Moment Spectra (WHOMS) are chosen as the basic representation. For every time instant t, the WHOMS express the time-varying Higher-Order Moment Spectrum (HOMS) in the same way the WD does for the instantaneous power spectrum [Fonollosa and Nikias, 1991a,b, 1993; Swami, 1991; Giannakis and Dantawate, 1991]. In the sequel, the continuous WHOMS are defined and studied for an arbitrary order. This definition is carefully conceived to preserve the properties of the WD or extend them to the higher-order spectra domains. In particular, the properties related to the instantaneous power and spectral density function in the WD are shown to be related to the instantaneous $(n + 1)^{th}$-order moment and $(n + 1)^{th}$-order HOMS.

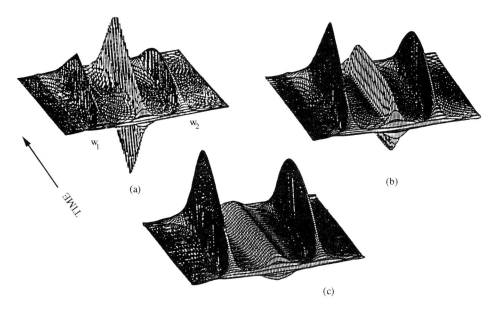

Figure 11.3 The Choi-Williams distribution for the sum of two sine waves of frequencies f_1 and f_2 for (a) $\sigma \to \infty$ (b) $\sigma = 10^6$ and (c) $\sigma = 10^5$.

11.4.1 Definition

Let $x(t)$ be a complex deterministic signal. The WHOMS of order n of $x(t)$ is defined as [Fonollosa and Nikias, 1991a]

$$W_{nx}(t, f_1, \ldots, f_n) = \int_{\tau_1} \cdots \int_{\tau_n} x^*(t - \frac{1}{n+1}\sum_{m=1}^{n}\tau_m)$$
$$\prod_{i=1}^{n} x(t + \frac{n}{n+1}\tau_i - \frac{1}{n+1}\sum_{j=1, j \neq i}^{n}\tau_j)\exp(-j2\pi f_i \tau_i)d\tau_i. \tag{11.9}$$

This definition was derived as follows. $W_{nx}(t, f_1, \ldots, f_n)$ is an n-dimensional Fourier transform of an n-dimensional local moment function $m_n^x(t, \tau_1, \ldots \tau_n)$; viz,

$$W_{nx}(t, f_1, \ldots, f_n) = \int_{\tau_1} \cdots \int_{\tau_n} m_n^x(t, \tau_1, \ldots, \tau_n) \prod_{i=1}^{n} \exp(-j2\pi f_i \tau_i)d\tau_i \quad (11.10.1)$$

where

$$m_n^x(t, \tau_1, \ldots \tau_n) = x^*(t - a)\prod_{i=1}^{n} x(t + \tau_i - a) \tag{11.10.2}$$

and a is an arbitrary delay. To fulfill the three basic properties of time-frequency distributions in a higher-order moment spectrum domain, the value of a should be chosen properly. In particular, to attain the instantaneous frequency as the mean frequency in the multifrequency space at a given time, $m_n^x(t, \tau_1, \ldots \tau_n)$ should be centered at time instant t in such a way that

$$\frac{1}{n+1}[(t-a) + \sum_{i=1}^{n}(t+\tau_i - a)] = t. \tag{11.11}$$

Consequently

$$a = \frac{1}{n+1}\sum_{i=1}^{n} \tau_i \tag{11.12}$$

and

$$m_n^x(t, \tau_1, \ldots \tau_n) = x^*(t - \frac{1}{n+1}\sum_{n=1}^{n} \tau_m)\prod_{i=1}^{n} x(t + \frac{n}{n+1}\tau_i - \frac{1}{n+1}\sum_{j=1, j\neq i}^{n} \tau_j) \tag{11.13}$$

which leads to the definition (11.9). Special cases of the WHOMS include the Wigner Bispectrum (WB) for $n = 2$, i.e.,

$$\begin{aligned}W_{2x}(t, f_1, f_2) &= \int_{\tau_1}\int_{\tau_2} x^*(t - \tfrac{1}{3}\tau_1 - \tfrac{1}{3}\tau_2)x(t + \tfrac{2}{3}\tau_1 - \tfrac{1}{3}\tau_2)x(t + \tfrac{2}{3}\tau_2 - \tfrac{1}{3}\tau_1)\\ &\quad \exp\{-j2\pi(f_1\tau_1 + f_2\tau_2)\}d\tau_1 d\tau_2,\end{aligned} \tag{11.14}$$

and the Wigner Trispectrum (WT) for $n = 3$, i.e.,

$$\begin{aligned}W_{3x}(t, f_1, f_2, f_3) &= \int_{\tau_1}\int_{\tau_2}\int_{\tau_3} x^*(t - \tfrac{1}{4}\tau_1 - \tfrac{1}{4}\tau_2 - \tfrac{1}{4}\tau_3)x(t + \tfrac{3}{4}\tau_1 - \tfrac{1}{4}\tau_2 - \tfrac{1}{4}\tau_3)\\ &\quad x(t + \tfrac{3}{4}\tau_2 - \tfrac{1}{4}\tau_1 - \tfrac{1}{4}\tau_3)x(t + \tfrac{3}{4}\tau_3 - \tfrac{1}{4}\tau_1 - \tfrac{1}{4}\tau_2)\\ &\quad \exp\{-j2\pi(f_1\tau_1 + f_2\tau_2 + f_3\tau_3)\}d\tau_1 d\tau_2 d\tau_3.\end{aligned} \tag{11.15}$$

Note that the WD can be drawn from (11.9) for $n = 1$.

11.4.2 Properties

The properties of the WHOMS are discussed in this section [Fonollosa and Nikias, 1991a; Swami 1991a,b; 1992].)

Time Shift of $x(t)$

A temporal shift of the signal $x(t)$ results in the corresponding time delay in the WHOMS. Hence,

$$y(t) = x(t-a) \tag{11.16}$$

implies that
$$W_{ny}(t, f_1, \ldots f_n) = W_{nx}(t - a, f_1, \ldots, f_n). \tag{11.17}$$

Projection in Time of the WHOMS

The projection in time of the WHOMS, or integration over all frequencies, equals the instantaneous $(n+1)^{th}$-order moment. That is

$$\int_{f_1} \cdots \int_{f_n} W_{nx}(t, f_1, \ldots f_n) \prod_{i=1}^{n} df_i = x^*(t) x^n(t) \tag{11.18}$$

Mean Conditional Frequency and Instantaneous Frequency

The mean frequency of the distribution over the multifrequency space, at each time instant, gives the instantaneous frequency of the signal

$$E\{f_m\}_t = \frac{\int_{f_1} \cdots \int_{f_n} f_m W_{nx}(t, f_1, \ldots, f_n) \prod_{i=1}^{n} df_i}{\int_{f_1} \cdots \int_{f_n} W_{nx}(t, f_1, \ldots, f_n) \prod_{i=1}^{n} df_i} = \frac{2\phi'(t)}{n+1} \tag{11.19}$$

for $x(t) = A(t) \exp\{j\phi(t)\}$ and $m = 1, \ldots, n$.

Projection in Frequency

Projection of an n-dimensional multifrequency point (integration of the distribution over time) equals the $(n+1)^{th}$-order moment spectrum. By considering inverse Fourier transforms in (11.9) we obtain the WHOMS distribution in terms of spectral values, i.e.,

$$W_{nx}(t, f_1, \ldots, f_n) = \int_{\Omega} X^*(\sum_{i=1}^{n} f_i + \frac{1}{n+1}\Omega) \prod_{j=1}^{n} X(f_j - \frac{1}{n+1}\Omega) \exp\{-j2\pi\Omega t\} d\Omega. \tag{11.21}$$

Consequently

$$\int_t W_{nx}(t, f_1, \ldots, f_n) dt = X^*(\sum_{i=1}^{n} f_i) \prod_{j=1}^{n} X(f_j). \tag{11.22}$$

Mean Time Calculation and Generalized Group Delay

From the point of view of classical time-frequency distribution, the group delay appears as the mean time at a given frequency. In the WHOMS the group delay becomes the mean time at a given multifrequency point. The group delay, defined as the derivative of the Fourier phase of the signal, expresses the mean temporal position of the energy of this signal (or its second-order moment) as a function of frequency. That is,

$$\int_t t|x(t)|^2 dt = \int_f \phi'(f)|X(f)|^2 df \qquad (11.23)$$

where $x(t) \leftrightarrow X(f) = A(f)\exp(j\phi(f))$ and $\phi'(f) = d\phi(f)/dt$. Similar expressions can be derived for the $(n+1)^{th}$-order moments. That is,

$$\int_t tx^*(t)x^n(t)dt = \int_{f_1}\cdots\int_{f_n} GD_n(f_1,\ldots f_n)X^*\left(\sum_{i=1}^n f_i\right)\prod_{j=1}^n X(f_j)df_j, \qquad (11.24)$$

where the generalized group delay, $GD_n(f_1,\ldots,f_n)$, can be calculated using the WHOMS as the mean time at a multifrequency point:

$$E\{t\}_{f_1\ldots,f_n} = \frac{\int_t tW_{nx}(t,f_1,\ldots,f_n)dt}{\int_t W_{nx}(t,f_1,\ldots f_n)dt}. \qquad (11.25)$$

In addition to the basic properties, the following properties are satisfied.

Time Support of $x(t)$

In analogy to the WD, the WHOMS do not have to be zero whenever the signal is zero. However,

$$x(t) = 0 \text{ for every } t \notin [T_1,T_2] \Rightarrow W_{nx}(t,f_1,\ldots,f_n) = 0 \text{ for every } t \notin [T_1,T_2]. \qquad (11.26)$$

This property is easily verified considering the lag-centering condition (11.11), i.e., that the mean value of the arguments in the right side of equation (11.10.2) should be t. If (11.11) holds, then it is clear that some of the arguments of $x(t)$ will be greater than t and some will be smaller. Consequently, if $x(t)$ is zero except for t in $[T_1,T_2]$, so will be $m_n^x(t,\tau_1,\ldots,\tau_n)$ *for any value of* τ_i, and thus (11.26) follows.

Frequency Support of $X(f)$

The region of support of WHOMS in the frequency domain can be related to the support of $X(f)$. For band-limited $x(t), X(f) = 0$ except for f in $[F_1,F_2]$. Consequently, necessary conditions for the existence of $W_{nx}(t, f_1, \ldots, f_n)$ are that inequalities

$$\sum_{i=1}^{n} f_i + \frac{1}{n+1}\Omega \geq F_1 \qquad (11.27.1)$$

$$\sum_{i=1}^{n} f_i + \frac{1}{n+1}\Omega \leq F_2 \qquad (11.27.2)$$

$$f_j - \frac{1}{n+1}\Omega \geq F_1 \qquad (11.27.3)$$

$$f_j - \frac{1}{n+1}\Omega \leq F_2. \qquad (11.27.4)$$

are satisfied at least for one common value of Ω ($-\infty < \Omega < +\infty$).

Substitution of (11.27.1), (11.27.2), in (11.27.3), (11.27.4) leads to:

$$2f_j + \sum_{i=1,\ i\neq j}^{n} f_i \geq 2F_1 \qquad (11.28.1)$$

$$2f_j + \sum_{i=1,\ i\neq j}^{n} f_i \leq 2F_2. \qquad (11.28.2)$$

And substitution of (11.27.3) into (11.27.4) results in:

$$-B_x \geq f_i - f_j \geq B_x \qquad \text{for } B_x = F_2 - F_1. \qquad (11.28.3)$$

Notice that this region contains the domain of support of the higher-order moment spectra of $x(t)$ (particular case of $\Omega = 0$). Figure 11.4 illustrates this region of support for both the WD and the bispectrum of an analytic deterministic signal with bandwidth B_x.

Considering (11.22) we see that integration over time of the WHOMS yields the higher-order moment spectra of the signal, which is a desirable property for time-frequency distribution. On the other hand, we do not observe any contradiction in the fact that the frequency support of the WHOMS at any particular time, which can be interpreted as the instantaneous higher-order moment spectra, exceeds the support of the HOS of the signal.

Convolution and Product Property

Denoting by $\otimes_f$ and $\otimes_t$ the frequency and time convolution respectively,

$$x(t) = f(t)g(t) \Rightarrow W_{nx}(t, f_1, \ldots, f_n) = W_{nf}(t, f_1, \ldots, f_n) \otimes_f W_{ng}(t, f_1, \ldots, f_n),$$
$$(11.29.1)$$

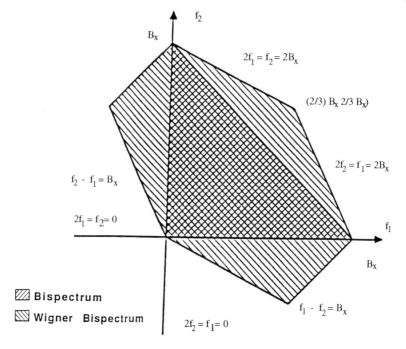

Figure 11.4 The region of support for WB and the bispectrum of an analytic deterministic signal with bandwidth B_x ([Fonollosa and Nikias, 1993] © 1993 IEEE).

$$x(t) = f(t) \otimes g(t) \Rightarrow W_{nx}(t, f_1, \ldots, f_n) = W_{nf}(t, f_1, \ldots, f_n) \otimes_t W_{ng}(t, f_1, \ldots, f_n), \quad (11.29.2)$$

Symmetry Property

WHOMS are symmetric with respect to the hyperplanes $f_i = f_j$.

$$W_{nx}(t, f_1, \ldots, f_i, \ldots, f_j, \ldots, f_n) = W_{nx}(t, f_1, \ldots, f_j, \ldots, f_i, \ldots, f_n) \quad (11.30)$$

Derivation of the Signal $x(t), X(f)$

The signal itself $x(t)$ and its Fourier transform $X(f)$ can be recovered from WHOMS except for a constant term, $x(t_0)$ and $X(f_0)$:

$$\int_{f_1} \cdots \int_{f_n} W_{nx}(\frac{1}{n+1}(t+nt_0), f_1, \ldots, f_n) \prod_{i=1}^{n} \exp(-j2\pi f_i(t-t_0))df_i = x^*(t)x^n(t_0) \quad (11.31.1)$$

$$\int_t W_{nx}(t, \frac{1}{n+1}(f+f_0), \ldots, \frac{1}{n+1}(f+f_0)) \exp(j2\pi(f - nf_0)t)dt = X^*(f)X^n(f_0)$$
(11.31.2)

Example 11.6

Complex exponential signal $x(t) = A_0 \exp(j2\pi f_0 t)$, and its Fourier transform $X(f) = A_0 \delta(f - f_0)$. Substitution of $X(f)$ in (11.21) leads to:

$$W_{nx}(t, f_1, \ldots f_n) = A_0^{(n+1)} \exp(j2\pi f_0(n-1)t) \prod_{i=1}^{n} \delta(f_i - \frac{2}{n+1}f_0). \quad (11.32)$$

The impulse function of the Fourier transform of the signal is mapped into a multidimensional impulse function at frequencies $2f_0/(n+1)$. Special cases include $n = 1$; i.e., the WD,

$$W_x(t, f) = A_0^2 \delta(f - f_0);$$

and $n = 2$ and $n = 3$, the WD and WT, respectively:

$$W_{2x}(t, f_1, f_2) = A_0^3 \exp(j2\pi f_0 t)\delta(f_1 - \frac{2}{3}f_0)\delta(f_2 - \frac{2}{3}f_0),$$

$$W_{3x}(t, f_1, f_2, f_3) = A_0^4 \exp(j4\pi f_0 t)\delta(f_1 - \frac{1}{2}f_0)\delta(f_2 - \frac{1}{2}f_0)\delta(f_3 - \frac{1}{2}f_0).$$

Observe that the WD is the only case where the impulse functions are centered at frequency f_0. □

Example 11.7

Consider the impulse function $x(t) = A_0 \delta(t - t_0)$ and its Fourier transform $X(f) = A_0 \exp(-j2\pi f t_0)$. Substitution of $x(t)$ in (11.9) yields:

$$W_{nx}(t, f_1, \ldots, f_n) = A_0^{n+1} \delta(t - t_0).$$

This is a somewhat expected result considering the projection in time (11.17) and the time support (11.26) properties. □

11.4.3 General Class of Time-Frequency Higher-Order Spectra-Based Distributions

The unified approach given by Cohen to most time-frequency representations was described in section 11.3. This formulation can be extended to the case of higher-order moment spectra as follows [Fonollosa and Nikias, 1993].

Let us define

$$\text{WHOMS}_{GEN}(\Omega, \tau_1, \ldots, \tau_n) = \phi_{HOS}(\Omega, \tau_1, \ldots, \tau_n) \int_{f_1} \cdots \int_{f_k} \int_n W_{nx}(u, f_1, \ldots, f_n) \exp(j2\pi u\Omega)$$
$$\prod_{i=1}^{n} \exp(-j2\pi f_i \tau_i) df_i du$$

(11.33)

or in terms of the local higher-order moment function

$$\text{WHOMS}_{GEN}(\Omega, \tau_1, \ldots, \tau_n) = \phi_{HOS}(\Omega, \tau_1, \ldots, \tau_n) \int_n m_n^x(u, \tau_1, \ldots, \tau_n)$$
$$\exp(j2\pi u\Omega) du,$$

(11.34)

where $\phi_{HOS}(\Omega, \tau_1, \ldots, \tau_n)$ is a multidimensional kernel analogous to $\phi(\theta, \tau)$ defined in (11.6.1). Hence, the general class of higher-order time-frequency distributions is formulated as

$$\text{PHOS}_{GEN}(t, f_1, \cdots, f_n) = \int_\Omega \int_{\tau_1} \cdots \int_{\tau_n} \text{WHOMS}_{GEN}(\Omega, \tau_1, \ldots, \tau_n) \exp(j2\pi t\Omega)$$
$$\prod_{i=1}^{n} \exp(-j2\pi f_i \tau_i) d\tau_i d\Omega$$

(11.35.1)

or alternatively based on (11.34),

$$\text{PHOS}_{GEN}(t, f_1, \ldots, f_n) = \int_\Omega \int_{\tau_1} \cdots \int_{\tau_n} \int_u \phi_{HOS}(\Omega, \tau_1, \ldots \tau_n) m_k^x(u, \tau_1, \ldots, \tau_n)$$
$$\exp(j2\pi u\Omega) \exp(-j2\pi t\Omega) \prod_{i=1}^{n} \exp(-j2\pi f_i \tau_i) d\tau_i d\Omega du.$$

(11.35.2)

The properties of any particular member of the general class can be obtained from the properties of the WHOMS and the particular kernel.

11.4.4 Reduced Interference Higher-Order Distribution (RIHOD)

As it was discussed in section 11.3.1, the main difficulty with the WD is the presence of interference between different components of the signal, or otherwise, the cross-term effect. This problem is actually more dominant in the WHOMS, since the bilinear expression of the WD is substituted with a multilinear form. For the WD, the Choi-Williams distribution has been used to reduce the spurious peaks caused by cross-terms, without sacrificing highly desirable properties of a time-frequency representation. The equivalent reduced interference distribution in the higher-order moment spectra domain is obtained using the kernel [Fonollosa and Nikias, 1991a,b]

$$\phi_{RIHOD}(\Omega, \tau_1, \cdots, \tau_n) = \prod_{i=1}^{n} \exp(-\tau_i^2 \Omega^2 / \sigma). \qquad (11.36)$$

The basic properties of WHOMS are preserved if (11.36) is used in (11.35.2) since

$$\phi_{RIHOD}(0, \tau_1, \ldots, \tau_n) = 1, \quad \text{for every } \tau_1, \ldots, \tau_n \qquad (11.37.1)$$

$$\frac{\partial}{\partial \Omega} \phi_{RIHOD}(\Omega, \tau_1, \ldots, \tau_n)|_{\Omega=0} = 0, \quad \text{for every } \tau_1, \ldots, \tau_n \qquad (11.37.2)$$

$$\phi_{RIHOD}(\Omega, 0, \ldots, 0) = 1; \quad \text{for every } \Omega \qquad (11.37.3)$$

$$\frac{\partial}{\partial \tau_i} \phi_{RIHOD}(\Omega, \tau_1, \ldots, \tau_n)|\tau_1 = \cdots = \tau_n = 0, \quad \text{for every } \Omega \text{ and every } \tau_i. \qquad (11.37.4)$$

Using the kernel $\phi_{RIHOD}(\Omega, \tau_1, \ldots, \tau_n)$, the Reduced Interference Higher-Order Distribution (RIHOD) is obtained as

$$RIHOD_{nx}(t, f_1, \ldots, f_n) = \int_\Omega \int_{\tau_1} \cdots \int_{\tau_n} WHOMS_{nx}(\Omega, \tau_1, \ldots, \tau_n) \qquad (11.38)$$
$$\prod_{i=1}^{n} \exp(-\tau_i^2 \Omega^2 / \sigma) d\tau_i \, d\Omega$$

which is an extension of the Choi-Williams distribution of (11.8) to a higher-order moment spectra domain.

11.5 WIGNER POLYSPECTRA: DISCRETE CASE

When it comes to practical signal processing applications of the WD, a discretization in time and frequency is necessary. A discrete version of the WD was initially defined by Claasen and Mecklenbrauker [1980a,b]. Later, Peyrin and Prost [1986] proposed a unified definition for the Discrete-Time, Discrete-Frequency, and Discrete-Time/Frequency Distribution.

The second aforementioned approach was followed by Fonollosa and Nikias [1993] to define the Discrete Time and Frequency WHOMS. Starting with the continuous definition applied on a discretized time signal, the Discrete Time WHOMS

(DT-WHOMS) is derived. Then, the definition of the WHOMS in the continuous frequency domain is used with a periodic signal, discrete in frequency, to derive the Discrete Frequency WHOMS (DF-WHOMS). Finally, a discretization in time and frequency is considered simultaneously, by applying the original continuous definition to a discrete-time and periodic signal, resulting in the DTF-WHOMS.

11.5.1 Discrete Time WHOMS (DT-WHOMS)

Let $x_c(t)$ be a continuous, complex in general signal and $x_s(t)$ the signal derived after uniform sampling of $x_c(t)$ by a train of ideal impulse functions of period T. The resultant discrete sequence is denoted by $x(k) = x_c(kT)$, i.e.,

$$x_S(t) = \sum_k x_c(kT)\delta(t - kT) = \sum_k x(k)\delta(t - kT). \qquad (11.39)$$

Applying expression (11.9) to $x_S(t)$, we obtain

$$W_{nx_S}(t, f_1, \ldots, f_n) = \sum_k W_{nx}(k, f_1, \ldots, f_n)\delta(t - k\frac{T}{n+1}), \qquad (11.40)$$

$$\begin{aligned} W_{nx}(k, f_1, \ldots, f_n) &= \sum_{m_1} \cdots \sum_{m_n} x^*(k - \sum_{q=1}^n m_q) \prod_{j=1}^n x(m_j) \\ &\exp\{-j2\pi f_j(2m_j + \sum_{i=1, i\neq j}^n m_i - k)T\}. \end{aligned} \qquad (11.41)$$

It is clear from (11.40) that the WHOMS of the discretized signal is sampled in time at a rate of $(n + 1)/T$, which is $(n + 1)$ times faster than the sampling rate of the original signal. In analogy with the definition of the discrete sequence $x(k)$, it is natural to define the DT-WHOMS as the weight of the impulse functions, $W_{nx}(k, f_1, \ldots, f_n)$, discrete in time but still a continuous function in the multifrequency space. The DT-WHOMS is periodic in each frequency, with the same periodicity, $1/T$, as the Fourier transform of the sampled signal.

11.5.2 Discrete Frequency WHOMS (DF-WHOMS)

Consider a continuous temporal signal, non-periodic in general. Its Fourier transform will be a continuous function in frequency. Let $X_p(f)$ be generated after uniform sampling of $X_{np}(f)$ by a train of ideal impulse functions of period Δ. Then, $X_p(f)$ becomes the Fourier transform of a periodic repetition of the original non-periodic signal with period $1/\Delta$. The resulting discrete frequency sequence is defined by $X(k) = X_{np}(k\Delta)$ and

$$X_p(t) = \sum_k X_{np}(k\Delta)\delta(f - k\Delta) = \sum_k X(k)\delta(f - k\Delta). \quad (11.42)$$

Substitution of $X_p(f)$ in the continuous frequency definition of WHOMS, given by (11.21), leads to

$$W_{nx_p}(t, f_1, \ldots, f_n) = \sum_{p_1} \cdots \sum_{p_n} W_{nx}(t, p_1, \ldots, p_n) \prod_{i=1}^n \delta(f_i - (np_i - \sum_{j=1, j\neq i}^n p_j)\frac{\Delta}{n+1}), \quad (11.43)$$

where

$$W_{nx}(t, p_1, \ldots, p_n) = \sum_m X^*(m) \prod_{i=1}^n X(p_i - m) \exp -j2\pi[(n+1)m - \sum_{i=1}^n p_i]\Delta t. \quad (11.44)$$

Notice that $W_{nx}(t, p_1, \ldots, p_n)$ is the weight of an impulse function situated in the multifrequency space of WHOMS at position:

$$f_i = (np_i - \sum_{j=1,\, j\neq i}^n p_j)\frac{\Delta}{n+1} = l_i \frac{\Delta}{n+1}, \qquad \text{for } i = 1, \ldots, n. \quad (11.45.1)$$

$$l_i = np_i - \sum_{j=1,\, j\neq i}^n p_j \quad (11.45.2)$$

where the index l_i is used to represent the position of $W_{nx}(t, p_1, \ldots, p_n)$ in the frequency component f_i as a function of all the indexes $p_1, \ldots, p_n$. Although all integer values are valid for p_i, the only possible values of l_i are the integers satisfying

$$\frac{1}{n+1}(2l_i + \sum_{j=1,\, j\neq i}^n l_j) = p_i = \text{Integer value, for } i = 1, \ldots, n. \quad (11.46)$$

Consequently, the WHOMS of a frequency sampled signal results in a *nonuniformly* sampled function in all frequency axes at some, *but not all*, integer multiples of $\Delta/(n+1)$. Following the same arguments given for the DT-WHOMS, it is natural to define the DF-WHOMS as the weight of the impulse functions, $W_{nx}(t, p_1, \ldots, p_n)$; a discrete function in each frequency component, but still continuous in time. Observe that, in analogy with the inverse Fourier transform of $X_p(f)$, $W_{nx}(t, p_1, \ldots, p_n)$ is a periodic function in time with periodicity $1/\Delta$.

11.5.3 Discrete Time and Frequency WHOMS (DTF-WHOMS)

The previous definitions for the DT-WHOMS and DF-WHOMS permit a formulation of the DTF-WHOMS. Consider a continuous signal that is uniformly sampled in both time and frequency domains. The sampling periods in time and frequency and T and $1/NT$, respectively.

$$x_{sp}(t) = \sum_m \sum_k x_c(t)\delta(t-(k+mN)T) = \sum_{k=0}^{N-1} x(k)\delta(t-kT) \otimes \sum_m \delta(t-mNT). \tag{11.47}$$

It follows from definitions in the time domain (11.9), or in the frequency domain (11.22), and the convolution and product property (11.29), that

$$W_{nx_{sp}}(t,f_1,\ldots,f_n) = \sum_k \sum_{p_1}\cdots\sum_{p_n} W_{nx}(k,p_1,\ldots,p_n)$$

$$\delta(t-k\frac{T}{n+1})\prod_{i=1}^n \delta(f_i - l_i\frac{1}{(n+1)NT}), \tag{11.48}$$

where l_i are given by (11.45) and (11.46). If the time domain definition (11.9) was applied to derive (11.48), the time domain expression would yield

$$W_{nx}(k,p_1,\ldots,p_n) = \frac{1}{N}\exp\{j\frac{2\pi}{N(n+1)}\sum_{l=1}^n p_l\}$$

$$\cdot \sum_{m_1=0}^{N-1}\cdots\sum_{m_k=0}^{N-1} x^*(k-\sum_{i=1}^n m_i)\prod_{\ell=1}^n x(m_\ell)\exp\{-j\frac{2\pi}{N}m_\ell p_\ell\}. \tag{11.49}$$

The equivalent expression in the frequency domain is obtained using (11.22)

$$W_{nx}(k,p_1,\cdots,p_n) = \frac{1}{N}\exp\{j\frac{2\pi k}{N(k+1)}\sum_{\ell=1}^n p_l\}$$

$$\sum_{m=1}^N X^*(m)\prod_{j=1}^n X(p_j-m)\exp(-j\frac{2\pi}{N}mk). \tag{11.50}$$

The WHOMS of $x_{sp}(t)$ becomes a sampled function at positions $kT/(n+1)$ and $l_i/(n+1)NT$ in time and for every frequency component, respectively. The resultant temporal sampling is uniform. However, the WHOMS is not uniformly sampled in the frequency domain because not all integers are permitted for l_i [but only the ones defined in (11.45) and (11.46)]. The DTF-WHOMS is then defined as the weight of the impulse functions in the WHOMS of the discrete and periodic signal, $W_{ns}(k,p_1,\ldots,p_n)$.

This is actually a periodic function with respect to the indexes n and p_i with period $(k+1)N$. Nonetheless, a careful analysis reveals that only knowledge of

$W_{nx}(k, p_1, \ldots, p_n)$ for $0 \leq k \leq N-1, 0 \leq p_i \leq N-1$ is necessary to reconstruct $W_{nx}(k, p_1, \ldots, p_n)$ elsewhere since:

$$W_{nx}(k + qN, p_1, \ldots, p_n) = \exp\{j\frac{2\pi q}{(n+1)}\sum_{i=1}^{n} p_i\} W_{nx}(k, p_1, \ldots, p_n), \quad (11.51.1)$$

$$W_{nx}(k, p_1, \ldots, p_i + qN, \ldots p_n) = \exp\{j\frac{2\pi q}{(n+1)}n\} W_{nx}(k, p_1, \ldots, p_n), \quad (11.51.2)$$

for any integer q. Figure 11.5 illustrates the nonredundant multifrequency domain of the DTF-WHOMS definition for $n = 2$.

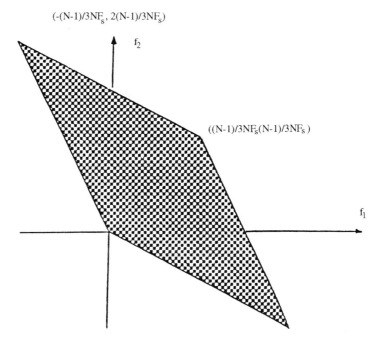

Figure 11.5 Nonredundant region of support in the frequency domain of DTF-WHOMS for $n = 2$ ([Fonollosa and Nikias, 1993] © 1993 IEEE).

11.5.4 Computation of DTF-WHOMS

Expressions (11.49) and (11.50) suggest a fast algorithm for the calculation of the DTF-WHOMS. Relation (11.49) can be expressed as:

$$W_{nx}(k, p_1, \ldots, p_n) = \frac{1}{N}\exp\{j\frac{2\pi}{N(n+1)}(k\sum_{l=1}^{n} p_l)\}DFT^n[f_k(m_1, \cdots, m_n)]_{p_1, \ldots, p_n},$$
$$(11.52.1)$$

where

$$f_k(m_1,\ldots,m_n) = x^*(k - \sum_{i=1}^{n} m_i) \prod_{j=1}^{n} x(m_j). \qquad (11.52.2)$$

For every temporal sample k, the function $f_k(m_1,\ldots,m_n)$ can be defined and an n-dimensional FFT can be utilized for the calculation of the DTF-WHOMS, except for a phase factor. Thus, $W_{nx}(k,p_1,\cdots,p_n)$ can be computed in the nonredundant region of support, $0 \le k \le N-1, 0 \le p_i \le N-1$, with N n-dimensional FFTs of order N.

If (11.50) is considered for the DTF-WHOMS,

$$W_{nx}(k,p_1,\ldots,p_n) = \frac{1}{N}\exp\{j\frac{2\pi}{N(n+1)}(k\sum_{l=1}^{n} p_l)\}DFT[f_{p_1,\ldots,p_n}(m)]_k, \qquad (11.53.1)$$

where

$$f_{p_1,\ldots,p_n}(m) = X^*(m)\prod_{j=1}^{n} X(p_j - m). \qquad (11.53.2)$$

For every multifrequency point $p_1,\ldots,p_n$, the function $f_{p_1,\ldots,p_n}(m)$ can be Fourier transformed to derive the DTF-WHOMS at that point and for any temporal index $0 \le k \le N-1$. Consequently, N^n one-dimensional FFTs of order N are necessary to compute $W_{nx}(k,p_1,\ldots,p_n)$ in its entire nonredundant domain of support.

Assuming that the n-dimensional FFTs are implemented by means of one-dimensional FFTs, both possible implementations require $O(N^{n+1}\log_2 N)$ operations.

11.5.5 Relation Between the WHOMS and the DTF-WHOMS After Sampling

Continuous signals are usually sampled to obtain discrete sequences in real data analysis applications. Thus, it becomes useful to relate the WHOMS to the DTF-WHOMS after sampling. In order to derive this relation, we will calculate initially the WHOMS of a periodic train of impulses.

WHOMS of a Periodic Train of Impulses

The impulse train function of period L and its Fourier transform are denoted by

$$dt_L(t) = \sum_{m} \delta(t - mL) \qquad (11.54)$$

$$DT_L(f) = \frac{1}{L} \sum_k \delta(f - k\frac{1}{L}). \tag{11.55}$$

The procedure used to define the expression of DT-WHOMS can be used to calculate the WHOMS of $dt_L(t)$ substituting $x(k) = 1, T = L$, in (11.39) and (11.40), since the impulse train can be viewed as a sampled signal with unity valued samples. That is,

$$W_{ndt_L}(t, f_1, \ldots, f_n) = \sum_k \sum_{p_1} \cdots \sum_{p_n} \prod_{j=1}^{n} \exp\{-j2\pi f_j (2p_j + \sum_{i=1, i \neq j}^{n} p_i - k)L\}$$

$$\delta(t - k\frac{L}{n+1}). \tag{11.56}$$

Another equivalent expression can be obtained considering the definition of DF-WHOMS, $\Delta = 1/L$ in (11.43), (11.44), since a train of impulses in the time domain is represented by another impulse train in the frequency domain:

$$W_{ndt_L}(t, f_1, \ldots, f_k) = \frac{1}{L^{(n+1)}} \sum_{p_1} \cdots \sum_{p_n} \sum_m \exp\{-j2\pi[(n+1)m - \sum_{i=1}^{n} p_i]\frac{1}{L}t\}$$

$$\cdot \prod_{i=1}^{k} \delta(f_i - (kp_i - \sum_{j=1, j \neq i}^{n} p_j)\frac{1}{L(n+1)}). \tag{11.57}$$

These expressions can be used to calculate the effect of sampling a continuous signal in relation with its WHOMS.

Nonaliasing Constraints for DT-WHOMS: WHOMS of Sampled Signals

A sampled signal can be represented as a continuous signal *multiplied* by an impulse train function,

$$x_S(t) = x_c(t) \sum_k \delta(t - kT). \tag{11.58}$$

Taking into account the product property of WHOMS in (11.29.1), the DT-WHOMS may be viewed as a result of a convolution in the multifrequency domain between the WHOMS of the continuous signal and the WHOMS of the impulse train of period $L = T$. Taking for $W_{ndt_L}(t, f_1, \ldots, f_n)$ the expression (11.57) and applying the product property (11.29.1), the following relation between the WHOMS of any signal prior and after sampling is obtained,

$$\sum_k W_{nx_s}(k,f_1,\ldots,f_n)\delta(t-k\frac{T}{n+1}) = \frac{1}{T(n+1)}\sum_{p_1}\cdots\sum_{p_n}\sum_m \exp\{-j2\pi[(n+1)m-\sum_{i=1}^m p_i]\frac{1}{T}t\}$$
$$\cdot W_{nx_c}(t,\ldots,f_i-l_i\frac{1}{T(n+1)},\ldots),$$
(11.59)

for $l_i = (np_i - \sum_{j=1,j\neq i}^n p_j)$.

We observe that the DT-WHOMS becomes a superposition of frequency delayed versions of the original WHOMS before sampling. Again, we have defined the index l_i as a function of all p_j, for $j = 1,\ldots,n$. Every integer is valid for p_j although only certain combinations of integers, the ones satisfying (11.46), are possible for l_i. This nonuniformity in the frequency aliasing due to sampling is of paramount importance since it determines the surprising non-aliasing constraints.

To avoid aliasing in the multifrequency domain, the regions of support of WHOMS of the continuous signal should not overlap when displaced in frequency by $l_i/T(n+1)$. It can be shown that aliasing is avoided if:

$$\frac{1}{T(n+1)} \geq \frac{2B_x}{(n+1)} \Rightarrow B_x \leq \frac{1}{2T},$$
(11.60)

i.e., the sampling frequency is at least two times the bandwidth of the continuous signal. This result is illustrated in Figure 11.6 for $n = 2$.

The aliasing issue is also discussed in [Swami, 1992].

Nonaliasing Constraints for DF-WHOMS: WHOMS of Periodic Signals

It is well known that a time-limited signal may be expressed uniquely by its Fourier coefficients, i.e., samples of its Fourier transform. This representation is equivalent to considering a periodic repetition of the signal in the time domain. Using the DF-WHOMS with the Fourier coefficients of the signal,

$$X_p(f) = X_{np}(f)\sum_k \delta(f-k\Delta),$$
(11.61)

considering for $W_{ndt_L}(t,f_1,\ldots,f_n)$ equation (3.18), and applying the convolution property (2.29.b), we obtain:

$$\sum_{p_1}\cdots\sum_{p_n} W_{nx_p}(t,p_1,\ldots,p_n)\prod_{i=1}^n \delta(f_1-l_i\frac{\Delta}{n+1}) =$$
$$\sum_k\sum_{p_1}\cdots\sum_{p_k}\prod_{j=1}^n \exp\{-j2\pi f_j\frac{1}{\Delta}(2p_j+\sum_{i=1,i\neq j}^n p_i-n)\}W_{nx_{np}}(t-k\frac{1}{(n+1)\Delta},f_1,\ldots,f_n),$$
(11.62)

Chapter 11   511

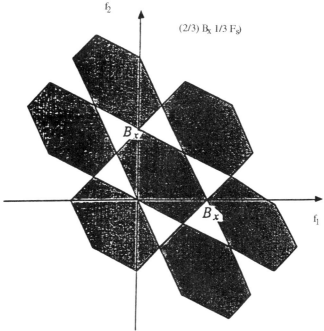

Figure 11.6   Illustration of the nonaliasing constraint in the Wigner Bispectrum of a deterministic analytic signal of bandwidth B_x i.e., $\frac{2}{3}B_x \leq \frac{1}{3T_s}$; $F_s = \frac{1}{T_s} \geq 2B_x$ ([Fonollosa and Nikias, 1993]. © 1993 IEEE).

where l_i is given by (3.7) and (3.8). The effect of a time-limited observation is that the DF-WHOMS turns out to be a superposition of time-delayed versions of the original WHOMS, before frequency sampling by Δ. Consequently, temporal aliasing is avoided when the temporal duration of WHOMS of the nonperiodic signal is shorter than $1/(n+1)\Delta$.

Nonaliasing Constraints for DTF-WHOMS: WHOMS of Periodic and Sampled Signals

If the nonaliasing constraints for DT-WHOMS and DF-WHOMS are combined, they define the constraints for DTF-WHOMS. In this case, $\Delta = 1/NT$ and both frequency and time aliasing have to be taken into account. As mentioned for the DT-WHOMS (due to the nonuniform aliasing in the multifrequency domain), the sampling frequency prior to the computation of the DTF-WHOMS should be at least twice the Nyquist frequency. If the signal can not be filtered, then interpolation by a factor of two becomes necessary. To avoid time-aliasing, the signal must be limited to one interval of length $NT/(n+1)$ in time. Considering property (11.26), the time support of WHOMS is the same independent of the duration of the signal, and therefore, the signal itself should be limited to $NT/(n+1)$. As such, zero padding of $nN/(n+1)$ samples is necessary.

Example 11.8

The Wigner Bispectrum (e.g., DTF-WHOMS for $n = 2$) of order $N = 256$ is calculated for the generalized discrete chirp signal,

$$x(k) = \exp\{j2\pi[(\frac{k - N/6}{N})^2\alpha + (\frac{k - N/6}{N})\beta]\},$$

$k = 0, \cdots, N/3 - 1$ and for different values of α and β.

For $\alpha = 0, x(k)$ is a complex exponential sequence and, considering (11.39), a sharp peak should appear in its DTF-WHOMS at every time instant. In particular, Figure 11.7 shows the magnitude of $W_{2x}(k, p_1, p_2)$ in the middle of the temporal interval, i.e., $k = N/2$. The discrete frequency of $x(k)$ is $\beta = 40$. For arbitrary values of α and β, $x(k)$ is a chirp signal and thus its frequency is linearly changing with time. This time-varying instantaneous frequency is also reflected in the DTF-WHOMS of the signal. Figure 11.8 shows the magnitude of $W_{2x}(k, p_1, p_2)$ at (a) $k = N/4$, (b) $k = N/2$ and (c) $k = 3N/4$. □

11.6 APPLICATIONS OF WIGNER HIGHER-ORDER SPECTRA

The detection of unknown deterministic transient signals in additive stationary noise is an important problem in underwater acoustics and radar. Different higher-order spectra-based methods have been reported for the detection and parameter estimation of transient signals [Papadopoulos and Nikias, 1990, Hinich, 1990]. Some additional methods have also been proposed for the detection of transients based on time-frequency representations [Friedlander and Porat, 1989, Probasco and Boudreaux-Bartels, 1990; Boashash and O'Shea, 1990]. In the first two papers, the Gabor coefficients have been employed for the detection of short duration signals. In the third paper, a methodology is developed for the detection and classification of cylinder firings of marine engines. The detection is based on the autocorrelation function, while the classification on the Wigner-Ville distribution (WD).

Frazer and Boashash [1991] demonstrated through real data analysis that the use of higher-order spectra reveals information that is hidden in conventional signal

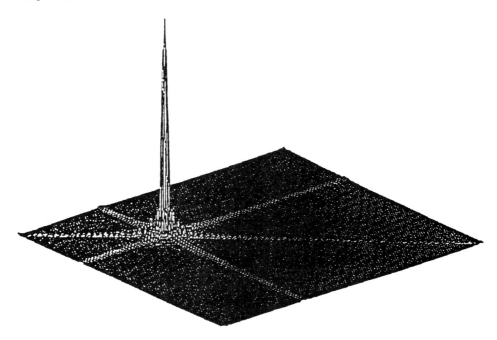

Figure 11.7 Illustration of the Wigner Bispectrum of order $N = 256$ of a complex exponential signal of frequency $\beta = 40$ at the temporal sample $k = N/2$ ([Fonollosa and Nikias, 1993]. © 1993 IEEE).

analysis procedures. The analyzed data were digitized recordings of underwater acoustic emissions of whales, and the analysis was performed using the Wigner Bispectrum defined by Gerr [1988] as

$$G(t, f_1, f_2) = \int_{-\infty}^{\infty} \int_{-\infty}^{\infty} x(t - \frac{2}{3}u_1 - \frac{1}{3}u_2) x(t + \frac{1}{3}u_1 - \frac{1}{3}u_2) \\ x(t + \frac{1}{3}u_1 + \frac{2}{3}u_2) e^{-j2\pi(f_1+f_2)u_1 - j2\pi f u_2} du_1 \, du_2, \quad (11.63)$$

and the time-varying bispectrum given by

$$B(t, f_1, f_2) = \int_{t-\Delta/2}^{t+\Delta/2} \int_{t-\Delta/2}^{t+\Delta/2} x(t) x(t+u_1) x(t+u_2) e^{-j2\pi f_1 u_1 - j2\pi f_2 u_2} du_1 \, du_2. \quad (11.64)$$

Note that (11.63) was derived by retaining the lag-centering property of the Wigner-Ville distribution and requires that if the signal is third-order stationary, then the expectation of $G(t, f_1, f_2)$ leads to the traditional bispectrum.

A simple and novel test for the detection of transient signals in noise using WHOMS was presented by Fonollosa and Nikias [1993], and is described in the sequel.

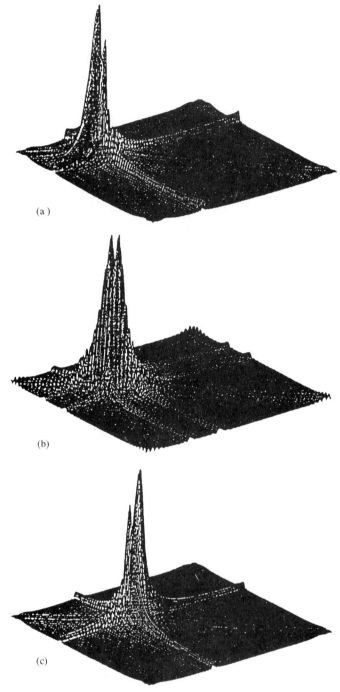

Figure 11.8 Illustration of the Wigner Bispectrum of order $N = 256$ of a complex chirp signal of mean frequency $\beta = 40$ and $\alpha = 60$, (a) [top] at the temporal sample $k = N/4$, (b) [center] at $k = N/2$ and (c) [bottom] at $k = 3N/4$ ([Fonollosa and Nikias, 1993] © 1993 IEEE).

Let $X(k)$ be a sequence consisting of a deterministic signal $s(k)$ embedded in additive white noise. In analyzing long data sequences, windowing becomes necessary in order to compute the WHOMS. Let a window of length N be used for the computation of DFT-WHOMS for $n = 1$ (WD) and $n = 2$ (WB):

$$W_{1x}(k,p) = \frac{1}{N} \exp(j\frac{\pi}{N}kp) \sum_{m=0}^{N-1} X^*(k-m)X(m)\exp(-j\frac{2\pi}{N}mp), \qquad (11.65)$$

and

$$W_{2x}(n,p_1,p_2) = \frac{1}{N} \exp\{j\frac{2\pi}{3N}n(p_1+p_2)\}.$$

$$\sum_{m_1=0}^{N-1} \sum_{m_2=0}^{N-1} X^*(n-m_1-m_2)X(m_1)X(m_2)\exp\{-j\frac{2\pi}{N}(m_1p_1+m_2p_2)\}. \qquad (11.66)$$

In addition, the output of a matched filter is calculated for comparison purposes. Assuming that the noise is white in the bandwidth of the transient signal, $s(k)$, the matched filter is designed with impulse response equal to

$$h(k) = s(T_1 - k), \qquad (11.67)$$

where T is the duration, in samples, of the deterministic transient $s(k)$. If $X(k)$ is the received signal (noise only or signal plus noise), the output of the matched filter is given by

$$y_{mf}(k) = X(k) * h(k). \qquad (11.68)$$

In the analysis of long data sequences, windowing becomes necessary. When the WD or WB are applied, the window length is determined by the order N used in the computation of (11.65) or (11.66). Time aliasing is avoided if $x(k)$ is limited from $k = 0$ to $N/2 - 1$ and $N/3 - 1$ in (11.65) and (11.66), respectively. Furthermore, to avoid frequency aliasing, an interpolation of order 2 of the data is necessary in both expressions.

Let's assume that the signal to be analyzed is passed through an anti-aliasing low-pass filter of cutoff frequency $F_s/4$. The output of the filter is sampled at rate F_s to obtain $X(k)$. The following hypotheses are considered for $X(k)$.

$$H_0 : X(k) = W(k) \qquad (11.69.1)$$

where $W(k)$ is a zero-mean stationary noise process and

$$H_1 : X(k) = s(k) + W(k), \qquad (11.69.2)$$

where $s(k)$ is a deterministic transient signal and, consequently, $X(k)$ is nonstationary.

For H_0, the expected value of both the WD and WB are independent of k; they correspond to the Power Spectrum and Bispectrum of the noise, respectively. For H_1, on the other hand, the nonstationarity of $X(k)$ will be reflected in the WD and in the WB. In other words, both the WD and WB functions will be changing with the temporal index k.

A decision variable of a function $f(k), DS\gamma\{f(k)\}$, may be defined as:

$$DS\gamma\{f(k)\} = \frac{(\prod_{i=0}^{M-1}|f(k-i)|^{\gamma^i})^{1/\Gamma}}{1/\Gamma \sum_{i=0}^{M-1} \gamma^i |f(k-i)|} \qquad (11.70.1)$$

for

$$\Gamma = \sum_{i=0}^{M-1} \gamma^i = \frac{1-\gamma^M}{1-\gamma} \qquad (11.70.2)$$

where M is the number of available data points of $f(k)$ and $0 < \gamma \leq 1$. $DS\gamma\{f(k)\}$ is the ratio of the geometric mean over the arithmetic mean of the sequence $|f(k)|$. It can be shown that

$$\begin{aligned} &\text{(i)} \quad & 0 \leq DS\gamma\{f(k)\} \leq 1 & \\ &\text{(ii)} \quad & DS\gamma\{f(k)\} = 1 \Leftrightarrow |f(i)| = |f(j)| \; \forall i,j. & \end{aligned} \qquad (11.71)$$

The forgetting factor γ emphasizes recent samples by means of one-sided exponential windowing and can be set to unity if the rectangular window is assumed. The decision variable (11.68) can be used to detect the transient signal as follows.

Under hypothesis H_0, $X(k)$ is a stationary process and thus

$$\begin{aligned} DS\gamma\{E\{W_{2x}(k,p_1,p_2)\}\} &= 1 \\ DS\gamma\{E\{W_{1x}(k,p)\}\} &= 1 \\ DS\gamma\{E\{y_{op}(k)\}\} &= 1. \end{aligned} \qquad (11.72)$$

Under H_1, however, $X(k)$ is nonstationary and

$$\begin{aligned} DS\gamma\{E\{W_{1x}(k,p)\}\} &< 1 \\ DS\gamma\{E\{W_{2x}(k,p_1,p_2)\}\} &< 1 \\ DS\gamma\{E\{y_{op}(k)\}\} &< 1. \end{aligned} \qquad (11.73)$$

The ability of each transformation to detect the nonstationarity will be reflected in the discriminating power of $DS\gamma$ between H_0 and H_1.

Simulation results are given for the transient signal $s(k)$ illustrated in Figure 11.9(a). White Gaussian noise was added to $s(k)$ to generate $X(k)$. The signal-to-noise ratio, (SNR) is defined as

$$\text{SNR} = 10 \log_{10} \max_k (s(k))/\gamma_2^w).$$

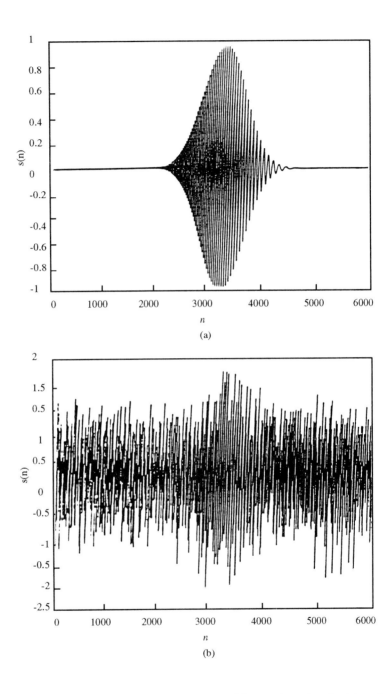

Figure 11.9 (a) Noise-free transient signal, and (b) transient signal with noise with SNR = 0 dB ([Fonollosa and Nikias, 1993] © 1993 IEEE).

Figure 11.9(b) illustrates $X(k)$ for SNR = 0 dB.

At every $L = 20$ samples, the following averaged functions are calculated:

$$AWB(k, p_1, p_2) = \sum_{i=0}^{L/2-1} W_{2x}(Lk - i, p_1, p_2) \qquad (11.74.1)$$

$$AWD(k, p) = \sum_{i=0}^{L/2-1} W_{1x}(Lk - i, p) \qquad (11.74.2)$$

$$AO(k) = \sum_{i=0}^{L/2-1} |y_{op}(Lk - i)|, \qquad (11.74.3)$$

and their corresponding detection statistics follow by combining (11.74) with (11.70.1); viz,

WB Detector:

$$DS - WB(k) = \sum_{p_1=0}^{N-1} \sum_{p_2=0}^{N-1} DF\gamma\{AWB(k, p_1, p_2)\} \qquad (11.75.1)$$

WD Detector:

$$DS - WD(k) = \sum_{p=0}^{N-1} DF\gamma\{AWD(k, p)\} \qquad (11.75.2)$$

Matched Filter Detector:

$$DS - MF(k) = DF\gamma\{AO(k)\}. \qquad (11.75.3)$$

Figures 11.10(a),(b), and (c) show $DS-WB(k), DS-WD(k)$, and $DS-MF(k)$ as functions of k, respectively, for SNR = 10,5, and 0 dB.

From these figures, it is apparent that detection statistics $DS - WB(k), DS - WD(k)$, and $DS - MF(k)$ attain initially a constant value of 0.7, 0.65 and 0.85, respectively, and then decrease in value to converge around a point in the data record that corresponds to the beginning of the transient signal. Note that a constant value in detection statistics over a time interval corresponds to stationary data (i.e., noise only) whereas dramatic change in their value corresponds to a presence of nonstationarity, i.e., a transient signal in stationary noise. Comparing Figures 11.10(a) and 11.10(b), we see that for SNR > 0 the Wigner Bispectrum (WB) detection statistic performs better than the detection statistic based on the WD. The function $DS-MF(k)$, shown in Figure 11.10(c), illustrates the detection

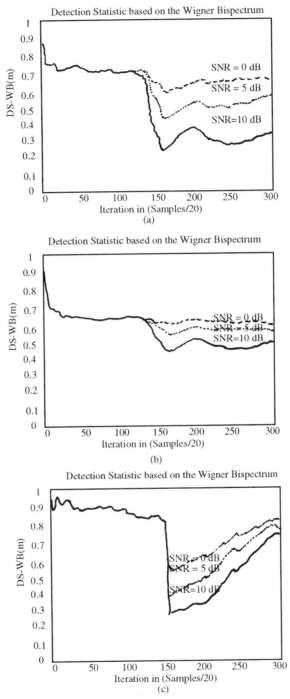

Figure 11.10 Detection statistics for SNR = 10, 5, and 0 dB; (a) DS-WB(k), (b) DS-WD(k), and (c) DS-MF(k) ([Fonollosa and Nikias, 1993] © 1993 IEEE).

performance obtained with the optimum method, *which requires exact knowledge of the shape of the transient signal*. Note that the detection statistics based on both the WB and WD do not require a priori knowledge of the shape of the transient signal. Similar performance for $DS-WB(k)$, $DS-WD(k)$ and $DS-MF(k)$ has been observed for other types of chirp-like transient signals.

11.7 SUMMARY

Time-frequency distributions have become one of the important areas of research and applications in digital signal processing. In this chapter, we described the time-frequency distributions based on higher-order moments and spectra, beginning with the continuous-time formulations and concluding with the discrete-time formulations, and some of their applications. Our discussion also included a brief presentation of the Wigner-Ville distribution, Cohen's general class of distributions, as well as reduced interference distributions.

The tutorial paper by Cohen [1989] covers time-frequency distributions based on second-order moments and spectrum, their properties, interrelationships, and applications. The papers by Fonollosa and Nikias [1991a,b, and 1993] are the main source of references of our treatment of the higher-order moment spectra-based time-frequency distributions. Additional important papers on this subject include those of Gerr [1988], Swami [1991], Giannakis and Dantawate [1991], and Frazer and Boashash [1991].

REFERENCES

Allen, J. B., "Short-Time Spectral Analysis, Synthesis and Modification by Discrete Fourier Transform," *IEEE Trans. Acoust., Speech, Signal Processing*, **ASSP-25**, pp. 235–238, June, 1977.

Boashash, B., and G. Frazer, "Time-varying Higher-Order Spectra, Generalized Wigner-Ville Distribution and the Analysis of Underwater Acoustic Data," *Proc. IEEE ICASSP-92*, **V**, pp. 193–196, San Francisco, 1992.

Boashash, B., and P. O'Shea, "A Methodology for Detection and Classification of Some Underwater Acoustic Signals Using Time-Frequency Analysis Techniques," *IEEE Trans. Acoust., Speech, Signal Processing*, **ASSP-38, (11)** pp. 1829–1841, November, 1990.

Boashash, B., and L. B. White, "Instantaneous Frequency Estimation and Automatic Time-Varying Filtering," *Proc. IEEE ICASSP-90*, pp. 1221–1224, April, 1990.

Bertrand, J., and P. Bertrand, "Affine Time-Frequency Distributions" in **Time-Frequency Signal Analysis – Methods and Applications**, B. Boashash, (ed.) Melbourne, Australia: Longman-Chesire, 1991.

Boudreaux-Bartels, G. F., and T. W. Parks, "Time-Varying Filtering and Signal Estimation Using Wigner Distribution Synthesis Techniques," *IEEE Trans. Acoust., Speech, Signal Processing*, **ASSP-34**, pp. 442–451, June, 1986.

Choi, H., and W. J. Williams, "Improved Time-Frequency Representation of Multicomponent Signals Using Exponential Kernels," *IEEE Trans. Acoust., Speech, Signal Processing*, **ASSP-37**, pp. 862–871, June, 1989.

Claasen, T.A.C.M., and W. F. G. Mecklenbrauker, "The Wigner Distribution. A Tool for Time-Frequency Signal Analysis; Part I: Discrete Time Signals," *Philips. J. Res.*, **35**, pp. 217–250, 1980a.

Claasen, T.A.C.M., and W. F. G. Mecklenbrauker, "The Wigner Distribution. A Tool for Time-Frequency Signal Analysis; Part II: Continuous Time Signals," *Philips. J. Res.*, **35**, pp. 276–300, 1980b.

Cohen, L., "Generalized Phase-Space Distribution Functions," *J. Math Phys.*, **7(7)**, pp. 781–786, 1986.

Cohen, L., "Time-Frequency Distributions. A Review," *Proc. of the IEEE*, **7(7)**, pp. 941–981, July, 1989.

Dantawate, A. V., and G. B. Giannakis, "Cyclic-cumulant Based Identification of Almost Periodically Time-Varying Systems: Parametric Methods," *Proc. IEEE ICASSP-92*, **V**, pp. 229–232, San Francisco, 1992.

Daubechies, I., "The Wavelet Transform, Time-Frequency Localization and Signal

Analysis," *IEEE Trans. Info. Th.*, **36(5)**, pp. 961–1005, Sept., 1990.

Daubechies, I., "The Wavelet Transform: A Method for Time-Frequency Localization," Chapter in **Advances in Spectrum Analysis and Array Processing, 1**, S. Haykin (ed.), pp. 366–417, Englewood Cliffs, N.J.: Prentice Hall, 1991.

Flandrin, P., "Some Aspects of Non-stationary Signal Processing with Emphasis on Time Frequency and Time-Scale Methods," in [Meyer, 1989], pp. 68–98. J. M. Combes, J. M. Grossman and P. Tchamitchian, eds., *Proc. of Int. Conf. on Wavelets, Time Frequency Methods and Phase Space: Inverse Problems and Theoretical Imaging*, Marseille, France, Dec., 1987, Berlin: Springer-Verlag, 1989.

Fonollosa, J. R., and C. L. Nikias, "Wigner Polyspectra: Higher-Order Spectra in Time Varying Signal Processing," *Proc. IEEE ICASSP-91*, pp. 3085–3088, Toronto, Canada, May, 1991a.

Fonollosa, J. R., and C. L. Nikias, "Transient Signal Detection Using the Wigner Bispectrum", *Proc. of Twenty-Fifth Annual Asilomar Conference on Signals, Systems and Computers*, Pacific Grove, CA, Nov., 1991b.

Fonollosa, J. R., and C. L. Nikias, "Wigner Higher-Order Moment Spectra: Definition, Properties, Computation and Applications to Transient Signal Analysis," *IEEE Trans. on Signal Processing*, January, 1993.

Frazer, G. J., and B. Boashash, "Detection of Underwater Transient Acoustic Signals Using Time-Frequency Distributions and Higher-Order Spectra," *Proc. of Twenty-Fifth Annual Asilomar Conference on Signals, Systems and Computers*, Pacific Grove, CA, Nov., 1991.

Friedlander, B., and B. Porat, "Detection of Transient Signals by the Gabor Representation," *IEEE Trans. Acoust. Speech, Signal Processing*, **37(2)**, pp. 169–180, Feb., 1989.

Gabor, D., "Theory of Communication," *J. IEE* (London), **93 (III)**, pp. 429–457, Nov., 1946.

Gardner, W. A., "Spectral Characterization of N-th Order Cyclostationarity," *Proc. IEEE-ASSP Workshop on Spectrum Estimation*, Rochester, N.Y., pp. 251–255,

1990.

Gerr, N. L., "Introducing a Third-Order Wigner Distribution," *Proc. of the IEEE*, **76 (3)**, pp. 290–292, March, 1988.

Giannakis, G. B., and A. V. Dantawate, "Polyspectral Analysis of Non-Stationary Signals: Bases, Consistency and HOS-WV," *Proc. International Signal Processing Workshop on Higher-Order Statistics*, Chamrouse, France, pp. 167–170, July, 1991.

Helstrom, C.W., "An Expansion of a Signal in Gaussian Elementary Signals," *IEEE Trans. Info. Th.*, **IT-12**, pp. 81–82, Jan., 1966.

Hinich, M. J., "Detecting a Transient Signal by Bispectral Analysis," *IEEE Trans. Acoust., Speech, Signal Processing*, **ASSP-38, (7)**, pp. 1277–1283, July, 1990.

Hlawatsch, F., and G. F. Boudreaux-Bartels, "Linear and Quadratic Time-Frequency Signal Representations," *IEEE Sig. Proc. Magazine*, pp. 21–67, April, 1992.

Mallat, S., "A Theory for Multiresolution Signal Decomposition: The Wavelet Representation," *IEEE Trans. Pat. Anal., Machine Intell.*, **11(7)**, pp. 674–693, July, 1989.

Marinovic, N. M., V. G. Oklobdzija, et al., "VLSI Architecture of a Real-Time Wigner Distribution Processor for Acoustic Signals," *Proc. IEEE ICASSP-88*, pp. 2112–2115, 1988.

Meyer, Y., "Orthonormal Wavelets," in **Wavelets Time-Frequency Methods and Phase-Space**, J. M. Combes, J. M. Grossman and P. Tchomitchian (eds.), *Proc. of Int. Conf. on Wavelets, Time Frequency Methods and Phase Space: Inverse Problems and Theoretical Imaging*, Marseille, France, Dec., 1987, Berlin: Springer-Verlag, 1989.

Nawab, S. N., and T. F. Quatieri, "Short-time Fourier Transform," chapter in **Advanced Topics in Signal Processing**, J. S. Lim, and A. V. Oppenheim (eds.), Englewood Cliffs, N.J.: Prentice Hall, 1988.

Papadopoulos, C. K., and C. L. Nikias, "Parameter Estimation of Exponentially Damped Sinusoids Using Higher Order Statistics," *IEEE Trans. Acoust., Speech,*

Signal Processing, **ASSP-38 (8)**, pp. 1424–1436, August, 1990.

Peyrin, F., and R. Prost, "A Unified Definition for the Discrete-Time, Discrete-Frequency and Discrete-Time/Frequency Wigner Distributions," *IEEE Trans. Acoust., Speech, Signal Processing*, **ASSP-34 (4)**, pp. 858–867, August, 1986.

Probasco, J. W., and G. F. Boudreaux-Bartels, "Detection of Transient Signals by the Gabor Representation and Arbitrary Elementary Signals," *Proc. of the IEEE 1990 Digital Signal Processing Workshop*, New Paltz, New York, pp. 310, Sept., 1990.

Rabiner, L. R., and R. W. Schafer, **Digital Processing of Speech Signals**, Englewood Cliffs, N.J.: Prentice Hall, 1978.

Rioul, O., and M. Vetterli, "Wavelets and Signal Processing," *IEEE Sig. Proc. Magazine*, pp. 14–38, Oct., 1991.

Swami, A., "Third-order Wigner Distribution: Definitions and Properties," *Proc. IEEE ICASSP-91*, pp. 3081–3084, Toronto, Canada, May, 1991a.

Swami, A., "Some New Results in Higher-Order Statistics," *Proc. International Signal Processing Workshop on Higher-Order Statistics*, Chamrouse, France, pp. 135–138, July, 1991b.

Swami, A., "Higher-Order Wigner Distributions," *Proceedings SPIE-92*, 1770-42, San Diego, July, 1992.

Ville, J., "Theorie et applications de la notion de signal analytique," *Cables et Transmission*, **2A**, pp. 61–74, 1948.

Wexler, J., and S. Raz, "Discrete Gabor Expansions," *Signal Processing*, **21**, pp. 207–220, 1990.

Wigner, E. P., "On the Quantum Correction for Thermodynamic Equilibrium," *Phys. Rev.*, **40**, pp. 749–759, 1932.

12

CURRENT AND FUTURE TRENDS

Rapid developments in the field of higher-order statistics (or spectra) naturally led to an expansion of its technical literature. As a consequence of such a rapid development, the first workshop on Higher-Order Spectral (HOS) Analysis was held in Vail, Colorado, in 1989, followed by an international workshop on Higher-Order Statistics in 1991, which was held in Chamrousse, France. Since the completion of the first draft of this manuscript, there have been many important new developments in the field with regard to theoretical approaches and applications. Very worthy of mentioning is the important work on array processing and direction finding by Cardoso [1991], Lacoume and Gaeta [1991], Lacoume and Ruiz [1991], and Lagunas and Vazquez [1991]; the results on nonlinear system modeling by Bondon, Benidir, and Picinbono [1991], Chevalier and Picinbono [1991], and Duvaut [1991]; the work on image processing and multidimensional signal processing by Anderson and Giannakis [1991], Chandran and Elgar [1991], Sadler and Giannakis [1992], Tsatsanis and Giannakis [1992], Ramponi and Carrato [1991], and Swami, et al. [1990]; the research work on detection by Giannakis and Tsatsanis [1992], Kletter and Messer [1990], and Dogan and Mendel [1992]; and, the application of the bispectrum to radar by Jouney and Walton [1991].

Current and future research in the field includes nonstationary and cyclostationary signal analysis with HOS-based time frequency representations, HOS-based wavelet representations as well as utilization of higher-order statistics for the analysis of cyclostationary and chaotic signals. Applications of HOS-based methods to telecommunications, sonar, radar, seismic, speech, image, biomedicine, and other areas of signal processing will continue to be an important trend. Finally, computationally efficient hardware and software implementations of HOS-based algorithms are becoming increasingly important.

A broad and increasingly important class of statistical models for non-Gaussian signal processing phenomena is that of the so-called heavy-tailed or *stable* distributions. As a result, they usually have infinite variances, in which case the second-order moment theory or higher-order spectral analysis methods no longer apply. In other words, the traditional autocorrelations or higher-order statistics are meaningless. Instead, the *covariations or fractional lower-order moments* (e.g., order less than second) are used as tools for signal analysis. Shao and Nikias [1992] summarize, in an engineering tutorial paper, the methodologies of processing non-Gaussian stable signals and point to future research directions.

REFERENCES

Anderson, J., and G. B. Giannakis, "Noise Insensitive Image Motion Estimation Using Cumulants," *Proc. 1991 ICAASP*, pp. 2721–2724, Toronto, Canada, May 1991.

Bondon, P., M. Benidir, M., and B. Picinbono, "Bispectrum Modeling Using Quadratic Filters," *Proc. 1991 ICASSP*, pp. 3113–3116, Toronto, Canada, May, 1991.

Cardoso, J-F., "Super-Symmetric Decomposition of the Fourth-Order Cumulant Tensor, Blind Identification of More Sources than Sensors," *Proc. 1991 ICASSP*, pp. 3109–3112, Toronto, Canada, May, 1991.

Chandran, V., and S. Elgar, "Shape Discrimination Using Invariants Defined from Higher-Order Spectra," *Proc. 1991 ICASSP*, pp. 3105–3108, Toronto, Canada, May, 1991.

Chevalier, P., and B. Picinbono, "Second Order Volterra Array Processor Mismatched to the Fourth Order Moments of the Jammers," *Proc. of Int'l Workshop on Higher-Order Statistics*, pp. 305–308, Chamrousse, France, July, 1991.

Common, P., "Independent Component Analysis," *Proc. of Int'l Workshop on Higher-Order Statistics*, pp. 111–120, Chamrousse, France, July, 1991.

Dogan, M. C., and J. M. Mendel, "Real-Time Robust Pitch Detector," *Proc. 1991 ICASSP*, pp. I-129–I-132, San Francisco, March, 1992.

Duvaut, P., "Non-Linear Filtering in Signal Processing," *Proc. of Int'l. Workshop on Higher-Order Statistics*, pp. 41–50, Chamrousse, France, July, 1991.

Giannakis, G. B., and M. K. Tsatsanis, "A Unifying Maximum-Likelihood View of Cumulant and Polyspectral Measures for Non-Gaussian Signal Classification and Estimation," *IEEE Trans. on Information Theory*, March, 1992.

Jouney, I., and E. K. Walton, "Applications of the Bispectrum in Radar Signatures Analysis and Target Identification," *Proc. of Int'l. Workshop on Higher-Order Statistics*, pp. 171–174, Chamrousse, France, July, 1991.

Kletter, D., and H. Messer, "Optimal Detection of a Random Multitone Signal and Its Relation to Bispectral Analysis," *Proc. 1990 ICASSP*, pp. 2391–2394, Albuquerque, NM, April, 1990.

Lacoume, J. L., and M. Gaeta, "Complex Random Variables; a Tensorial Approach," *Proc. of Int'l. Workshop on Higher-Order Statistics*, pp. 25–28, Chamrousse, France, July, 1991.

Lacoume, J. L., and P. Ruiz, "Source Identification: A Solution Based on the Cumulants," *Proc. 4th ASSP Workshop on Spectral Estimation and Modeling*, pp. 199–203, August, 1991.

Lagunas, M. A., and G. Vazquez, "Array Processing from Third Order Functions," *Proc. of Int'l. Workshop on Higher-Order Statistics*, pp. 21720, Chamrousse, France, July, 1991.

Ramponi, G., and S. Carrato, "Modified Fourth-Order Moments in Texture Recognition," *Proc. of Int'l. Workshop on Higher-Order Statistics*, pp. 139–142, Chamrousse, France, July, 1991.

Sadler, B., and G. B. Giannakis, "Shift and Rotation Invariant Object Reconstruction Using the Bispectrum," *J. of the Optical Society of America A*, Jan., 1992.

Shao, M., and C. L. Nikias, "Signal Processing with Fractional Lower Order Moments-Stable Processes and their Applications," *Proceedings of IEEE*, under review, 1992.

Swami, A., G. B. Giannakis, and J. M. Mendel, "Modeling of Multi-Dimensional Non-Gaussian Processes Using Cumulants," *J. of Multidimensional Signals and*

Systems, **1**, pp. 11–37, March, 1990.

Tsatsanis, M. K., and G. B. Giannakis, "Object and Texture Classification Using Higher Order Statistics," *IEEE Transactions on Pattern Analysis and Machine Intelligence*, **14**(7), pp. 733–750, July, 1992.

INDEX

absolutely summable 37
adaptive algorithms for ARMA
    Processes 400
adaptive blind equalization
    algorithms 418
adaptive filtering 383-384,418
adaptive IIR algorithms 384,397,
    444
adaptive lattice algorithms 406
adaptive lattice linear
    algorithm 406
adaptive lattice linear prediction
    algorithm 408
adaptive lattice linear prediction
    using cumulants 402
adaptive lattice method 442
adaptive relaxation
    technique 247,249-250
adaptive time delay
    estimation 414-415,
    417,420,442
Akaike modified criterion 281
aliasing 4,147,232, 509-511,515
alternating maximization 358
anticausal 274,295-296
aperiodic sequence 77,79
applications 1,2,4,7,27,60,61,66,
    67,69,71,75,121,147,161,
    163,177,191,201,205,216,
    217,218,237,238,242,253,
    261,264,275,310,322,
    362,383,445,448,460,
    484,485,503,508,512,520,
    521,522,524
AR 271-272,274-275,278-279,281,
    284-287,289-290,292-296,
    298,303-304,306-308, 311,
    401-402,406-407, 442,
    461,465-467, 470-472,
    475-477,479, 483-485

ARMA 266,271-272,274-275,
    282-284,286-288,290-291,
    293,295-296,301-302,
    307-308,310-311,378,
    401,442,444-445,484
ARMA model 272,274-276,282,
    284-285,287-290,293,
    295-296,307-309,311,445
ARMA orders 306
ARMA processes 265,308-309,
    397,400
array processing 339-340,342-344,
    346-348,377,380,522
atmospheric turbulence 238,384,
    386,388,390,393,406,425
autocorrelation sequences 15-16,48,
    60,62,80,83,96,297,300,
    323,449,478,480
autocorrelations 173-178,190,194
    199-200,206,266,268-269,
    271, 276,280-281,284,295,
    297, 304,307,311,318,367,
    414, 437,467,476,478,481
autoregressive (AR) 347,361,367,
    407,461,465
average power 71-74,77,101,110,478
Bartelt-Lohman-Wirnitzer
    algorithm 223-224,227-228,
    232
beamforming 313,339,342,
    347-352,355,377
bearing calculation 313,315-317
bearing estimation 313,315,337,
    339-340,342,344,346,
    349,351-353,355,357,
    360,378-379
bearing formula 315
Benveniste-Goursat (BG) 422,424-426,
    433-434,437,443
bibliography 4-5

bicepstrum 177,179-188,192,
    200,203-208,210-211,
    214-216,230-233,240,
    242,257,380,477
bicoherency 31,64,187-192,216,230,
    462-464,470,478,480
Bispectrum Iterative Reconstruction
    Algorithm (BIRA)240
BIRA 240,242-243,245-250,
    253,255,257,259
bispectral 211,218,357,379,381
bispectral estimates 142,161,282
bispectral phase unwrapping 225
bispectrum2,5,61,63,68,88,89,92,95,
    110,112,114,116,118,119,
    121,126,127,129,130,131,
    132,134,135,137,138,139,
    140,142,143, 144, 145,146,
    147,148,149,150,151,152,
    153,154,155,156,157,158,
    159,160,161, 177,219,
    220,221,228,231,232,236,
    237,239,240,245,249,
    252,261,262, 263,264,
    496,499,500,511,512,
    513,514,516,518,522
bispectrum bias supremum 131
bispectrum domains 324
bispectrum estimation 445
bispectrum factorization 192-193
bispectrum phase 220-221,
    227,240,249,263
bispectrum principal domain 147
Bispectrum Signal Reconstruction
    (BSR) 231,235
bispectrum variance 129-143
Blackman-Tuckey method 124
Blind deconvolution BIRA 253,418,443
blind equalization 384,418-419,
    425,427,433-434,443,445-446
blind parametric equalization using
    higher-order moments 437
Brillinger 232,378,455,456,483-484
Brillinger algorithm 221,222
BSR 233-235
Bussgang 424,426,433,443
Bussgang algorithms 418,
    422,425-426,442-443
Bussgang techniques
Bussgang-type algorithms 437

c(q,k) 267
c(q,k) formula 267,292
causal AR 279
cepstral coefficients 174,177,180-182,
    184-185,187,191
cepstral parameters 166,169-172,
    174, 189,193-194,
    197,206-207,429
cepstrum 163,167,169-175,
    177-178,181,189-192,
    204-205,216,431,433,437
    443,445
cepstrum analysis 163
cepstrum of bicoherency 187
cepstrum of higher-order
    spectrum 163
characteristic function 8-9,13,294
chirp signals 487
Choi-Williams 490,492,494-495,
    502-503,521
classification map 3
closed form solution 267
Cohen's general class 487,491,494,520
coherence index 31,64
complex cepstrum 163-167,169-170,
    173,180-181,191-192,
    204-205,209,214-216,218,445
complex demodulates
    class 123,141-142,159
complex processes 34
complex regression coefficients 34,57
constant modulus algorithm (CMA) 424
constrained third-order mean
    (CTOM) 465
constrained third-order mean
    method (CTOM) 471
conventional 1,123-124,126,132,
    137-145,147-148,159
conventional methods 123-124,
    132,138,142-143,
    145,147-148,159,163,324-327,
    330-331,333-336,342,461,472-
    474,476-477,479,483
conventional TDE methods 322,325
cost functions 280,384-385,
    388,391,396,426-427,437,440
covariance sequence 15,21,25,
    28,314,375,449
Cramer spectral representation 21
Cramer-Rao bound 320,359

Index

CRIMNO 427,443
criterion with memory
    nonlinearity 427
cross bispectra 326
cross bispectrum 201-205,
    207-208,210, 216,324-
    326,331,339, 352,354,356-
    357,360,414, 451,455,443
cross spectra 317-320,352,357,360
cross- bispectrum 201,202,
    203,205,207,326,331,
    352,358,361,414,451,455
cross-bispectrum
    beamforming 352,358,361
cross-cumulant spectra 31,34
cross-cumulants 17,31,33-34,272,414,437
cross-energy 78
cross-moments 83-84,86,98,106
cross-spectral matrix (CSM) 352,357
cross-spectra 200-201,319
cross-spectrum 450,462
cross-tricepstrum 205,216,414,437,443
crystallography 238,263
CTOM 472-474
cubic phase coupling 448,482-484
cumulant spectra 2,7,20-21,24,26,
    30-31,33-34,37,123,138,140
cumulant spectrum 20-21,
    31,34,39,53,125,138-139,145
cumulant spectrum estimates 125,145
cumulant-based beamforming
    methods 347,349,351
cumulant-based ESPRIT Method 350
cumulant-based noise
    subspace method 347-348
cumulant-based solutions 279
cumulant-based signal subspaced
    method
cumulants 7-9,10-17,19,22-23,25-29,
    31-32,37,39,47-48,55-56,
    59-61,64,68,71,81,124-126,
    145,177,179,182,184,
    187,196,199-201,206-207,
    217,267-269,275,277-278,
    284-285,289,293-295,
    303-304,306,309-311
cumulative energy transfer 98
cyclostationarity 488,523
damped sinusoidal signals 366
Daniell 127,130-131

deconvolution 4,67,163,193-194,
    216,219,253-254,257-264,
    309,380,406-407,409,
    411-412,415,418,422,
    443,445
delta functions 81,91,419
detection of deterministic signals 376
detection of transient
    signals 371,373,488,
    512-513,522-524
detection statistic 376
deterministic energy signals 164,
    172,175,177,179-181,
    185,187-188,191,200-202,209
deterministic signals 2,37,71-72,78,84,88,
    495,100,121,124,159,163,
    172,181,216,228,242,253-254,
    314,374-376,488-499-500,511-
    512,515
deterministic transient
    signal 90,92,94,100,132
deterministic transients 2,
    68,71-72,75,78,95
DF-WHOMS 509-511
Dianat-Raghuveer 232
differential bicepstrum 180-182,205,216
differential cepstrum 169-172,181
direct approach 142,149,151,326
direct class 123,132,138
discrete frequency WHOMS
    (DF-WHOMS) 504
Discrete Time and Frequency
    WHOMS 505
Discrete Time WHOMS
    (DT-WHOMS) 504
dispersion relation 453
double $c(q,k)$ algorithm 286,289,
    291
DT-WHOMS 504-509-510,512
duality magnitude recovery 234
duality phase recovery 234
Eckart 314
Eckart window 319
eigenvector 341,343-344,
    346-347,361,366,370
energy 72-74,77-78,80,88-89,
    92-99,126,143,146,230,
    243,247,249,250-253,315,
    374,453-454,487-489,490
energy cascading 453,454

energy signals 71-73,75,77-78,
        80,82-85,87,89-90,98,
        100-101, 109,121,123-124,
        132 135,155,163,230,367
energy spectrum 88-89,92-93,95,99,487
energy transfer 29,98,160,454
equalization 4,383-384,391,418-419,
        422-425,427-428,433-
        434,437,441-442
ergodicity 19,68,472
error bicoherence 143
ESPRIT 1,344,-346,350-353,378,380-381
estimation of higher-order statistics 124
exponential 9-11,72,74,82-83,
        92-93,138,145,212,224,
        245,277,290,293,304,403,
        501,512-513,516,521
exponentially 245,366,379,380
fast Fourier transform (FFT) 184
FFT 326,476,508
finite-duration sequences 75-76,
        82,92,96,100,120
finite-duration signals 71-72
first-order moment 80
fluctuation field 453-454
Fourier analysis 72-73
Fourier magnitude 219-220,232,
        238-240,245,247,
        437,457
Fourier phase 219-223,226-228,
        230,233,237-240,249,259,
        437,458,498
Fourier phase retrieval algorithms 220
Fourier series 74,123
Fourier-Stieltjes representation 21
Fourier- transforms123,126, 128,
        129,132,147,151,160, 165,
        167,169,172,175, 177,
        184,189,190,193, 197,
        201,206,209,215, 219,
        220,221,228,229, 230,
        232,233,238,239, 241,
        251,259,262,282, 283,
        317,331,332,355, 359,
        375, 419, 449,450, 451,
        453,454,455,457, 461,
        462,463,476,583, 487,
        488,490, 259,262,
fourth-order cumulant
        matrices 377

fourth-order cumulant matrix 339,
        346-347
fourth-order cumulants 349,
        400,406,427-429,433-434,437-
        438,440, 445,452,482
fourth-order moments 16,35,
        82-83,88,92,103-104,110-111,
        362,370-371, 429,437-438
fourth-order statistics 270, 367,
        427,433,442
frequency relations 461-462
frequency response 441
frequency response function 266
Gabor coefficients 512
Gabor representation 487,522,524
Gaussian 9-10,14-17,24,31,39-40,
        47,53,55,57,61,63,64-65,
        87,100,163,177,179-180,182,
        184-185,187-188,191,193-199
        201,207,209,211,214,216,
        218,233,235,247-249,254,256,
        259,265-266,268,290,293,295,
        303-306,308,313-314,320,
        322-323,328-336,346-354,
        356,359-361,366-367,369,
        371-377,380-381,391,
        394-395,414-415,417-419,
        424,428,433-434,439,447,
        449-451,453,45-459,
        465-466,472,474,476,
        483-484,516,523
Gaussian noise 267,476
Gaussian signals 443
GCC 317,323,330-332,337-338
Generalized Cross-Correlation
        Method 317
generalized group delay 498
GM method 276,277
group delay 489,498
Hamming 127,130-131
Hannan-Thompson 314,320
harmonic decomposition 1,380
harmonic retrieval 339,486
higher-order periodogram 132-134
higher-order spectra 2-5,25,66-67,
        123-125,144,147,149-150,159,
        163,165,167,193,215-216,
        219-220,237,259,261,263,
        265-266,307,311
higher-order spectra (HOS) 447

Index

higher-order spectral 66,311
higher-order spectral
    analysis 124,308,485
higher-order spectrum 21,67,220,261,310
higher-order spectrum factorization 56
higher-order statistics(HOS)
    2,4-5,7,27,67,123-126,
    266,282,308-311,313,
    339-340,366-367,375,
    377-378,380-381,383,397,
    427,441-445,457
higher-order whiteness279
high-resolution techniques 344,361,465
homomorphic deconvolution 163
Huzii method278,280
identifiability 266,309,311,406
index of efficiency 129
indirect approach 142
indirect class 123-124,138
instantaneous frequency 489-490,
    496-497,512,521
instantaneous power 72,487,489,494
integrated bispectrum 27
integrated trispectrum 30
interferometric techniques 238
inverse filter 193-194,419,422
inverse filter reconstruction 194
inverse transforms 165,175,177,
    179,189,197,205
K-distribution 11
Kronecker delta function 25,81,91
kurtosis 16,23-25,29,
    39,53,89,111,332,337,
    347,350,432,438
Kurtosis Measures 16
Laplace 9-10,337
Least Mean 393
Least-mean fourth (LMF)
    384,387,391,414
least-mean fourth-order
    algorithm 394
Lii and Rosenblatt 214
Lii-Rosenblatt 217,232,380
Lii-Rosenblatt algorithm 222,223,227
line array configuration 313,339
linear growth rate 453
linear least estimation algorithm 439
linear least-squares algorithm 440
linear least-squares estimation
    algorithm 438

linear non-Gaussian 63-64
linear non-Gaussian
    processes 7,66,87,138,163,
    179,180-181,185,187-188,191,
    193,201,216,228,230
linear phase shifts 33,34
linear processes 53,57,64
linear programming 270-271
LMF 384,390-392,
    394-397,429,433,442,446
local autocorrelation function 488-489
localization accuracy 316
log-likelihood function357
LTI 249,252,256
LTI systems 37-39,41,
    50-52,56-59,90-91,116,295
MA 266-268,270-271,
    275-279,284-308,311,
    397,400-401,406,411-412,
    442,444
MA methods 266,397
magnitude recovery 234-235
magnitude recovery
    algorithms 219-220,230,261
magnitude retrieval 238
magnitude-squared coherence 320
matched filter 375,377-378,515,518
Matsuoka-Ulrych 232
Matsuoka-Ulrych Algorithm 224
maximum likelihood 266,313-314,
    320,343,347,352,366-367
maximum likelihood capon
    method348
maximum phase 40,45,49,88,94,96,
    255-257,274-275,296,299,
    428-429,433
maximum phase components 164
mean conditional frequency 497
mean fourth-cumulant criterion 332
mean frequency 489,496-497,514
mean time calculation 498
mean value 16
mean-square error
    (MSE) 383-385,423,434,441
measure the non-Gaussian nature 14
minimum norm method 366,377
minimum phase 40,45,47-49,239,
    255-257,263,265-266,271-273,
    275,278-279,286,288,295-296,
    299,428-429,433,458-459

minimum phase components 164
minimum phase signal 94
minimum variance
    distortionless technique 343
mixed phase 46,94,256
mixed spectrum 32-33
model order determination 275,295
moment spectra 2,30-31,
    71,84-85,87-93,95,
    98-100,107-112,114-121,
    123,136,140-141
moment spectrum estimate 125,141
moments 7-9,10-15,18-19,22-23,
    27,31,33,37,39,62-
    64,68,124,126,
    132,145,149,241,259,323,362,
    367,369,456,460-461,465-467,
    467-468,471,494,498,520
MSE 321
MSE optimal window 131
multilinear kernel 488
multilinear representations 488
multipath 253-254,257,446
MUSIC 1,339,344,348-351,
    361,363-364,366,377
n-th order coherence
    function 31,64,99,100,119
neural network 271,310,419
noise subspace 313,343-344,346,348
nonaliasing constraints 509-511
noncausal AR 266,275,278-279,
    407,411-412
noncausal AR methods 278
non-Gaussian 401,406-407,
    412,414-415,417,419,423-
    424,428
non-Gaussian linear processes
    31,37,39,64, 262,309-311,
non-Gaussian processes 24,27,
    37,228,230,242,309,
    384,397,443,444
nonlinear 3-5,61,66,261,271,281,366,
    384,418,423-427,437,440,
    445,447-448
nonlinear filter 419
nonlinear (N/L) time series 448
nonlinear least squares 270,441
nonlinear least squares algorithm 440
nonlinear least-squares estimation
    algorithm 439,440

nonlinear least-squares 440
nonlinear processes 7,58,63-64
nonminimum phase 311
Nonparametric BSR 232
nonstationary signal analysis 487
nonlinear quadratic system 461
nonlinear system identification 447
nonlinear systems 58-60,61,63
nonminimum phase 3-4,38-40,45-46,
    48-49,69,94,96,144,163-164,
    180,184,257,266,271,
    282,288,290,295,297,
    299,309-310
nonminimum phase LTI systems 45
nonminimum phase signals 163,181,216
nonminimum phase system
    identification 182,184,307
nonminimum phase
    systems 3,202,265-266
nonnormality 14
Nonparametric BSR 232
nonparametric methods for signal
    reconstruction 219
nonrecursive phase recovery 261
normalized bispectrum 31, 230,467,470
normalized cumulant spectrum 31
normalized kurtosis 16
nth-order cumulants 15,145,147
null hypothesis 374
OARM 475-476
optimization methods 270
optimized AR Method
    (OARM) 465,475,477,479
optimum receiver 375
Padé rational approximation 282
Page 487,492
parameter estimation 2-4,37,
    267,271,275,275,282,
    290,293-295,310-311,313,
    367,369,371,401,414,429,445,
    445,470,524
parametric approach 123,161
parametric bispectrum estimation 466
parametric estimation 465,484-485,512
parametric methods 265,
    266,307,311,407,461,465,483
parametric TDE method 327
Parseval's identity 76
Parseval's relation 73,75
Parzen 127,130-135,145-146,467

periodic 79
periodic sequences 73-77,103-106,
    110,112,115,118,120-121
periodic signals 71-72,
    74,100-101,106-107,
    108-111,114,117-119,
    123,504,506
periodic train of impulses 508
phase 458,487,498,508
phase conditions 94
phase coupling 448, 459-462,465,467,
    470-472,476,482-485
phase information 37-38,45,49,
    88-89,174,181,231-232,
    237,240,274
phase recovery 219,224,
    228,230,232-235,261,263
phase recovery
    algorithms 219-220,225,230,
    232,261
phase relations 31, 60-61,67,160,
    459-460
phase retrieval algorithms 220-221
phase retrieval problem 238
Phase transform 319
phase unwrapping 167-169,172,174,
    177, 181,183-184,205,
    218,221,224-225,227-228,
    231-232,261
PHAT 314,319-320
Pisarenko 1,344,348,476,480
POCS 219,249,252,261
POCS methods 249
Poisson triggered processes 56-57
polar rasters 149,159-160
polycepstra 2-4,7,21,45,57-58,63,66,
    68,132,147,149,160,163,
    193-194,215-216,230,
    235,261
polyspectra algorithms 419
polycepstra-based BSR 232
POTEA 443
power cepstrum and tricoherence
    equalization algorithm
    (POTEA) 437
power cepstrum 172-175,
    177,181,235,241-242,
    256-257
power signals 71-75,77, 100-101,
    110,121

power spectra 28,131, 133-134,
    160,487
power spectrum 1-3,21,24-26,
    34,36,38-44,47-52,54,58,
    60-64,172,174,187-188,
    190-191,200,265,282,318,
    330,494,516
power spectrum
    estimation 2
power spectrum estimation
    techniques 1,278
prediction error filter 402-405
prefilters 317
Priestley 127,130-131,303,310
projection in frequency 497
Prony 1,313,366-367
Prony's method 366
q-slice algorithm 291-293
quadratic nonlinearity 452-453
quadratic phase coupling 4, 59-60,
    62,66,147,278,459-462,
    465,467,470-472,476,482-485
quadratic system 59,62-63
range calculation 313,315-317
rank determination 295,306,370
Rayleigh 9,11
real cepstrum 172,174
recursive least squares
    (RLS) 384,414,433
recursive phase recovery
    algorithms 261
reduced interference
    distributions 492,502,520
reduced interference high-order
    distribution (RIHOD) 502
regression coefficients 34,57-58
residual time series 284,285-286,307
Rihaczek 487,492
RIHOD 503
RLF 384,390,396-398,442-443
Roth window 318
ROTH 314,318,330-331,333-336
Sasaki 126,127,129-135,145-146,161,322-
    323,325,381
SCOT 314,319,330-331,333-336
second characteristic function 8,13
second-order moments 8,16,
    18,35,63,101,118-119
second-order statistics 340,380,
    383-384,387,406,442

second-order Volterra filters 448
semi-invariants 8
short-time Fourier
    transform (STFT) 487,523
signal analysis map 87
signal detection 375
signal reconstruction 163,216,
    18-291,231,235,237-239,
    249,260-264
signal recovery 219,235,237-238,261
signal recovery from partial
    information 237
signal subspace 313,341,
    343-344,347-348,381
singular value decomposition 273,
    285-286,306,328,365,379
singularities 235
sinusoidal 480
sinusoidal processes 58
skewness 16,23-27,29,39-
    40,48,53,55,81,89,102,
    138,145,279,289,323,
    325,330,352,359-360,
    380,397,414-415
Slices of Cumulants 26,406,409
Smoothed Coherence Transform 319
smoothed cross-correlation 317,321
smoothing windows 317,321
spatial covariance matrix 339-346,
    348,353,360-361
spatial fourth-order cumulant
    matrix 346
special covariance
    matrix 364
spectral factorization 56
spectrally equivalent 47,279-280
Spectrogram 487
spectrum factorization 191
spike-array 4,122
square-law device 63-65
statistical properties 123,142,
    145,159,369
statistical test 303-304
statistics 337
steepest descent method 385-386
stop-and-go 424,434
stop-and-go algorithm 437,445
stop-and-go (SG) 433
super-FFT approach 149,153
symmetry regions 23,150,156

system identification 265-266,
    275,292,310-311,383,
    406,412,442,444-447
Taylor expansion 8
TDE 313,321-323,325,327-332,337
TEA 429,430,433-434,437
test for aliasing 147-149,159-160
test for linearity 193,381
test statistics 373,377
thermal noise 347
third-order cumulant
    estimate 199-201,207
third-order cumulants 207,267-272,
    275-276,280-281,284,
    286-291,293,295,303-306,
    311,397,400,406-407,411,
    414-415,417,447,450-452,457,
    461,465-466,471,475-476,
    478,480-483
third-order moment sequence 199-202
third-order moment window 232,236,247
third-order moments 16,18,64,81-84,
    86,88-89,92-94,96,
    98-99,102-105,108,111,
    113-114,116,132,149,177,
    179,190,233,236-237,245,
    247,258-259,367,369,
    466-468,471
third-order recursion 475
third-order recursion (TOR) 471
third-order recursion (TOR)
    method 465
third-order recursion
    equation 367,407
third-order recursion
    method 465,468,470
third-order statistics 181-184,
    191,200,322-323,
    329,331,367,377,380
third-order Wigner distribution 488,
    523-524
three-wave coupling equation 453-454
three-wave energy transfer function 454
time delay estimation 34,58,313-315,
    317,321,323-324,330-331,
    337,377-381,414
time-average operator 19,74
time delay 84,99,313-317,320-325,
    327-328,330-337,
    378-381,414

Index 537

time reversible 80
time shift 488,496
time-frequency
    distributions 487-488,491-492,
    494,496,498-499,502,520-522
time-frequency energy distributions 487
Time-Reversible Process 17
time-varying bispectrum 513
Toeplitz 466
Toeplitz approximation
    method (TAM) 322,340,344
TOR 470,472-474,476-477,479
transient signals 488,512-513,520,522
tricepstrum 2,21,23,25,39,54-56,
    88-89,92,112-114,132,145-146,
    149-150,155-159,177,179,
    185-186,194-198,205,216,
    219-220,259,263,284,310,
    339,427-428,443,445

tricepstrum equalization
    algorithm (TEA) 427-428
two-dimensional phase
    unwrapping 221,233
uncertainty principle 461
uniform 9-10,127,131,249,
    351,394-395,504,506
uniform window 127
variance 16,23-25,38,40,48,53,64,89,
    123,125,129,133-134,138,
    142-147,160,170,172,184-185,
    200-201,209,211,215,233,268,
    279,282,290,293-294,303-304,
    316- 320,340,343,345,350,
    370-371,374,385,397,
    401,403-404,417-
    419,424,438,440-
    441,451-453,456,488
Volterra 123,419,447-448,
    450-451,454,484-485
Volterra filters 59,448,455,483
Volterra series expansions 448
Volterra type 58
Wavelet transform (WT) 487,522
WB 496,500,515-516,518,520
WB Detector 518
WD 487-492,494,496,498-499,
    501-503,512,515-516,518,520
Wiener 321,328,485
Wiener filter theory 383

Wiener methods 414
Wiener Processor 321
Wiener-Khintchine identity 21
Wigner bispectrum 36,37,496,
    512-514,518,522
Wigner distribution 36,67
Wigner Higher-Order Moment
    Spectra (WHOMS) 488,
    494-500,502-513,515,522
Wigner Polyspectra 503,522
Wigner time-frequency 36
Wigner Trispectrum 496
Wigner-Ville 487
Wigner-Ville distribution 488,
    512-513,520
window bandwidth 144
Yule-Walker 343
Yule-Walker methods 279-280,467,470
zero forcing (ZF) constraint 433
zero-padding approach 149,151,154
zero-phase signals 73,96
Zurbenko method 138-140,159,161

MATLAB® – HIGH-PERFORMANCE NUMERIC COMPUTATION

MATLAB, the companion software to *Higher Order Spectral Analysis: A Nonlinear Signal Processing Framework* by Chrysostomos L. Nikias and Athina P. Petropulu (Prentice-Hall, 1993), is used extensively in this textbook's presentation of the theory, and for problem-solving. In addition, the *Hi-Spec Toolbox*, for use with MATLAB, has been created to complement the book. MATLAB is widely used in applied math, electrical engineering, mechanical engineering, economics, physics, and other disciplines.

- MATLAB is available for platforms including PCs, Macintosh, workstations, and supercomputers.
- Educational and quantity discounts are available.
- PC and Mac **Classroom Kits** provide a cost-effective way to use MATLAB for teaching a course with unlimited enrollment, while **The Student Edition of MATLAB** lets students obtain software for their own PC or Mac.
- **Application Toolboxes** add specific functions to MATLAB for control system design, system identification, signal processing, optimization, and other areas.

The Hi-Spec Toolbox

The *Hi-Spec Toolbox* for use with MATLAB includes functions for higher-order spectrum estimation (conventional and parametric approaches), adaptive linear prediction, magnitude and phase retrieval, quadratic phase coupling, harmonic retrieval, time-delay estimation, and array processing using higher-order statistics.

For more information, call The MathWorks, Inc. at (508) 653-1415 or return this card today.

☑ *Yes!* I want information on:
- ☐ Control system design tools
- ☐ Digital signal processing tools
- ☐ System identification tools
- ☐ Classroom Kits and The Student Edition
- ☐ Hi-Spec Toolbox

Name _____
Company _____
Dept/MS _____
Address _____
City/State/Zip _____
Country _____ Tele _____
Computer(s) _____

Nikias

BUSINESS REPLY MAIL
FIRST CLASS   PERMIT NO. 82   NATICK, MA

POSTAGE WILL BE PAID BY ADDRESSEE

THE MATHWORKS, INC.
24 Prime Park Way
Natick, MA 01760-9889

NO POSTAGE
NECESSARY IF
MAILED IN THE
UNITED STATES